Principles of Digital Communication

The renowned communication theorist Robert Gallager brings his lucid writing style to this first-year graduate textbook on the fundamental system aspects of digital communication. With the clarity and insight that have characterized his teaching and earlier textbooks he develops a simple framework and then combines this with careful proofs to help the reader understand modern systems and simplified models in an intuitive yet precise way. Although many features of various modern digital communication systems are discussed, the focus is always on principles explained using a hierarchy of simple models.

A major simplifying principle of digital communication is to separate source coding and channel coding by a standard binary interface. Data compression, i.e., source coding, is then treated as the conversion of arbitrary communication sources into binary data streams. Similarly digital modulation, i.e., channel coding, becomes the conversion of binary data into waveforms suitable for transmission over communication channels. These waveforms are viewed as vectors in signal space, modeled mathematically as Hilbert space.

A self-contained introduction to random processes is used to model the noise and interference in communication channels. The principles of detection and decoding are then developed to extract the transmitted data from noisy received waveforms. An introduction to coding and coded modulation then leads to Shannon's noisy-channel coding theorem. The final topic is wireless communication. After developing models to explain various aspects of fading, there is a case study of cellular CDMA communication which illustrates the major principles of digital communication.

Throughout, principles are developed with both mathematical precision and intuitive explanations, allowing readers to choose their own mix of mathematics and engineering. An extensive set of exercises ranges from confidence-building examples to more challenging problems. Instructor solutions and other resources are available at www.cambridge.org/9780521879071.

'Prof. Gallager is a legendary figure . . . known for his insights and excellent style of exposition'
Professor Lang Tong, Cornell University

'a compelling read'
Professor Emre Telatar, EPFL

'It is surely going to be a classic in the field'
Professor Hideki Imai, University of Tokyo

Robert G. Gallager has had a profound influence on the development of modern digital communication systems through his research and teaching. As a Professor at M.I.T. since 1960 in the areas of information theory, communication technology, and data networks. He is a member of the U.S. National Academy of Engineering, the U.S. National Academy of Sciences, and, among many honors, received the IEEE Medal of Honor in 1990 and the Marconi prize in 2003. This text has been his central academic passion over recent years.

Principles of Digital Communication

ROBERT G. GALLAGER
Massachusetts Institute of Technology

CAMBRIDGE
UNIVERSITY PRESS

University Printing House, Cambridge CB2 8BS, United Kingdom

Published in the United States of America by Cambridge University Press, New York

Cambridge University Press is part of the University of Cambridge.

It furthers the University's mission by disseminating knowledge in the pursuit of education, learning and research at the highest international levels of excellence.

www.cambridge.org
Information on this title: www.cambridge.org/9780521879071

First published 2008

A catalogue record for this publication is available from the British Library

ISBN 978-0-521-87907-1 Hardback

Contents

Preface

Digital communication is an enormous and rapidly growing industry, roughly comparable in size to the computer industry. The objective of this text is to study those aspects of digital communication systems that are unique. That is, rather than focusing on hardware and software for these systems (which is much like that in many other fields), we focus on the fundamental system aspects of modern digital communication.

Digital communication is a field in which theoretical ideas have had an unusually powerful impact on system design and practice. The basis of the theory was developed in 1948 by Claude Shannon, and is called information theory. For the first 25 years or so of its existence, information theory served as a rich source of academic research problems and as a tantalizing suggestion that communication systems could be made more efficient and more reliable by using these approaches. Other than small experiments and a few highly specialized military systems, the theory had little interaction with practice. By the mid 1970s, however, mainstream systems using information-theoretic ideas began to be widely implemented. The first reason for this was the increasing number of engineers who understood both information theory and communication system practice. The second reason was that the low cost and increasing processing power of digital hardware made it possible to implement the sophisticated algorithms suggested by information theory. The third reason was that the increasing complexity of communication systems required the architectural principles of information theory.

The theoretical principles here fall roughly into two categories – the first provides analytical tools for determining the performance of particular systems, and the second puts fundamental limits on the performance of any system. Much of the first category can be understood by engineering undergraduates, while the second category is distinctly graduate in nature. It is not that graduate students know so much more than undergraduates, but rather that undergraduate engineering students are trained to master enormous amounts of detail and the equations that deal with that detail. They are not used to the patience and deep thinking required to understand abstract performance limits. This patience comes later with thesis research.

My original purpose was to write an undergraduate text on digital communication, but experience teaching this material over a number of years convinced me that I could not write an honest exposition of principles, including both what is possible and what is not possible, without losing most undergraduates. There are many excellent undergraduate texts on digital communication describing a wide variety of systems, and I did not see the need for another. Thus this text is now aimed at graduate students, but is accessible to patient undergraduates.

The relationship between theory, problem sets, and engineering/design in an academic subject is rather complex. The theory deals with relationships and analysis for *models* of real systems. A good theory (and information theory is one of the best) allows for simple analysis of simplified models. It also provides structural principles that allow insights from these simple models to be applied to more complex and realistic models. Problem sets provide students with an opportunity to analyze these highly simplified models, and, with patience, to start to understand the general principles. Engineering deals with making the approximations and judgment calls to create simple models that focus on the critical elements of a situation, and from there to design workable systems.

The important point here is that engineering (at this level) cannot really be separated from theory. Engineering is necessary to choose appropriate theoretical models, and theory is necessary to find the general properties of those models. To oversimplify, engineering determines what the reality is and theory determines the consequences and structure of that reality. At a deeper level, however, the engineering perception of reality heavily depends on the perceived structure (all of us carry oversimplified models around in our heads). Similarly, the structures created by theory depend on engineering common sense to focus on important issues. Engineering sometimes becomes overly concerned with detail, and theory becomes overly concerned with mathematical niceties, but we shall try to avoid both these excesses here.

Each topic in the text is introduced with highly oversimplified toy models. The results about these toy models are then related to actual communication systems, and these are used to generalize the models. We then iterate back and forth between analysis of models and creation of models. Understanding the performance limits on classes of models is essential in this process.

There are many exercises designed to help the reader understand each topic. Some give examples showing how an analysis breaks down if the restrictions are violated. Since analysis always treats models rather than reality, these examples build insight into how the results about models apply to real systems. Other exercises apply the text results to very simple cases, and others generalize the results to more complex systems. Yet more explore the sense in which theoretical models apply to particular practical problems.

It is important to understand that the purpose of the exercises is not so much to get the "answer" as to acquire understanding. Thus students using this text will learn much more if they discuss the exercises with others and think about what they have learned after completing the exercise. The point is not to manipulate equations (which computers can now do better than students), but rather to understand the equations (which computers cannot do).

As pointed out above, the material here is primarily graduate in terms of abstraction and patience, but requires only a knowledge of elementary probability, linear systems, and simple mathematical abstraction, so it can be understood at the undergraduate level. For both undergraduates and graduates, I feel strongly that learning to reason about engineering material is more important, both in the workplace and in further education, than learning to pattern match and manipulate equations.

Most undergraduate communication texts aim at familiarity with a large variety of different systems that have been implemented historically. This is certainly valuable in the workplace, at least for the near term, and provides a rich set of examples that are valuable for further study. The digital communication field is so vast, however, that learning from examples is limited, and in the long term it is necessary to learn the underlying principles. The examples from undergraduate courses provide a useful background for studying these principles, but the ability to reason abstractly that comes from elementary pure mathematics courses is equally valuable.

Most graduate communication texts focus more on the analysis of problems, with less focus on the modeling, approximation, and insight needed to see how these problems arise. Our objective here is to use simple models and approximations as a way to understand the general principles. We will use quite a bit of mathematics in the process, but the mathematics will be used to establish general results precisely, rather than to carry out detailed analyses of special cases.

Acknowledgements

This book has evolved from lecture notes for a one-semester course on digital communication given at MIT for the past ten years. I am particularly grateful for the feedback I have received from the other faculty members, namely Professors Amos Lapidoth, Dave Forney, Greg Wornell, and Lizhong Zheng, who have used these notes in the MIT course. Their comments, on both tutorial and technical aspects, have been critically helpful. The notes in the early years were written jointly with Amos Lapidoth and Dave Forney. The notes have been rewritten and edited countless times since then, but I am very grateful for their ideas and wording, which, even after many modifications, have been an enormous help. I am doubly indebted to Dave Forney for reading the entire text a number of times and saving me from many errors, ranging from conceptual to grammatical and stylistic.

I am indebted to a number of others, including Randy Berry, Sanjoy Mitter, Baris Nakiboglu, Emre Telatar, David Tse, Edmund Yeh, and some anonymous reviewers for important help with both content and tutorial presentation.

Emre Koksal, Tengo Saengudomlert, Shan-Yuan Ho, Manish Bhardwaj, Ashish Khisti, Etty Lee, and Emmanuel Abbe have all made major contributions to the text as teaching assistants for the MIT course. They have not only suggested new exercises and prepared solutions for others, but have also given me many insights about why certain material is difficult for some students, and suggested how to explain it better to avoid this confusion. The final test for clarity, of course, comes from the three or four hundred students who have taken the course over the last ten years, and I am grateful to them for looking puzzled when my explanations have failed and asking questions when I have been almost successful.

Finally, I am particularly grateful to my wife, Marie, for making our life a delight, even during the worst moments of writing yet another book.

1 Introduction to digital communication

Communication has been one of the deepest needs of the human race throughout recorded history. It is essential to forming social unions, to educating the young, and to expressing a myriad of emotions and needs. Good communication is central to a civilized society.

The various communication disciplines in engineering have the purpose of providing technological aids to human communication. One could view the smoke signals and drum rolls of primitive societies as being technological aids to communication, but communication technology as we view it today became important with telegraphy, then telephony, then video, then computer communication, and today the amazing mixture of all of these in inexpensive, small portable devices.

Initially these technologies were developed as separate networks and were viewed as having little in common. As these networks grew, however, the fact that all parts of a given network had to work together, coupled with the fact that different components were developed at different times using different design methodologies, caused an increased focus on the underlying principles and architectural understanding required for continued system evolution.

This need for basic principles was probably best understood at American Telephone and Telegraph (AT&T), where Bell Laboratories was created as the research and development arm of AT&T. The Math Center at Bell Labs became the predominant center for communication research in the world, and held that position until quite recently. The central core of the principles of communication technology were developed at that center.

Perhaps the greatest contribution from the Math Center was the creation of Information Theory [27] by Claude Shannon (Shannon, 1948). For perhaps the first 25 years of its existence, Information Theory was regarded as a beautiful theory but not as a central guide to the architecture and design of communication systems. After that time, however, both the device technology and the engineering understanding of the theory were sufficient to enable system development to follow information theoretic principles.

A number of information theoretic ideas and how they affect communication system design will be explained carefully in subsequent chapters. One pair of ideas, however, is central to almost every topic. The first is to view all communication sources, e.g., speech waveforms, image waveforms, and text files, as being representable by binary sequences. The second is to design communication systems that first convert the

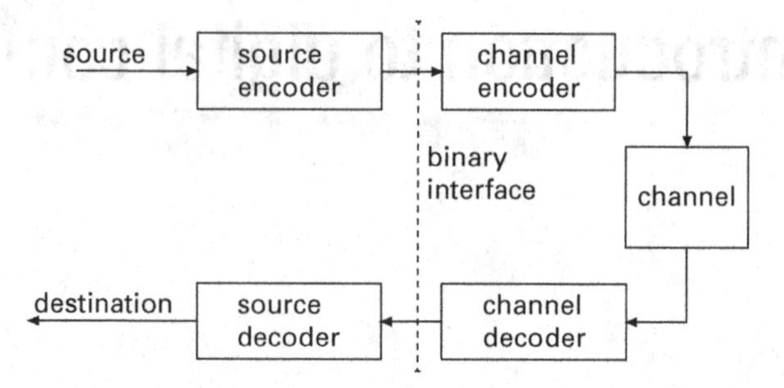

Figure 1.1. Placing a binary interface between source and channel. The source encoder converts the source output to a binary sequence and the channel encoder (often called a modulator) processes the binary sequence for transmission over the channel. The channel decoder (demodulator) recreates the incoming binary sequence (hopefully reliably), and the source decoder recreates the source output.

source output into a binary sequence and then convert that binary sequence into a form suitable for transmission over particular physical media such as cable, twisted wire pair, optical fiber, or electromagnetic radiation through space.

Digital communication systems, by definition, are communication systems that use such a digital[1] sequence as an interface between the source and the channel input (and similarly between the channel output and final destination) (see Figure 1.1).

The idea of converting an analog source output to a binary sequence was quite revolutionary in 1948, and the notion that this should be done before channel processing was even more revolutionary. Today, with digital cameras, digital video, digital voice, etc., the idea of digitizing any kind of source is commonplace even among the most technophobic. The notion of a binary interface before channel transmission is almost as commonplace. For example, we all refer to the speed of our Internet connection in bits per second.

There are a number of reasons why communication systems now usually contain a binary interface between source and channel (i.e., why digital communication systems are now standard). These will be explained with the necessary qualifications later, but briefly they are as follows.

- Digital hardware has become so cheap, reliable, and miniaturized that digital interfaces are eminently practical.
- A standardized binary interface between source and channel simplifies implementation and understanding, since source coding/decoding can be done independently of the channel, and, similarly, channel coding/decoding can be done independently of the source.

[1] A digital sequence is a sequence made up of elements from a finite alphabet (e.g. the binary digits $\{0, 1\}$, the decimal digits $\{0, 1, \ldots, 9\}$, or the letters of the English alphabet). The binary digits are almost universally used for digital communication and storage, so we only distinguish digital from binary in those few places where the difference is significant.

- A standardized binary interface between source and channel simplifies networking, which now reduces to sending binary sequences through the network.
- One of the most important of Shannon's information theoretic results is that if a source can be transmitted over a channel in any way at all, it can be transmitted using a binary interface between source and channel. This is known as the *source/channel separation theorem.*

In the remainder of this chapter, the problems of source coding and decoding and channel coding and decoding are briefly introduced. First, however, the notion of layering in a communication system is introduced. One particularly important example of layering was introduced in Figure 1.1, where source coding and decoding are viewed as one layer and channel coding and decoding are viewed as another layer.

1.1 Standardized interfaces and layering

Large communication systems such as the Public Switched Telephone Network (PSTN) and the Internet have incredible complexity, made up of an enormous variety of equipment made by different manufacturers at different times following different design principles. Such complex networks need to be based on some simple architectural principles in order to be understood, managed, and maintained. Two such fundamental architectural principles are *standardized interfaces* and *layering*.

A standardized interface allows the user or equipment on one side of the interface to ignore all details about the other side of the interface except for certain specified interface characteristics. For example, the binary interface[2] in Figure 1.1 allows the source coding/decoding to be done independently of the channel coding/decoding.

The idea of layering in communication systems is to break up communication functions into a string of separate layers, as illustrated in Figure 1.2.

Each layer consists of an input module at the input end of a communcation system and a "peer" output module at the other end. The input module at layer i processes the information received from layer $i+1$ and sends the processed information on to layer $i-1$. The peer output module at layer i works in the opposite direction, processing the received information from layer $i-1$ and sending it on to layer i.

As an example, an input module might receive a voice waveform from the next higher layer and convert the waveform into a binary data sequence that is passed on to the next lower layer. The output peer module would receive a binary sequence from the next lower layer at the output and convert it back to a speech waveform.

As another example, a *modem* consists of an input module (a modulator) and an output module (a demodulator). The modulator receives a binary sequence from the next higher input layer and generates a corresponding modulated waveform for transmission over a channel. The peer module is the remote demodulator at the other end of the channel. It receives a more or less faithful replica of the transmitted

[2] The use of a binary sequence at the interface is not quite enough to specify it, as will be discussed later.

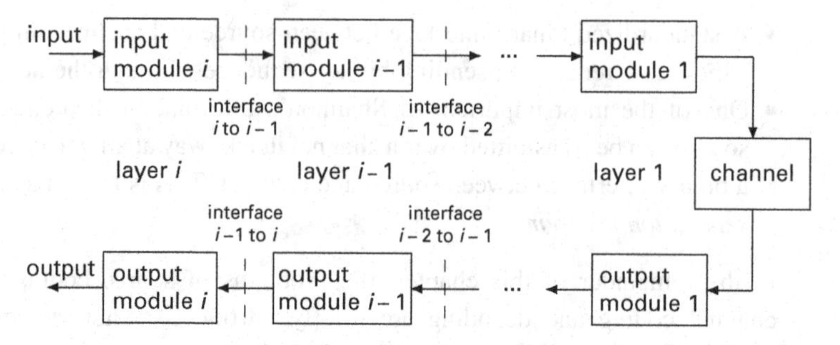

Figure 1.2. Layers and interfaces. The specification of the interface between layers i and $i-1$ should specify how input module i communicates with input module $i-1$, how the corresponding output modules communicate, and, most important, the input/output behavior of the system to the right of the interface. The designer of layer $i-1$ uses the input/output behavior of the layers to the right of $i-1$ to produce the required input/output performance to the right of layer i. Later examples will show how this multilayer process can simplify the overall system design.

waveform and reconstructs a typically faithful replica of the binary sequence. Similarly, the local demodulator is the peer to a remote modulator (often collocated with the remote demodulator above). Thus a modem is an input module for communication in one direction and an output module for independent communication in the opposite direction. Later chapters consider modems in much greater depth, including how noise affects the channel waveform and how that affects the reliability of the recovered binary sequence at the output. For now, however, it is enough simply to view the modulator as converting a binary sequence to a waveform, with the peer demodulator converting the waveform back to the binary sequence.

As another example, the source coding/decoding layer for a waveform source can be split into three layers, as shown in Figure 1.3. One of the advantages of this layering is that discrete sources are an important topic in their own right (discussed in Chapter 2) and correspond to the inner layer of Figure 1.3. Quantization is also an important topic in its own right (discussed in Chapter 3). After both of these are understood, waveform sources become quite simple to understand.

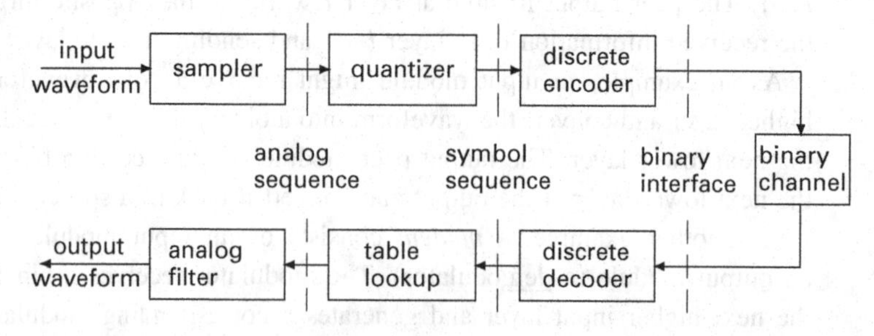

Figure 1.3. Breaking the source coding/decoding layer into three layers for a waveform source. The input side of the outermost layer converts the waveform into a sequence of samples and the output side converts the recovered samples back to the waveform. The quantizer then converts each sample into one of a finite set of symbols, and the peer module recreates the sample (with some distortion). Finally the inner layer encodes the sequence of symbols into binary digits.

The channel coding/decoding layer can also be split into several layers, but there are a number of ways to do this which will be discussed later. For example, binary error-correction coding/decoding can be used as an outer layer with modulation and demodulation as an inner layer, but it will be seen later that there are a number of advantages in combining these layers into what is called coded modulation.[3] Even here, however, layering is important, but the layers are defined differently for different purposes.

It should be emphasized that layering is much more than simply breaking a system into components. The input and peer output in each layer encapsulate all the lower layers, and all these lower layers can be viewed in aggregate as a communication channel. Similarly, the higher layers can be viewed in aggregate as a simple source and destination.

The above discussion of layering implicitly assumed a point-to-point communication system with one source, one channel, and one destination. Network situations can be considerably more complex. With broadcasting, an input module at one layer may have multiple peer output modules. Similarly, in multiaccess communication a multiplicity of input modules have a single peer output module. It is also possible in network situations for a single module at one level to interface with multiple modules at the next lower layer or the next higher layer. The use of layering is at least as important for networks as it is for point-to-point communications systems. The physical layer for networks is essentially the channel encoding/decoding layer discussed here, but textbooks on networks rarely discuss these physical layer issues in depth. The network control issues at other layers are largely separable from the physical layer communication issues stressed here. The reader is referred to Bertsekas and Gallager (1992), for example, for a treatment of these control issues.

The following three sections provide a fuller discussion of the components of Figure 1.1, i.e. of the fundamental two layers (source coding/decoding and channel coding/decoding) of a point-to-point digital communication system, and finally of the interface between them.

1.2 Communication sources

The source might be discrete, i.e. it might produce a sequence of discrete symbols, such as letters from the English or Chinese alphabet, binary symbols from a computer file, etc. Alternatively, the source might produce an analog waveform, such as a voice signal from a microphone, the output of a sensor, a video waveform, etc. Or, it might be a sequence of images such as X-rays, photographs, etc.

Whatever the nature of the source, the output from the source will be modeled as a sample function of a random process. It is not obvious why the inputs to communication

[3] Terminology is nonstandard here. A channel coder (including both coding and modulation) is often referred to (both here and elsewhere) as a modulator. It is also often referred to as a modem, although a modem is really a device that contains both modulator for communication in one direction and demodulator for communication in the other.

systems should be modeled as random, and in fact this was not appreciated before Shannon developed information theory in 1948.

The study of communication before 1948 (and much of it well after 1948) was based on Fourier analysis; basically one studied the effect of passing sine waves through various kinds of systems and components and viewed the source signal as a superposition of sine waves. Our study of channels will begin with this kind of analysis (often called Nyquist theory) to develop basic results about sampling, intersymbol interference, and bandwidth.

Shannon's view, however, was that if the recipient knows that a sine wave of a given frequency is to be communicated, why not simply regenerate it at the output rather than send it over a long distance? Or, if the recipient knows that a sine wave of unknown frequency is to be communicated, why not simply send the frequency rather than the entire waveform?

The essence of Shannon's viewpoint is that the set of possible source outputs, rather than any particular output, is of primary interest. The reason is that the communication system must be designed to communicate whichever one of these possible source outputs actually occurs. The objective of the communication system then is to transform each possible source output into a transmitted signal in such a way that these possible transmitted signals can be best distinguished at the channel output. A probability measure is needed on this set of possible source outputs to distinguish the typical from the atypical. This point of view drives the discussion of all components of communication systems throughout this text.

1.2.1 Source coding

The source encoder in Figure 1.1 has the function of converting the input from its original form into a sequence of bits. As discussed before, the major reasons for this almost universal conversion to a bit sequence are as follows: inexpensive digital hardware, standardized interfaces, layering, and the source/channel separation theorem.

The simplest source coding techniques apply to discrete sources and simply involve representing each successive source symbol by a sequence of binary digits. For example, letters from the 27-symbol English alphabet (including a SPACE symbol) may be encoded into 5-bit blocks. Since there are 32 distinct 5-bit blocks, each letter may be mapped into a distinct 5-bit block with a few blocks left over for control or other symbols. Similarly, upper-case letters, lower-case letters, and a great many special symbols may be converted into 8-bit blocks ("bytes") using the standard ASCII code.

Chapter 2 treats coding for discrete sources and generalizes the above techniques in many ways. For example, the input symbols might first be segmented into m-tuples, which are then mapped into blocks of binary digits. More generally, the blocks of binary digits can be generalized into variable-length sequences of binary digits. We shall find that any given discrete source, characterized by its alphabet and probabilistic description, has a quantity called *entropy* associated with it. Shannon showed that this source entropy is equal to the minimum number of binary digits per source symbol required to map the source output into binary digits in such a way that the source symbols may be retrieved from the encoded sequence.

Some discrete sources generate finite segments of symbols, such as email messages, that are statistically unrelated to other finite segments that might be generated at other times. Other discrete sources, such as the output from a digital sensor, generate a virtually unending sequence of symbols with a given statistical characterization. The simpler models of Chapter 2 will correspond to the latter type of source, but the discussion of universal source coding in Section 2.9 is sufficiently general to cover both types of sources and virtually any other kind of source.

The most straightforward approach to analog source coding is called analog to digital (A/D) conversion. The source waveform is first sampled at a sufficiently high rate (called the "Nyquist rate"). Each sample is then quantized sufficiently finely for adequate reproduction. For example, in standard voice telephony, the voice waveform is sampled 8000 times per second; each sample is then quantized into one of 256 levels and represented by an 8-bit byte. This yields a source coding bit rate of 64 kilobits per second (kbps).

Beyond the basic objective of conversion to bits, the source encoder often has the further objective of doing this as efficiently as possible – i.e. transmitting as few bits as possible, subject to the need to reconstruct the input adequately at the output. In this case source encoding is often called data compression. For example, modern speech coders can encode telephone-quality speech at bit rates of the order of 6–16 kbps rather than 64 kbps.

The problems of sampling and quantization are largely separable. Chapter 3 develops the basic principles of quantization. As with discrete source coding, it is possible to quantize each sample separately, but it is frequently preferable to segment the samples into blocks of n and then quantize the resulting n-tuples. As will be shown later, it is also often preferable to view the quantizer output as a discrete source output and then to use the principles of Chapter 2 to encode the quantized symbols. This is another example of layering.

Sampling is one of the topics in Chapter 4. The purpose of sampling is to convert the analog source into a sequence of real-valued numbers, i.e. into a discrete-time, analog-amplitude source. There are many other ways, beyond sampling, of converting an analog source to a discrete-time source. A general approach, which includes sampling as a special case, is to expand the source waveform into an orthonormal expansion and use the coefficients of that expansion to represent the source output. The theory of orthonormal expansions is a major topic of Chapter 4. It forms the basis for the signal space approach to channel encoding/decoding. Thus Chapter 4 provides us with the basis for dealing with waveforms for both sources and channels.

1.3 Communication channels

Next we discuss the channel and channel coding in a generic digital communication system.

In general, a channel is viewed as that part of the communication system between source and destination that is given and not under the control of the designer. Thus, to a source-code designer, the channel might be a digital channel with binary input and output; to a telephone-line modem designer, it might be a 4 kHz voice channel; to a cable modem designer, it might be a physical coaxial cable of up to a certain length, with certain bandwidth restrictions.

When the channel is taken to be the physical medium, the amplifiers, antennas, lasers, etc. that couple the encoded waveform to the physical medium might be regarded as part of the channel or as as part of the channel encoder. It is more common to view these coupling devices as part of the channel, since their design is quite separable from that of the rest of the channel encoder. This, of course, is another example of layering.

Channel encoding and decoding when the channel is the physical medium (either with or without amplifiers, antennas, lasers, etc.) is usually called *(digital) modulation* and *demodulation*, respectively. The terminology comes from the days of analog communication where modulation referred to the process of combining a lowpass signal waveform with a high-frequency sinusoid, thus placing the signal waveform in a frequency band appropriate for transmission and regulatory requirements. The analog signal waveform could modulate the amplitude, frequency, or phase, for example, of the sinusoid, but, in any case, the original waveform (in the absence of noise) could be retrieved by demodulation.

As digital communication has increasingly replaced analog communication, the modulation/demodulation terminology has remained, but now refers to the entire process of digital encoding and decoding. In most cases, the binary sequence is first converted to a baseband waveform and the resulting baseband waveform is converted to bandpass by the same type of procedure used for analog modulation. As will be seen, the challenging part of this problem is the conversion of binary data to baseband waveforms. Nonetheless, this entire process will be referred to as modulation and demodulation, and the conversion of baseband to passband and back will be referred to as frequency conversion.

As in the study of any type of system, a channel is usually viewed in terms of its possible inputs, its possible outputs, and a description of how the input affects the output. This description is usually probabilistic. If a channel were simply a linear time-invariant system (e.g. a filter), it could be completely characterized by its impulse response or frequency response. However, the channels here (and channels in practice) always have an extra ingredient – noise.

Suppose that there were no noise and a single input voltage level could be communicated exactly. Then, representing that voltage level by its infinite binary expansion, it would be possible in principle to transmit an infinite number of binary digits by transmitting a single real number. This is ridiculous in practice, of course, precisely because noise limits the number of bits that can be reliably distinguished. Again, it was Shannon, in 1948, who realized that noise provides the fundamental limitation to performance in communication systems.

The most common channel model involves a waveform input $X(t)$, an added noise waveform $Z(t)$, and a waveform output $Y(t) = X(t) + Z(t)$ that is the sum of the input

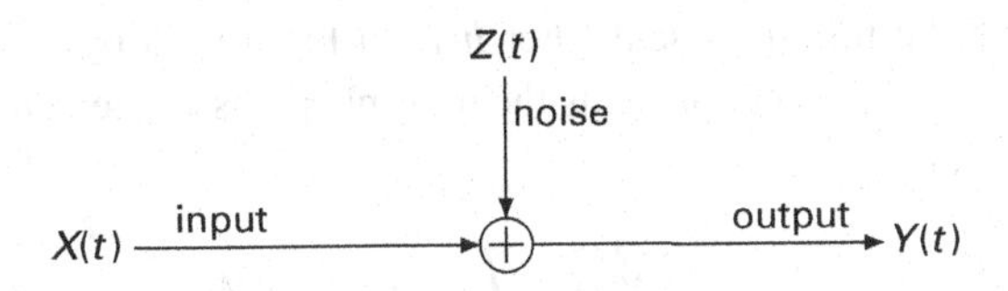

Figure 1.4. Additive white Gaussian noise (AWGN) channel.

and the noise, as shown in Figure 1.4. Each of these waveforms are viewed as random processes. Random processes are studied in Chapter 7, but for now they can be viewed intuitively as waveforms selected in some probabilitistic way. The noise $Z(t)$ is often modeled as white Gaussian noise (also to be studied and explained later). The input is usually constrained in power and bandwidth.

Observe that for any channel with input $X(t)$ and output $Y(t)$, the noise could be defined to be $Z(t) = Y(t) - X(t)$. Thus there must be something more to an additive-noise channel model than what is expressed in Figure 1.4. The additional required ingredient for noise to be called additive is that its probabilistic characterization does not depend on the input.

In a somewhat more general model, called a *linear Gaussian channel*, the input waveform $X(t)$ is first filtered in a linear filter with impulse response $h(t)$, and then independent white Gaussian noise $Z(t)$ is added, as shown in Figure 1.5, so that the channel output is given by

$$Y(t) = X(t) * h(t) + Z(t),$$

where "*" denotes convolution. Note that Y at time t is a function of X over a range of times, i.e.

$$Y(t) = \int_{-\infty}^{\infty} X(t-\tau)h(\tau)\mathrm{d}\tau + Z(t).$$

The linear Gaussian channel is often a good model for wireline communication and for line-of-sight wireless communication. When engineers, journals, or texts fail to describe the channel of interest, this model is a good bet.

The linear Gaussian channel is a rather poor model for non-line-of-sight mobile communication. Here, multiple paths usually exist from source to destination. Mobility of the source, destination, or reflecting bodies can cause these paths to change in time in a way best modeled as random. A better model for mobile communication is to

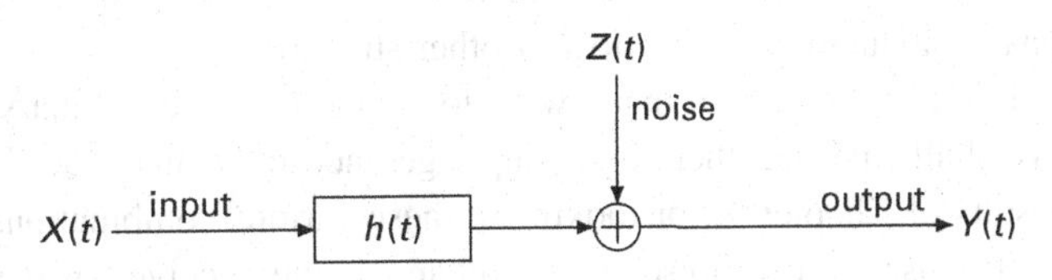

Figure 1.5. Linear Gaussian channel model.

replace the time-invariant filter $h(t)$ in Figure 1.5 by a randomly time varying linear filter, $H(t, \tau)$, that represents the multiple paths as they change in time. Here the output is given by

$$Y(t) = \int_{-\infty}^{\infty} X(t-u)H(u, t)\mathrm{d}u + Z(t).$$

These randomly varying channels will be studied in Chapter 9.

1.3.1 Channel encoding (modulation)

The channel encoder box in Figure 1.1 has the function of mapping the binary sequence at the source/channel interface into a channel waveform. A particularly simple approach to this is called binary pulse amplitude modulation (2-PAM). Let $\{u_1, u_2, \ldots, \}$ denote the incoming binary sequence, and let each $u_n = \pm 1$ (rather than the traditional 0/1). Let $p(t)$ be a given elementary waveform such as a rectangular pulse or a $\sin(\omega t)/\omega t$ function. Assuming that the binary digits enter at R bps, the sequence $u_1, u_2, \ldots$ is mapped into the waveform $\sum_n u_n p(t - n/R)$.

Even with this trivially simple modulation scheme, there are a number of interesting questions, such as how to choose the elementary waveform $p(t)$ so as to satisfy frequency constraints and reliably detect the binary digits from the received waveform in the presence of noise and intersymbol interference.

Chapter 6 develops the principles of modulation and demodulation. The simple 2-PAM scheme is generalized in many ways. For example, multilevel modulation first segments the incoming bits into m-tuples. There are $M = 2^m$ distinct m-tuples, and in M-PAM, each m-tuple is mapped into a different numerical value (such as $\pm 1, \pm 3, \pm 5, \pm 7$ for $M = 8$). The sequence $u_1, u_2, \ldots$ of these values is then mapped into the waveform $\sum_n u_n p(t - mn/R)$. Note that the rate at which pulses are sent is now m times smaller than before, but there are 2^m different values to be distinguished at the receiver for each elementary pulse.

The modulated waveform can also be a complex baseband waveform (which is then modulated up to an appropriate passband as a real waveform). In a scheme called quadrature amplitude modulation (QAM), the bit sequence is again segmented into m-tuples, but now there is a mapping from binary m-tuples to a set of $M = 2^m$ complex numbers. The sequence $u_1, u_2, \ldots$ of outputs from this mapping is then converted to the complex waveform $\sum_n u_n p(t - mn/R)$.

Finally, instead of using a fixed signal pulse $p(t)$ multiplied by a selection from M real or complex values, it is possible to choose M different signal pulses, $p_1(t), \ldots, p_M(t)$. This includes frequency shift keying, pulse position modulation, phase modulation, and a host of other strategies.

It is easy to think of many ways to map a sequence of binary digits into a waveform. We shall find that there is a simple geometric "signal-space" approach, based on the results of Chapter 4, for looking at these various combinations in an integrated way.

Because of the noise on the channel, the received waveform is different from the transmitted waveform. A major function of the demodulator is that of detection.

The detector attempts to choose which possible input sequence is most likely to have given rise to the given received waveform. Chapter 7 develops the background in random processes necessary to understand this problem, and Chapter 8 uses the geometric signal-space approach to analyze and understand the detection problem.

1.3.2 Error correction

Frequently the error probability incurred with simple modulation and demodulation techniques is too high. One possible solution is to separate the channel encoder into two layers: first an error-correcting code, then a simple modulator.

As a very simple example, the bit rate into the channel encoder could be reduced by a factor of three, and then each binary input could be repeated three times before entering the modulator. If at most one of the three binary digits coming out of the demodulator were incorrect, it could be corrected by majority rule at the decoder, thus reducing the error probability of the system at a considerable cost in data rate.

The scheme above (repetition encoding followed by majority-rule decoding) is a very simple example of error-correction coding. Unfortunately, with this scheme, small error probabilities are achieved only at the cost of very small transmission rates.

What Shannon showed was the very unintuitive fact that more sophisticated coding schemes can achieve arbitrarily low error probability at any data rate below a value known as the *channel capacity*. The channel capacity is a function of the probabilistic description of the output conditional on each possible input. Conversely, it is not possible to achieve low error probability at rates above the channel capacity. A brief proof of this *channel coding theorem* is given in Chapter 8, but readers should refer to texts on Information Theory such as Gallager (1968) and Cover and Thomas (2006) for detailed coverage.

The channel capacity for a bandlimited additive white Gaussian noise channel is perhaps the most famous result in information theory. If the input power is limited to P, the bandwidth limited to W, and the noise power per unit bandwidth is N_0, then the capacity (in bits per second) is given by

$$C = \mathsf{W}\log_2\left(1+\frac{P}{N_0\mathsf{W}}\right).$$

Only in the past few years have channel coding schemes been developed that can closely approach this channel capacity.

Early uses of error-correcting codes were usually part of a two-layer system similar to that above, where a digital error-correcting encoder is followed by a modulator. At the receiver, the waveform is first demodulated into a noisy version of the encoded sequence, and then this noisy version is decoded by the error-correcting decoder. Current practice frequently achieves better performance by combining error correction coding and modulation together in coded modulation schemes. Whether the error correction and traditional modulation are separate layers or combined, the combination is generally referred to as a modulator, and a device that does this modulation on data in one direction and demodulation in the other direction is referred to as a modem.

The subject of error correction has grown over the last 50 years to the point where complex and lengthy textbooks are dedicated to this single topic (see, for example, Lin and Costello (2004) and Forney (2005)). This text provides only an introduction to error-correcting codes.

Chapter 9, the final topic of the text, considers channel encoding and decoding for wireless channels. Considerable attention is paid here to modeling physical wireless media. Wireless channels are subject not only to additive noise, but also random fluctuations in the strength of multiple paths between transmitter and receiver. The interaction of these paths causes fading, and we study how this affects coding, signal selection, modulation, and detection. Wireless communication is also used to discuss issues such as channel measurement, and how these measurements can be used at input and output. Finally, there is a brief case study of CDMA (code division multiple access), which ties together many of the topics in the text.

1.4 Digital interface

The interface between the source coding layer and the channel coding layer is a sequence of bits. However, this simple characterization does not tell the whole story. The major complicating factors are as follows.

- Unequal rates: the rate at which bits leave the source encoder is often not perfectly matched to the rate at which bits enter the channel encoder.
- Errors: source decoders are usually designed to decode an exact replica of the encoded sequence, but the channel decoder makes occasional errors.
- Networks: encoded source outputs are often sent over networks, traveling serially over several channels; each channel in the network typically also carries the output from a number of different source encoders.

The first two factors above appear both in point-to-point communication systems and in networks. They are often treated in an *ad hoc* way in point-to-point systems, whereas they must be treated in a standardized way in networks. The third factor, of course, must also be treated in a standardized way in networks.

The usual approach to these problems in networks is to convert the superficially simple binary interface into multiple layers, as illustrated in Figure 1.6

How the layers in Figure 1.6 operate and work together is a central topic in the study of networks and is treated in detail in network texts such as Bertsekas and Gallager (1992). These topics are not considered in detail here, except for the very brief introduction to follow and a few comments as required later.

1.4.1 Network aspects of the digital interface

The output of the source encoder is usually segmented into packets (and in many cases, such as email and data files, is already segmented in this way). Each of the network layers then adds some overhead to these packets, adding a header in the case of TCP

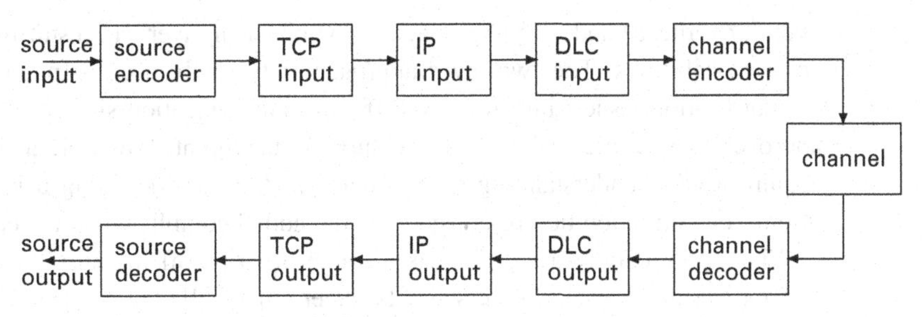

Figure 1.6. The replacement of the binary interface in Figure 1.5 with three layers in an oversimplified view of the internet. There is a TCP (transport control protocol) module associated with each source/destination pair; this is responsible for end-to-end error recovery and for slowing down the source when the network becomes congested. There is an IP (Internet protocol) module associated with each node in the network; these modules work together to route data through the network and to reduce congestion. Finally there is a DLC (data link control) module associated with each channel; this accomplishes rate matching and error recovery on the channel. In network terminology, the channel, with its encoder and decoder, is called the *physical layer.*

(transmission control protocol) and IP (internet protocol) and adding both a header and trailer in the case of DLC (data link control). Thus, what enters the channel encoder is a sequence of frames, where each frame has the structure illustrated in Figure 1.7.

These data frames, interspersed as needed by idle-fill, are strung together, and the resulting bit stream enters the channel encoder at its synchronous bit rate. The header and trailer supplied by the DLC must contain the information needed for the receiving DLC to parse the received bit stream into frames and eliminate the idle-fill.

The DLC also provides protection against decoding errors made by the channel decoder. Typically this is done by using a set of 16 or 32 parity checks in the frame trailer. Each parity check specifies whether a given subset of bits in the frame contains an even or odd number of 1s. Thus if errors occur in transmission, it is highly likely that at least one of these parity checks will fail in the receiving DLC. This type of DLC is used on channels that permit transmission in both directions. Thus, when an erroneous frame is detected, it is rejected and a frame in the opposite direction requests a retransmission of the erroneous frame. Thus the DLC header must contain information about frames traveling in both directions. For details about such protocols, see, for example, Bertsekas and Gallager (1992).

An obvious question at this point is why error correction is typically done both at the physical layer and at the DLC layer. Also, why is feedback (i.e. error detection and retransmission) used at the DLC layer and not at the physical layer? A partial answer is that, if the error correction is omitted at one of the layers, the error probability is increased. At the same time, combining both procedures (with the same

DLC header	IP header	TCP header	source encoded packet	DLC trailer

Figure 1.7. Structure of a data frame using the layers of Figure 1.6.

overall overhead) and using feedback at the physical layer can result in much smaller error probabilities. The two-layer approach is typically used in practice because of standardization issues, but, in very difficult communication situations, the combined approach can be preferable. From a tutorial standpoint, however, it is preferable to acquire a good understanding of channel encoding and decoding using transmission in only one direction before considering the added complications of feedback.

When the receiving DLC accepts a frame, it strips off the DLC header and trailer and the resulting packet enters the IP layer. In the IP layer, the address in the IP header is inspected to determine whether the packet is at its destination or must be forwarded through another channel. Thus the IP layer handles routing decisions, and also sometimes the decision to drop a packet if the queues at that node are too long.

When the packet finally reaches its destination, the IP layer strips off the IP header and passes the resulting packet with its TCP header to the TCP layer. The TCP module then goes through another error recovery phase,[4] much like that in the DLC module, and passes the accepted packets, without the TCP header, on to the destination decoder. The TCP and IP layers are also jointly responsible for congestion control, which ultimately requires the ability either to reduce the rate from sources as required or simply to drop sources that cannot be handled (witness dropped cell-phone calls).

In terms of sources and channels, these extra layers simply provide a sharper understanding of the digital interface between source and channel. That is, source encoding still maps the source output into a sequence of bits, and, from the source viewpoint, all these layers can simply be viewed as a channel to send that bit sequence reliably to the destination.

In a similar way, the input to a channel is a sequence of bits at the channel's synchronous input rate. The output is the same sequence, somewhat delayed and with occasional errors.

Thus both source and channel have digital interfaces, and the fact that these are slightly different because of the layering is, in fact, an advantage. The source encoding can focus solely on minimizing the output bit rate (perhaps with distortion and delay constraints) but can ignore the physical channel or channels to be used in transmission. Similarly the channel encoding can ignore the source and focus solely on maximizing the transmission bit rate (perhaps with delay and error rate constraints).

1.5 Supplementary reading

An excellent text that treats much of the material here with more detailed coverage but less depth is Proakis (2000). Another good general text is Wilson (1996). The classic work that introduced the signal space point of view in digital communication is

[4] Even after all these layered attempts to prevent errors, occasional errors are inevitable. Some are caught by human intervention, many do not make any real difference, and a final few have consequences. C'est la vie. The purpose of communication engineers and network engineers is not to eliminate all errors, which is not possible, but rather to reduce their probability as much as practically possible.

Wozencraft and Jacobs (1965). Good undergraduate treatments are provided in Proakis and Salehi (1994), Haykin (2002), and Pursley (2005).

Readers who lack the necessary background in probability should consult Ross (1994) or Bertsekas and Tsitsiklis (2002). More advanced treatments of probability are given in Ross (1996) and Gallager (1996). Feller (1968, 1971) still remains the classic text on probability for the serious student.

Further material on information theory can be found, for example, in Gallager (1968) and Cover and Thomas (2006). The original work by Shannon (1948) is fascinating and surprisingly accessible.

The field of channel coding and decoding has become an important but specialized part of most communication systems. We introduce coding and decoding in Chapter 8, but a separate treatment is required to develop the subject in depth. At MIT, the text here is used for the first of a two-term sequence, and the second term uses a polished set of notes by D. Forney (2005), available on the web. Alternatively, Lin and Costello (2004) is a good choice among many texts on coding and decoding.

Wireless communication is probably the major research topic in current digital communication work. Chapter 9 provides a substantial introduction to this topic, but a number of texts develop wireless communcation in much greater depth. Tse and Viswanath (2005) and Goldsmith (2005) are recommended, and Viterbi (1995) is a good reference for spread spectrum techniques.

2 Coding for discrete sources

2.1 Introduction

A general block diagram of a point-to-point digital communication system was given in Figure 1.1. The source encoder converts the sequence of symbols from the source to a sequence of binary digits, preferably using as few binary digits per symbol as possible. The source decoder performs the inverse operation. Initially, in the spirit of source/channel separation, we ignore the possibility that errors are made in the channel decoder and assume that the source decoder operates on the source encoder output.

We first distinguish between three important classes of sources.

- **Discrete sources** The output of a discrete source is a sequence of symbols from a known discrete alphabet $\mathcal{X}$. This alphabet could be the alphanumeric characters, the characters on a computer keyboard, English letters, Chinese characters, the symbols in sheet music (arranged in some systematic fashion), binary digits, etc. The discrete alphabets in this chapter are assumed to contain a finite set of symbols.[1]

  It is often convenient to view the sequence of symbols as occurring at some fixed rate in time, but there is no need to bring time into the picture (for example, the source sequence might reside in a computer file and the encoding can be done off-line).

  This chapter focuses on source coding and decoding for discrete sources. Supplementary references for source coding are given in Gallager (1968, chap. 3) and Cover and Thomas (2006, chap. 5). A more elementary partial treatment is given in Proakis and Salehi (1994, sect. 4.1–4.3).
- **Analog waveform sources** The output of an analog source, in the simplest case, is an analog real waveform, representing, for example, a speech waveform. The word analog is used to emphasize that the waveform can be arbitrary and is not restricted to taking on amplitudes from some discrete set of values.

  It is also useful to consider analog waveform sources with outputs that are complex functions of time; both real and complex waveform sources are discussed later.

  More generally, the output of an analog source might be an image (represented as an intensity function of horizontal/vertical location) or video (represented as

[1] A set is usually defined to be discrete if it includes either a finite or countably infinite number of members. The countably infinite case does not extend the basic theory of source coding in any important way, but it is occasionally useful in looking at limiting cases, which will be discussed as they arise.

an intensity function of horizontal/vertical location and time). For simplicity, we restrict our attention to analog waveforms, mapping a single real variable, time, into a real or complex-valued intensity.

- **Discrete-time sources with analog values (analog sequence sources)** These sources are halfway between discrete and analog sources. The source output is a sequence of real numbers (or perhaps complex numbers). Encoding such a source is of interest in its own right, but is of interest primarily as a subproblem in encoding analog sources. That is, analog waveform sources are almost invariably encoded by first either sampling the analog waveform or representing it by the coefficients in a series expansion. Either way, the result is a sequence of numbers, which is then encoded.

There are many differences between discrete sources and the latter two types of analog sources. The most important is that a discrete source can be, and almost always is, encoded in such a way that the source output can be uniquely retrieved from the encoded string of binary digits. Such codes are called *uniquely decodable*.[2] On the other hand, for analog sources, there is usually no way to map the source values to a bit sequence such that the source values are uniquely decodable. For example, an infinite number of binary digits is required for the exact specification of an arbitrary real number between 0 and 1. Thus, some sort of quantization is necessary for these analog values, and this introduces distortion. Source encoding for analog sources thus involves a trade-off between the bit rate and the amount of distortion.

Analog sequence sources are almost invariably encoded by first quantizing each element of the sequence (or more generally each successive n-tuple of sequence elements) into one of a finite set of symbols. This symbol sequence is a discrete sequence which can then be encoded into a binary sequence.

Figure 2.1 summarizes this layered view of analog and discrete source coding. As illustrated, discrete source coding is both an important subject in its own right

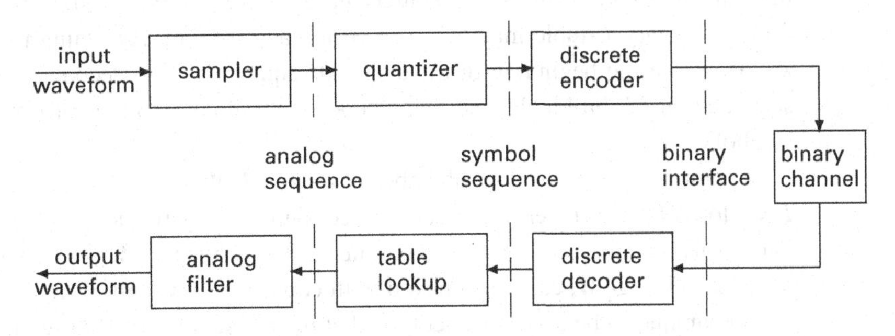

Figure 2.1. Discrete sources require only the inner layer above, whereas the inner two layers are used for analog sequences, and all three layers are used for waveform sources.

[2] Uniquely decodable codes are sometimes called noiseless codes in elementary treatments. *Uniquely decodable* captures both the intuition and the precise meaning far better than *noiseless*. Unique decodability is defined in Section 2.3.1.

for encoding text-like sources, but is also the inner layer in the encoding of analog sequences and waveforms.

The remainder of this chapter discusses source coding for discrete sources. The following chapter treats source coding for analog sequences, and Chapter 4 discusses waveform sources.

2.2 Fixed-length codes for discrete sources

The simplest approach to encoding a discrete source into binary digits is to create a code $\mathcal{C}$ that maps each symbol x of the alphabet $\mathcal{X}$ into a distinct codeword $\mathcal{C}(x)$, where $\mathcal{C}(x)$ is a block of binary digits. Each such block is restricted to have the same block length L, which is why such a code is called a *fixed-length code*.

For example, if the alphabet $\mathcal{X}$ consists of the seven symbols $\{a, b, c, d, e, f, g\}$, then the following fixed-length code of block length $L = 3$ could be used:

$$\begin{aligned}
\mathcal{C}(a) &= 000,\\
\mathcal{C}(b) &= 001,\\
\mathcal{C}(c) &= 010,\\
\mathcal{C}(d) &= 011,\\
\mathcal{C}(e) &= 100,\\
\mathcal{C}(f) &= 101,\\
\mathcal{C}(g) &= 110.
\end{aligned}$$

The source output, $x_1, x_2, \ldots,$ would then be encoded into the encoded output $\mathcal{C}(x_1)\mathcal{C}(x_2)\ldots$ and thus the encoded output contains L bits per source symbol. For the above example the source sequence *bad*... would be encoded into 001000011... Note that the output bits are simply run together (or, more technically, concatenated).

There are 2^L different combinations of values for a block of L bits. Thus, if the number of symbols in the source alphabet, $M = |\mathcal{X}|$, satisfies $M \leq 2^L$, then a different binary L-tuple may be assigned to each symbol. Assuming that the decoder knows where the beginning of the encoded sequence is, the decoder can segment the sequence into L-bit blocks and then decode each block into the corresponding source symbol.

In summary, if the source alphabet has size M, then this coding method requires $L = \lceil \log_2 M \rceil$ bits to encode each source symbol, where $\lceil w \rceil$ denotes the smallest integer greater than or equal to the real number w. Thus $\log_2 M \leq L < \log_2 M + 1$. The lowerbound, $\log_2 M$, can be achieved with equality if and only if M is a power of 2.

A technique to be used repeatedly is that of first segmenting the sequence of source symbols into successive blocks of n source symbols at a time. Given an alphabet $\mathcal{X}$ of M symbols, there are M^n possible n-tuples. These M^n n-tuples are regarded as the elements of a super-alphabet. Each n-tuple can be encoded rather than encoding the original symbols. Using fixed-length source coding on these n-tuples, each source n-tuple can be encoded into $L = \lceil \log_2 M^n \rceil$ bits. The rate $\overline{L} = L/n$ of encoded bits per original source symbol is then bounded by

$$\overline{L} = \frac{\lceil \log_2 M^n \rceil}{n} \geq \frac{n \log_2 M}{n} = \log_2 M;$$
$$\overline{L} = \frac{\lceil \log_2 M^n \rceil}{n} < \frac{n(\log_2 M) + 1}{n} = \log_2 M + \frac{1}{n}.$$

Thus $\log_2 M \leq \overline{L} < \log_2 M + 1/n$, and by letting n become sufficiently large the average number of coded bits per source symbol can be made arbitrarily close to $\log_2 M$, regardless of whether M is a power of 2.

Some remarks are necessary.

- This simple scheme to make $\overline{L}$ arbitrarily close to $\log_2 M$ is of greater theoretical interest than practical interest. As shown later, $\log_2 M$ is the minimum possible binary rate for uniquely decodable source coding if the source symbols are independent and equiprobable. Thus this scheme asymptotically approaches this minimum.
- This result begins to hint at why measures of information are logarithmic in the alphabet size.[3] The logarithm is usually taken to base 2 in discussions of binary codes. Henceforth $\log n$ means "$\log_2 n$."
- This method is nonprobabilistic; it takes no account of whether some symbols occur more frequently than others, and it works robustly regardless of the symbol frequencies. But if it is known that some symbols occur more frequently than others, then the rate $\overline{L}$ of coded bits per source symbol can be reduced by assigning shorter bit sequences to more common symbols in a *variable-length source code*. This will be our next topic.

2.3 Variable-length codes for discrete sources

The motivation for using variable-length encoding on discrete sources is the intuition that data compression can be achieved by mapping more probable symbols into shorter bit sequences and less likely symbols into longer bit sequences. This intuition was used in the Morse code of old-time telegraphy in which letters were mapped into strings of dots and dashes, using shorter strings for common letters and longer strings for less common letters.

A *variable-length code* $\mathcal{C}$ maps each source symbol a_j in a source alphabet $\mathcal{X} = \{a_1, \ldots, a_M\}$ to a binary string $\mathcal{C}(a_j)$, called a *codeword*. The number of bits in $\mathcal{C}(a_j)$ is called the *length* $l(a_j)$ of $\mathcal{C}(a_j)$. For example, a variable-length code for the alphabet $\mathcal{X} = \{a, b, c\}$ and its lengths might be given by

$$\begin{aligned} \mathcal{C}(a) &= 0, & l(a) &= 1; \\ \mathcal{C}(b) &= 10, & l(b) &= 2; \\ \mathcal{C}(c) &= 11, & l(c) &= 2. \end{aligned}$$

[3] The notion that information can be viewed as a logarithm of a number of possibilities was first suggested by Hartley (1928).

Successive codewords of a variable-length code are assumed to be transmitted as a continuing sequence of bits, with no demarcations of codeword boundaries (i.e., no commas or spaces). The source decoder, given an original starting point, must determine where the codeword boundaries are; this is called *parsing*.

A potential system issue with variable-length coding is the requirement for buffering. If source symbols arrive at a fixed rate and the encoded bit sequence must be transmitted at a fixed bit rate, then a buffer must be provided between input and output. This requires some sort of recognizable "fill" to be transmitted when the buffer is empty and the possibility of lost data when the buffer is full. There are many similar system issues, including occasional errors on the channel, initial synchronization, terminal synchronization, etc. Many of these issues are discussed later, but they are more easily understood after the more fundamental issues are discussed.

2.3.1 Unique decodability

The major property that is usually required from any variable-length code is that of *unique decodability*. This essentially means that, for any sequence of source symbols, that sequence can be reconstructed unambiguously from the encoded bit sequence. Here initial synchronization is assumed: the source decoder knows which is the first bit in the coded bit sequence. Note that without initial synchronization not even a fixed-length code can be uniquely decoded.

Clearly, unique decodability requires that $\mathcal{C}(a_j) \neq \mathcal{C}(a_i)$ for each $i \neq j$. More than that, however, it requires that strings[4] of encoded symbols be distinguishable. The following definition states this precisely.

Definition 2.3.1 A code $\mathcal{C}$ for a discrete source is uniquely decodable if, for any string of source symbols, say $x_1, x_2, \ldots, x_n$, the concatenation[5] of the corresponding codewords, $\mathcal{C}(x_1)\mathcal{C}(x_2)\cdots\mathcal{C}(x_n)$, differs from the concatenation of the codewords $\mathcal{C}(x'_1)\mathcal{C}(x'_2)\cdots\mathcal{C}(x'_m)$ for any other string $x'_1, x'_2, \ldots, x'_m$ of source symbols.

In other words, $\mathcal{C}$ is uniquely decodable if all concatenations of codewords are distinct.

Remember that there are no commas or spaces between codewords; the source decoder has to determine the codeword boundaries from the received sequence of bits. (If commas were inserted, the code would be ternary rather than binary.)

For example, the above code $\mathcal{C}$ for the alphabet $\mathcal{X} = \{a, b, c\}$ is soon shown to be uniquely decodable. However, the code $\mathcal{C}'$ defined by

$$\mathcal{C}'(a) = 0,$$
$$\mathcal{C}'(b) = 1,$$
$$\mathcal{C}'(c) = 01,$$

[4] A *string* of symbols is an n-tuple of symbols for any finite n. A *sequence* of symbols is an n-tuple in the limit $n \to \infty$, although the word sequence is also used when the length might be either finite or infinite.

[5] The concatenation of two strings, say $u_1 \cdots u_l$ and $v_1 \cdots v_{l'}$ is the combined string $u_1 \cdots u_l v_1 \cdots v_{l'}$.

is not uniquely decodable, even though the codewords are all different. If the source decoder observes 01, it cannot determine whether the source emitted $(a\,b)$ or (c).

Note that the property of unique decodability depends only on the set of codewords and not on the mapping from symbols to codewords. Thus we can refer interchangeably to uniquely decodable codes and uniquely decodable codeword sets.

2.3.2 Prefix-free codes for discrete sources

Decoding the output from a uniquely decodable code, and even determining whether it is uniquely decodable, can be quite complicated. However, there is a simple class of uniquely decodable codes called *prefix-free codes*. As shown later, these have the following advantages over other uniquely decodable codes.[6]

- If a uniquely decodable code exists with a certain set of codeword lengths, then a prefix-free code can easily be constructed with the same set of lengths.
- The decoder can decode each codeword of a prefix-free code immediately on the arrival of the last bit in that codeword.
- Given a probability distribution on the source symbols, it is easy to construct a prefix-free code of minimum expected length.

Definition 2.3.2 A *prefix* of a string $y_1 \cdots y_l$ is any initial substring $y_1 \cdots y_{l'}$, $l' \leq l$, of that string. The prefix is *proper* if $l' < l$. A code is *prefix-free* if no codeword is a prefix of any other codeword.

For example, the code $\mathcal{C}$ with codewords 0, 10, and 11 is prefix-free, but the code $\mathcal{C}'$ with codewords 0, 1, and 01 is not. Every fixed-length code with distinct codewords is prefix-free.

We will now show that every prefix-free code is uniquely decodable. The proof is constructive, and shows how the decoder can uniquely determine the codeword boundaries.

Given a prefix-free code $\mathcal{C}$, a corresponding *binary code tree* can be defined which grows from a root on the left to leaves on the right representing codewords. Each branch is labeled 0 or 1 and each node represents the binary string corresponding to the branch labels from the root to that node. The tree is extended just enough to include each codeword. That is, each node in the tree is either a codeword or proper prefix of a codeword (see Figure 2.2).

The prefix-free condition ensures that each codeword corresponds to a leaf node (i.e., a node with no adjoining branches going to the right). Each intermediate node (i.e., node having one or more adjoining branches going to the right) is a prefix of some codeword reached by traveling right from the intermediate node.

[6] With all the advantages of prefix-free codes, it is difficult to understand why the more general class is even discussed. This will become clearer much later.

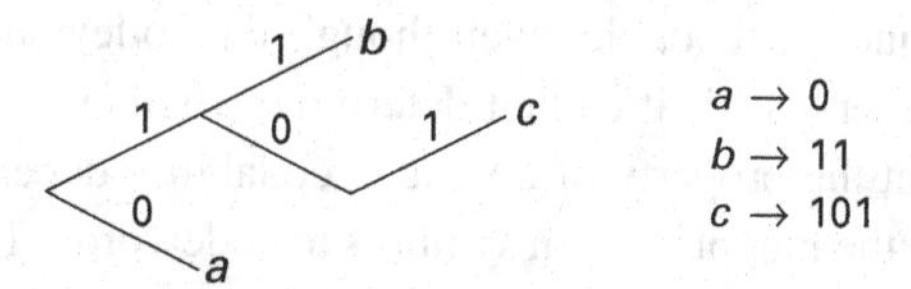

Figure 2.2. Binary code tree for a prefix-free code.

The tree in Figure 2.2 has an intermediate node, 10, with only one right-going branch. This shows that the codeword for c could be shortened to 10 without destroying the prefix-free property. This is shown in Figure 2.3.

A prefix-free code will be called *full* if no new codeword can be added without destroying the prefix-free property. As just seen, a prefix-free code is also full if no codeword can be shortened without destroying the prefix-free property. Thus the code of Figure 2.2 is not full, but that of Figure 2.3 is.

To see why the prefix-free condition guarantees unique decodability, consider the tree for the concatenation of two codewords. This is illustrated in Figure 2.4 for the code of Figure 2.3. This new tree has been formed simply by grafting a copy of the original tree onto each of the leaves of the original tree. Each concatenation of two codewords thus lies on a different node of the tree and also differs from each single codeword. One can imagine grafting further trees onto the leaves of Figure 2.4 to obtain a tree representing still more codewords concatenated together. Again all concatenations of codewords lie on distinct nodes, and thus correspond to distinct binary strings.

An alternative way to see that prefix-free codes are uniquely decodable is to look at the codeword parsing problem from the viewpoint of the source decoder. Given the encoded binary string for any string of source symbols, the source decoder can decode

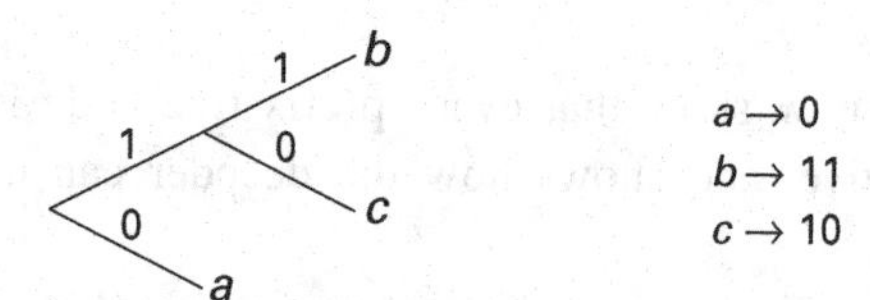

Figure 2.3. Code with shorter lengths than that of Figure 2.2.

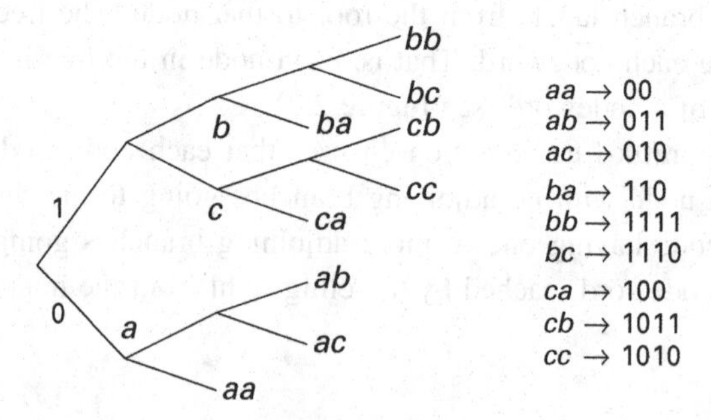

Figure 2.4. Binary code tree for two codewords; upward branches represent 1s.

the first symbol simply by reading the string from left to right and following the corresponding path in the code tree until it reaches a leaf, which must correspond to the first codeword by the prefix-free property. After stripping off the first codeword, the remaining binary string is again a string of codewords, so the source decoder can find the second codeword in the same way, and so on *ad infinitum*.

For example, suppose a source decoder for the code of Figure 2.3 decodes the sequence $1010011\cdots$. Proceeding through the tree from the left, it finds that 1 is not a codeword, but that 10 is the codeword for c. Thus c is decoded as the first symbol of the source output, leaving the string $10011\cdots$. Then c is decoded as the next symbol, leaving $011\cdots$, which is decoded into a and then b, and so forth.

This proof also shows that prefix-free codes can be decoded with no delay. As soon as the final bit of a codeword is received at the decoder, the codeword can be recognized and decoded without waiting for additional bits. For this reason, prefix-free codes are sometimes called instantaneous codes.

It has been shown that all prefix-free codes are uniquely decodable. The converse is not true, as shown by the following code:

$$\mathcal{C}(a) = 0,$$
$$\mathcal{C}(b) = 01,$$
$$\mathcal{C}(c) = 011.$$

An encoded sequence for this code can be uniquely parsed by recognizing 0 as the beginning of each new code word. A different type of example is given in Exercise 2.6.

With variable-length codes, if there are errors in data transmission, then the source decoder may lose codeword boundary synchronization and may make more than one symbol error. It is therefore important to study the synchronization properties of variable-length codes. For example, the prefix-free code $\{0, 10, 110, 1110, 11110\}$ is instantaneously self-synchronizing, because every 0 occurs at the end of a codeword. The shorter prefix-free code $\{0, 10, 110, 1110, 1111\}$ is probabilistically self-synchronizing; again, any observed 0 occurs at the end of a codeword, but since there may be a sequence of 1111 codewords of unlimited length, the length of time before resynchronization is a random variable. These questions are not pursued further here.

2.3.3 The Kraft inequality for prefix-free codes

The Kraft inequality (Kraft, 1949) is a condition determining whether it is possible to construct a prefix-free code for a given discrete source alphabet $\mathcal{X} = \{a_1, \ldots, a_M\}$ with a given set of codeword lengths $\{l(a_j); 1 \le j \le M\}$.

Theorem 2.3.1 (Kraft inequality for prefix-free codes) *Every prefix-free code for an alphabet $\mathcal{X} = \{a_1, \ldots, a_M\}$ with codeword lengths $\{l(a_j); 1 \le j \le M\}$ satisfies the following:*

$$\sum_{j=1}^{M} 2^{-l(a_j)} \leq 1. \tag{2.1}$$

Conversely, if (2.1) is satisfied, then a prefix-free code with lengths $\{l(a_j); 1 \leq j \leq M\}$ *exists. Moreover, every full prefix-free code satisfies (2.1) with equality and every nonfull prefix-free code satisfies it with strict inequality.*

For example, this theorem implies that there exists a full prefix-free code with codeword lengths $\{1, 2, 2\}$ (two such examples have already been given), but there exists no prefix-free code with codeword lengths $\{1, 1, 2\}$.

Before proving Theorem 2.3.1, we show how to represent codewords as base 2 expansions (the base 2 analog of base 10 decimals) in the binary number system. After understanding this representation, the theorem will be almost obvious. The base 2 expansion $.y_1 y_2 \cdots y_l$ represents the rational number $\sum_{m=1}^{l} y_m 2^{-m}$. For example, .011 represents $1/4 + 1/8$.

Ordinary decimals with l digits are frequently used to indicate an approximation of a real number to l places of accuracy. Here, in the same way, the base 2 expansion $.y_1 y_2 \cdots y_l$ is viewed as "covering" the interval[7] $[\sum_{m=1}^{l} y_m 2^{-m}, \sum_{m=1}^{l} y_m 2^{-m} + 2^{-l})$. This interval has size 2^{-l} and includes all numbers whose base 2 expansions start with $.y_1 \ldots y_l$.

In this way, any codeword $\mathcal{C}(a_j)$ of length l is represented by a rational number in the interval [0, 1) and covers an interval of size 2^{-l}, which includes all strings that contain $\mathcal{C}(a_j)$ as a prefix (see Figure 2.5). The proof of Theorem 2.1 follows.

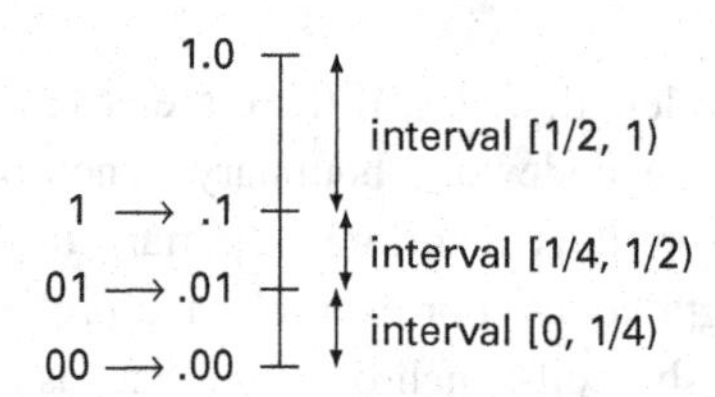

Figure 2.5. Base 2 expansion numbers and intervals representing codewords. The codewords represented above are (00, 01, and 1).

Proof First, assume that $\mathcal{C}$ is a prefix-free code with codeword lengths $\{l(a_j), 1 \leq j \leq M\}$. For any distinct a_j and a_i in $\mathcal{X}$, it was shown above that the base 2 expansion corresponding to $\mathcal{C}(a_j)$ cannot lie in the interval corresponding to $\mathcal{C}(a_i)$ since $\mathcal{C}(a_i)$ is not a prefix of $\mathcal{C}(a_j)$. Thus the lower end of the interval corresponding to any codeword $\mathcal{C}(a_j)$ cannot lie in the interval corresponding to any other codeword. Now, if two of these intervals intersect, then the lower end of one of them must lie

[7] Brackets and parentheses, respectively, are used to indicate closed and open boundaries; thus the interval $[a, b)$ means the set of real numbers u such that $a \leq u < b$.

in the other, which is impossible. Thus the two intervals must be disjoint and thus the set of all intervals associated with the codewords are disjoint. Since all these intervals are contained in the interval $[0, 1)$, and the size of the interval corresponding to $\mathcal{C}(a_j)$ is $2^{-l(a_j)}$, (2.1) is established.

Next note that if (2.1) is satisfied with strict inequality, then some interval exists in $[0, 1)$ that does not intersect any codeword interval; thus another codeword can be "placed" in this interval and the code is not full. If (2.1) is satisfied with equality, then the intervals fill up $[0, 1)$. In this case no additional codeword can be added and the code is full.

Finally we show that a prefix-free code can be constructed from any desired set of codeword lengths $\{l(a_j), 1 \le j \le M\}$ for which (2.1) is satisfied. Put the set of lengths in nondecreasing order, $l_1 \le l_2 \le \cdots \le l_M$, and let $u_1, \ldots, u_M$ be the real numbers corresponding to the codewords in the construction to be described. The construction is quite simple: $u_1 = 0$, and for all j, $1 < j \le M$:

$$u_j = \sum_{i=1}^{j-1} 2^{-l_i}. \tag{2.2}$$

Each term on the right is an integer multiple of 2^{-l_j}, so u_j is also an integer multiple of 2^{-l_j}. From (2.1), $u_j < 1$, so u_j can be represented by a base 2 expansion with l_j places. The corresponding codeword of length l_j can be added to the code while preserving prefix-freedom (see Figure 2.6). □

Some final remarks on the Kraft inequality are in order.

- Just because a code has lengths that satisfy (2.1), it does not follow that the code is prefix-free, or even uniquely decodable.
- Exercise 2.11 shows that Theorem 2.3.1 also holds for all uniquely decodable codes; i.e., there exists a uniquely decodable code with codeword lengths $\{l(a_j), 1 \le j \le M\}$ if and only if (2.1) holds. This will imply that if a uniquely decodable code exists with a certain set of codeword lengths, then a prefix-free code exists with the same set of lengths. So why use any code other than a prefix-free code?

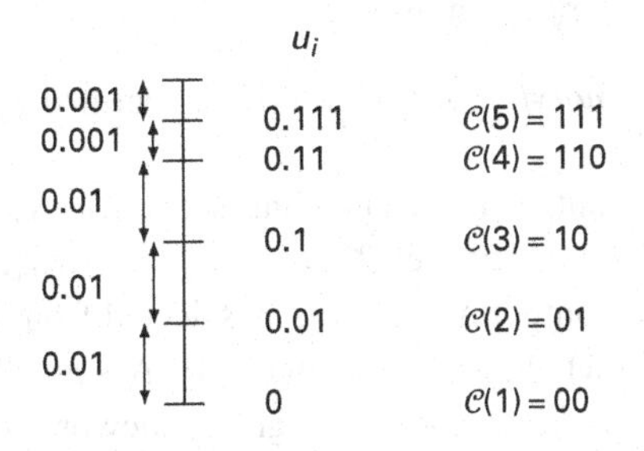

Figure 2.6. Construction of codewords for the set of lengths $\{2, 2, 2, 3, 3\}$; $\mathcal{C}(i)$ is formed from u_i by representing u_i to l_i places.

2.4 Probability models for discrete sources

It was shown above that prefix-free codes exist for any set of codeword lengths satisfying the Kraft inequality. When is it desirable to use one of these codes; i.e., when is the expected number of coded bits per source symbol less than $\log M$ and why is the expected number of coded bits per source symbol the primary parameter of importance?

These questions cannot be answered without a probabilistic model for the source. For example, the $M = 4$ prefix-free set of codewords $\{0, 10, 110, 111\}$ has an expected length $2.25 > 2 = \log M$ if the source symbols are equiprobable, but if the source symbol probabilities are $\{1/2, 1/4, 1/8, 1/8\}$, then the expected length is $1.75 < 2$.

The discrete sources that one meets in applications usually have very complex statistics. For example, consider trying to compress email messages. In typical English text, some letters, such as e and o, occur far more frequently than q, x, and z. Moreover, the letters are not independent; for example, h is often preceded by t, and q is almost always followed by u. Next, some strings of letters are words, while others are not; those that are not have probability near 0 (if in fact the text is correct English). Over longer intervals, English has grammatical and semantic constraints, and over still longer intervals, such as over multiple email messages, there are still further constraints.

It should be clear therefore that trying to find an accurate probabilistic model of a real-world discrete source is not going to be a productive use of our time. An alternative approach, which has turned out to be very productive, is to start out by trying to understand the encoding of "toy" sources with very simple probabilistic models. After studying such toy sources, it will be shown how to generalize to source models with more and more general structure, until, presto, real sources can be largely understood even without good stochastic models. This is a good example of a problem where having the patience to look carefully at simple and perhaps unrealistic models pays off handsomely in the end.

The type of toy source that will now be analyzed in some detail is called a discrete memoryless source.

2.4.1 Discrete memoryless sources

A *discrete memoryless source* (DMS) is defined by the following properties.

- The source output is an unending sequence, $X_1, X_2, X_3, \ldots$, of randomly selected symbols from a *finite* set $\mathcal{X} = \{a_1, a_2, \ldots, a_M\}$, called the *source alphabet.*
- Each source output $X_1, X_2, \ldots$ is selected from $\mathcal{X}$ using the same probability mass function (pmf) $\{p_X(a_1), \ldots, p_X(a_M)\}$. Assume that $p_X(a_j) > 0$ for all j, $1 \le j \le M$, since there is no reason to assign a codeword to a symbol of zero probability and no reason to model a discrete source as containing impossible symbols.
- Each source output X_k is statistically independent of the previous outputs $X_1, \ldots, X_{k-1}$.

The randomly chosen symbols coming out of the source are called *random symbols*. They are very much like random variables except that they may take on nonnumeric values. Thus, if X denotes the result of a fair coin toss, then it can be modeled as a random symbol that takes values in the set {HEADS, TAILS} with equal probability. Note that if X is a nonnumeric random symbol, then it makes no sense to talk about its expected value. However, the notion of statistical independence between random symbols is the same as that for random variables, i.e. the event that X_i is any given element of $\mathcal{X}$ is independent of the events corresponding to the values of the other random symbols.

The word memoryless in the definition refers to the statistical independence between different random symbols, i.e. each variable is chosen with no memory of how the previous random symbols were chosen. In other words, the source symbol sequence is independent and identically distributed (iid).[8]

In summary, a DMS is a semi-infinite iid sequence of random symbols,

$$X_1, X_2, X_3, \ldots,$$

each drawn from the finite set $\mathcal{X}$, each element of which has positive probability.

A sequence of independent tosses of a biased coin is one example of a DMS. The sequence of symbols drawn (with replacement) in a Scrabble™ game is another. The reason for studying these sources is that they provide the tools for studying more realistic sources.

2.5 Minimum $\overline{L}$ for prefix-free codes

The Kraft inequality determines which sets of codeword lengths are possible for prefix-free codes. Given a discrete memoryless source (DMS), we want to determine what set of codeword lengths can be used to *minimize* the expected length of a prefix-free code for that DMS. That is, we want to minimize the expected length subject to the Kraft inequality.

Suppose a set of lengths $l(a_1), \ldots, l(a_M)$ (subject to the Kraft inequality) is chosen for encoding each symbol into a prefix-free codeword. Define $L(X)$ (or more briefly L) as a random variable representing the codeword length for the randomly selected source symbol. The expected value of L for the given code is then given by

$$\overline{L} = \mathsf{E}[L] = \sum_{j=1}^{M} l(a_j) p_X(a_j).$$

We want to find $\overline{L}_{\min}$, which is defined as the minimum value of $\overline{L}$ over all sets of codeword lengths satisfying the Kraft inequality.

Before finding $\overline{L}_{\min}$, we explain why this quantity is of interest. The number of bits resulting from using the above code to encode a long block $\boldsymbol{X} = (X_1, X_2, \ldots, X_n)$ of

[8] Do not confuse this notion of memorylessness with any nonprobabalistic notion in system theory.

symbols is $S_n = L(X_1)+L(X_2)+\cdots+L(X_n)$. This is a sum of n iid random variables (rvs), and the law of large numbers, which is discussed in Section 2.7.1, implies that S_n/n, the number of bits per symbol in this long block, is very close to $\overline{L}$ with probability very close to 1. In other words, $\overline{L}$ is essentially the rate (in bits per source symbol) at which bits come out of the source encoder. This motivates the objective of finding $\overline{L}_{\min}$ and later of finding codes that achieve the minimum.

Before proceeding further, we simplify our notation. We have been carrying along a completely arbitrary finite alphabet $\mathcal{X} = \{a_1, \ldots, a_M\}$ of size $M = |\mathcal{X}|$, but this problem (along with most source coding problems) involves only the probabilities of the M symbols and not their names. Thus we define the source alphabet to be $\{1, 2, \ldots, M\}$, we denote the symbol probabilities by $p_1, \ldots, p_M$, and we denote the corresponding codeword lengths by $l_1, \ldots, l_M$. The expected length of a code is then given by

$$\overline{L} = \sum_{j=1}^{M} l_j p_j.$$

Mathematically, the problem of finding $\overline{L}_{\min}$ is that of minimizing $\overline{L}$ over all sets of integer lengths $l_1, \ldots, l_M$ subject to the Kraft inequality:

$$\overline{L}_{\min} = \min_{l_1,\ldots,l_M:\sum_j 2^{-l_j}\le 1} \left\{ \sum_{j=1}^{M} p_j l_j \right\}. \tag{2.3}$$

2.5.1 Lagrange multiplier solution for the minimum $\overline{L}$

The minimization in (2.3) is over a function of M variables, $l_1, \ldots, l_M$, subject to constraints on those variables. Initially, consider a simpler problem where there are no integer constraint on the l_j. This simpler problem is then to minimize $\sum_j p_j l_j$ over all real values of $l_1, \ldots, l_M$ subject to $\sum_j 2^{-l_j} \le 1$. The resulting minimum is called $\overline{L}_{\min}$(noninteger).

Since the allowed values for the lengths in this minimization include integer lengths, it is clear that $\overline{L}_{\min}(\text{noninteger}) \le \overline{L}_{\min}$. This noninteger minimization will provide a number of important insights about the problem, so its usefulness extends beyond just providing a lowerbound on $\overline{L}_{\min}$.

Note first that the minimum of $\sum_j l_j p_j$ subject to $\sum_j 2^{-l_j} \le 1$ must occur when the constraint is satisfied with equality; otherwise, one of the l_j could be reduced, thus reducing $\sum_j p_j l_j$ without violating the constraint. Thus the problem is to minimize $\sum_j p_j l_j$ subject to $\sum_j 2^{-l_j} = 1$.

Problems of this type are often solved by using a Lagrange multiplier. The idea is to replace the minimization of one function, subject to a constraint on another function, by the minimization of a linear combination of the two functions, in this case the minimization of

$$\sum_j p_j l_j + \lambda \sum_j 2^{-l_j}. \tag{2.4}$$

If the method works, the expression can be minimized for each choice of λ (called a *Lagrange multiplier*); λ can then be chosen so that the optimizing choice of $l_1, \ldots, l_M$

satisfies the constraint. The minimizing value of (2.4) is then $\sum_j p_j l_j + \lambda$. This choice of $l_1, \ldots, l_M$ minimizes the original constrained optimization, since for any $l'_1, \ldots, l'_M$ that satisfies the constraint $\sum_j 2^{-l'_j} = 1$, the expression in (2.4) is $\sum_j p_j l'_j + \lambda$, which must be greater than or equal to $\sum_j p_j l_j + \lambda$.

We can attempt[9] to minimize (2.4) simply by setting the derivitive with respect to each l_j equal to 0. This yields

$$p_j - \lambda(\ln 2)2^{-l_j} = 0; \quad 1 \le j \le M. \tag{2.5}$$

Thus, $2^{-l_j} = p_j/(\lambda \ln 2)$. Since $\sum_j p_j = 1$, λ must be equal to $1/\ln 2$ in order to satisfy the constraint $\sum_j 2^{-l_j} = 1$. Then $2^{-l_j} = p_j$, or equivalently $l_j = -\log p_j$. It will be shown shortly that this stationary point actually achieves a minimum. Substituting this solution into (2.3), we obtain

$$\overline{L}_{\min}(\text{noninteger}) = -\sum_{j=1}^{M} p_j \log p_j. \tag{2.6}$$

The quantity on the right side of (2.6) is called the *entropy*[10] of X, and is denoted $\mathsf{H}[X]$. Thus

$$\mathsf{H}[X] = -\sum_j p_j \log p_j.$$

In summary, the entropy $\mathsf{H}[X]$ is a lowerbound to $\overline{L}$ for prefix-free codes and this lowerbound is achieved when $l_j = -\log p_j$ for each j. The bound was derived by ignoring the integer constraint, and can be met only if $-\log p_j$ is an integer for each j; i.e., if each p_j is a power of 2.

2.5.2 Entropy bounds on $\overline{L}$

We now return to the problem of minimizing $\overline{L}$ with an integer constraint on lengths. The following theorem both establishes the correctness of the previous noninteger optimization and provides an upperbound on $\overline{L}_{\min}$.

Theorem 2.5.1 (Entropy bounds for prefix-free codes) *Let X be a discrete random symbol with symbol probabilities $p_1, \ldots, p_M$. Let $\overline{L}_{\min}$ be the minimum expected codeword length over all prefix-free codes for X. Then*

$$\mathsf{H}[X] \le \overline{L}_{\min} < \mathsf{H}[X] + 1 \quad \text{bit/symbol}. \tag{2.7}$$

Furthermore, $\overline{L}_{\min} = \mathsf{H}[X]$ if and only if each probability p_j is an integer power of 2.

---

[9] There are well known rules for when the Lagrange multiplier method works and when it can be solved simply by finding a stationary point. The present problem is so simple, however, that this machinery is unnecessary.

[10] Note that X is a random symbol and carries with it all of the accompanying baggage, including a pmf. The entropy $\mathsf{H}[X]$ is a numerical function of the random symbol including that pmf; in the same way, $\mathsf{E}[L]$ is a numerical function of the rv L. Both $\mathsf{H}[X]$ and $\mathsf{E}[L]$ are expected values of particular rvs, and braces are used as a mnemonic reminder of this. In distinction, $L(X)$ above is a rv in its own right; it is based on some function $l(x)$ mapping $\mathcal{X} \to \mathbb{R}$ and takes the sample value $l(x)$ for all sample points such that $X = x$.

Proof It is first shown that $\mathsf{H}[X] \le \overline{L}$ for all prefix-free codes. Let $l_1, \ldots, l_M$ be the codeword lengths of an arbitrary prefix-free code. Then

$$\mathsf{H}[X] - \overline{L} = \sum_{j=1}^{M} p_j \log\frac{1}{p_j} - \sum_{j=1}^{M} p_j l_j = \sum_{j=1}^{M} p_j \log\frac{2^{-l_j}}{p_j}, \tag{2.8}$$

where $\log 2^{-l_j}$ has been substituted for $-l_j$.

We now use the very useful inequality $\ln u \le u - 1$, or equivalently $\log u \le (\log e)(u-1)$, which is illustrated in Figure 2.7. Note that equality holds only at the point $u = 1$.

Substituting this inequality in (2.8) yields

$$\mathsf{H}[X] - \overline{L} \le (\log e) \sum_{j=1}^{M} p_j \left(\frac{2^{-l_j}}{p_j} - 1\right) = (\log e)\left(\sum_{j=1}^{M} 2^{-l_j} - \sum_{j=1}^{M} p_j\right) \le 0, \tag{2.9}$$

where the Kraft inequality and $\sum_j p_j = 1$ have been used. This establishes the left side of (2.7). The inequality in (2.9) is strict unless $2^{-l_j}/p_j = 1$, or equivalently $l_j = -\log p_j$, for all j. For integer l_j, this can be satisfied with equality if and only if p_j is an integer power of 2 for all j. For arbitrary real values of l_j, this proves that (2.5) minimizes (2.3) without the integer constraint, thus verifying (2.6).

To complete the proof, it will be shown that a prefix-free code exists with $\overline{L} < \mathsf{H}[X] + 1$. Choose the codeword lengths to be

$$l_j = \lceil -\log p_j \rceil,$$

where the ceiling notation $\lceil u \rceil$ denotes the smallest integer less than or equal to u. With this choice,

$$-\log p_j \le l_j < -\log p_j + 1. \tag{2.10}$$

Since the left side of (2.10) is equivalent to $2^{-l_j} \le p_j$, the Kraft inequality is satisfied:

$$\sum_j 2^{-l_j} \le \sum_j p_j = 1.$$

Thus a prefix-free code exists with the above lengths. From the right side of (2.10), the expected codeword length of this code is upperbounded by

$$\overline{L} = \sum_j p_j l_j < \sum_j p_j(-\log p_j + 1) = \mathsf{H}[X] + 1.$$

Since $\overline{L}_{\min} \le \overline{L}$, $\overline{L}_{\min} < \mathsf{H}[X] + 1$, completing the proof. □

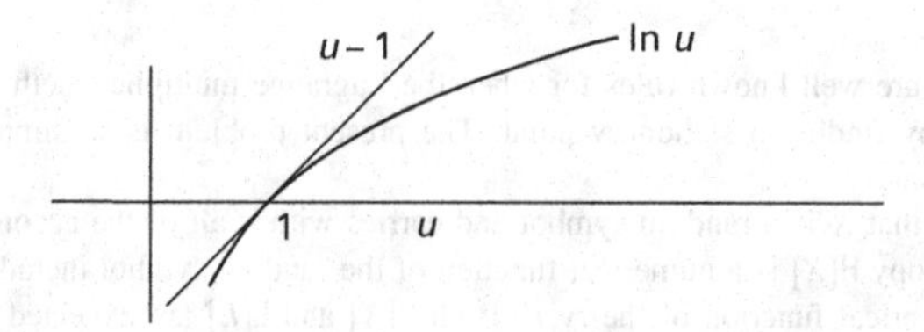

Figure 2.7. Inequality $\ln u \le u - 1$. The inequality is strict except at $u = 1$.

Both the proof above and the noninteger minimization in (2.6) suggest that the optimal length of a codeword for a source symbol of probability p_j should be approximately $-\log p_j$. This is not quite true, because, for example, if $M = 2$ and $p_1 = 2^{-20}$, $p_2 = 1 - 2^{-20}$, then $-\log p_1 = 20$, but the optimal l_1 is 1. However, the last part of the above proof shows that if each l_i is chosen as an integer approximation to $-\log p_i$, then $\overline{L}$ is at worst within one bit of $\mathsf{H}[X]$.

For sources with a small number of symbols, the upperbound in the theorem appears to be too loose to have any value. When these same arguments are applied later to long blocks of source symbols, however, the theorem leads directly to the source coding theorem.

2.5.3 Huffman's algorithm for optimal source codes

In the very early days of information theory, a number of heuristic algorithms were suggested for choosing codeword lengths l_j to approximate $-\log p_j$. Both Claude Shannon and Robert Fano had suggested such heuristic algorithms by 1948. It was conjectured at that time that, since this was an integer optimization problem, its optimal solution would be quite difficult. It was quite a surprise therefore when David Huffman came up with a very simple and straightforward algorithm for constructing optimal (in the sense of minimal $\overline{L}$) prefix-free codes (Huffman, 1952). Huffman developed the algorithm in 1950 as a term paper in Robert Fano's information theory class at MIT.

Huffman's trick, in today's jargon, was to "think outside the box." He ignored the Kraft inequality and looked at the binary code tree to establish properties that an optimal prefix-free code should have. After discovering a few simple properties, he realized that they led to a simple recursive procedure for constructing an optimal code.

The simple examples in Figure 2.8 illustrate some key properties of optimal codes. After stating these properties precisely, the Huffman algorithm will be almost obvious.

The property of the length assignments in the three-word example above can be generalized as follows: the longer the codeword, the less probable the corresponding symbol must be. We state this more precisely in Lemma 2.5.1.

Lemma 2.5.1 *Optimal codes have the property that if $p_i > p_j$, then $l_i \le l_j$.*

Proof Assume to the contrary that a code has $p_i > p_j$ and $l_i > l_j$. The terms involving symbols i and j in $\overline{L}$ are $p_i l_i + p_j l_j$. If the two codewords are interchanged, thus interchanging l_i and l_j, this sum decreases, i.e.

$$(p_i l_i + p_j l_j) - (p_i l_j + p_j l_i) = (p_i - p_j)(l_i - l_j) > 0.$$

Thus $\overline{L}$ decreases, so any code with $p_i > p_j$ and $l_i > l_j$ is nonoptimal. □

An even simpler property of an optimal code is given in Lemma 2.5.2.

Lemma 2.5.2 *Optimal prefix-free codes have the property that the associated code tree is full.*

Proof If the tree is not full, then a codeword length could be reduced (see Figures 2.2 and 2.3). □

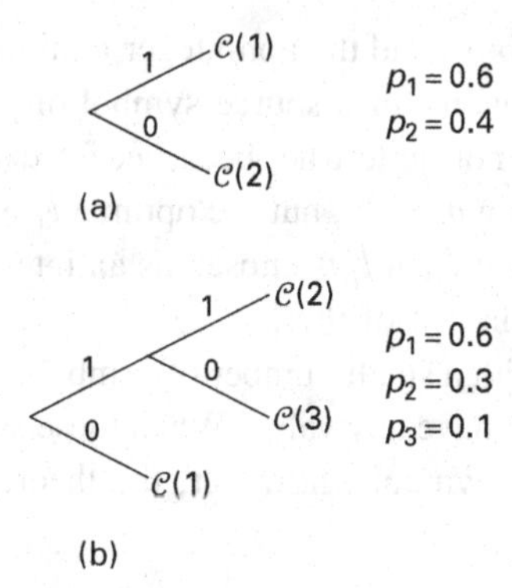

Figure 2.8. Some simple optimal codes. (a) With two symbols, the optimal codeword lengths are 1 and 1. (b) With three symbols, the optimal lengths are 1, 2, and 2. The least likely symbols are assigned words of length 2.

Define the *sibling* of a codeword as the binary string that differs from the codeword in only the final digit. A sibling in a full code tree can be either a codeword or an intermediate node of the tree.

Lemma 2.5.3 *Optimal prefix-free codes have the property that, for each of the longest codewords in the code, the sibling of that codeword is another longest codeword.*

Proof A sibling of a codeword of maximal length cannot be a prefix of a longer codeword. Since it cannot be an intermediate node of the tree, it must be a codeword. □

For notational convenience, assume that the $M = |\mathcal{X}|$ symbols in the alphabet are ordered so that $p_1 \geq p_2 \geq \cdots \geq p_M$.

Lemma 2.5.4 *Let X be a random symbol with a pmf satisfying $p_1 \geq p_2 \geq \cdots \geq p_M$. There is an optimal prefix-free code for X in which the codewords for $M-1$ and M are siblings and have maximal length within the code.*

Proof There are finitely many codes satisfying the Kraft inequality with equality,[11] so consider a particular one that is optimal. If $p_M < p_j$ for each $j < M$, then, from Lemma 2.5.1, $l_M \geq l_j$ for each and l_M has maximal length. If $p_M = p_j$ for one or more $j < M$, then l_j must be maximal for at least one such j. Then if l_M is not maximal, $\mathcal{C}(j)$ and $\mathcal{C}(M)$ can be interchanged with no loss of optimality, after which l_M is maximal. Now if $\mathcal{C}(k)$ is the sibling of $\mathcal{C}(M)$ in this optimal code, then l_k also has maximal length. By the argument above, $\mathcal{C}(M-1)$ can then be exchanged with $\mathcal{C}(k)$ with no loss of optimality. □

The Huffman algorithm chooses an optimal code tree by starting with the two least likely symbols, specifically M and $M-1$, and constraining them to be siblings

[11] Exercise 2.10 proves this for those who enjoy such things.

in the as yet unknown code tree. It makes no difference which sibling ends in 1 and which in 0. How is the rest of the tree to be chosen?

If the above pair of siblings is removed from the as yet unknown tree, the rest of the tree must contain $M-1$ leaves, namely the $M-2$ leaves for the original first $M-2$ symbols and the parent node of the removed siblings. The probability p'_{M-1} associated with this new leaf is taken as $p_{M-1}+p_M$. This tree of $M-1$ leaves is viewed as a code for a reduced random symbol X' with a reduced set of probabilities given as $p_1, \ldots, p_{M-2}$ for the original first $M-2$ symbols and p'_{M-1} for the new symbol $M-1$.

To complete the algorithm, an optimal code is constructed for X'. It will be shown that an optimal code for X can be generated by constructing an optimal code for X', and then grafting siblings onto the leaf corresponding to symbol $M-1$. Assuming this fact for the moment, the problem of constructing an optimal M-ary code has been replaced with constructing an optimal $(M-1)$-ary code. This can be further reduced by applying the same procedure to the $(M-1)$-ary random symbol, and so forth, down to a binary symbol for which the optimal code is obvious.

The following example in Figures 2.9 to 2.11 will make the entire procedure obvious. It starts with a random symbol X with probabilities $\{0.4, 0.2, 0.15, 0.15, 0.1\}$ and generates the reduced random symbol X' in Figure 2.9. The subsequent reductions are shown in Figures 2.10 and 2.11.

Another example, using a different set of probabilities and leading to a different set of codeword lengths, is given in Figure 2.12.

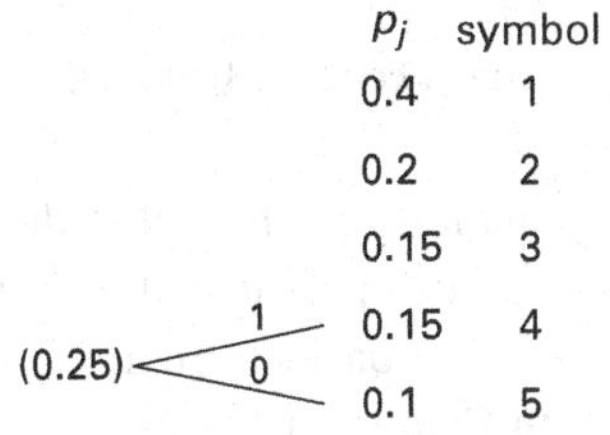

Figure 2.9. Step 1 of the Huffman algorithm; finding X' from X. The two least likely symbols, 4 and 5, have been continued as siblings. The reduced set of probabilities then becomes $\{0.4, 0.2, 0.15, 0.25\}$.

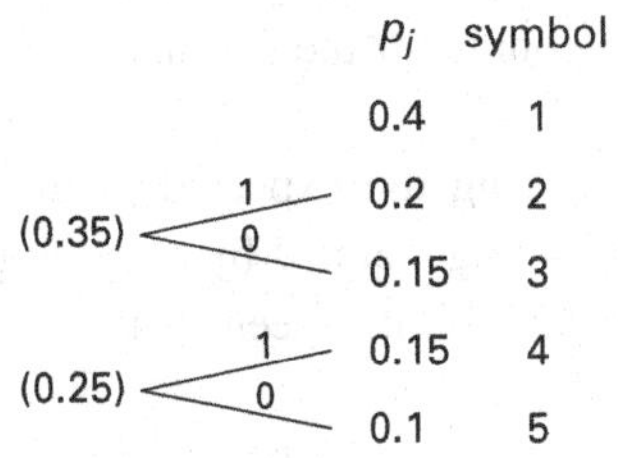

Figure 2.10. Finding X'' from X'. The two least likely symbols in the reduced set, with probabilities 0.15 and 0.2, have been combined as siblings. The reduced set of probabilities then becomes $\{0.4, 0.35, 0.25\}$.

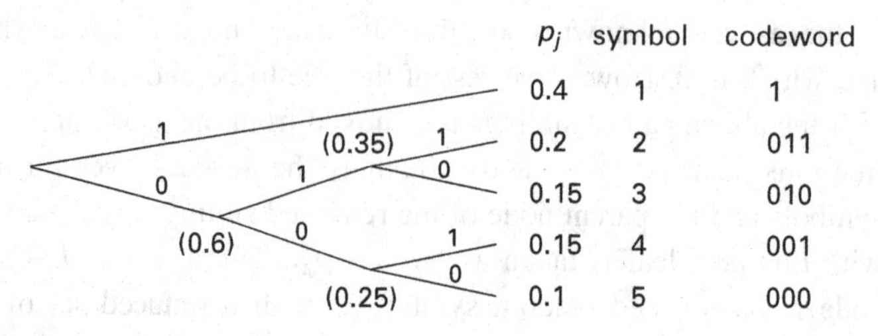

Figure 2.11. Completed Huffman code.

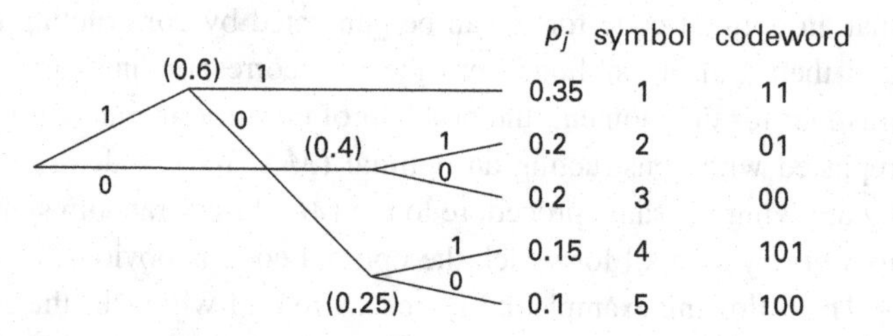

Figure 2.12. Completed Huffman code for a different set of probabilities.

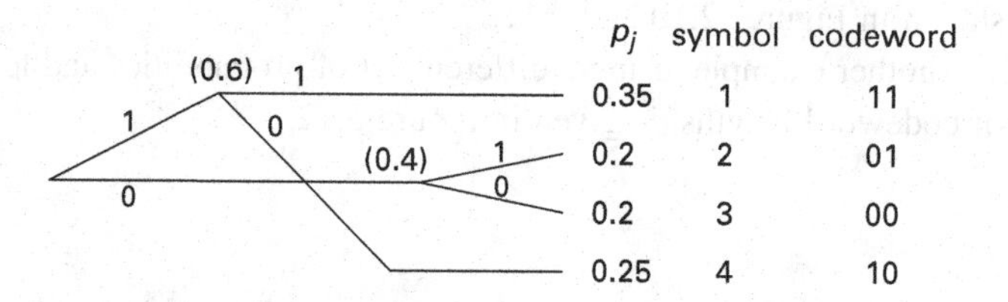

Figure 2.13. Completed reduced Huffman code for Figure 2.12.

The only thing remaining to show that the Huffman algorithm constructs optimal codes is to show that an optimal code for the reduced random symbol X' yields an optimal code for X. Consider Figure 2.13, which shows the code tree for X' corresponding to X in Figure 2.12.

Note that Figures 2.12 and 2.13 differ in that $\mathcal{C}(4)$ and $\mathcal{C}(5)$, each of length 3 in Figure 2.12, have been replaced by a single codeword of length 2 in Figure 2.13. The probability of that single symbol is the sum of the two probabilities in Figure 2.12. Thus the expected codeword length for Figure 2.12 is that for Figure 2.13, increased by $p_4 + p_5$. This accounts for the fact that $\mathcal{C}(4)$ and $\mathcal{C}(5)$ have lengths one greater than their parent node.

In general, comparing the expected length $\overline{L}'$ of *any* code for X' and the corresponding $\overline{L}$ of the code generated by extending $\mathcal{C}'(M-1)$ in the code for X' into two siblings for $M-1$ and M, it is seen that

$$\overline{L} = \overline{L}' + p_{M-1} + p_M.$$

This relationship holds for all codes for X in which $\mathcal{C}(M-1)$ and $\mathcal{C}(M)$ are siblings (which includes at least one optimal code). This proves that $\overline{L}$ is minimized by minimizing $\overline{L}'$, and also shows that $\overline{L}_{\min} = \overline{L}'_{\min} + p_{M-1} + p_M$. This completes the proof of the optimality of the Huffman algorithm.

It is curious that neither the Huffman algorithm nor its proof of optimality gives any indication of the entropy bounds, $\mathsf{H}[X] \leq \overline{L}_{\min} < \mathsf{H}[X]+1$. Similarly, the entropy bounds do not suggest the Huffman algorithm. One is useful in finding an optimal code; the other provides insightful performance bounds.

As an example of the extent to which the optimal lengths approximate $-\log p_j$, the source probabilities in Figure 2.11 are $\{0.40, 0.20, 0.15, 0.15, 0.10\}$, so $-\log p_j$ takes the set of values $\{1.32, 2.32, 2.74, 2.74, 3.32\}$ bits; this approximates the lengths $\{1, 3, 3, 3, 3\}$ of the optimal code quite well. Similarly, the entropy is $\mathsf{H}[X] = 2.15$ bits/symbol and $\overline{L}_{\min} = 2.2$ bits/symbol, quite close to $\mathsf{H}[X]$. However, it would be difficult to guess these optimal lengths, even in such a simple case, without the algorithm.

For the example of Figure 2.12, the source probabilities are $\{0.35, 0.20, 0.20, 0.15, 0.10\}$, the values of $-\log p_i$ are $\{1.51, 2.32, 2.32, 2.74, 3.32\}$, and the entropy is $\mathsf{H}[X] = 2.20$. This is not very different from Figure 2.11. However, the Huffman code now has lengths $\{2, 2, 2, 3, 3\}$ and average length $\overline{L} = 2.25$ bits/symbol. (The code of Figure 2.11 has average length $\overline{L} = 2.30$ for these source probabilities.) It would be hard to predict these perturbations without carrying out the algorithm.

2.6 Entropy and fixed-to-variable-length codes

Entropy is now studied in more detail, both to understand the entropy bounds better and to understand the entropy of n-tuples of successive source letters.

The entropy $\mathsf{H}[X]$ is a fundamental measure of the randomness of a random symbol X. It has many important properties. The property of greatest interest here is that it is the smallest expected number $\overline{L}$ of bits per source symbol required to map the sequence of source symbols into a bit sequence in a uniquely decodable way. This will soon be demonstrated by generalizing the variable-length codes of the preceding few sections to codes in which multiple-source symbols are encoded together. First, however, several other properties of entropy are derived.

Definition 2.6.1 The *entropy* of a discrete random symbol[12] X with alphabet $\mathcal{X}$ is given by

$$\mathsf{H}[X] = \sum_{x\in\mathcal{X}} p_X(x) \log \frac{1}{p_X(x)} = -\sum_{x\in\mathcal{X}} p_X(x) \log p_X(x). \tag{2.11}$$

[12] If one wishes to consider discrete random symbols with one or more symbols of zero probability, one can still use this formula by recognizing that $\lim_{p\to 0} p\log(1/p) = 0$ and then defining $0 \log 1/0$ as 0 in (2.11). Exercise 2.18 illustrates the effect of zero probability symbols in a variable-length prefix code.

Using logarithms to base 2, the units of $\mathsf{H}[X]$ are *bits/symbol*. If the base of the logarithm is e, then the units of $\mathsf{H}[X]$ are called nats/symbol. Conversion is easy; just remember that $\log y = (\ln y)/(\ln 2)$ or $\ln y = (\log y)/(\log e)$, both of which follow from $y = e^{\ln y} = 2^{\log y}$ by taking logarithms. Thus using another base for the logarithm just changes the numerical units of entropy by a scale factor.

Note that the entropy $\mathsf{H}[X]$ of a discrete random symbol X depends on the probabilities of the different outcomes of X, but not on the names of the outcomes. Thus, for example, the entropy of a random symbol taking the values GREEN, BLUE, and RED with probabilities 0.2, 0.3, 0.5, respectively, is the same as the entropy of a random symbol taking on the values SUNDAY, MONDAY, FRIDAY with the same probabilities 0.2, 0.3, 0.5.

The entropy $\mathsf{H}[X]$ is also called the *uncertainty* of X, meaning that it is a measure of the randomness of X. Note that entropy is the expected value of the rv $\log(1/p_X(X))$. This random variable is called the *log pmf* rv.[13] Thus the entropy is the expected value of the log pmf rv.

Some properties of entropy are as follows.

- For any discrete random symbol X, $\mathsf{H}[X] \geq 0$. This follows because $p_X(x) \leq 1$, so $\log(1/p_X(x)) \geq 0$. The result follows from (2.11).
- $\mathsf{H}[X] = 0$ if and only if X is deterministic. This follows since $p_X(x)\log(1/p_X(x)) = 0$ if and only if $p_X(x)$ equals 0 or 1.
- The entropy of an equiprobable random symbol X with an alphabet $\mathcal{X}$ of size M is $\mathsf{H}[X] = \log M$. This follows because, if $p_X(x) = 1/M$ for all $x \in \mathcal{X}$, then

$$\mathsf{H}[X] = \sum_{x\in\mathcal{X}} \frac{1}{M}\log M = \log M.$$

  In this case, the rv $-\log(p_X(X))$ has the constant value $\log M$.
- More generally, the entropy $\mathsf{H}[X]$ of a random symbol X defined on an alphabet $\mathcal{X}$ of size M satisfies $\mathsf{H}[X] \leq \log M$, with equality only in the equiprobable case. To see this, note that

$$\begin{aligned}\mathsf{H}[X] - \log M &= \sum_{x\in\mathcal{X}} p_X(x)\left[\log\frac{1}{p_X(x)} - \log M\right] = \sum_{x\in\mathcal{X}} p_X(x)\left[\log\frac{1}{Mp_X(x)}\right] \\ &\leq (\log e)\sum_{x\in\mathcal{X}} p_X(x)\left[\frac{1}{Mp_X(x)} - 1\right] = 0.\end{aligned}$$

  This uses the inequality $\log u \leq (\log e)(u-1)$ (after omitting any terms for which $p_X(x) = 0$). For equality, it is necessary that $p_X(x) = 1/M$ for all $x \in \mathcal{X}$.

In summary, of all random symbols X defined on a given finite alphabet $\mathcal{X}$, the highest entropy occurs in the equiprobable case, namely $\mathsf{H}[X] = \log M$, and the lowest occurs

[13] This rv is often called *self-information* or *surprise*, or *uncertainty*. It bears some resemblance to the ordinary meaning of these terms, but historically this has caused much more confusion than enlightenment; log pmf, on the other hand, emphasizes what is useful here.

in the deterministic case, namely $\mathsf{H}[X] = 0$. This supports the intuition that the entropy of a random symbol X is a measure of its randomness.

For any pair of discrete random symbols X and Y, XY is another random symbol. The sample values of XY are the set of all pairs xy, $x \in \mathcal{X}$, $y \in \mathcal{Y}$, and the probability of each sample value xy is $p_{XY}(x, y)$. An important property of entropy is that if X and Y are independent discrete random symbols, then $\mathsf{H}[XY] = \mathsf{H}[X] + \mathsf{H}[Y]$. This follows from:

$$\begin{aligned} \mathsf{H}[XY] &= -\sum_{\mathcal{X}\times\mathcal{Y}} p_{XY}(x, y) \log p_{XY}(x, y) \\ &= -\sum_{\mathcal{X}\times\mathcal{Y}} p_X(x) p_Y(y) (\log p_X(x) + \log p_Y(y)) = \mathsf{H}[X] + \mathsf{H}[Y]. \end{aligned} \tag{2.12}$$

Extending this to n random symbols, the entropy of a random symbol X^n corresponding to a block of n iid outputs from a discrete memoryless source is $\mathsf{H}[X^n] = n\mathsf{H}[X]$; i.e., each symbol increments the entropy of the block by $\mathsf{H}[X]$ bits.

2.6.1 Fixed-to-variable-length codes

Recall that in Section 2.2 the sequence of symbols from the source was segmented into successive blocks of n symbols which were then encoded. Each such block was a discrete random symbol in its own right, and thus could be encoded as in the single-symbol case. It was seen that by making n large, fixed-length codes could be constructed in which the number $\overline{L}$ of encoded bits per source symbol approached $\log M$ as closely as desired.

The same approach is now taken for variable-length coding of discrete memoryless sources. A block of n source symbols, $X_1, X_2, \ldots, X_n$ has entropy $\mathsf{H}[X^n] = n\mathsf{H}[X]$. Such a block is a random symbol in its own right and can be encoded using a variable-length prefix-free code. This provides a fixed-to-variable-length code, mapping n-tuples of source symbols to variable-length binary sequences. It will be shown that the expected number $\overline{L}$ of encoded bits per source symbol can be made as close to $\mathsf{H}[X]$ as desired.

Surprisingly, this result is very simple. Let $\mathsf{E}[L(X^n)]$ be the expected length of a variable-length prefix-free code for X^n. Denote the minimum expected length of any prefix-free code for X^n by $\mathsf{E}[L(X^n)]_{\min}$. Theorem 2.5.1 then applies. Using (2.7),

$$\mathsf{H}[X^n] \le \mathsf{E}[L(X^n)]_{\min} < \mathsf{H}[X^n] + 1. \tag{2.13}$$

Define $\overline{L}_{\min,n} = \mathsf{E}[L(X^n)]_{\min}/n$; i.e., $\overline{L}_{\min,n}$ is the minimum number of bits per source symbol over all prefix-free codes for X^n. From (2.13), we have

$$\mathsf{H}[X] \le \overline{L}_{\min,n} < \mathsf{H}[X] + \frac{1}{n}. \tag{2.14}$$

This simple result establishes the following important theorem.

Theorem 2.6.1 (Prefix-free source coding theorem) *For any discrete memoryless source with entropy* $\mathsf{H}[X]$*, and any integer* $n \geq 1$*, there exists a prefix-free encoding of source n-tuples for which the expected codeword length per source symbol* $\overline{L}$ *is at most* $\mathsf{H}[X] + 1/n$*. Furthermore, no prefix-free encoding of fixed-length source blocks of any length n results in an expected codeword length* $\overline{L}$ *less than* $\mathsf{H}[X]$.

This theorem gives considerable significance to the entropy $\mathsf{H}[X]$ of a discrete memoryless source: $\mathsf{H}[X]$ is the minimum expected number $\overline{L}$ of bits per source symbol that can be achieved by fixed-to-variable-length prefix-free codes.

There are two potential questions about the significance of the theorem. First, is it possible to find uniquely decodable codes other than prefix-free codes for which $\overline{L}$ is less than $\mathsf{H}[X]$? Second, is it possible to reduce $\overline{L}$ further by using variable-to-variable-length codes?

For example, if a binary source has $p_1 = 10^{-6}$ and $p_0 = 1 - 10^{-6}$, fixed-to-variable-length codes must use remarkably long n-tuples of source symbols to approach the entropy bound. Run-length coding, which is an example of variable-to-variable-length coding, is a more sensible approach in this case: the source is first encoded into a sequence representing the number of source 0s between each 1, and then this sequence of integers is encoded. This coding technique is further developed in Exercise 2.23.

Section 2.7 strengthens Theorem 2.6.1, showing that $\mathsf{H}[X]$ is indeed a lowerbound to $\overline{L}$ over all uniquely decodable encoding techniques.

2.7 The AEP and the source coding theorems

We first review the weak[14] law of large numbers (WLLN) for sequences of iid rvs. Applying the WLLN to a particular iid sequence, we will establish a form of the remarkable asymptotic equipartition property (AEP).

Crudely, the AEP says that, given a very long string of n iid discrete random symbols $X_1, \ldots, X_n$, there exists a "typical set" of sample strings $(x_1, \ldots, x_n)$ whose aggregate probability is almost 1. There are roughly $2^{n\mathsf{H}[X]}$ typical strings of length n, and each has a probability roughly equal to $2^{-n\mathsf{H}[X]}$. We will have to be careful about what the words "almost" and "roughly" mean here.

The AEP will give us a fundamental understanding not only of source coding for discrete memoryless sources, but also of the probabilistic structure of such sources and the meaning of entropy. The AEP will show us why general types of source encoders, such as variable-to-variable-length encoders, cannot have a strictly smaller expected length per source symbol than the best fixed-to-variable-length prefix-free codes for discrete memoryless sources.

[14] The word *weak* is something of a misnomer, since this is one of the most useful results in probability theory. There is also a strong law of large numbers; the difference lies in the limiting behavior of an infinite sequence of rvs, but this difference is not relevant here. The weak law applies in some cases where the strong law does not, but this also is not relevant here.

2.7.1 The weak law of large numbers

Let $Y_1, Y_2, \ldots,$ be a sequence of iid rvs. Let $\overline{Y}$ and σ_Y^2 be the mean and variance of each Y_j. Define the *sample average* A_Y^n of $Y_1, \ldots, Y_n$ as follows:

$$A_Y^n = \frac{S_Y^n}{n}, \quad \text{where} \quad S_Y^n = Y_1 + \cdots + Y_n.$$

The sample average A_Y^n is itself an rv, whereas, of course, the mean $\overline{Y}$ is simply a real number. Since the sum S_Y^n has mean $n\overline{Y}$ and variance $n\sigma_Y^2$, the sample average A_Y^n has mean $\mathsf{E}[A_Y^n] = \overline{Y}$ and variance $\sigma_{A_Y^n}^2 = \sigma_{S_Y^n}^2/n^2 = \sigma_Y^2/n$. It is important to understand that the variance of the sum *increases* with n and that the variance of the normalized sum (the sample average, A_Y^n) *decreases* with n.

The Chebyshev inequality states that if $\sigma_X^2 < \infty$ for an rv X, then $\Pr(|X - \overline{X}| \geq \varepsilon) \leq \sigma_X^2/\varepsilon^2$ for any $\varepsilon > 0$ (see Exercise 2.3 or any text on probability such as Ross (1994) or Bertsekas and Tsitsiklis (2002)). Applying this inequality to A_Y^n yields the simplest form of the WLLN: for any $\varepsilon > 0$,

$$\Pr(|A_Y^n - \overline{Y}| \geq \varepsilon) \leq \frac{\sigma_Y^2}{n\varepsilon^2}. \tag{2.15}$$

This is illustrated in Figure 2.14.

Since the right side of (2.15) approaches 0 with increasing n for any fixed $\varepsilon > 0$,

$$\lim_{n\to\infty} \Pr(|A_Y^n - \overline{Y}| \geq \varepsilon) = 0. \tag{2.16}$$

For large n, (2.16) says that $A_Y^n - \overline{Y}$ is small with high probability. It does not say that $A_Y^n = \overline{Y}$ with high probability (or even nonzero probability), and it does not say that $\Pr(|A_Y^n - \overline{Y}| \geq \varepsilon) = 0$. As illustrated in Figure 2.14, both a nonzero ε and a nonzero probability are required here, even though they can be made simultaneously as small as desired by increasing n.

In summary, the sample average A_Y^n is a rv whose mean $\overline{Y}$ is independent of n, but whose standard deviation $\sigma_Y/\sqrt{n}$ approaches 0 as $n \to \infty$. Therefore the distribution

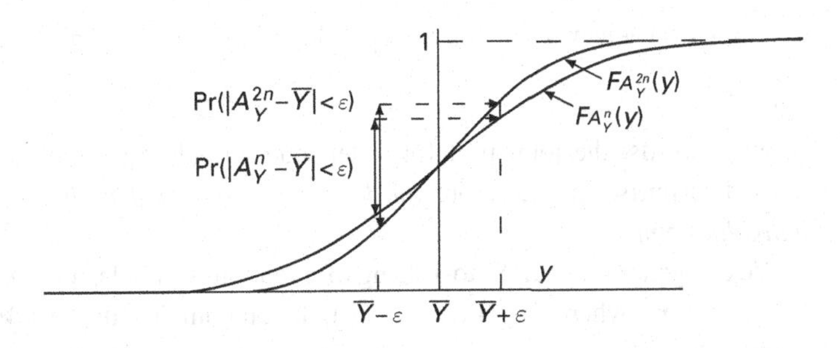

Figure 2.14. Distribution function of the sample average for different n. As n increases, the distribution function approaches a unit step at $\overline{Y}$. The closeness to a step within $\overline{Y} \pm \varepsilon$ is upperbounded by (2.15).

of the sample average becomes concentrated near $\overline{Y}$ as n increases. The WLLN is simply this concentration property, stated more precisely by either (2.15) or (2.16).

The WLLN, in the form of (2.16), applies much more generally than the simple case of iid rvs. In fact, (2.16) provides the central link between probability models and the real phenomena being modeled. One can observe the outcomes both for the model and reality, but probabilities are assigned only for the model. The WLLN, applied to a sequence of rvs in the model, and the concentration property (if it exists), applied to the corresponding real phenomenon, provide the basic check on whether the model corresponds reasonably to reality.

2.7.2 The asymptotic equipartition property

This section starts with a sequence of iid random symbols and defines a sequence of rvs as functions of those symbols. The WLLN, applied to these rvs, will permit the classification of sample sequences of symbols as being "typical" or not, and then lead to the results alluded to earlier.

Let $X_1, X_2, \ldots$ be a sequence of iid discrete random symbols with a common pmf $p_X(x)>0$, $x\in\mathcal{X}$. For each symbol x in the alphabet $\mathcal{X}$, let $w(x) = -\log p_X(x)$. For each X_k in the sequence, define $W(X_k)$ to be the rv that takes the value $w(x)$ for $X_k = x$. Then $W(X_1), W(X_2), \ldots$ is a sequence of iid discrete rvs, each with mean given by

$$\mathsf{E}[W(X_k)] = -\sum_{x\in\mathcal{X}} p_X(x) \log p_X(x) = \mathsf{H}[X], \tag{2.17}$$

where $\mathsf{H}[X]$ is the entropy of the random symbol X.

The rv $W(X_k)$ is the *log pmf* of X_k and the entropy of X_k is the mean of $W(X_k)$.

The most important property of the log pmf for iid random symbols comes from observing, for example, that for the event $X_1 = x_1, X_2 = x_2$, the outcome for $W(X_1) + W(X_2)$ is given by

$$w(x_1) + w(x_2) = -\log p_X(x_1) - \log p_X(x_2) = -\log\{p_{X_1X_2}(x_1x_2)\}. \tag{2.18}$$

In other words, the joint pmf for independent random symbols is the product of the individual pmfs, and therefore *the log of the joint pmf is the sum of the logs of the individual pmfs.*

We can generalize (2.18) to a string of n random symbols, $\boldsymbol{X}^n = (X_1, \ldots, X_n)$. For an event $\boldsymbol{X}^n = \boldsymbol{x}^n$, where $\boldsymbol{x}^n = (x_1, \ldots, x_n)$, the outcome for the sum $W(X_1) + \cdots + W(X_n)$ is given by

$$\sum_{k=1}^{n} w(x_k) = -\sum_{k=1}^{n} \log p_X(x_k) = -\log p_{\boldsymbol{X}^n}(\boldsymbol{x}^n). \tag{2.19}$$

The WLLN can now be applied to the sample average of the log pmfs. Let

$$A_W^n = \frac{W(X_1)+\cdots+W(X_n)}{n} = \frac{-\log p_{X^n}(X^n)}{n} \tag{2.20}$$

be the sample average of the log pmf. From (2.15), it follows that

$$\Pr\left(\left|A_W^n - \mathsf{E}[W(X)]\right| \geq \varepsilon\right) \leq \frac{\sigma_W^2}{n\varepsilon^2}. \tag{2.21}$$

Substituting (2.17) and (2.20) into (2.21) yields

$$\Pr\left(\left|\frac{-\log p_{X^n}(X^n)}{n} - \mathsf{H}[X]\right| \geq \varepsilon\right) \leq \frac{\sigma_W^2}{n\varepsilon^2}. \tag{2.22}$$

In order to interpret this result, define the *typical set* T_ε^n for any $\varepsilon > 0$ as follows:

$$T_\varepsilon^n = \left\{x^n : \left|\frac{-\log p_{X^n}(x^n)}{n} - \mathsf{H}[X]\right| < \varepsilon\right\}. \tag{2.23}$$

Thus T_ε^n is the set of source strings of length n for which the sample average of the log pmf is within ε of its mean $\mathsf{H}[X]$. Equation (2.22) then states that the aggregrate probability of all strings of length n not in T_ε^n is at most $\sigma_W^2/(n\varepsilon^2)$. Thus,

$$\Pr(X^n \in T_\varepsilon^n) \geq 1 - \frac{\sigma_W^2}{n\varepsilon^2}. \tag{2.24}$$

As n increases, the aggregate probability of T_ε^n approaches 1 for any given $\varepsilon > 0$, so T_ε^n is certainly a typical set of source strings. This is illustrated in Figure 2.15.

Rewrite (2.23) in the following form:

$$T_\varepsilon^n = \left\{x^n : n(\mathsf{H}[X]-\varepsilon) < -\log p_{X^n}(x^n) < n(\mathsf{H}[X]+\varepsilon)\right\}.$$

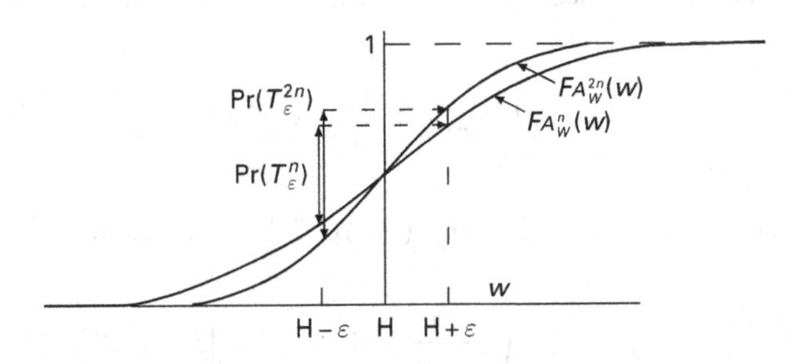

Figure 2.15. Distribution function of the sample average log pmf. As n increases, the distribution function approaches a unit step at H. The typical set is the set of sample strings of length n for which the sample average log pmf stays within ε of H; as illustrated, its probability approaches 1 as $n \to \infty$.

Multiplying by -1 and exponentiating, we obtain

$$T_\varepsilon^n = \{x^n : 2^{-n(\mathsf{H}[X]+\varepsilon)} < p_{X^n}(x^n) < 2^{-n(\mathsf{H}[X]-\varepsilon)}\}. \tag{2.25}$$

Equation (2.25) has the intuitive connotation that the n-strings in T_ε^n are approximately equiprobable. This is the same kind of approximation that one would use in saying that $10^{-1001} \approx 10^{-1000}$; these numbers differ by a factor of 10, but for such small numbers it makes sense to compare the exponents rather than the numbers themselves. In the same way, the ratio of the upperbound to lowerbound in (2.25) is $2^{2\varepsilon n}$, which grows unboundedly with n for fixed ε. However, as may be seen from (2.23), $-(1/n)\log p_{X^n}(x^n)$ is approximately equal to $\mathsf{H}[X]$ for all $x^n \in T_\varepsilon^n$. This is the important notion, and it does no harm to think of the n-strings in T_ε^n as being approximately equiprobable.

The set of all n-strings of source symbols is thus separated into the typical set T_ε^n and the complementary atypical set $(T_\varepsilon^n)^c$. The atypical set has aggregate probability no greater than $\sigma_W^2/(n\varepsilon^2)$, and the elements of the typical set are approximately equiprobable (in this peculiar sense), each with probability about $2^{-n\mathsf{H}[X]}$.

The typical set T_ε^n depends on the choice of ε. As ε decreases, the equiprobable approximation (2.25) becomes tighter, but the bound (2.24) on the probability of the typical set is further from 1. As n increases, however, ε can be slowly decreased, thus bringing the probability of the typical set closer to 1 and simultaneously tightening the bounds on equiprobable strings.

Let us now estimate the number of elements $|T_\varepsilon^n|$ in the typical set. Since $p_{X^n}(x^n) > 2^{-n(\mathsf{H}[X]+\varepsilon)}$ for each $x^n \in T_\varepsilon^n$,

$$1 \geq \sum_{x^n \in T_\varepsilon^n} p_{X^n}(x^n) > |T_\varepsilon^n| 2^{-n(\mathsf{H}[X]+\varepsilon)}.$$

This implies that $|T_\varepsilon^n| < 2^{n(\mathsf{H}[X]+\varepsilon)}$. In other words, since each $x^n \in T_\varepsilon^n$ contributes at least $2^{-n(\mathsf{H}[X]+\varepsilon)}$ to the probability of T_ε^n, the number of these contributions can be no greater than $2^{n(\mathsf{H}[X]+\varepsilon)}$.

Conversely, since $\Pr(T_\varepsilon^n) \geq 1 - \sigma_W^2/(n\varepsilon^2)$, $|T_\varepsilon^n|$ can be lowerbounded by

$$1 - \frac{\sigma_W^2}{n\varepsilon^2} \leq \sum_{x^n \in T_\varepsilon^n} p_{X^n}(x^n) < |T_\varepsilon^n| 2^{-n(\mathsf{H}[X]-\varepsilon)},$$

which implies $|T_\varepsilon^n| > [1 - \sigma_W^2/(n\varepsilon^2)]2^{n(\mathsf{H}[X]-\varepsilon)}$. In summary,

$$\left(1 - \frac{\sigma_W^2}{n\varepsilon^2}\right) 2^{n(\mathsf{H}[X]-\varepsilon)} < |T_\varepsilon^n| < 2^{n(\mathsf{H}[X]+\varepsilon)}. \tag{2.26}$$

For large n, then, the typical set T_ε^n has aggregate probability approximately 1 and contains approximately $2^{n\mathsf{H}[X]}$ elements, each of which has probability approximately

$2^{-n\mathsf{H}[X]}$. That is, asymptotically for very large n, the random symbol $\boldsymbol{X}^n$ resembles an equiprobable source with alphabet size $2^{n\mathsf{H}[X]}$.

The quantity $\sigma_W^2/(n\varepsilon^2)$ in many of the equations above is simply a particular upperbound to the probability of the atypical set. It becomes arbitrarily small as n increases for any fixed $\varepsilon > 0$. Thus it is insightful simply to replace this quantity with a real number δ; for any such $\delta > 0$ and any $\varepsilon > 0$, $\sigma_W^2/(n\varepsilon^2) \leq \delta$ for large enough n.

This set of results, summarized in the following theorem, is known as the *asymptotic equipartition property* (AEP).

Theorem 2.7.1 (Asymptotic equipartition property) *Let X^n be a string of n iid discrete random symbols $\{X_k; 1 \leq k \leq n\}$, each with entropy* $\mathsf{H}[X]$. *For all $\delta > 0$ and all sufficiently large n,* $\Pr(T_\varepsilon^n) \geq 1-\delta$, *and $|T_\varepsilon^n|$ is bounded by*

$$(1-\delta)2^{n(\mathsf{H}[X]-\varepsilon)} < |T_\varepsilon^n| < 2^{n(\mathsf{H}[X]+\varepsilon)}. \tag{2.27}$$

Finally, note that the total number of different strings of length n from a source with alphabet size M is M^n. For nonequiprobable sources, namely sources with $\mathsf{H}[X] < \log M$, the ratio of the number of typical strings to total strings is approximately $2^{-n(\log M - \mathsf{H}[X])}$, which approaches 0 exponentially with n. Thus, for large n, the great majority of n-strings are atypical. It may be somewhat surprising that this great majority counts for so little in probabilistic terms. As shown in Exercise 2.26, the most probable of the individual sequences are also atypical. There are too few of them, however, to have any significance.

We next consider source coding in the light of the AEP.

2.7.3 Source coding theorems

Motivated by the AEP, we can take the approach that an encoder operating on strings of n source symbols need only provide a codeword for each string $\boldsymbol{x}^n$ in the typical set T_ε^n. If a sequence $\boldsymbol{x}^n$ occurs that is not in T_ε^n, then a source coding failure is declared. Since the probability of $\boldsymbol{x}^n \notin T_\varepsilon^n$ can be made arbitrarily small by choosing n large enough, this situation is tolerable.

In this approach, since there are less than $2^{n(\mathsf{H}[X]+\varepsilon)}$ strings of length n in T_ε^n, the number of source codewords that need to be provided is fewer than $2^{n(\mathsf{H}[X]+\varepsilon)}$. Choosing fixed-length codewords of length $\lceil n(\mathsf{H}[X]+\varepsilon)\rceil$ is more than sufficient and even allows for an extra codeword, if desired, to indicate that a coding failure has occurred. In bits per source symbol, taking the ceiling function into account, $\overline{L} \leq \mathsf{H}[X]+\varepsilon+1/n$. Note that $\varepsilon > 0$ is arbitrary, and for any such ε, $\Pr(\text{failure}) \to 0$ as $n \to \infty$. This proves Theorem 2.7.2.

Theorem 2.7.2 (Fixed-to-fixed-length source coding theorem) *For any discrete memoryless source with entropy* $\mathsf{H}[X]$, *any $\varepsilon > 0$, any $\delta > 0$, and any sufficiently large n, there is a fixed-to-fixed-length source code with* $\Pr(\text{failure}) \leq \delta$ *that maps blocks of n source symbols into fixed-length codewords of length $\overline{L} \leq \mathsf{H}[X]+\varepsilon+1/n$ bits per source symbol.*

We saw in Section 2.2 that the use of fixed-to-fixed-length source coding requires $\log M$ bits per source symbol if unique decodability is required (i.e. no failures are allowed), and now we see that this is reduced to arbitrarily little more than $\mathsf{H}[X]$ bits per source symbol if arbitrarily rare failures are allowed. This is a good example of a situation where "arbitrarily small $\delta > 0$" and 0 behave very differently.

There is also a converse to this theorem following from the other side of the AEP theorem. This says that the error probability approaches 1 for large n if strictly fewer than $\mathsf{H}[X]$ bits per source symbol are provided.

Theorem 2.7.3 (Converse for fixed-to-fixed-length codes) *Let X^n be a string of n iid discrete random symbols $\{X_k; 1 \le k \le n\}$, with entropy $\mathsf{H}[X]$ each. For any $\nu > 0$, let X^n be encoded into fixed-length codewords of length $\lfloor n(\mathsf{H}[X] - \nu) \rfloor$ bits. For every $\delta > 0$, and for all sufficiently large n given δ,*

$$\Pr(\text{failure}) > 1 - \delta - 2^{-\nu n/2}. \tag{2.28}$$

Proof Apply the AEP, Theorem 2.7.1, with $\varepsilon = \nu/2$. Codewords can be provided for at most $2^{n(\mathsf{H}[X]-\nu)}$ typical source n-sequences, and from (2.25) each of these has a probability at most $2^{-n(\mathsf{H}[X]-\nu/2)}$. Thus the aggregate probability of typical sequences for which codewords are provided is at most $2^{-n\nu/2}$. From the AEP theorem, $\Pr(T_\varepsilon^n) \ge 1 - \delta$ is satisfied for large enough n. Codewords[15] can be provided for at most a subset of T_ε^n of probability $2^{-n\nu/2}$, and the remaining elements of T_ε^n must all lead to errors, thus yielding (2.28). □

In going from fixed-length codes of slightly more than $\mathsf{H}[X]$ bits per source symbol to codes of slightly less than $\mathsf{H}[X]$ bits per source symbol, the probability of failure goes from almost 0 to almost 1, and as n increases those limits are approached more and more closely.

2.7.4 The entropy bound for general classes of codes

We have seen that the expected number of encoded bits per source symbol is lowerbounded by $\mathsf{H}[X]$ for iid sources using either fixed-to-fixed-length or fixed-to-variable-length codes. The details differ in the sense that very improbable sequences are simply dropped in fixed-length schemes but have abnormally long encodings, leading to buffer overflows, in variable-length schemes.

We now show that other types of codes, such as variable-to-fixed, variable-to-variable, and even more general codes are also subject to the entropy limit. Rather than describing the highly varied possible nature of these source codes, this will be shown by simply defining certain properties that the associated decoders must have. By doing this, it is also shown that as yet undiscovered coding schemes must also be subject to the same limits. The fixed-to-fixed-length converse in Section 2.7.3 is the key to this.

[15] Note that the proof allows codewords to be provided for atypical sequences; it simply says that a large portion of the typical set cannot be encoded.

For any encoder, there must be a decoder that maps the encoded bit sequence back into the source symbol sequence. For prefix-free codes on k-tuples of source symbols, the decoder waits for each variable-length codeword to arrive, maps it into the corresponding k-tuple of source symbols, and then starts decoding for the next k-tuple. For fixed-to-fixed-length schemes, the decoder waits for a block of code symbols and then decodes the corresponding block of source symbols.

In general, the source produces a nonending sequence $X_1, X_2, \ldots$ of source letters which are encoded into a nonending sequence of encoded binary digits. The decoder observes this encoded sequence and decodes source symbol X_n when enough bits have arrived to make a decision on it.

For any given coding and decoding scheme for a given iid source, define the rv D_n as the number of received bits that permit a decision on $\boldsymbol{X}^n = X_1, \ldots, X_n$. This includes the possibility of coders and decoders for which some sample source strings $\boldsymbol{x}^n$ are decoded incorrectly or postponed infinitely. For these $\boldsymbol{x}^n$, the sample value of D_n is taken to be infinite. It is assumed that all decisions are final in the sense that the decoder cannot decide on a particular $\boldsymbol{x}^n$ after observing an initial string of the encoded sequence and then change that decision after observing more of the encoded sequence. What we would like is a scheme in which decoding is correct with high probability and the sample value of the rate, D_n/n, is small with high probability. What the following theorem shows is that for large n, the sample rate can be strictly below the entropy only with vanishingly small probability. This then shows that the entropy lowerbounds the data rate in this strong sense.

Theorem 2.7.4 (Converse for general coders/decoders for iid sources) *Let $\boldsymbol{X}^\infty$ be a sequence of discrete random symbols $\{X_k; 1 \le k \le \infty\}$. For each integer $n \ge 1$, let $\boldsymbol{X}^n$ be the first n of those symbols. For any given encoder and decoder, let D_n be the number of received bits at which the decoder can correctly decode $\boldsymbol{X}^n$. Then for any $\nu > 0$ and $\delta > 0$, and for any sufficiently large n given ν and δ,*

$$\Pr\{D_n \le n(\mathsf{H}[X] - \nu)\} < \delta + 2^{-\nu n/2}. \tag{2.29}$$

Proof For any sample value $\boldsymbol{x}^\infty$ of the source sequence, let $\boldsymbol{y}^\infty$ denote the encoded sequence. For any given integer $n \ge 1$, let $m = \lfloor n[\mathsf{H}[X] - \nu]\rfloor$. Suppose that $\boldsymbol{x}^n$ is decoded upon observation of $\boldsymbol{y}^j$ for some $j \le m$. Since decisions are final, there is only one source n-string, namely $\boldsymbol{x}^n$, that can be decoded by the time $\boldsymbol{y}^m$ is observed. This means that out of the 2^m possible initial m-strings from the encoder, there can be at most[16] 2^m n-strings from the source that can be decoded from the observation of the first m encoded outputs. The aggregate probability of any set of 2^m source n-strings is bounded in Theorem 2.7.3, and (2.29) simply repeats that bound. □

[16] There are two reasons why the number of decoded n-strings of source symbols by time m can be less than 2^m. The first is that the first n source symbols might not be decodable until after the mth encoded bit is received. The second is that multiple m-strings of encoded bits might lead to decoded strings with the same first n source symbols.

2.8 Markov sources

The basic coding results for discrete memoryless sources have now been derived. Many of the results, in particular the Kraft inequality, the entropy bounds on expected length for uniquely decodable codes, and the Huffman algorithm, do not depend on the independence of successive source symbols.

In this section, these results are extended to sources defined in terms of finite-state Markov chains. The state of the Markov chain[17] is used to represent the "memory" of the source. Labels on the transitions between states are used to represent the next symbol out of the source. Thus, for example, the state could be the previous symbol from the source, or it could be the previous 300 symbols. It is possible to model as much memory as desired while staying in the regime of finite-state Markov chains.

Example 2.8.1 Consider a binary source with outputs $X_1, X_2, \ldots$ Assume that the symbol probabilities for X_m are conditioned on X_{k-2} and X_{k-1} but are independent of all previous symbols given the preceding two symbols. This pair of previous symbols is modeled by a *state* S_{k-1}. The alphabet of possible states is then the set of binary pairs, $\mathcal{S} = \{[00], [01], [10], [11]\}$. In Figure 2.16, the states are represented as the nodes of the graph representing the Markov chain, and the source outputs are labels on the graph transitions. Note, for example, that from state $S_{k-1} = [01]$ (representing $X_{k-2}=0, X_{k-1}=1$), the output $X_k=1$ causes a transition to $S_k = [11]$ (representing $X_{k-1}=1, X_k=1$). The chain is assumed to start at time 0 in a state S_0 given by some arbitrary pmf.

Note that this particular source is characterized by long strings of 0s and long strings of 1s interspersed with short transition regions. For example, starting in state 00, a representative output would be

$$00000000101111111111111011111111010100000000\cdots$$

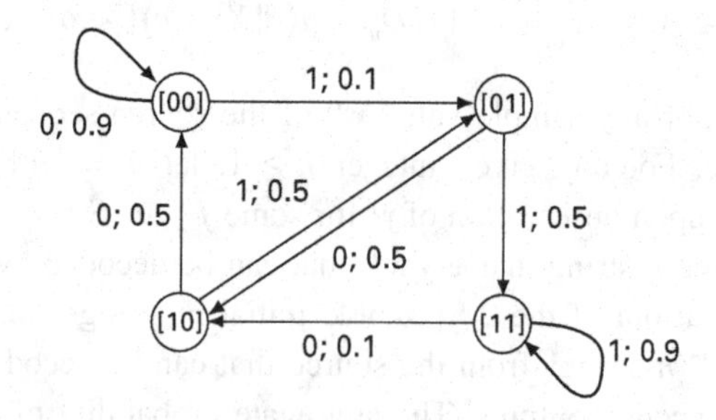

Figure 2.16. Markov source: each transition $s' \to s$ is labeled by the corresponding source output and the transition probability $\Pr(S_k = s | S_{k-1} = s')$.

[17] The basic results about finite-state Markov chains, including those used here, are established in many texts such as Gallager (1996) and Ross (1996). These results are important in the further study of digital communcation, but are not essential here.

Note that if $s_k = [x_{k-1}x_k]$ then the next state must be either $s_{k+1} = [x_k 0]$ or $s_{k+1} = [x_k 1]$; i.e., each state has only two transitions coming out of it.

Example 2.8.1 is now generalized to an arbitrary discrete Markov source.

Definition 2.8.1 A *finite-state Markov chain* is a sequence $S_0, S_1, \ldots$ of discrete random symbols from a finite alphabet, $\mathcal{S}$. There is a pmf $q_0(s), s \in \mathcal{S}$ on S_0, and there is a conditional pmf $Q(s|s')$ such that, for all $m \geq 1$, all $s \in \mathcal{S}$, and all $s' \in \mathcal{S}$,

$$\Pr(S_k = s|S_{k-1} = s') = \Pr(S_k = s|S_{k-1} = s', \ldots, S_0 = s_0) = Q(s|s'). \tag{2.30}$$

There is said to be a *transition* from s' to s, denoted $s' \to s$, if $Q(s|s') > 0$.

Note that (2.30) says, first, that the conditional probability of a state, given the past, depends only on the previous state, and, second, that these transition probabilities $Q(s|s')$ do not change with time.

Definition 2.8.2 A *Markov source* is a sequence of discrete random symbols $\mathcal{X}_1, \mathcal{X}_2, \ldots$ with a common alphabet $\mathcal{X}$ which is based on a finite-state Markov chain $S_0, S_1, \ldots$ Each transition $(s' \to s)$ in the Markov chain is labeled with a symbol from $\mathcal{X}$; each symbol from $\mathcal{X}$ can appear on at most one outgoing transition from each state.

Note that the state alphabet $\mathcal{S}$ and the source alphabet $\mathcal{X}$ are, in general, different. Since each source symbol appears on at most one transition from each state, the initial state $S_0 = s_0$, combined with the source output, $X_1 = x_1, X_2 = x_2, \ldots$, uniquely identifies the state sequence, and, of course, the state sequence uniquely specifies the source output sequence. If $x \in \mathcal{X}$ labels the transition $s' \to s$, then the conditional probability of that x is given by $P(x|s') = Q(s|s')$. Thus, for example, in the transition $[00] \to [0]1$ in Figure 2.16, $Q([01]|[00]) = P(1|[00])$.

The reason for distinguishing the Markov chain alphabet from the source output alphabet is to allow the state to represent an arbitrary combination of past events rather than just the previous source output. This feature permits Markov sources to provide reasonable models for surprisingly complex forms of memory.

A state s is *accessible* from state s' in a Markov chain if there is a path in the corresponding graph from $s' \to s$, i.e. if $\Pr(S_k = s|S_0 = s') > 0$ for some $k > 0$. The period of a state s is the greatest common divisor of the set of integers $k \geq 1$ for which $\Pr(S_k = s|S_0 = s) > 0$. A finite-state Markov chain is *ergodic* if all states are accessible from all other states and if all states are aperiodic, i.e. have period 1.

We will consider only Markov sources for which the Markov chain is ergodic. An important fact about ergodic Markov chains is that the chain has steady-state probabilities $q(s)$ for all $s \in \mathcal{S}$, given by the following unique solution to the linear equations:

$$q(s) = \sum_{s' \in \mathcal{S}} q(s')Q(s|s'); \quad s \in \mathcal{S}, \tag{2.31}$$

$$\sum_{s \in \mathcal{S}} q(s) = 1.$$

These steady-state probabilities are approached asymptotically from any starting state, i.e.

$$\lim_{k\to\infty} \Pr(S_k = s|S_0 = s') = q(s) \qquad \text{for all } s, s' \in \mathcal{S}. \tag{2.32}$$

2.8.1 Coding for Markov sources

The simplest approach to coding for Markov sources is that of using a separate prefix-free code for each state in the underlying Markov chain. That is, for each $s \in \mathcal{S}$, select a prefix-free code whose lengths $l(x, s)$ are appropriate for the conditional pmf $P(x|s) > 0$. The codeword lengths for the code used in state s must of course satisfy the Kraft inequality $\sum_x 2^{-l(x,s)} \le 1$. The minimum expected length, $\overline{L}_{\min}(s)$, for each such code can be generated by the Huffman algorithm and satisfies

$$\mathsf{H}[X|s] \le \overline{L}_{\min}(s) < \mathsf{H}[X|s] + 1, \tag{2.33}$$

where, for each $s \in \mathcal{S}$, $\mathsf{H}[X|s] = \sum_x -P(x|s)\log P(x|s)$.

If the initial state S_0 is chosen according to the steady-state pmf $\{q(s); s \in \mathcal{S}\}$, then, from (2.31), the Markov chain remains in steady state and the overall expected codeword length is given by

$$\mathsf{H}[X|S] \le \overline{L}_{\min} < \mathsf{H}[X|S] + 1, \tag{2.34}$$

where

$$\overline{L}_{\min} = \sum_{s\in\mathcal{S}} q(s)\overline{L}_{\min}(s) \tag{2.35}$$

and

$$\mathsf{H}[X|S] = \sum_{s\in\mathcal{S}} q(s)\mathsf{H}[X|s]. \tag{2.36}$$

Assume that the encoder transmits the initial state s_0 at time 0. If M' is the number of elements in the state space, then this can be done with $\lceil \log M' \rceil$ bits, but this can be ignored since it is done only at the beginning of transmission and does not affect the long term expected number of bits per source symbol. The encoder then successively encodes each source symbol x_k using the code for the state at time $k-1$. The decoder, after decoding the initial state s_0, can decode x_1 using the code based on state s_0. After determining s_1 from s_0 and x_1, the decoder can decode x_2 using the code based on s_1. The decoder can continue decoding each source symbol, and thus the overall code is uniquely decodable. We next must understand the meaning of the conditional entropy in (2.36).

2.8.2 Conditional entropy

It turns out that the conditional entropy $\mathsf{H}[X|S]$ plays the same role in coding for Markov sources as the ordinary entropy $\mathsf{H}[X]$ plays for the memoryless case. Rewriting (2.36), we obtain

$$\mathsf{H}[X|S] = \sum_{s\in\mathcal{S}} \sum_{x\in\mathcal{X}} q(s)P(x|s) \log \frac{1}{P(x|s)}.$$

This is the expected value of the rv $\log[1/P(X|S)]$.

An important entropy relation, for arbitrary discrete rvs, is given by

$$\mathsf{H}[XS] = \mathsf{H}[S] + \mathsf{H}[X|S]. \tag{2.37}$$

To see this,

$$\begin{aligned}\mathsf{H}[XS] &= \sum_{s,x} q(s)P(x|s) \log \frac{1}{q(s)P(x|s)} \\ &= \sum_{s,x} q(s)P(x|s) \log \frac{1}{q(s)} + \sum_{s,x} q(s)P(x|s) \log \frac{1}{P(x|s)} \\ &= \mathsf{H}[S] + \mathsf{H}[X|S].\end{aligned}$$

Exercise 2.19 demonstrates that

$$\mathsf{H}[XS] \le \mathsf{H}[S] + \mathsf{H}[X].$$

Comparing this and (2.37), it follows that

$$\mathsf{H}[X|S] \le \mathsf{H}[X]. \tag{2.38}$$

This is an important inequality in information theory. If the entropy $\mathsf{H}[X]$ is a measure of mean uncertainty, then the conditional entropy $\mathsf{H}[X|S]$ should be viewed as a measure of mean uncertainty after the observation of the outcome of S. If X and S are not statistically independent, then intuition suggests that the observation of S should reduce the mean uncertainty in X; this equation indeed verifies this.

Example 2.8.2 Consider Figure 2.16 again. It is clear from symmetry that, in steady state, $p_X(0) = p_X(1) = 1/2$. Thus $\mathsf{H}[X] = 1$ bit. Conditional on $S = 00$, X is binary with pmf $\{0.1, 0.9\}$, so $\mathsf{H}[X|[00]] = -0.1\log 0.1 - 0.9\log 0.9 = 0.47$ bits. Similarly, $\mathsf{H}[X|[11]] = 0.47$ bits, and $\mathsf{H}[X|[01]] = \mathsf{H}[X|[10]] = 1$ bit. The solution to the steady-state equations in (2.31) is $q([00]) = q([11]) = 5/12$ and $q([01]) = q([10]) = 1/12$. Thus, the conditional entropy, averaged over the states, is $\mathsf{H}[X|S] = 0.558$ bits.

For this example, it is particularly silly to use a different prefix-free code for the source output for each prior state. The problem is that the source is binary, and thus the prefix-free code will have length 1 for each symbol no matter what the state. As with the memoryless case, however, the use of fixed-to-variable-length codes is a solution to these problems of small alphabet sizes and integer constraints on codeword lengths.

Let $\mathsf{E}[L(\boldsymbol{X}^n)]_{\min,s}$ be the minimum expected length of a prefix-free code for $\boldsymbol{X}^n$ conditional on starting in state s. Then, applying (2.13) to the situation here,

$$\mathsf{H}[\boldsymbol{X}^n|s] \le \mathsf{E}[L(\boldsymbol{X}^n)]_{\min,s} < \mathsf{H}[\boldsymbol{X}^n|s] + 1.$$

Assume as before that the Markov chain starts in steady state S_0. Thus it remains in steady state at each future time. Furthermore, assume that the initial *sample state* is known at the decoder. Then the sample state continues to be known at each future time. Using a minimum expected length code for each initial sample state, we obtain

$$\mathsf{H}[\boldsymbol{X}^n|S_0] \le \mathsf{E}[L(\boldsymbol{X}^n)]_{\min,S_0} < \mathsf{H}[\boldsymbol{X}^n|S_0] + 1. \tag{2.39}$$

Since the Markov source remains in steady state, the average entropy of each source symbol given the state is $H(X|S_0)$, so intuition suggests (and Exercise 2.32 verifies) that

$$\mathsf{H}[\boldsymbol{X}^n|S_0] = n\mathsf{H}[X|S_0]. \tag{2.40}$$

Defining $\overline{L}_{\min,n} = \mathsf{E}[L(\boldsymbol{X}^n)]_{\min,S_0}/n$ as the minimum expected codeword length per input symbol when starting in steady state, we obtain

$$\mathsf{H}[X|S_0] \le \overline{L}_{\min,n} < \mathsf{H}[X|S_0] + 1/n. \tag{2.41}$$

The asymptotic equipartition property (AEP) also holds for Markov sources. Here, however, there are[18] approximately $2^{n\mathsf{H}[X|S]}$ typical strings of length n, each with probability approximately equal to $2^{-n\mathsf{H}[X|S]}$. It follows, as in the memoryless case, that $\mathsf{H}[X|S]$ is the minimum possible rate at which source symbols can be encoded subject either to unique decodability or to fixed-to-fixed-length encoding with small probability of failure. The arguments are essentially the same as in the memoryless case.

The analysis of Markov sources will not be carried further here, since the additional required ideas are minor modifications of the memoryless case. Curiously, most of our insights and understanding about source coding come from memoryless sources. At the same time, however, most sources of practical importance can be insightfully modeled as Markov and hardly any can be reasonably modeled as memoryless. In dealing with practical sources, we combine the insights from the memoryless case with modifications suggested by Markov memory.

The AEP can be generalized to a still more general class of discrete sources called *ergodic sources*. These are essentially sources for which sample time averages converge in some probabilistic sense to ensemble averages. We do not have the machinery to define ergodicity, and the additional insight that would arise from studying the AEP for this class would consist primarily of mathematical refinements.

---

[18] There are additional details here about whether the typical sequences include the initial state or not, but these differences become unimportant as n becomes large.

2.9 Lempel–Ziv universal data compression

The Lempel–Ziv data compression algorithms differ from the source coding algorithms studied in previous sections in the following ways.

- They use variable-to-variable-length codes in which both the number of source symbols encoded and the number of encoded bits per codeword are variable. Moreover, the codes are time-varying.
- They do not require prior knowledge of the source statistics, yet over time they adapt so that the average codeword length $\overline{L}$ per source symbol is minimized in some sense to be discussed later. Such algorithms are called *universal*.
- They have been widely used in practice; they provide a simple approach to understanding universal data compression even though newer schemes now exist.

The Lempel–Ziv compression algorithms were developed in 1977–1978. The first, LZ77 (Ziv and Lempel, 1977), uses string-matching on a sliding window; the second, LZ78 (Ziv and Lempel, 1978), uses an adaptive dictionary. The LZ78 algorithm was implemented many years ago in UNIX `compress`, and in many other places. Implementations of LZ77 appeared somewhat later (Stac Stacker, Microsoft Windows), and is still widely used.

In this section, the LZ77 algorithm is described, accompanied by a high-level description of why it works. Finally, an approximate analysis of its performance on Markov sources is given, showing that it is effectively optimal.[19] In other words, although this algorithm operates in ignorance of the source statistics, it compresses substantially as well as the best algorithm designed to work with those statistics.

2.9.1 The LZ77 algorithm

The LZ77 algorithm compresses a sequence $\boldsymbol{x} = x_1, x_2, \ldots$ from some given discrete alphabet $\mathcal{X}$ of size $M = |\mathcal{X}|$. At this point, no probabilistic model is assumed for the source, so $\boldsymbol{x}$ is simply a sequence of symbols, not a sequence of random symbols. A subsequence $(x_m, x_{m+1}, \ldots, x_n)$ of $\boldsymbol{x}$ is represented by $\boldsymbol{x}_m^n$.

The algorithm keeps the w most recently encoded source symbols in memory. This is called a sliding window of size w. The number w is large, and can be thought of as being in the range of 2^{10} to 2^{20}, say. The parameter w is chosen to be a power of 2. Both complexity and, typically, performance increase with w.

Briefly, the algorithm operates as follows. Suppose that at some time the source symbols $\boldsymbol{x}_1^P$ have been encoded. The encoder looks for the longest match, say of length n, between the not-yet-encoded n-string $\boldsymbol{x}_{P+1}^{P+n}$ and a stored string $\boldsymbol{x}_{P+1-u}^{P+n-u}$ starting in the window of length w. The clever algorithmic idea in LZ77 is to encode this string of n symbols simply by encoding the integers n and u; i.e., by pointing to the previous

[19] A proof of this optimality for discrete ergodic sources has been given by Wyner and Ziv (1994).

occurrence of this string in the sliding window. If the decoder maintains an identical window, then it can look up the string x_{P+1-u}^{P+n-u}, decode it, and keep up with the encoder.

More precisely, the LZ77 algorithm operates as follows.

(1) Encode the first w symbols in a fixed-length code without compression, using $\lceil \log M \rceil$ bits per symbol. (Since $w\lceil \log M \rceil$ will be a vanishing fraction of the total number of encoded bits, the efficiency of encoding this preamble is unimportant, at least in theory.)

(2) Set the pointer $P = w$. (This indicates that all symbols up to and including x_P have been encoded.)

(3) Find the largest $n \geq 2$ such that $x_{P+1}^{P+n} = x_{P+1-u}^{P+n-u}$ for some u in the range $1 \leq u \leq w$. (Find the longest match between the not-yet-encoded symbols starting at $P+1$ and a string of symbols starting in the window; let n be the length of that longest match and u the distance back into the window to the start of that match.) The string x_{P+1}^{P+n} is encoded by encoding the integers n and u.

Here are two examples of finding this longest match. In the first, the length of the match is $n = 3$ and the match starts $u = 7$ symbols before the pointer. In the second, the length of the match is 4 and it starts $u = 2$ symbols before the pointer. This illustrates that the string and its match can overlap.

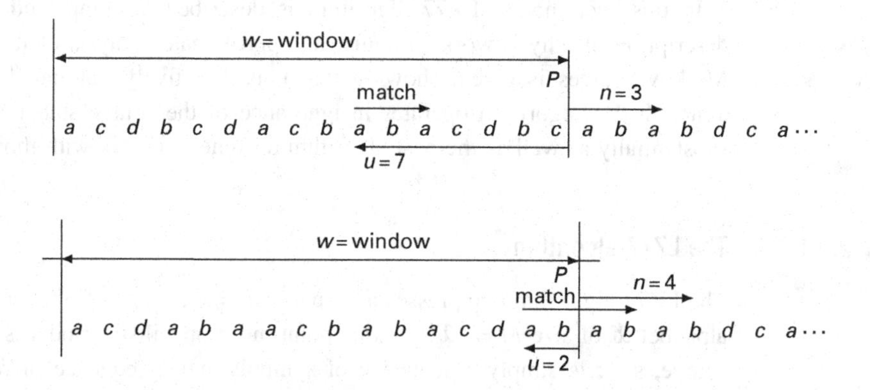

If no match exists for $n \geq 2$, then, independently of whether a match exists for $n = 1$, set $n = 1$ and directly encode the single source symbol x_{P+1} without compression.

(4) Encode the integer n into a codeword from the unary–binary code. In the unary–binary code, as illustrated in Table 2.1, a positive integer n is encoded into the binary representation of n, preceded by a prefix of $\lfloor \log_2 n \rfloor$ zeros.

Thus the codewords starting with 0^k1 correspond to the set of 2^k integers in the range $2^k \leq n \leq 2^{k+1} - 1$. This code is prefix-free (picture the corresponding binary tree). It can be seen that the codeword for integer n has length $2\lfloor \log n \rfloor + 1$; it is seen later that this is negligible compared with the length of the encoding for u.

(5) If $n > 1$, encode the positive integer $u \leq w$ using a fixed-length code of length $\log w$ bits. (At this point the decoder knows n, and can simply count back by u

Table 2.1. The unary–binary code

n	Prefix	Base 2 expansion	Codeword
1		1	1
2	0	10	010
3	0	11	011
4	00	100	00100
5	00	101	00101
6	00	110	00110
7	00	111	00111
8	000	1000	0001000

in the previously decoded string to find the appropriate n-tuple, even if there is overlap as above.)

(6) Set the pointer P to $P+n$ and go to step (3). (Iterate forever.)

2.9.2 Why LZ77 works

The motivation behind LZ77 is information-theoretic. The underlying idea is that if the unknown source happens to be, say, a Markov source of entropy $\mathsf{H}[X|S]$, then the AEP says that, for any large n, there are roughly $2^{n\mathsf{H}[X|S]}$ typical source strings of length n. On the other hand, a window of size w contains w source strings of length n, counting duplications. This means that if $w \ll 2^{n\mathsf{H}[X|S]}$, then most typical sequences of length n cannot be found in the window, suggesting that matches of length n are unlikely. Similarly, if $w \gg 2^{n\mathsf{H}[X|S]}$, then it is reasonable to suspect that most typical sequences will be in the window, suggesting that matches of length n or more are likely.

The above argument, approximate and vague as it is, suggests that, in order to achieve large typical match sizes n_t, the window w should be exponentially large, on the order of $w \approx 2^{n_t \mathsf{H}[X|S]}$, which means

$$n_t \approx \frac{\log w}{\mathsf{H}[X|S]}; \qquad \text{typical match size.} \tag{2.42}$$

The encoding for a match requires $\log w$ bits for the match location and $2\lfloor \log n_t \rfloor + 1$ for the match size n_t. Since n_t is proportional to $\log w$, $\log n_t$ is negligible compared with $\log w$ for very large w. Thus, for the typical case, about $\log w$ bits are used to encode about n_t source symbols. Thus, from (2.42), the required rate, in bits per source symbol, is about $\overline{L} \approx \mathsf{H}[X|S]$.

The above argument is very imprecise, but the conclusion is that, for very large window size, $\overline{L}$ is reduced to the value required when the source is known and an optimal fixed-to-variable prefix-free code is used.

The imprecision above involves more than simply ignoring the approximation factors in the AEP. More conceptual issues, resolved in Wyner and Ziv (1994), are, first, that the strings of source symbols that must be encoded are somewhat special since they start at the end of previous matches, and, second, duplications of typical sequences within the window have been ignored.

2.9.3 Discussion

Let us recapitulate the basic ideas behind the LZ77 algorithm.

(1) Let N_x be the number of occurrences of symbol x in a window of very large size w. If the source satisfies the WLLN, then the relative frequency N_x/w of appearances of x in the window will satisfy $N_x/w \approx p_X(x)$ with high probability. Similarly, let $N_{\boldsymbol{x}^n}$ be the number of occurrences of $\boldsymbol{x}^n$ which start in the window. The relative frequency $N_{\boldsymbol{x}^n}/w$ will then satisfy $N_{\boldsymbol{x}^n}/w \approx p_{X^n}(\boldsymbol{x}^n)$ with high probability for very large w. This association of relative frequencies with probabilities is what makes LZ77 a universal algorithm which needs no prior knowledge of source statistics.[20]

(2) Next, as explained in Section 2.8, the probability of a typical source string $\boldsymbol{x}^n$ for a Markov source is approximately $2^{-n\mathsf{H}[X|S]}$. If $w \gg 2^{n\mathsf{H}[X|S]}$, then, according to (1) above, $N_{\boldsymbol{x}^n} \approx w p_{X^n}(\boldsymbol{x}^n)$ should be large and $\boldsymbol{x}^n$ should occur in the window with high probability. Alternatively, if $w \ll 2^{n\mathsf{H}[X|S]}$, then $\boldsymbol{x}^n$ will probably not occur. Consequently, the match will usually occur for $n \approx (\log w)/\mathsf{H}[X|S]$ as w becomes very large.

(3) Finally, it takes about $\log w$ bits to point to the best match in the window. The unary–binary code uses $2\lfloor \log n \rfloor + 1$ bits to encode the length n of the match. For typical n, this is on the order of $2\log(\log w/\mathsf{H}[X|S])$, which is negligible for large enough w compared with $\log w$.

Consequently, LZ77 requires about $\log w$ encoded bits for each group of about $(\log w)/\mathsf{H}[X|S]$ source symbols, so it nearly achieves the optimal efficiency of $\overline{L} = \mathsf{H}[X|S]$ bits/symbol, as w becomes very large.

Discrete sources, as they appear in practice, often can be viewed over different time scales. Over very long time scales, or over the sequences presented to different physical encoders running the same algorithm, there is often very little common structure, sometimes varying from one language to another, or varying from text in a language to data from something else.

Over shorter time frames, corresponding to a single file or a single application type, there is often more structure, such as that in similar types of documents from the same language. Here it is more reasonable to view the source output as a finite length segment of, say, the output of an ergodic Markov source.

---

[20] As Yogi Berra said, "You can observe a whole lot just by watchin'."

What this means is that universal data compression algorithms must be tested in practice. The fact that they behave optimally for unknown sources that can be modeled to satisfy the AEP is an important guide, but not the whole story.

The above view of different time scales also indicates that a larger window need not always improve the performance of the LZ77 algorithm. It suggests that long matches will be more likely in recent portions of the window, so that fixed-length encoding of the window position is not the best approach. If shorter codewords are used for more recent matches, then it requires a shorter time for efficient coding to start to occur when the source statistics abruptly change. It also then makes sense to start coding from some arbitrary window known to both encoder and decoder rather than filling the entire window with data before starting to use the LZ77 alogorithm.

2.10 Summary of discrete source coding

Discrete source coding is important both for discrete sources, such as text and computer files, and also as an inner layer for discrete-time analog sequences and fully analog sources. It is essential to focus on the range of possible outputs from the source rather than any one particular output. It is also important to focus on *probabilistic* models so as to achieve the best compression for the most common outputs with less care for very rare outputs. Even universal coding techniques, such as LZ77, which are designed to work well in the absence of a probability model, require probability models to understand and evaluate how they work.

Variable-length source coding is the simplest way to provide good compression for common source outputs at the expense of rare outputs. The necessity to concatenate successive variable-length codewords leads to the nonprobabilistic concept of unique decodability. Prefix-free codes provide a simple class of uniquely decodable codes. Both prefix-free codes and the more general class of uniquely decodable codes satisfy the Kraft inequality on the number of possible codewords of each length. Moreover, for any set of lengths satisfying the Kraft inequality, there is a simple procedure for constructing a prefix-free code with those lengths. Since the expected length, and other important properties of codes, depend only on the codeword lengths (and how they are assigned to source symbols), there is usually little reason to use variable-length codes that are not also prefix-free.

For a DMS with given probabilities on the symbols of a source code, the entropy is a lowerbound on the expected length of uniquely decodable codes. The Huffman algorithm provides a simple procedure for finding an optimal (in the sense of minimum expected codeword length) variable-length prefix-free code. The Huffman algorithm is also useful for deriving properties about optimal variable-length source codes (see Exercises 2.12 to 2.18).

All the properties of variable-length codes extend immediately to fixed-to-variable-length codes. Here, the source output sequence is segmented into blocks of n symbols, each of which is then encoded as a single symbol from the alphabet of source n-tuples. For a DMS the minimum expected codeword length per source symbol then lies between $\mathsf{H}[U]$ and $\mathsf{H}[U]+1/n$. Thus prefix-free fixed-to-variable-length codes can approach the entropy bound as closely as desired.

One of the disadvantages of fixed-to-variable-length codes is that bits leave the encoder at a variable rate relative to incoming symbols. Thus, if the incoming symbols have a fixed rate and the bits must be fed into a channel at a fixed rate (perhaps with some idle periods), then the encoded bits must be queued and there is a positive probability that any finite-length queue will overflow.

An alternative point of view is to consider fixed-length-to-fixed-length codes. Here, for a DMS, the set of possible n-tuples of symbols from the source can be partitioned into a typical set and an atypical set. For large n, the AEP says that there are essentially $2^{n\mathsf{H}[U]}$ typical n-tuples with an aggregate probability approaching 1 with increasing n. Encoding just the typical n-tuples requires about $\mathsf{H}[U]$ bits per symbol, thus approaching the entropy bound without the above queueing problem, but, of course, with occasional errors.

As detailed in the text, the AEP can be used to look at the long-term behavior of arbitrary source coding algorithms to show that the entropy bound cannot be exceeded without a failure rate that approaches 1.

The above results for discrete memoryless sources extend easily to ergodic Markov sources. The text does not carry out this analysis in detail since readers are not assumed to have the requisite knowledge about Markov chains (see Gallager (1968) for the detailed analysis). The important thing here is to see that Markov sources can model n-gram statistics for any desired n and thus can model fairly general sources (at the cost of very complex models). From a practical standpoint, universal source codes such as LZ77 are usually a more reasonable approach to complex and partly unknown sources.

2.11 Exercises

2.1 Chapter 1 pointed out that voice waveforms could be converted to binary data by sampling at 8000 times per second and quantizing to 8 bits per sample, yielding 64 kbps. It then said that modern speech coders can yield telephone-quality speech at 6–16 kbps. If your objective were simply to reproduce the words in speech recognizably without concern for speaker recognition, intonation, etc., make an estimate of how many kbps would be required. Explain your reasoning. (Note: there is clearly no "correct answer" here; the question is too vague for that. The point of the question is to get used to questioning objectives and approaches.)

2.2 Let V and W be discrete rvs defined on some probability space with a joint pmf $p_{VW}(v, w)$.

(a) Prove that $\mathsf{E}[V+W] = \mathsf{E}[V]+\mathsf{E}[W]$. Do not assume independence.

(b) Prove that if V and W are independent rvs, then $\mathsf{E}[V \cdot W] = \mathsf{E}[V] \cdot \mathsf{E}[W]$.

(c) Assume that V and W are not independent. Find an example where $\mathsf{E}[V \cdot W] \neq \mathsf{E}[V] \cdot \mathsf{E}[W]$ and another example where $\mathsf{E}[V \cdot W] = \mathsf{E}[V] \cdot \mathsf{E}[W]$.

(d) Assume that V and W are independent and let σ_V^2 and σ_W^2 be the variances of V and W, respectively. Show that the variance of $V+W$ is given by $\sigma_{V+W}^2 = \sigma_V^2 + \sigma_W^2$.

2.3 (a) For a nonnegative integer-valued rv N, show that $\mathsf{E}[N] = \sum_{n>0} \Pr(N \geq n)$.

(b) Show, with whatever mathematical care you feel comfortable with, that for an arbitrary nonnegative rv X, $\mathsf{E}(X) = \int_0^\infty \Pr(X \geq a)da$.

(c) Derive the Markov inequality, which says that for any $a>0$ and any nonnegative rv X, $\Pr(X \geq a) \leq \mathsf{E}[X]/a$. [Hint. Sketch $\Pr(X > a)$ as a function of a and compare the area of the rectangle with horizontal length a and vertical length $\Pr(X \geq a)$ in your sketch with the area corresponding to $\mathsf{E}[X]$.]

(d) Derive the Chebyshev inequality, which says that $\Pr(|Y - \mathsf{E}[Y]| \geq b) \leq \sigma_Y^2/b^2$ for any rv Y with finite mean $\mathsf{E}[Y]$ and finite variance σ_Y^2. [Hint. Use part (c) with $(Y - \mathsf{E}[Y])^2 = X$.]

2.4 Let $X_1, X_2, \ldots, X_n, \ldots$ be a sequence of independent identically distributed (iid) analog rvs with the common probability density function $f_X(x)$. Note that $\Pr(X_n = \alpha) = 0$ for all α and that $\Pr(X_n = X_m) = 0$ for $m \neq n$.

(a) Find $\Pr(X_1 \leq X_2)$. (Give a numerical answer, not an expression; no computation is required and a one- or two-line explanation should be adequate.)

(b) Find $\Pr(X_1 \leq X_2; X_1 \leq X_3)$; in other words, find the probability that X_1 is the smallest of $\{X_1, X_2, X_3\}$. (Again, think – don't compute.)

(c) Let the rv N be the index of the first rv in the sequence to be less than X_1; i.e., $\Pr(N = n) = \Pr(X_1 \leq X_2; X_1 \leq X_3; \cdots; X_1 \leq X_{n-1}; X_1 > X_n)$. Find $\Pr(N \geq n)$ as a function of n. [Hint. Generalize part (b).]

(d) Show that $\mathsf{E}[N] = \infty$. [Hint. Use part (a) of Exercise 2.3.]

(e) Now assume that $X_1, X_2, \ldots$ is a sequence of iid rvs each drawn from a finite set of values. Explain why you can't find $\Pr(X_1 \leq X_2)$ without knowing the pmf. Explain why $\mathsf{E}[N] = \infty$.

2.5 Let $X_1, X_2, \ldots, X_n$ be a sequence of n binary iid rvs. Assume that $\Pr(X_m = 1) = \Pr(X_m = 0) = 1/2$. Let Z be a parity check on $X_1, \ldots, X_n$; i.e., $Z = X_1 \oplus X_2 \oplus \cdots \oplus X_n$ (where $0 \oplus 0 = 1 \oplus 1 = 0$ and $0 \oplus 1 = 1 \oplus 0 = 1$).

(a) Is Z independent of X_1? (Assume $n > 1$.)

(b) Are $Z, X_1, \ldots, X_{n-1}$ independent?

(c) Are $Z, X_1, \ldots, X_n$ independent?

(d) Is Z independent of X_1 if $\Pr(X_i = 1) \neq 1/2$? (You may take $n = 2$ here.)

2.6 Define a suffix-free code as a code in which no codeword is a suffix of any other codeword.

(a) Show that suffix-free codes are uniquely decodable. Use the definition of unique decodability in Section 2.3.1, rather than the intuitive but vague idea of decodability with initial synchronization.

(b) Find an example of a suffix-free code with codeword lengths (1, 2, 2) that is not a prefix-free code. Can a codeword be decoded as soon as its last bit arrives at the decoder? Show that a decoder might have to wait for an arbitrarily long time before decoding (this is why a careful definition of unique decodability is required).

(c) Is there a code with codeword lengths (1, 2, 2) that is both prefix-free and suffix-free? Explain your answer.

2.7 The algorithm given in essence by (2.2) for constructing prefix-free codes from a set of codeword lengths uses the assumption that the lengths have been ordered first. Give an example in which the algorithm fails if the lengths are not ordered first.

2.8 Suppose that, for some reason, you wish to encode a source into symbols from a D-ary alphabet (where D is some integer greater than 2) rather than into a binary alphabet. The development of Section 2.3 can be easily extended to the D-ary case, using D-ary trees rather than binary trees to represent prefix-free codes. Generalize the Kraft inequality, (2.1), to the D-ary case and outline why it is still valid.

2.9 Suppose a prefix-free code has symbol probabilities $p_1, p_2, \ldots, p_M$ and lengths $l_1, \ldots, l_M$. Suppose also that the expected length $\overline{L}$ satisfies $\overline{L} = \mathsf{H}[X]$.

(a) Explain why $p_i = 2^{-l_i}$ for each i.

(b) Explain why the sequence of encoded binary digits is a sequence of iid equiprobable binary digits. [Hint. Use Figure 2.4 to illustrate this phenomenon and explain in words why the result is true in general. Do not attempt a general proof.]

2.10 (a) Show that in a code of M codewords satisfying the Kraft inequality with equality, the maximum length is at most $M-1$. Explain why this ensures that the number of distinct such codes is finite.

(b) Consider the number $S(M)$ of distinct full code trees with M terminal nodes. Count two trees as being different if the corresponding set of codewords is different. That is, ignore the set of source symbols and the mapping between source symbols and codewords. Show that $S(2) = 1$ and show that, for $M > 2$, $S(M) = \sum_{j=1}^{M-1} S(j)S(M-j)$, where $S(1) = 1$ by convention.

2.11 Proof of the Kraft inequality for uniquely decodable codes.

(a) Assume a uniquely decodable code has lengths $l_1, \ldots, l_M$. In order to show that $\sum_j 2^{-l_j} \leq 1$, demonstrate the following identity for each integer $n \geq 1$:

$$\left[\sum_{j=1}^{M} 2^{-l_j}\right]^n = \sum_{j_1=1}^{M}\sum_{j_2=1}^{M}\cdots\sum_{j_n=1}^{M} 2^{-(l_{j_1}+l_{j_2}+\cdots+l_{j_n})}.$$

(b) Show that there is one term on the right for each concatenation of n codewords (i.e. for the encoding of one n-tuple x^n) where $l_{j_1} + l_{j_2} + \cdots + l_{j_n}$ is the aggregate length of that concatenation.

(c) Let A_i be the number of concatenations which have overall length i and show that

$$\left[\sum_{j=1}^{M} 2^{-l_j}\right]^n = \sum_{i=1}^{nl_{\max}} A_i 2^{-i}.$$

(d) Using the unique decodability, upperbound each A_i and show that

$$\left[\sum_{j=1}^{M} 2^{-l_j}\right]^n \le nl_{\max}.$$

(e) By taking the nth root and letting $n \to \infty$, demonstrate the Kraft inequality.

2.12 A source with an alphabet size of $M = |\mathcal{X}| = 4$ has symbol probabilities $\{1/3, 1/3, 2/9, 1/9\}$.

(a) Use the Huffman algorithm to find an optimal prefix-free code for this source.
(b) Use the Huffman algorithm to find another optimal prefix-free code with a different set of lengths.
(c) Find another prefix-free code that is optimal but cannot result from using the Huffman algorithm.

2.13 An alphabet of $M = 4$ symbols has probabilities $p_1 \ge p_2 \ge p_3 \ge p_4 > 0$.

(a) Show that if $p_1 = p_3 + p_4$, then a Huffman code exists with all lengths equal and that another exists with a codeword of length 1, one of length 2, and two of length 3.
(b) Find the largest value of p_1, say $p_{\max}$, for which $p_1 = p_3 + p_4$ is possible.
(c) Find the smallest value of p_1, say $p_{\min}$, for which $p_1 = p_3 + p_4$ is possible.
(d) Show that if $p_1 > p_{\max}$, then every Huffman code has a length 1 codeword.
(e) Show that if $p_1 > p_{\max}$, then every optimal prefix-free code has a length 1 codeword.
(f) Show that if $p_1 < p_{\min}$, then all codewords have length 2 in every Huffman code.
(g) Suppose $M > 4$. Find the smallest value of $p'_{\max}$ such that $p_1 > p'_{\max}$ guarantees that a Huffman code will have a length 1 codeword.

2.14 Consider a source with M equiprobable symbols.

(a) Let $k = \lceil \log M \rceil$. Show that, for a Huffman code, the only possible codeword lengths are k and $k-1$.
(b) As a function of M, find how many codewords have length $k = \lceil \log M \rceil$. What is the expected codeword length $\overline{L}$ in bits per source symbol?
(c) Define $y = M/2^k$. Express $\overline{L} - \log M$ as a function of y. Find the maximum value of this function over $1/2 < y \le 1$. This illustrates that the entropy bound, $\overline{L} < \mathsf{H}[X] + 1$, is rather loose in this equiprobable case.

2.15 Let a discrete memoryless source have M symbols with alphabet $\{1, 2, \ldots, M\}$ and ordered probabilities $p_1 > p_2 > \cdots > p_M > 0$. Assume also that $p_1 < p_{M-1} + p_M$. Let $l_1, l_2, \ldots, l_M$ be the lengths of a prefix-free code of minimum expected length for such a source.

(a) Show that $l_1 \leq l_2 \leq \cdots \leq l_M$.
(b) Show that if the Huffman algorithm is used to generate the above code, then $l_M \leq l_1 + 1$. [Hint. Look only at the first step of the algorithm.]
(c) Show that $l_M \leq l_1 + 1$ whether or not the Huffman algorithm is used to generate a minimum expected length prefix-free code.
(d) Suppose $M = 2^k$ for integer k. Determine $l_1, \ldots, l_M$.
(e) Suppose $2^k < M < 2^{k+1}$ for integer k. Determine $l_1, \ldots, l_M$.

2.16 (a) Consider extending the Huffman procedure to codes with ternary symbols $\{0, 1, 2\}$. Think in terms of codewords as leaves of ternary trees. Assume an alphabet with $M = 4$ symbols. Note that you cannot draw a full ternary tree with four leaves. By starting with a tree of three leaves and extending the tree by converting leaves into intermediate nodes, show for what values of M it is possible to have a complete ternary tree.
(b) Explain how to generalize the Huffman procedure to ternary symbols, bearing in mind your result in part (a).
(c) Use your algorithm for the set of probabilities $\{0.3, 0.2, 0.2, 0.1, 0.1, 0.1\}$.

2.17 Let X have M symbols, $\{1, 2, \ldots, M\}$ with ordered probabilities $p_1 \geq p_2 \geq \cdots \geq p_M > 0$. Let X' be the reduced source after the first step of the Huffman algorithm.

(a) Express the entropy $\mathsf{H}[X]$ for the original source in terms of the entropy $\mathsf{H}[X']$ of the reduced source as follows:

$$\mathsf{H}[X] = \mathsf{H}[X'] + (p_M + p_{M-1})H(\gamma), \tag{2.43}$$

where $H(\gamma)$ is the binary entropy function, $H(\gamma) = -\gamma \log \gamma - (1-\gamma) \log(1-\gamma)$. Find the required value of γ to satisfy (2.43).
(b) In the code tree generated by the Huffman algorithm, let v_1 denote the intermediate node that is the parent of the leaf nodes for symbols M and $M-1$. Let $q_1 = p_M + p_{M-1}$ be the probability of reaching v_1 in the code tree. Similarly, let $v_2, v_3, \ldots,$ denote the subsequent intermediate nodes generated by the Huffman algorithm. How many intermediate nodes are there, including the root node of the entire tree?
(c) Let $q_1, q_2, \ldots,$ be the probabilities of reaching the intermediate nodes $v_1, v_2, \ldots,$ (note that the probability of reaching the root node is 1). Show that $\overline{L} = \sum_i q_i$. [Hint. Note that $\overline{L} = \overline{L}' + q_1$.]
(d) Express $\mathsf{H}[X]$ as a sum over the intermediate nodes. The ith term in the sum should involve q_i and the binary entropy $H(\gamma_i)$ for some γ_i to be determined. You may find it helpful to define α_i as the probability of moving upward from intermediate node v_i, conditional on reaching v_i. [Hint. Look at part (a).]
(e) Find the conditions (in terms of the probabilities and binary entropies above) under which $\overline{L} = \mathsf{H}[X]$.
(f) Are the formulas for $\overline{L}$ and $\mathsf{H}[X]$ above specific to Huffman codes alone, or do they apply (with the modified intermediate node probabilities and entropies) to arbitrary full prefix-free codes?

2.18 Consider a discrete random symbol X with $M+1$ symbols for which $p_1 \geq p_2 \geq \cdots \geq p_M > 0$ and $p_{M+1} = 0$. Suppose that a prefix-free code is generated for X and that, for some reason, this code contains a codeword for $M+1$ (suppose, for example, that p_{M+1} is actually positive but so small that it is approximated as 0).

(a) Find $\overline{L}$ for the Huffman code including symbol $M+1$ in terms of $\overline{L}$ for the Huffman code omitting a codeword for symbol $M+1$.

(b) Suppose now that instead of one symbol of zero probability, there are n such symbols. Repeat part (a) for this case.

2.19 In (2.12), it is shown that if X and Y are independent discrete random symbols, then the entropy for the random symbol XY satisfies $\mathsf{H}[XY] = \mathsf{H}[X] + \mathsf{H}[Y]$. Here we want to show that, without the assumption of independence, we have $\mathsf{H}[XY] \leq \mathsf{H}[X] + \mathsf{H}[Y]$.

(a) Show that

$$\mathsf{H}[XY] - \mathsf{H}[X] - \mathsf{H}[Y] = \sum_{x \in \mathcal{X}, y \in \mathcal{Y}} p_{XY}(x, y) \log \frac{p_X(x) p_Y(y)}{p_{X,Y}(x, y)}.$$

(b) Show that $\mathsf{H}[XY] - \mathsf{H}[X] - \mathsf{H}[Y] \leq 0$, i.e. that $\mathsf{H}[XY] \leq \mathsf{H}[X] + \mathsf{H}[Y]$.

(c) Let $X_1, X_2, \ldots, X_n$ be discrete random symbols, not necessarily independent. Use your answer to part (b) to show that

$$H[X_1 X_2 \cdots X_n] \leq \sum_{j=1}^{n} H[X_j].$$

2.20 Consider a random symbol X with the symbol alphabet $\{1, 2, \ldots, M\}$ and a pmf $\{p_1, p_2, \ldots, p_M\}$. This exercise derives a relationship called *Fano's inequality* between the entropy $\mathsf{H}[X]$ and the probability p_1 of the first symbol. This relationship is used to prove the converse to the noisy channel coding theorem. Let Y be a random symbol that is 1 if $X = 1$ and 0 otherwise. For parts (a) through (d), consider M and p_1 to be fixed.

(a) Express $\mathsf{H}[Y]$ in terms of the binary entropy function, $H_b(\alpha) = -\alpha \log(\alpha) - (1-\alpha)\log(1-\alpha)$.

(b) What is the conditional entropy $\mathsf{H}[X \mid Y=1]$?

(c) Show that $\mathsf{H}[X \mid Y=0] \leq \log(M-1)$ and show how this bound can be met with equality by appropriate choice of $p_2, \ldots, p_M$. Combine this with part (b) to upperbound $\mathsf{H}[X \mid Y]$.

(d) Find the relationship between $\mathsf{H}[X]$ and $\mathsf{H}[XY]$

(e) Use $\mathsf{H}[Y]$ and $\mathsf{H}[X \mid Y]$ to upperbound $\mathsf{H}[X]$ and show that the bound can be met with equality by appropriate choice of $p_2, \ldots, p_M$.

(f) For the same value of M as before, let $p_1, \ldots, p_M$ be arbitrary and let $p_{\max}$ be $\max\{p_1, \ldots, p_M\}$. Is your upperbound in (e) still valid if you replace p_1 by $p_{\max}$? Explain your answer.

2.21 A discrete memoryless source emits iid random symbols $X_1, X_2, \ldots$ Each random symbol X has the symbols $\{a, b, c\}$ with probabilities $\{0.5, 0.4, 0.1\}$, respectively.

(a) Find the expected length $\overline{L}_{\min}$ of the best variable-length prefix-free code for X.

(b) Find the expected length $\overline{L}_{\min,2}$, normalized to bits per symbol, of the best variable-length prefix-free code for X^2.

(c) Is it true that for any DMS, $\overline{L}_{\min} \geq \overline{L}_{\min,2}$? Explain your answer.

2.22 For a DMS X with alphabet $\mathcal{X} = \{1, 2, \ldots, M\}$, let $L_{\min,1}$, $L_{\min,2}$, and $L_{\min,3}$ be the normalized average lengths in bits per source symbol for a Huffman code over $\mathcal{X}$, $\mathcal{X}^2$ and $\mathcal{X}^3$, respectively. Show that $L_{\min,3} \leq (2/3)L_{\min,2} + (1/3)L_{\min,1}$.

2.23 (Run-length coding) Suppose $X_1, X_2, \ldots,$ is a sequence of binary random symbols with $p_X(a) = 0.9$ and $p_X(b) = 0.1$. We encode this source by a variable-to-variable-length encoding technique known as run-length coding. The source output is first mapped into intermediate digits by counting the number of occurrences of a between each b. Thus an intermediate output occurs on each occurence of the symbol b. Since we do not want the intermediate digits to get too large, however, the intermediate digit 8 is used on the eighth consecutive a, and the counting restarts at this point. Thus, outputs appear on each b and on each eighth a. For example, the first two lines below illustrate a string of source outputs and the corresponding intermediate outputs:

b	*a*	*a*	*a*	*b*	*a*	*a*	*a*	*a*	*a*	*a*	*a*	*a*	*a*	*a*	*b*	*b*	*a*	*a*	*a*	*a*	*b*
0				3								8			2	0					4
0000				0011								1			0010	0000					0100

The final stage of encoding assigns the codeword 1 to the intermediate integer 8, and assigns a 4 bit codeword consisting of 0 followed by the 3 bit binary representation for each integer 0 to 7. This is illustrated in the third line above.

(a) Show why the overall code is uniquely decodable.

(b) Find the expected total number of output bits corresponding to each occurrence of the letter b. This total number includes the 4 bit encoding of the letter b *and* the 1 bit encodings for each consecutive string of eight occurrences of a preceding that letter b.

(c) By considering a string of 10^{20} binary symbols into the encoder, show that the number of occurrences of b per input symbol is, with very high probability, very close to 0.1.

(d) Combine parts (b) and (c) to find $\overline{L}$, the expected number of output bits per input symbol.

2.24 (a) Suppose a DMS emits h and t with probability 1/2 each. For $\varepsilon = 0.01$, what is T_ε^5?

(b) Find T_ε^1 for $\Pr(h) = 0.1$, $\Pr(t) = 0.9$, and $\varepsilon = 0.001$.

2.25 Consider a DMS with a two-symbol alphabet $\{a, b\}$, where $p_X(a) = 2/3$ and $p_X(b) = 1/3$. Let $X^n = X_1, \ldots, X_n$ be a string of random symbols from the source with $n = 100\,000$.

(a) Let $W(X_j)$ be the log pmf rv for the jth source output, i.e. $W(X_j) = -\log 2/3$ for $X_j = a$ and $-\log 1/3$ for $X_j = b$. Find the variance of $W(X_j)$.

(b) For $\varepsilon = 0.01$, evaluate the bound on the probability of the typical set given in (2.24).

(c) Let N_a be the number of occurrences of a in the string $X^n = X_1, \ldots, X_n$. The rv N_a is the sum of n iid rvs. Show what these rvs are.

(d) Express the rv $W(X^n)$ as a function of the rv N_a. Note how this depends on n.

(e) Express the typical set in terms of bounds on N_a (i.e. $T_\varepsilon^n = \{x^n : \alpha < N_a < \beta\}$ and calculate α and β).

(f) Find the mean and variance of N_a. Approximate $\Pr(T_\varepsilon^n)$ by the central limit theorem approximation. The central limit theorem approximation is to evaluate $\Pr(T_\varepsilon^n)$ assuming that N_a is Gaussian with the mean and variance of the actual N_a.

One point of this exercise is to illustrate that the Chebyshev inequality used in bounding $\Pr(T_\varepsilon)$ in the text is very weak (although it is a strict bound, whereas the Gaussian approximation here is relatively accurate but not a bound). Another point is to show that n must be very large for the typical set to look typical.

2.26 For the rvs in Exercise 2.25, find $\Pr(N_a = i)$ for $i = 0, 1, 2$. Find the probability of each individual string x^n for those values of i. Find the particular string x^n that has maximum probability over all sample values of X^n. What are the next most probable n-strings? Give a brief discussion of why the most probable n-strings are not regarded as typical strings.

2.27 Let $X_1, X_2, \ldots$ be a sequence of iid symbols from a finite alphabet. For any block length n and any small number $\varepsilon > 0$, define the *good* set of n-tuples x^n as the set given by

$$G_\varepsilon^n = \left\{x^n : \; p_{X^n}(x^n) > 2^{-n[\mathsf{H}[X]+\varepsilon]}\right\}.$$

(a) Explain how G_ε^n differs from the typical set T_ε^n.

(b) Show that $\Pr(G_\varepsilon^n) \geq 1 - \sigma_W^2/n\varepsilon^2$, where W is the log pmf rv for X. Nothing elaborate is expected here.

(c) Derive an upperbound on the number of elements in G_ε^n of the form $|G_\varepsilon^n| < 2^{n(\mathsf{H}[X]+\alpha)}$ and determine the value of α. (You are expected to find the smallest such α that you can, but not to prove that no smaller value can be used in an upperbound.)

(d) Let $G_\varepsilon^n - T_\varepsilon^n$ be the set of n-tuples x^n that lie in G_ε^n but not in T_ε^n. Find an upperbound to $|G_\varepsilon^n - T_\varepsilon^n|$ of the form $|G_\varepsilon^n - T_\varepsilon^n| \leq 2^{n(\mathsf{H}[X]+\beta)}$. Again find the smallest β that you can.

(e) Find the limit of $|G_\varepsilon^n - T_\varepsilon^n|/|T_\varepsilon^n|$ as $n \to \infty$.

2.28 The typical set T_ε^n defined in the text is often called a weakly typical set, in contrast to another kind of typical set called a strongly typical set. Assume a discrete memoryless source and let $N_j(\boldsymbol{x}^n)$ be the number of symbols in an n-string $\boldsymbol{x}^n$ taking on the value j. Then the strongly typical set S_ε^n is defined as follows:

$$S_\varepsilon^n = \left\{ \boldsymbol{x}^n : \; p_j(1-\varepsilon) < \frac{N_j(\boldsymbol{x}^n)}{n} < p_j(1+\varepsilon); \quad \text{for all } j \in \mathcal{X} \right\}.$$

(a) Show that $p_{X^n}(\boldsymbol{x}^n) = \prod_j p_j^{N_j(\boldsymbol{x}^n)}$.

(b) Show that every $\boldsymbol{x}^n$ in S_ε^n has the following property:

$$\mathsf{H}[X](1-\varepsilon) < \frac{-\log p_{X^n}(\boldsymbol{x}^n)}{n} < \mathsf{H}[X](1+\varepsilon).$$

(c) Show that if $\boldsymbol{x}^n \in S_\varepsilon^n$, then $\boldsymbol{x}^n \in T_{\varepsilon'}^n$ with $\varepsilon' = \mathsf{H}[X]\varepsilon$, i.e. that $S_\varepsilon^n \subseteq T_{\varepsilon'}^n$.

(d) Show that for any $\delta > 0$ and all sufficiently large n,

$$\Pr(X^n \notin S_\varepsilon^n) \le \delta.$$

[Hint. Taking each letter j separately, $1 \le j \le M$, show that for all sufficiently large n, $\Pr\left(\left|N_j/n - p_j\right| \ge \varepsilon\right) \le \delta/M$.]

(e) Show that for all $\delta > 0$ and all sufficiently large n,

$$(1-\delta)2^{n(\mathsf{H}[X]-\varepsilon)} < |S_\varepsilon^n| < 2^{n(\mathsf{H}[X]+\varepsilon)}. \tag{2.44}$$

Note that parts (d) and (e) constitute the same theorem for the strongly typical set as Theorem 2.7.1 establishes for the weakly typical set. Typically the n required for (2.44) to hold (with the correspondence in part (c) between ε and ε') is considerably larger than than that for (2.27) to hold. We will use strong typicality later in proving the noisy channel coding theorem.

2.29 (a) The random variable D_n in Section 2.7.4 was defined as the initial string length of encoded bits required to decode the first n symbols of the source input. For the run-length coding example in Exercise 2.23, list the input strings and corresponding encoded output strings that must be inspected to decode the first source letter, and from this find the pmf function of D_1. [Hint. As many as eight source letters must be encoded before X_1 can be decoded.]

(b) Find the pmf of D_2. One point of this exercise is to convince you that D_n is a useful rv for proving theorems, but not an rv that is useful for detailed computation. It also shows clearly that D_n can depend on more than the first n source letters.

2.30 The Markov chain $S_0, S_1, \ldots$ defined by Figure 2.17 starts in steady state at time 0 and has four states, $\mathcal{S} = \{1, 2, 3, 4\}$. The corresponding Markov source $X_1, X_2, \ldots$ has a source alphabet $\mathcal{X} = \{a, b, c\}$ of size 3.

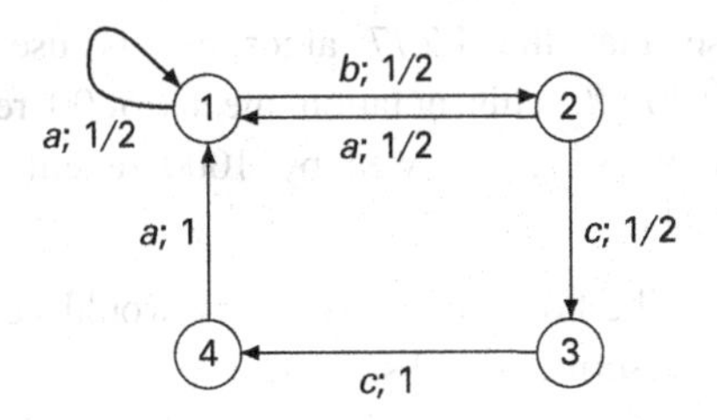

Figure 2.17.

(a) Find the steady-state probabilities $\{q(s)\}$ of the Markov chain.
(b) Find $\mathsf{H}[X_1]$.
(c) Find $\mathsf{H}[X_1|S_0]$.
(d) Describe a uniquely decodable encoder for which $\overline{L} = \mathsf{H}[X_1|S_0)$. Assume that the initial state is known to the decoder. Explain why the decoder can track the state after time 0.
(e) Suppose you observe the source output without knowing the state. What is the maximum number of source symbols you must observe before knowing the state?

2.31 Let $X_1, X_2, \ldots, X_n$ be discrete random symbols. Derive the following chain rule:

$$\mathsf{H}[X_1, \ldots, X_n] = \mathsf{H}[X_1] + \sum_{k=2}^{n} \mathsf{H}[X_k|X_1, \ldots, X_{k-1}].$$

[Hint. Use the chain rule for $n = 2$ in (2.37) and ask yourself whether a k-tuple of random symbols is itself a random symbol.]

2.32 Consider a discrete ergodic Markov chain $S_0, S_1, \ldots$ with an arbitrary initial state distribution.

(a) Show that $\mathsf{H}[S_2|S_1S_0] = \mathsf{H}[S_2|S_1]$ (use the basic definition of conditional entropy).
(b) Show with the help of Exercise 2.31 that, for any $n \geq 2$,

$$\mathsf{H}[S_1S_2 \cdots S_n|S_0] = \sum_{k=1}^{n} \mathsf{H}[S_k|S_{k-1}].$$

(c) Simplify this for the case where S_0 is in steady state.
(d) For a Markov source with outputs $X_1X_2 \cdots$, explain why $\mathsf{H}[X_1 \cdots X_n|S_0] = \mathsf{H}[S_1 \cdots S_n|S_0]$. You may restrict this to $n = 2$ if you desire.
(e) Verify (2.40).

2.33 Perform an LZ77 parsing of the string $\underline{00011101}0010101100$. Assume a window of length $W = 8$; the initial window is underlined above. You should parse the rest of the string using the Lempel–Ziv algorithm.

2.34 Suppose that the LZ77 algorithm is used on the binary string $x_1^{10\,000} = 0^{5000}1^{4000}0^{1000}$. This notation means 5000 repetitions of 0, followed by 4000 repetitions of 1, followed by 1000 repetitions of 0. Assume a window size $w = 1024$.

(a) Describe how the above string would be encoded. Give the encoded string and describe its substrings.

(b) How long is the encoded string?

(c) Suppose that the window size is reduced to $w = 8$. How long would the encoded string be in this case? (Note that such a small window size would only work well for really simple examples like this one.)

(d) Create a Markov source model with two states that is a reasonably good model for this source output. You are not expected to do anything very elaborate here; just use common sense.

(e) Find the entropy in bits per source symbol for your source model.

2.35 (a) Show that if an optimum (in the sense of minimum expected length) prefix-free code is chosen for any given pmf (subject to the condition $p_i > p_j$ for $i < j$), the codeword lengths satisfy $l_i \leq l_j$ for all $i < j$. Use this to show that, for all $j \geq 1$,

$$l_j \geq \lfloor \log j \rfloor + 1.$$

(b) The asymptotic efficiency of a prefix-free code for the positive integers is defined to be $\lim_{j\to\infty} l_j / \log j$. What is the asymptotic efficiency of the unary–binary code?

(c) Explain how to construct a prefix-free code for the positive integers where the asymptotic efficiency is 1. [Hint. Replace the unary code for the integers $n = \lfloor \log j \rfloor + 1$ in the unary–binary code with a code whose length grows more slowly with increasing n.]

3 Quantization

3.1 Introduction to quantization

Chapter 2 discussed coding and decoding for discrete sources. Discrete sources are a subject of interest in their own right (for text, computer files, etc.) and also serve as the inner layer for encoding analog source sequences and waveform sources (see Figure 3.1). This chapter treats coding and decoding for a sequence of analog values. Source coding for analog values is usually called *quantization*. Note that this is also the middle layer for waveform encoding/decoding.

The input to the quantizer will be modeled as a sequence $U_1, U_2, \dots,$ of analog random variables (rvs). The motivation for this is much the same as that for modeling the input to a discrete source encoder as a sequence of random symbols. That is, the design of a quantizer should be responsive to the set of possible inputs rather than being designed for only a single sequence of numerical inputs. Also, it is desirable to treat very rare inputs differently from very common inputs, and a probability density is an ideal approach for this. Initially, $U_1, U_2, \dots$ will be taken as independent identically distributed (iid) analog rvs with some given probability density function (pdf) $f_U(u)$.

A quantizer, by definition, maps the incoming sequence $U_1, U_2, \dots,$ into a sequence of discrete rvs $V_1, V_2, \dots,$ where the objective is that V_m, for each m in the sequence, should represent U_m with as little distortion as possible. Assuming that the discrete encoder/decoder at the inner layer of Figure 3.1 is uniquely decodable, the sequence $V_1, V_2, \dots$ will appear at the output of the discrete encoder and will be passed through the middle layer (denoted "table lookup") to represent the input $U_1, U_2, \dots$. The output side of the quantizer layer is called a "table lookup" because the alphabet for each discrete random variable V_m is a finite set of real numbers, and these are usually mapped into another set of symbols such as the integers 1 to M for an M-symbol alphabet. Thus on the output side a lookup function is required to convert back to the numerical value V_m.

As discussed in Section 2.1, the quantizer output V_m, if restricted to an alphabet of M possible values, cannot represent the analog input U_m perfectly. Increasing M, i.e. quantizing more finely, typically reduces the distortion, but cannot eliminate it.

When an analog rv U is quantized into a discrete rv V, the mean-squared distortion is defined to be $\mathsf{E}[(U-V)^2]$. Mean-squared distortion (often called mean-squared error) is almost invariably used in this text to measure distortion. When studying the conversion of waveforms into sequences in Chapter 4, it will be seen that mean-squared distortion

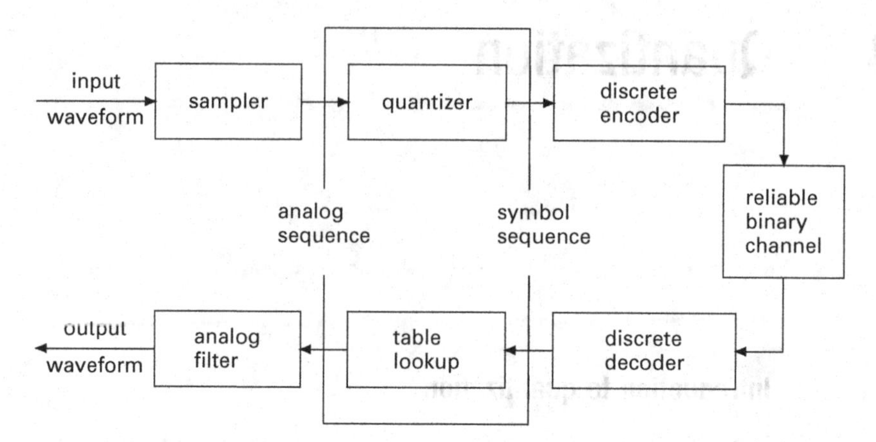

Figure 3.1. Encoding and decoding of discrete sources, analog sequence sources, and waveform sources. Quantization, the topic of this chapter, is the middle layer and should be understood before trying to understand the outer layer, which deals with waveform sources.

is a particularly convenient measure for converting the distortion for the sequence into the distortion for the waveform.

There are some disadvantages to measuring distortion only in a mean-squared sense. For example, efficient speech coders are based on models of human speech. They make use of the fact that human listeners are more sensitive to some kinds of reconstruction error than others, so as, for example, to permit larger errors when the signal is loud than when it is soft. Speech coding is a specialized topic which we do not have time to explore (see, for example, Gray (1990)). However, understanding compression relative to a mean-squared distortion measure will develop many of the underlying principles needed in such more specialized studies.

In what follows, scalar quantization is considered first. Here each analog rv in the sequence is quantized independently of the other rvs. Next, vector quantization is considered. Here the analog sequence is first segmented into blocks of n rvs each; then each n-tuple is quantized as a unit.

Our initial approach to both scalar and vector quantization will be to minimize mean-squared distortion subject to a constraint on the size of the quantization alphabet. Later, we consider minimizing mean-squared distortion subject to a constraint on the *entropy* of the quantized output. This is the relevant approach to quantization if the quantized output sequence is to be source-encoded in an efficient manner, i.e. to reduce the number of encoded bits per quantized symbol to little more than the corresponding entropy.

3.2 Scalar quantization

A *scalar quantizer* partitions the set $\mathbb{R}$ of real numbers into M subsets $\mathcal{R}_1, \ldots, \mathcal{R}_M$, called *quantization regions*. Assume that each quantization region is an interval; it will soon be seen why this assumption makes sense. Each region $\mathcal{R}_j$ is then represented

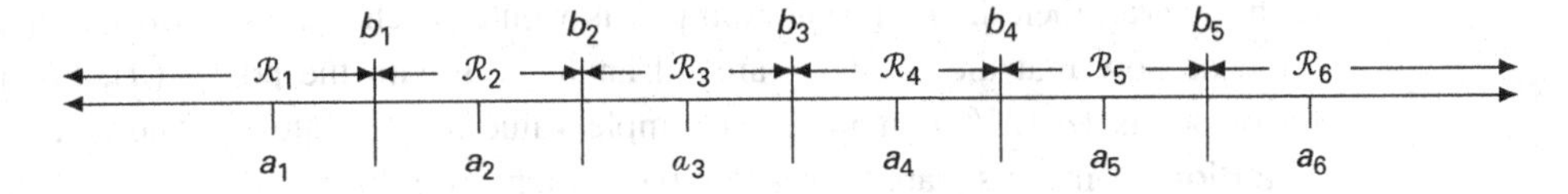

Figure 3.2. Quantization regions and representation points.

by a *representation point* $a_j \in \mathbb{R}$. When the source produces a number $u \in \mathcal{R}_j$, that number is quantized into the point a_j. A scalar quantizer can be viewed as a function $\{v(u) : \mathbb{R} \to \mathbb{R}\}$ that maps analog real values u into discrete real values $v(u)$, where $v(u) = a_j$ for $u \in \mathcal{R}_j$.

An analog sequence $u_1, u_2, \ldots$ of real-valued symbols is mapped by such a quantizer into the discrete sequence $v(u_1), v(u_2), \ldots$ Taking $u_1, u_2, \ldots$ as sample values of a random sequence $U_1, U_2, \ldots$, the map $v(u)$ generates an rv V_k for each U_k; V_k takes the value a_j if $U_k \in \mathcal{R}_j$. Thus each quantized output V_k is a discrete rv with the alphabet $\{a_1, \ldots, a_M\}$. The discrete random sequence $V_1, V_2, \ldots$ is encoded into binary digits, transmitted, and then decoded back into the same discrete sequence. For now, assume that transmission is error-free.

We first investigate how to choose the quantization regions $\mathcal{R}_1, \ldots, \mathcal{R}_M$ and how to choose the corresponding representation points. Initially assume that the regions are intervals, ordered as in Figure 3.2, with $\mathcal{R}_1 = (-\infty, b_1]$, $\mathcal{R}_2 = (b_1, b_2], \ldots, \mathcal{R}_M = (b_{M-1}, \infty)$. Thus an M-level quantizer is specified by $M-1$ interval endpoints, $b_1, \ldots, b_{M-1}$, and M representation points, $a_1, \ldots, a_M$.

For a given value of M, how can the regions and representation points be chosen to minimize mean-squared error? This question is explored in two ways as follows.

- Given a set of representation points $\{a_j\}$, how should the intervals $\{\mathcal{R}_j\}$ be chosen?
- Given a set of intervals $\{\mathcal{R}_j\}$, how should the representation points $\{a_j\}$ be chosen?

3.2.1 Choice of intervals for given representation points

The choice of intervals for given representation points, $\{a_j; 1 \le j \le M\}$, is easy: given any $u \in \mathbb{R}$, the squared error to a_j is $(u - a_j)^2$. This is minimized (over the fixed set of representation points $\{a_j\}$) by representing u by the closest representation point a_j. This means, for example, that if u is between a_j and a_{j+1}, then u is mapped into the closer of the two. Thus the boundary b_j between $\mathcal{R}_j$ and $\mathcal{R}_{j+1}$ must lie halfway between the representation points a_j and a_{j+1}, $1 \le j \le M-1$. That is, $b_j = a_j + a_{j+1}/2$. This specifies each quantization region, and also shows why each region should be an interval. Note that this minimization of mean-squared distortion does not depend on the probabilistic model for $U_1, U_2, \ldots$

3.2.2 Choice of representation points for given intervals

For the second question, the probabilistic model for $U_1, U_2, \ldots$ is important. For example, if it is known that each U_k is discrete and has only one sample value in each interval, then the representation points would be chosen as those sample values. Suppose now that the rvs $\{U_k\}$ are iid analog rvs with the pdf $f_U(u)$. For a given set of points $\{a_j\}$, $V(U)$ maps each sample value $u \in \mathcal{R}_j$ into a_j. The mean-squared distortion, or mean-squared error (MSE), is then given by

$$\mathrm{MSE} = \mathsf{E}[(U - V(U))^2] = \int_{-\infty}^{\infty} f_U(u)(u - v(u))^2 \, \mathrm{d}u = \sum_{j=1}^{M} \int_{\mathcal{R}_j} f_U(u)\left(u - a_j\right)^2 \mathrm{d}u. \quad (3.1)$$

In order to minimize (3.1) over the set of a_j, it is simply necessary to choose each a_j to minimize the corresponding integral (remember that the regions are considered fixed here). Let $f_j(u)$ denote the conditional pdf of U given that $\{u \in \mathcal{R}_j\}$; i.e.,

$$f_j(u) = \begin{cases} \frac{f_U(u)}{Q_j}, & \text{if } \; u \in \mathcal{R}_j; \\ 0, & \text{otherwise,} \end{cases} \quad (3.2)$$

where $Q_j = \Pr(U \in \mathcal{R}_j)$. Then, for the interval $\mathcal{R}_j$,

$$\int_{\mathcal{R}_j} f_U(u)\left(u - a_j\right)^2 \mathrm{d}u = Q_j \int_{\mathcal{R}_j} f_j(u)\left(u - a_j\right)^2 \mathrm{d}u. \quad (3.3)$$

Now (3.3) is minimized by choosing a_j to be the mean of a random variable with the pdf $f_j(u)$. To see this, note that for any rv Y and real number a,

$$\overline{(Y - a)^2} = \overline{Y^2} - 2a\overline{Y} + a^2,$$

which is minimized over a when $a = \overline{Y}$.

This provides a set of conditions that the endpoints $\{b_j\}$ and the points $\{a_j\}$ must satisfy to achieve the MSE – namely, each b_j must be the midpoint between a_j and a_{j+1} and each a_j must be the mean of an rv U_j with pdf $f_j(u)$. In other words, a_j must be the conditional mean of U conditional on $U \in \mathcal{R}_j$.

These conditions are necessary to minimize the MSE for a given number M of representation points. They are not sufficient, as shown by Example 3.2.1. Nonetheless, these necessary conditions provide some insight into the minimization of the MSE.

3.2.3 The Lloyd–Max algorithm

The *Lloyd–Max algorithm*[1] is an algorithm for finding the endpoints $\{b_j\}$ and the representation points $\{a_j\}$ to meet the above necessary conditions. The algorithm is

[1] This algorithm was developed independently by S. P. Lloyd in 1957 and J. Max in 1960. Lloyd's work was performed in the Bell Laboratories research department and became widely circulated, although it was not published until 1982. Max's work was published in 1960. See Lloyd (1982) and Max (1960).

almost obvious given the necessary conditions; the contribution of Lloyd and Max was to define the problem and develop the necessary conditions. The algorithm simply alternates between the optimizations of the previous subsections, namely optimizing the endpoints $\{b_j\}$ for a given set of $\{a_j\}$, and then optimizing the points $\{a_j\}$ for the new endpoints.

The Lloyd–Max algorithm is as follows. Assume that the number M of quantizer levels and the pdf $f_U(u)$ are given.

(1) Choose an arbitrary initial set of M representation points $a_1 < a_2 < \cdots < a_M$.
(2) For each j; $1 \le j \le M-1$, set $b_j = (1/2)(a_{j+1} + a_j)$.
(3) For each j; $1 \le j \le M$, set a_j equal to the conditional mean of U given $U \in (b_{j-1}, b_j]$ (where b_0 and b_M are taken to be $-\infty$ and $+\infty$, respectively).
(4) Repeat steps (2) and (3) until further improvement in MSE is negligible; then stop.

The MSE decreases (or remains the same) for each execution of step (2) and step (3). Since the MSE is nonnegative, it approaches some limit. Thus if the algorithm terminates when the MSE improvement is less than some given $\varepsilon > 0$, then the algorithm must terminate after a finite number of iterations.

Example 3.2.1 This example shows that the algorithm might reach a local minimum of MSE instead of the global minimum. Consider a quantizer with $M = 2$ representation points, and an rv U whose pdf $f_U(u)$ has three peaks, as shown in Figure 3.3.

It can be seen that one region must cover two of the peaks, yielding quite a bit of distortion, while the other will represent the remaining peak, yielding little distortion. In the figure, the two rightmost peaks are both covered by $\mathcal{R}_2$, with the point a_2 between them. Both the points and the regions satisfy the necessary conditions and cannot be locally improved. However, it can be seen that the rightmost peak is more probable than the other peaks. It follows that the MSE would be lower if $\mathcal{R}_1$ covered the two leftmost peaks.

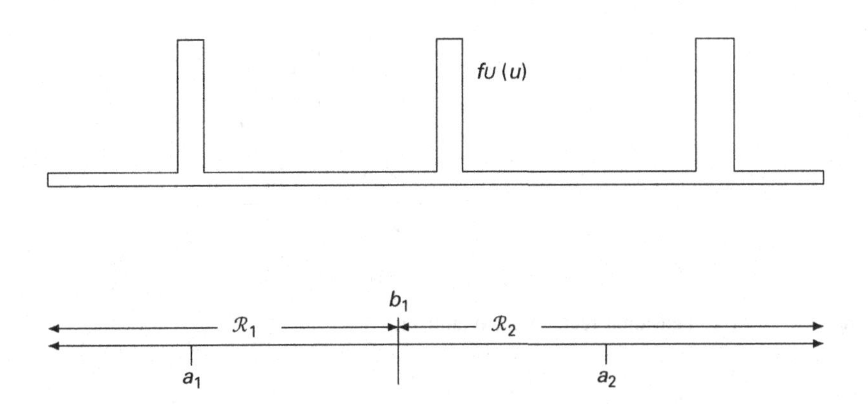

Figure 3.3. Example of regions and representation points that satisfy the Lloyd–Max conditions without minimizing mean-squared distortion.

The Lloyd–Max algorithm is a type of hill-climbing algorithm; starting with an arbitrary set of values, these values are modified until reaching the top of a hill where no more local improvements are possible.[2] A reasonable approach in this sort of situation is to try many randomly chosen starting points, perform the Lloyd–Max algorithm on each, and then take the best solution. This is somewhat unsatisfying since there is no general technique for determining when the optimal solution has been found.

3.3 Vector quantization

As with source coding of discrete sources, we next consider quantizing n source variables at a time. This is called *vector quantization*, since an n-tuple of rvs may be regarded as a vector rv in an n-dimensional vector space. We will concentrate on the case $n = 2$ so that illustrative pictures can be drawn.

One possible approach is to quantize each dimension independently with a scalar (one-dimensional) quantizer. This results in a rectangular grid of quantization regions, as shown in Figure 3.4. The MSE per dimension is the same as for the scalar quantizer using the same number of bits per dimension. Thus the best two-dimensional (2D) vector quantizer has an MSE per dimension at least as small as that of the best scalar quantizer.

To search for the minimum-MSE 2D vector quantizer with a given number M of representation points, the same approach is used as with scalar quantization.

Let (U, U') be the two rvs being jointly quantized. Suppose a set of M 2D representation points $\{(a_j, a'_j)\}$, $1 \le j \le M$, is chosen. For example, in Figure 3.4, there are 16 representation points, represented by small dots. Given a sample pair (u, u') and given the M representation points, which representation point should be chosen for the given (u, u')? Again, the answer is easy. Since mapping (u, u') into (a_j, a'_j) generates

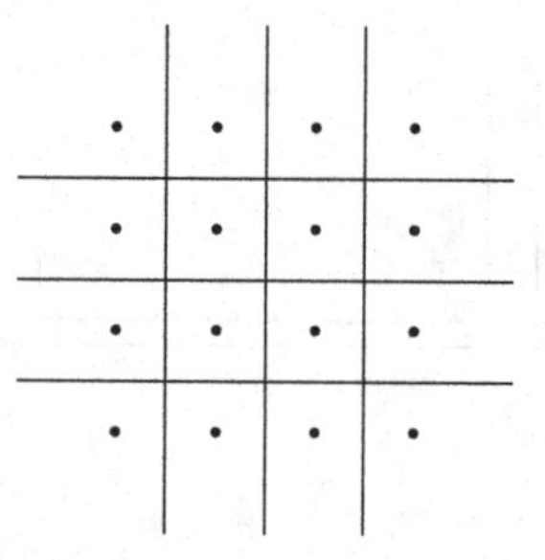

Figure 3.4. Two-dimensional rectangular quantizer.

[2] It would be better to call this a valley-descending algorithm, both because a minimum is desired and also because binoculars cannot be used at the bottom of a valley to find a distant lower valley.

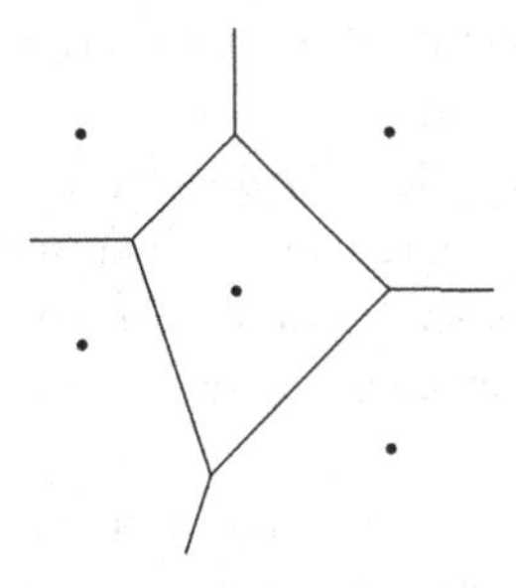

Figure 3.5. Voronoi regions for a given set of representation points.

a squared error equal to $(u - a_j)^2 + (u' - a'_j)^2$, the point (a_j, a'_j) which is closest to (u, u') in Euclidean distance should be chosen.

Consequently, the region $\mathcal{R}_j$ must be the set of points (u, u') that are closer to (a_j, a'_j) than to any other representation point. Thus the regions $\{\mathcal{R}_j\}$ are minimum-distance regions; these regions are called the *Voronoi* regions for the given representation points. The boundaries of the Voronoi regions are perpendicular bisectors between neighboring representation points. The minimum-distance regions are thus, in general, convex polygonal regions, as illustrated in Figure 3.5.

As in the scalar case, the MSE can be minimized for a given set of regions by choosing the representation points to be the conditional means within those regions. Then, given this new set of representation points, the MSE can be further reduced by using the Voronoi regions for the new points. This gives us a 2D version of the Lloyd–Max algorithm, which must converge to a local minimum of the MSE. This can be generalized straightforwardly to any dimension n.

As already seen, the Lloyd–Max algorithm only finds local minima to the MSE for scalar quantizers. For vector quantizers, the problem of local minima becomes even worse. For example, when $U_1, U_2, \ldots$ are iid, it is easy to see that the rectangular quantizer in Figure 3.4 satisfies the Lloyd–Max conditions if the corresponding scalar quantizer does (see Exercise 3.10). It will soon be seen, however, that this is not necessarily the minimum MSE.

Vector quantization was a popular research topic for many years. The problem is that quantizing complexity goes up exponentially with n, and the reduction in MSE with increasing n is quite modest, unless the samples are statistically highly dependent.

3.4 Entropy-coded quantization

We must now ask if minimizing the MSE for a given number M of representation points is the right problem. The minimum expected number of bits per symbol, $\overline{L}_{\min}$, required to encode the quantizer output was shown in Chapter 2 to be governed by the entropy $\mathsf{H}[V]$ of the quantizer output, not by the size M of the quantization alphabet. Therefore, anticipating efficient source coding of the quantized outputs, we should really try to minimize the MSE for a given entropy $\mathsf{H}[V]$ rather than a given number of representation points.

This approach is called *entropy-coded quantization* and is almost implicit in the layered approach to source coding represented in Figure 3.1. Discrete source coding close to the entropy bound is similarly often called entropy coding. Thus entropy-coded quantization refers to quantization techniques that are designed to be followed by entropy coding.

The entropy $\mathsf{H}[V]$ of the quantizer output is determined only by the probabilities of the quantization regions. Therefore, given a set of regions, choosing the representation points as conditional means minimizes their distortion without changing the entropy. However, given a set of representation points, the optimal regions are not necessarily Voronoi regions (e.g., in a scalar quantizer, the point separating two adjacent regions is not necessarily equidistant from the two representation points).

For example, for a scalar quantizer with a constraint $\mathsf{H}[V] \leq 1/2$ and a Gaussian pdf for U, a reasonable choice is three regions, the center one having high probability $1-2p$ and the outer ones having small, equal probability p, such that $\mathsf{H}[V] = 1/2$.

Even for scalar quantizers, minimizing MSE subject to an entropy constraint is a rather messy problem. Considerable insight into the problem can be obtained by looking at the case where the target entropy is large – i.e. when a large number of points can be used to achieve a small MSE. Fortunately this is the case of greatest practical interest.

Example 3.4.1 For the pdf illustrated in Figure 3.6, consider the minimum-MSE quantizer using a constraint on the number of representation points M compared to that using a constraint on the entropy $\mathsf{H}[V]$.

The pdf $f_U(u)$ takes on only two positive values, say $f_U(u) = f_1$ over an interval of size L_1 and $f_U(u) = f_2$ over a second interval of size L_2. Assume that $f_U(u) = 0$ elsewhere. Because of the wide separation between the two intervals, they can be quantized separately without providing any representation point in the region between the intervals. Let M_1 and M_2 be the number of representation points in each interval. In Figure 3.6, $M_1 = 9$ and $M_2 = 7$. Let $\Delta_1 = L_1/M_1$ and $\Delta_2 = L_2/M_2$ be the lengths of the quantization regions in the two ranges (by symmetry, each quantization region in a given interval should have the same length). The representation points are at the center of each quantization interval. The MSE, conditional on being in a quantization region of length Δ_i, is the MSE of a uniform distribution over an interval of length Δ_i,

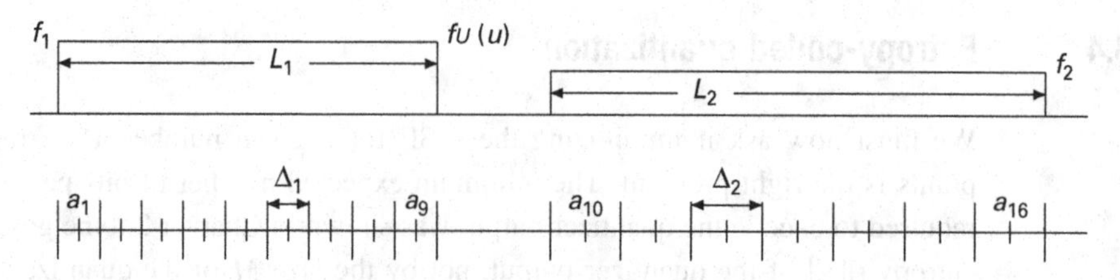

Figure 3.6. Comparison of constraint on M to constraint on $\mathsf{H}[U]$.

which is easily computed to be $\Delta_i^2/12$. The probability of being in a given quantization region of size Δ_i is $f_i\Delta_i$, so the overall MSE is given by

$$\text{MSE} = M_1\frac{\Delta_1^2}{12}f_1\Delta_1 + M_2\frac{\Delta_2^2}{12}f_2\Delta_2 = \frac{1}{12}\Delta_1^2 f_1 L_1 + \frac{1}{12}\Delta_2^2 f_2 L_2. \tag{3.4}$$

This can be minimized over Δ_1 and Δ_2 subject to the constraint that $M = M_1 + M_2 = L_1/\Delta_1 + L_2/\Delta_2$. Ignoring the constraint that M_1 and M_2 are integers (which makes sense for M large), Exercise 3.4 shows that the minimum MSE occurs when Δ_i is chosen inversely proportional to the cube root of f_i. In other words,

$$\frac{\Delta_1}{\Delta_2} = \left(\frac{f_2}{f_1}\right)^{1/3}. \tag{3.5}$$

This says that the size of a quantization region decreases with increasing probability density. This is reasonable, putting the greatest effort where there is the most probability. What is perhaps surprising is that this effect is so small, proportional only to a cube root.

Perhaps even more surprisingly, if the MSE is minimized subject to a constraint on entropy for this pdf, then Exercise 3.4 shows that, in the limit of high rate, the quantization intervals all have the same length! A scalar quantizer in which all intervals have the same length is called a *uniform scalar quantizer*. The following sections will show that uniform scalar quantizers have remarkable properties for high-rate quantization.

3.5 High-rate entropy-coded quantization

This section focuses on high-rate quantizers where the quantization regions can be made sufficiently small so that the probability density is approximately constant within each region. It will be shown that under these conditions the combination of a uniform scalar quantizer followed by discrete entropy coding is nearly optimum (in terms of mean-squared distortion) within the class of scalar quantizers. This means that a uniform quantizer can be used as a universal quantizer with very little loss of optimality. The probability distribution of the rvs to be quantized can be exploited at the level of discrete source coding. Note, however, that this essential optimality of uniform quantizers relies heavily on the assumption that mean-squared distortion is an appropriate distortion measure. With voice coding, for example, a given distortion at low signal levels is far more harmful than the same distortion at high signal levels.

In the following sections, it is assumed that the source output is a sequence $U_1, U_2, \ldots$ of iid real analog-valued rvs, each with a probability density $f_U(u)$. It is further assumed that the probability density function (pdf) $f_U(u)$ is smooth enough and the quantization fine enough that $f_U(u)$ is almost constant over each quantization region.

The analog of the entropy $\mathsf{H}[X]$ of a discrete rv X is the differential entropy $\mathsf{h}[U]$ of an analog rv U. After defining $\mathsf{h}[U]$, the properties of $\mathsf{H}[X]$ and $\mathsf{h}[U]$ will be compared.

The performance of a uniform scalar quantizer followed by entropy coding will then be analyzed. It will be seen that there is a tradeoff between the rate of the quantizer and the mean-squared error (MSE) between source and quantized output. It is also shown that the uniform quantizer is essentially optimum among scalar quantizers at high rate.

The performance of uniform vector quantizers followed by entropy coding will then be analyzed and similar tradeoffs will be found. A major result is that vector quantizers can achieve a gain over scalar quantizers (i.e. a reduction of MSE for given quantizer rate), but that the reduction in MSE is at most a factor of $\pi e/6 = 1.42$.

The changes in MSE for different quantization methods, and, similarly, changes in power levels on channels, are invariably calculated by communication engineers in decibels (dB). The number of decibels corresponding to a reduction of α in the mean-squared error is defined to be $10\log_{10}\alpha$. The use of a logarithmic measure allows the various components of mean-squared error or power gain to be added rather than multiplied.

The use of decibels rather than some other logarithmic measure, such as natural logs or logs to the base 2, is partly motivated by the ease of doing rough mental calculations. A factor of 2 is $10\log_{10}2 = 3.010\ldots$ dB, approximated as 3 dB. Thus $4 = 2^2$ is 6 dB and 8 is 9 dB. Since 10 is 10 dB, we also see that 5 is 10/2 or 7 dB. We can just as easily see that 20 is 13 dB and so forth. The limiting factor of 1.42 in the MSE above is then a reduction of 1.53 dB.

As in the discrete case, generalizations to analog sources with memory are possible, but not discussed here.

3.6 Differential entropy

The differential entropy $\mathsf{h}[U]$ of an analog random variable (rv) U is analogous to the entropy $\mathsf{H}[X]$ of a discrete random symbol X. It has many similarities, but also some important differences.

Definition 3.6.1 The *differential entropy* of an analog real rv U with pdf $f_U(u)$ is given by

$$\mathsf{h}[U] = \int_{-\infty}^{\infty} -f_U(u)\log f_U(u)\mathrm{d}u.$$

The integral may be restricted to the region where $f_U(u) > 0$, since $0\log 0$ is interpreted as 0. Assume that $f_U(u)$ is smooth and that the integral exists with a finite value. Exercise 3.7 gives an example where $\mathsf{h}(U)$ is infinite. As before, the logarithms are base 2 and the units of $\mathsf{h}[U]$ are bits per source symbol.

Like $\mathsf{H}[X]$, the differential entropy $\mathsf{h}[U]$ is the expected value of the rv $-\log f_U(U)$. The log of the joint density of several independent rvs is the sum of the logs of the individual pdfs, and this can be used to derive an AEP similar to the discrete case.

Unlike $\mathsf{H}[X]$, the differential entropy $\mathsf{h}[U]$ can be negative and depends on the scaling of the outcomes. This can be seen from the following two examples.

Example 3.6.1 (Uniform distributions) Let $f_U(u)$ be a uniform distribution over an interval $[a, a+\Delta]$ of length Δ; i.e., $f_U(u) = 1/\Delta$ for $u \in [a, a+\Delta]$, and $f_U(u) = 0$ elsewhere. Then $-\log f_U(u) = \log \Delta$, where $f_U(u) > 0$, and

$$\mathsf{h}[U] = \mathsf{E}[-\log f_U(U)] = \log \Delta.$$

Example 3.6.2 (Gaussian distribution) Let $f_U(u)$ be a Gaussian distribution with mean m and variance σ^2; i.e.,

$$f_U(u) = \sqrt{\frac{1}{2\pi\sigma^2}} \exp\left\{-\frac{(u-m)^2}{2\sigma^2}\right\}.$$

Then $-\log f_U(u) = (1/2)\log 2\pi\sigma^2 + (\log \mathrm{e})(u-m)^2/(2\sigma^2)$. Since $\mathsf{E}[(U-m)^2] = \sigma^2$, we have

$$\mathsf{h}[U] = \mathsf{E}[-\log f_U(U)] = \frac{1}{2}\log(2\pi\sigma^2) + \frac{1}{2}\log \mathrm{e} = \frac{1}{2}\log(2\pi \mathrm{e}\sigma^2).$$

It can be seen from these expressions that by making Δ or σ^2 arbitrarily small, the differential entropy can be made arbitrarily negative, while by making Δ or σ^2 arbitrarily large, the differential entropy can be made arbitrarily positive.

If the rv U is rescaled to αU for some scale factor $\alpha > 0$, then the differential entropy is increased by $\log \alpha$, both in these examples and in general. In other words, $\mathsf{h}[U]$ is not invariant to scaling. Note, however, that differential entropy is invariant to translation of the pdf, i.e. an rv and its fluctuation around the mean have the same differential entropy.

One of the important properties of entropy is that it does not depend on the labeling of the elements of the alphabet, i.e. it is invariant to invertible transformations. Differential entropy is very different in this respect, and, as just illustrated, it is modified by even such a trivial transformation as a change of scale. The reason for this is that the probability density is a probability per unit length, and therefore depends on the measure of length. In fact, as seen more clearly later, this fits in very well with the fact that source coding for analog sources also depends on an error term per unit length.

Definition 3.6.2 The *differential entropy* of an n-tuple of rvs $\boldsymbol{U}^n = (U_1, \ldots, U_n)$ with joint pdf $f_{\boldsymbol{U}^n}(\boldsymbol{u}^n)$ is given by

$$\mathsf{h}[\boldsymbol{U}^n] = \mathsf{E}[-\log f_{\boldsymbol{U}^n}(\boldsymbol{U}^n)].$$

Like entropy, differential entropy has the property that if U and V are independent rvs, then the entropy of the joint variable UV with pdf $f_{UV}(u, v) = f_U(u)f_V(v)$ is $\mathsf{h}[UV] = \mathsf{h}[U] + \mathsf{h}[V]$. Again, this follows from the fact that the log of the joint probability density of independent rvs is additive, i.e. $-\log f_{UV}(u, v) = -\log f_U(u) - \log f_V(v)$.

Thus the differential entropy of a vector rv $\boldsymbol{U}^n$, corresponding to a string of n iid rvs $U_1, U_2, \ldots, U_n$, each with the density $f_U(u)$, is $\mathsf{h}[\boldsymbol{U}^n] = n\mathsf{h}[U]$.

3.7 Performance of uniform high-rate scalar quantizers

This section analyzes the performance of uniform scalar quantizers in the limit of high rate. Appendix 3.10.1 continues the analysis for the nonuniform case and shows that uniform quantizers are effectively optimal in the high-rate limit.

For a uniform scalar quantizer, every quantization interval $\mathcal{R}_j$ has the same length $|\mathcal{R}_j| = \Delta$. In other words, $\mathbb{R}$ (or the portion of $\mathbb{R}$ over which $f_U(u) > 0$), is partitioned into equal intervals, each of length Δ (see Figure 3.7).

Assume there are enough quantization regions to cover the region where $f_U(u) > 0$. For the Gaussian distribution, for example, this requires an infinite number of representation points, $-\infty < j < \infty$. Thus, in this example the quantized discrete rv V has a countably infinite alphabet. Obviously, practical quantizers limit the number of points to a finite region $\mathcal{R}$ such that $\int_{\mathcal{R}} f_U(u)\mathrm{d}u \approx 1$.

Assume that Δ is small enough that the pdf $f_U(u)$ is approximately constant over any one quantization interval. More precisely, define $\overline{f}(u)$ (see Figure 3.8) as the average value of $f_U(u)$ over the quantization interval containing u:

$$\overline{f}(u) = \frac{\int_{\mathcal{R}_j} f_U(u)\mathrm{d}u}{\Delta} \qquad \text{for } u \in \mathcal{R}_j. \tag{3.6}$$

From (3.6) it is seen that $\Delta\overline{f}(u) = \Pr(\mathcal{R}_j)$ for all integers j and all $u \in \mathcal{R}_j$.

The *high-rate assumption* is that $f_U(u) \approx \overline{f}(u)$ for all $u \in \mathbb{R}$. This means that $f_U(u) \approx \Pr(\mathcal{R}_j)/\Delta$ for $u \in \mathcal{R}_j$. It also means that the conditional pdf $f_{U|\mathcal{R}_j}(u)$ of U conditional on $u \in \mathcal{R}_j$ is approximated by

$$f_{U|\mathcal{R}_j}(u) \approx \begin{cases} 1/\Delta, & u \in \mathcal{R}_j; \\ 0, & u \notin \mathcal{R}_j. \end{cases}$$

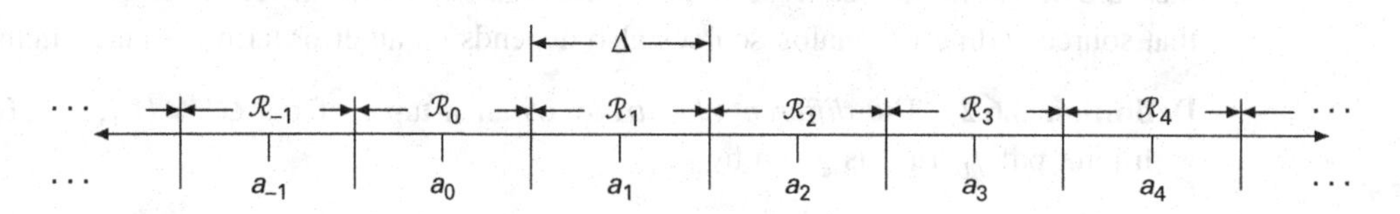

Figure 3.7. Uniform scalar quantizer.

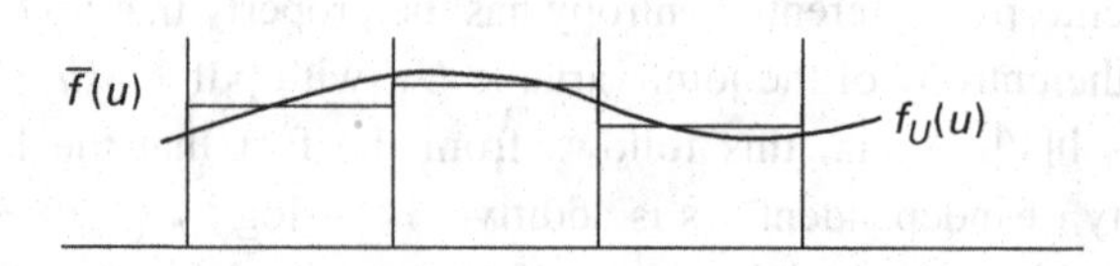

Figure 3.8. Average density over each $\mathcal{R}_j$.

Consequently, the conditional mean a_j is approximately in the center of the interval $\mathcal{R}_j$, and the mean-squared error is approximately given by

$$\text{MSE} \approx \int_{-\Delta/2}^{\Delta/2} \frac{1}{\Delta} u^2 \, \mathrm{d}u = \frac{\Delta^2}{12} \tag{3.7}$$

for each quantization interval $\mathcal{R}_j$. Consequently, this is also the overall MSE.

Next consider the entropy of the quantizer output V. The probability p_j that $V = a_j$ is given by both

$$p_j = \int_{\mathcal{R}_j} f_U(u)\mathrm{d}u \quad \text{and, for all } u \in \mathcal{R}_j, \quad p_j = \overline{f}(u)\Delta. \tag{3.8}$$

Therefore the entropy of the discrete rv V is given by

$$\begin{aligned} \mathsf{H}[V] &= \sum_j -p_j \log p_j = \sum_j \int_{\mathcal{R}_j} -f_U(u) \log[\overline{f}(u)\Delta]\mathrm{d}u \\ &= \int_{-\infty}^{\infty} -f_U(u) \log[\overline{f}(u)\Delta]\mathrm{d}u \qquad (3.9) \\ &= \int_{-\infty}^{\infty} -f_U(u) \log[\overline{f}(u)]\mathrm{d}u - \log \Delta, \qquad (3.10) \end{aligned}$$

where the sum of disjoint integrals was combined into a single integral.

Finally, using the high-rate approximation[3] $f_U(u) \approx \overline{f}(u)$, this becomes

$$\begin{aligned} \mathsf{H}[V] &\approx \int_{-\infty}^{\infty} -f_U(u) \log[f_U(u)\Delta]\mathrm{d}u \\ &= \mathsf{h}[U] - \log \Delta. \qquad (3.11) \end{aligned}$$

Since the sequence $U_1, U_2, \ldots$ of inputs to the quantizer is memoryless (iid), the quantizer output sequence $V_1, V_2, \ldots$ is an iid sequence of discrete random symbols representing quantization points, i.e. a discrete memoryless source. A uniquely decodable source code can therefore be used to encode this output sequence into a bit sequence at an average rate of $\overline{L} \approx \mathsf{H}[V] \approx \mathsf{h}[U] - \log \Delta$ bits/symbol. At the receiver, the mean-squared quantization error in reconstructing the original sequence is approximately $\text{MSE} \approx \Delta^2/12$.

The important conclusions from this analysis are illustrated in Figure 3.9 and are summarized as follows.

[3] Exercise 3.6 provides some insight into the nature of the approximation here. In particular, the difference between $\mathsf{h}[U] - \log \Delta$ and $\mathsf{H}[V]$ is $\int f_U(u) \log[\overline{f}(u)/f_U(u)]\mathrm{d}u$. This quantity is always nonpositive and goes to zero with Δ as Δ^2. Similarly, the approximation error on MSE goes to 0 as Δ^4.

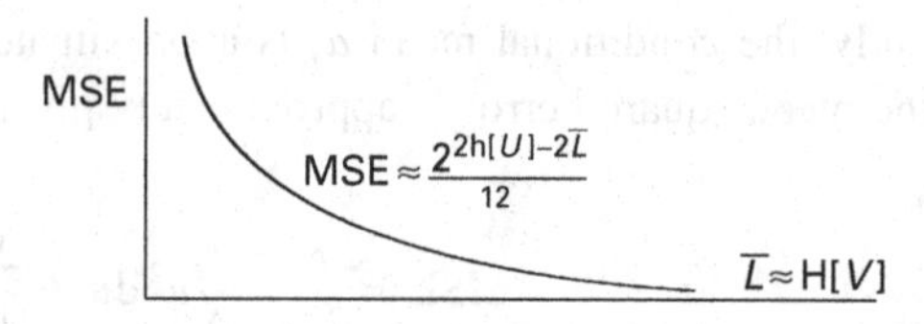

Figure 3.9. MSE as a function of $\overline{L}$ for a scalar quantizer with the high-rate approximation. Note that changing the source entropy $\mathsf{h}(U)$ simply shifts the figure to the right or left. Note also that log MSE is linear, with a slope of -2, as a function of $\overline{L}$.

- Under the high-rate assumption, the rate $\overline{L}$ for a uniform quantizer followed by discrete entropy coding depends only on the differential entropy $\mathsf{h}[U]$ of the source and the spacing Δ of the quantizer. It does not depend on any other feature of the source pdf $f_U(u)$, nor on any other feature of the quantizer, such as the number M of points, so long as the quantizer intervals cover $f_U(u)$ sufficiently completely and finely.
- The rate $\overline{L} \approx \mathsf{H}[V]$ and the MSE are parametrically related by Δ, i.e.

$$\overline{L} \approx \mathsf{h}(U) - \log \Delta; \qquad \text{MSE} \approx \frac{\Delta^2}{12}. \tag{3.12}$$

Note that each reduction in Δ by a factor of 2 will reduce the MSE by a factor of 4 and increase the required transmission rate $\overline{L} \approx \mathsf{H}[V]$ by 1 bit/symbol. Communication engineers express this by saying that each additional bit per symbol decreases the mean-squared distortion[4] by 6 dB. Figure 3.9 sketches the MSE as a function of $\overline{L}$.

Conventional b-bit analog to digital (A/D) converters are uniform scalar 2^b-level quantizers that cover a certain range $\mathcal{R}$ with a quantizer spacing $\Delta = 2^{-b}|\mathcal{R}|$. The input samples must be scaled so that the probability that $u \notin \mathcal{R}$ (the "overflow probability") is small. For a fixed scaling of the input, the tradeoff is again that increasing b by 1 bit reduces the MSE by a factor of 4.

Conventional A/D converters are not usually directly followed by entropy coding. The more conventional approach is to use A/D conversion to produce a very-high-rate digital signal that can be further processed by digital signal processing (DSP). This digital signal is then later compressed using algorithms specialized to the particular application (voice, images, etc.). In other words, the clean layers of Figure 3.1 oversimplify what is done in practice. On the other hand, it is often best to view compression in terms of the Figure 3.1 layers, and then use DSP as a way of implementing the resulting algorithms.

The relation $\mathsf{H}[V] \approx \mathsf{h}[u] - \log \Delta$ provides an elegant interpretation of differential entropy. It is obvious that there must be some kind of tradeoff between the MSE and the entropy of the representation, and the differential entropy specifies this tradeoff

[4] A quantity x expressed in dB is given by $10 \log_{10} x$. This very useful and common logarithmic measure is discussed in detail in Chapter 6.

in a very simple way for high-rate uniform scalar quantizers. Note that $\mathsf{H}[V]$ is the entropy of a finely quantized version of U, and the additional term $\log\Delta$ relates to the "uncertainty" within an individual quantized interval. It shows explicitly how the scale used to measure U affects $\mathsf{h}[U]$.

Appendix 3.10.1 considers nonuniform scalar quantizers under the high-rate assumption and shows that nothing is gained in the high-rate limit by the use of nonuniformity.

3.8 High-rate two-dimensional quantizers

The performance of uniform two-dimensional (2D) quantizers are now analyzed in the limit of high rate. Appendix 3.10.2 considers the nonuniform case and shows that uniform quantizers are again effectively optimal in the high-rate limit.

A 2D quantizer operates on two source samples $\boldsymbol{u} = (u_1, u_2)$ at a time; i.e. the source alphabet is $\boldsymbol{U} = \mathbb{R}^2$. Assuming iid source symbols, the joint pdf is then $f_{\boldsymbol{U}}(\boldsymbol{u}) = f_U(u_1)f_U(u_2)$, and the joint differential entropy is $\mathsf{h}[\boldsymbol{U}] = 2\mathsf{h}[U]$.

Like a uniform scalar quantizer, a uniform 2D quantizer is based on a fundamental quantization region $\mathcal{R}$ ("quantization cell") whose translates tile[5] the 2D plane. In the one-dimensional case, there is really only one sensible choice for $\mathcal{R}$, namely an interval of length Δ, but in higher dimensions there are many possible choices. For two dimensions, the most important choices are squares and hexagons, but in higher dimensions many more choices are available.

Note that if a region $\mathcal{R}$ tiles $\mathbb{R}^2$, then any scaled version $\alpha\mathcal{R}$ of $\mathcal{R}$ will also tile $\mathbb{R}^2$, and so will any rotation or translation of $\mathcal{R}$.

Consider the performance of a uniform 2D quantizer with a basic cell $\mathcal{R}$ which is centered at the origin $\boldsymbol{0}$. The set of cells, which are assumed to tile the region, are denoted by[6] $\{\mathcal{R}_j;\ j \in \mathbb{Z}^+\}$, where $\mathcal{R}_j = \boldsymbol{a}_j + \mathcal{R}$ and $\boldsymbol{a}_j$ is the center of the cell $\mathcal{R}_j$. Let $A(\mathcal{R}) = \int_{\mathcal{R}} d\boldsymbol{u}$ be the area of the basic cell. The average pdf in a cell $\mathcal{R}_j$ is given by $\Pr(\mathcal{R}_j)/A(\mathcal{R}_j)$. As before, define $\overline{f}(\boldsymbol{u})$ to be the average pdf over the region $\mathcal{R}_j$ containing $\boldsymbol{u}$. The high-rate assumption is again made, i.e. assume that the region $\mathcal{R}$ is small enough that $f_{\boldsymbol{U}}(\boldsymbol{u}) \approx \overline{f}(\boldsymbol{u})$ for all $\boldsymbol{u}$.

The assumption $f_{\boldsymbol{U}}(\boldsymbol{u}) \approx \overline{f}(\boldsymbol{u})$ implies that the conditional pdf, conditional on $\boldsymbol{u} \in \mathcal{R}_j$, is approximated by

$$f_{\boldsymbol{U}|\mathcal{R}_j}(\boldsymbol{u}) \approx \begin{cases} 1/A(\mathcal{R}), & \boldsymbol{u} \in \mathcal{R}_j; \\ 0, & \boldsymbol{u} \notin \mathcal{R}_j. \end{cases} \tag{3.13}$$

[5] A region of the 2D plane is said to *tile* the plane if the region, plus translates and rotations of the region, fill the plane without overlap. For example, the square and the hexagon tile the plane. Also, rectangles tile the plane, and equilateral triangles with rotations tile the plane.

[6] $\mathbb{Z}^+$ denotes the set of positive integers, so $\{\mathcal{R}_j;\ j \in \mathbb{Z}^+\}$ denotes the set of regions in the tiling, numbered in some arbitrary way of no particular interest here.

The conditional mean is approximately equal to the center $\boldsymbol{a}_j$ of the region $\mathcal{R}_j$. The mean-squared error per dimension for the basic quantization cell $\mathcal{R}$ centered on $\boldsymbol{0}$ is then approximately equal to

$$\text{MSE} \approx \frac{1}{2}\int_{\mathcal{R}} \|\boldsymbol{u}\|^2 \frac{1}{A(\mathcal{R})}\,\mathrm{d}\boldsymbol{u}. \tag{3.14}$$

The right side of (3.14) is the MSE for the quantization area $\mathcal{R}$ using a pdf equal to a constant; it will be denoted MSE_c. The quantity $\|\boldsymbol{u}\|$ is the length of the vector u_1, u_2, so that $\|\boldsymbol{u}\|^2 = u_1^2 + u_2^2$. Thus MSE_c can be rewritten as follows:

$$\text{MSE} \approx \text{MSE}_\text{c} = \frac{1}{2}\int_{\mathcal{R}} (u_1^2 + u_2^2)\frac{1}{A(\mathcal{R})}\mathrm{d}u_1\mathrm{d}u_2; \tag{3.15}$$

MSE_c is measured in units of squared length, just like $A(\mathcal{R})$. Thus the ratio $G(\mathcal{R}) = \text{MSE}_\text{c}/A(\mathcal{R})$ is a dimensionless quantity called the normalized second moment. With a little effort, it can be seen that $G(\mathcal{R})$ is invariant to scaling, translation, and rotation. Further, $G(\mathcal{R})$ does depend on the shape of the region $\mathcal{R}$, and, as seen below, it is $G(\mathcal{R})$ that determines how well a given shape performs as a quantization region. By expressing MSE_c as follows:

$$\text{MSE}_\text{c} = G(\mathcal{R})A(\mathcal{R}),$$

it is seen that the MSE is the product of a shape term and an area term, and these can be chosen independently.

As examples, $G(\mathcal{R})$ is given below for some common shapes.

- Square: for a square Δ on a side, $A(\mathcal{R}) = \Delta^2$. Breaking (3.15) into two terms, we see that each is identical to the scalar case and $\text{MSE}_\text{c} = \Delta^2/12$. Thus $G(\text{square}) = 1/12$.
- Hexagon: view the hexagon as the union of six equilateral triangles Δ on a side. Then $A(\mathcal{R}) = 3\sqrt{3}\Delta^2/2$ and $\text{MSE}_\text{c} = 5\Delta^2/24$. Thus $G(\text{hexagon}) = 5/(36\sqrt{3})$.
- Circle: for a circle of radius r, $A(\mathcal{R}) = \pi r^2$ and $\text{MSE}_\text{c} = r^2/4$ so $G(\text{circle}) = 1/(4\pi)$.

The circle is not an allowable quantization region, since it does not tile the plane. On the other hand, for a given area, this is the shape that minimizes MSE_c. To see this, note that for any other shape, differential areas further from the origin can be moved closer to the origin with a reduction in MSE_c. That is, the circle is the 2D shape that minimizes $G(\mathcal{R})$. This also suggests why $G(\text{hexagon}) < G(\text{square})$, since the hexagon is more concentrated around the origin than the square.

Using the high-rate approximation for any given tiling, each quantization cell $\mathcal{R}_j$ has the same shape and area and has a conditional pdf which is approximately uniform. Thus MSE_c approximates the MSE for each quantization region and thus approximates the overall MSE.

Next consider the entropy of the quantizer output. The probability that $\boldsymbol{U}$ falls in the region $\mathcal{R}_j$ is given by

$$p_j = \int_{\mathcal{R}_j} f_{\boldsymbol{U}}(\boldsymbol{u})\mathrm{d}\boldsymbol{u} \qquad \text{and, for all } \boldsymbol{u} \in \mathcal{R}_j, \quad p_j = \overline{f}(\boldsymbol{u})A(\mathcal{R}).$$

The output of the quantizer is the discrete random symbol V with the pmf p_j for each symbol j. As before, the entropy of V is given by

$$\begin{aligned}
\mathsf{H}[V] &= -\sum_j p_j \log p_j \\
&= -\sum_j \int_{\mathcal{R}_j} f_U(\boldsymbol{u}) \log[\bar{f}(\boldsymbol{u}) A(\mathcal{R})] \mathrm{d}\boldsymbol{u} \\
&= -\int f_U(\boldsymbol{u}) [\log \bar{f}(\boldsymbol{u}) + \log A(\mathcal{R})] \mathrm{d}\boldsymbol{u} \\
&\approx -\int f_U(\boldsymbol{u}) [\log f_U(\boldsymbol{u})] \mathrm{d}\boldsymbol{u} + \log A(\mathcal{R})] \\
&= 2\mathsf{h}[U] - \log A(\mathcal{R}),
\end{aligned}$$

where the high-rate approximation $f_U(\boldsymbol{u}) \approx \bar{f}(\boldsymbol{u})$ was used. Note that, since $\boldsymbol{U} = U_1 U_2$ for iid variables U_1 and U_2, the differential entropy of $\boldsymbol{U}$ is $2\mathsf{h}[U]$.

Again, an efficient uniquely decodable source code can be used to encode the quantizer output sequence into a bit sequence at an average rate per source symbol of

$$\overline{L} \approx \frac{\mathsf{H}[V]}{2} \approx \mathsf{h}[U] - \frac{1}{2} \log A(\mathcal{R}) \text{ bits/symbol.} \tag{3.16}$$

At the receiver, the mean-squared quantization error in reconstructing the original sequence will be approximately equal to the MSE given in (3.14).

We have the following important conclusions for a uniform 2D quantizer under the high-rate approximation.

- Under the high-rate assumption, the rate $\overline{L}$ depends only on the differential entropy $\mathsf{h}[U]$ of the source and the area $A(\mathcal{R})$ of the basic quantization cell $\mathcal{R}$. It does not depend on any other feature of the source pdf $f_U(u)$, and does not depend on the shape of the quantizer region, i.e. it does not depend on the normalized second moment $G(\mathcal{R})$.
- There is a tradeoff between the rate $\overline{L}$ and the MSE that is governed by the area $A(\mathcal{R})$. From (3.16), an increase of 1 bit/symbol in rate corresponds to a decrease in $A(\mathcal{R})$ by a factor of 4. From (3.14), this decreases the MSE by a factor of 4, i.e. by 6 dB.
- The ratio $G(\text{square})/G(\text{hexagon})$ is equal to $3\sqrt{3}/5 = 1.0392$ (0.17 dB) This is called the *quantizing gain* of the hexagon over the square. For a given $A(\mathcal{R})$ (and thus a given $\overline{L}$), the MSE for a hexagonal quantizer is smaller than that for a square quantizer (and thus also for a scalar quantizer) by a factor of 1.0392 (0.17 dB). This is a disappointingly small gain given the added complexity of 2D and hexagonal regions, and suggests that uniform scalar quantizers are good choices at high rates.

3.9 Summary of quantization

Quantization is important both for digitizing a sequence of analog signals and as the middle layer in digitizing analog waveform sources. Uniform scalar quantization is the simplest and often most practical approach to quantization. Before reaching this conclusion, two approaches to optimal scalar quantizers were taken. The first attempted to minimize the expected distortion subject to a fixed number M of quantization regions, and the second attempted to minimize the expected distortion subject to a fixed entropy of the quantized output. Each approach was followed by the extension to vector quantization.

In both approaches, and for both scalar and vector quantization, the emphasis was on minimizing mean-squared distortion or error (MSE), as opposed to some other distortion measure. As will be seen later, the MSE is the natural distortion measure in going from waveforms to sequences of analog values. For specific sources, such as speech, however, the MSE is not appropriate. For an introduction to quantization, however, focusing on the MSE seems appropriate in building intuition; again, our approach is building understanding through the use of simple models.

The first approach, minimizing the MSE with a fixed number of regions, leads to the Lloyd–Max algorithm, which finds a local minimum of the MSE. Unfortunately, the local minimum is not necessarily a global minimum, as seen by several examples. For vector quantization, the problem of local (but not global) minima arising from the Lloyd–Max algorithm appears to be the typical case.

The second approach, minimizing the MSE with a constraint on the output entropy is also a difficult problem analytically. This is the appropriate approach in a two-layer solution where the quantizer is followed by discrete encoding. On the other hand, the first approach is more appropriate when vector quantization is to be used but cannot be followed by fixed-to-variable-length discrete source coding.

High-rate scalar quantization, where the quantization regions can be made sufficiently small so that the probability density is almost constant over each region, leads to a much simpler result when followed by entropy coding. In the limit of high rate, a uniform scalar quantizer minimizes the MSE for a given entropy constraint. Moreover, the tradeoff between the minimum MSE and the output entropy is the simple universal curve of Figure 3.9. The source is completely characterized by its differential entropy in this tradeoff. The approximations in this result are analyzed in Exercise 3.6.

Two-dimensional (2D) vector quantization under the high-rate approximation with entropy coding leads to a similar result. Using a square quantization region to tile the plane, the tradeoff between the MSE per symbol and the entropy per symbol is the same as with scalar quantization. Using a hexagonal quantization region to tile the plane reduces the MSE by a factor of 1.0392, which seems hardly worth the trouble. It is possible that nonuniform 2D quantizers might achieve a smaller MSE than a hexagonal tiling, but this gain is still limited by the circular shaping gain, which is $\pi/3 = 1.0472$ (0.2 dB). Using nonuniform quantization regions at high rate leads to a lowerbound on the MSE which is lower than that for the scalar uniform quantizer by a factor of 1.0472, which, even if achievable, is scarcely worth the trouble.

The use of high-dimensional quantizers can achieve slightly higher gains over the uniform scalar quantizer, but the gain is still limited by a fundamental information-theoretic result to $\pi e/6 = 1.423$ (1.53 dB)

3.10 Appendixes

3.10.1 Nonuniform scalar quantizers

This appendix shows that the approximate MSE for uniform high-rate scalar quantizers in Section 3.7 provides an approximate lowerbound on the MSE for any nonuniform scalar quantizer, again using the high-rate approximation that the pdf of U is constant within each quantization region. This shows that, in the high-rate region, there is little reason to consider nonuniform scalar quantizers further.

Consider an arbitrary scalar quantizer for an rv U with a pdf $f_U(u)$. Let Δ_j be the width of the jth quantization interval, i.e. $\Delta_j = |\mathcal{R}_j|$. As before, let $\overline{f}(u)$ be the average pdf within each quantization interval, i.e.

$$\overline{f}(u) = \frac{\int_{\mathcal{R}_j} f_U(u)\mathrm{d}u}{\Delta_j} \qquad \text{for} \quad u \in \mathcal{R}_j.$$

The high-rate approximation is that $f_U(u)$ is approximately constant over each quantization region. Equivalently, $f_U(u) \approx \overline{f}(u)$ for all u. Thus, if region $\mathcal{R}_j$ has width Δ_j, the conditional mean a_j of U over $\mathcal{R}_j$ is approximately the midpoint of the region, and the conditional mean-squared error, MSE_j, given $U \in \mathcal{R}_j$, is approximately $\Delta_j^2/12$.

Let V be the quantizer output, i.e. the discrete rv such that $V = a_j$ whenever $U \in \mathcal{R}_j$. The probability p_j that $V = a_j$ is $p_j = \int_{\mathcal{R}_j} f_U(u)\mathrm{d}u$. The unconditional mean-squared error, i.e. $\mathsf{E}[(U - V)^2]$, is then given by

$$\mathrm{MSE} \approx \sum_j p_j \frac{\Delta_j^2}{12} = \sum_j \int_{\mathcal{R}_j} f_U(u) \frac{\Delta_j^2}{12}\,\mathrm{d}u. \tag{3.17}$$

This can be simplified by defining $\Delta(u) = \Delta_j$ for $u \in \mathcal{R}_j$. Since each u is in $\mathcal{R}_j$ for some j, this defines $\Delta(u)$ for all $u \in \mathbb{R}$. Substituting this in (3.17), we have

$$\mathrm{MSE} \approx \sum_j \int_{\mathcal{R}_j} f_U(u) \frac{\Delta(u)^2}{12}\,\mathrm{d}u \tag{3.18}$$

$$= \int_{-\infty}^{\infty} f_U(u) \frac{\Delta(u)^2}{12}\,\mathrm{d}u\,. \tag{3.19}$$

Next consider the entropy of V. As in (3.8), the following relations are used for p_j:

$$p_j = \int_{\mathcal{R}_j} f_U(u)\mathrm{d}u \quad \text{and, for all } u \in \mathcal{R}_j, \quad p_j = \overline{f}(u)\Delta(u);$$

$$\begin{aligned}\mathsf{H}[V] &= \sum_j -p_j \log p_j \\ &= \sum_j \int_{\mathcal{R}_j} -f_U(u) \log[\overline{f}(u)\Delta(u)] \mathrm{d}u \quad (3.20) \\ &= \int_{-\infty}^{\infty} -f_U(u) \log[\overline{f}(u)\Delta(u)] \mathrm{d}u, \quad (3.21)\end{aligned}$$

where the multiple integrals over disjoint regions have been combined into a single integral. The high-rate approximation $f_U(u) \approx \overline{f}(u)$ is next substituted into (3.21) as follows:

$$\begin{aligned}\mathsf{H}[V] &\approx \int_{-\infty}^{\infty} -f_U(u) \log[f_U(u)\Delta(u)] \mathrm{d}u \\ &= \mathsf{h}[U] - \int_{-\infty}^{\infty} f_U(u) \log \Delta(u) \mathrm{d}u. \quad (3.22)\end{aligned}$$

Note the similarity of this equation to (3.11).

The next step is to minimize the MSE subject to a constraint on the entropy $\mathsf{H}[V]$. This is done approximately by minimizing the approximation to the MSE in (3.22) subject to the approximation to $\mathsf{H}[V]$ in (3.19). Exercise 3.6 provides some insight into the accuracy of these approximations and their effect on this minimization.

Consider using a Lagrange multiplier to perform the minimization. Since the MSE decreases as $\mathsf{H}[V]$ increases, consider minimizing $\text{MSE} + \lambda\mathsf{H}[V]$. As λ increases, the MSE will increase and $\mathsf{H}[V]$ will decrease in the minimizing solution.

In principle, the minimization should be constrained by the fact that $\Delta(u)$ is constrained to represent the interval sizes for a realizable set of quantization regions. The minimum of $\text{MSE} + \lambda\mathsf{H}[V]$ will be lowerbounded by ignoring this constraint. The very nice thing that happens is that this unconstrained lowerbound occurs where $\Delta(u)$ is constant. This corresponds to a uniform quantizer, which is clearly realizable. In other words, subject to the high-rate approximation, the lowerbound on the MSE over all scalar quantizers is equal to the MSE for the uniform scalar quantizer. To see this, use (3.19) and (3.22):

$$\begin{aligned}\text{MSE} + \lambda\mathsf{H}[V] &\approx \int_{-\infty}^{\infty} f_U(u) \frac{\Delta(u)^2}{12} \, du + \lambda\mathsf{h}[U] - \lambda \int_{-\infty}^{\infty} f_U(u) \log \Delta(u) \mathrm{d}u \\ &= \lambda\mathsf{h}[U] + \int_{-\infty}^{\infty} f_U(u) \left\{ \frac{\Delta(u)^2}{12} - \lambda \log \Delta(u) \right\} \mathrm{d}u. \quad (3.23)\end{aligned}$$

This is minimized over all choices of $\Delta(u) > 0$ by simply minimizing the expression inside the braces for each real value of u. That is, for each u, differentiate the quantity inside the braces with respect to $\Delta(u)$, yielding $\Delta(u)/6 - \lambda(\log \mathrm{e})/\Delta(u)$. Setting the derivative equal to 0, it is seen that $\Delta(u) = \sqrt{\lambda(\log \mathrm{e})/6}$. By taking the second derivative, it can be seen that this solution actually minimizes the integrand for each u. The only important thing here is that the minimizing $\Delta(u)$ is independent of u.

This means that the approximation of the MSE is minimized, subject to a constraint on the approximation of $\mathsf{H}[V]$, by the use of a uniform quantizer.

The next question is the meaning of minimizing an approximation to something subject to a constraint which itself is an approximation. From Exercise 3.6, it is seen that both the approximation to the MSE and that to $\mathsf{H}[V]$ are good approximations for small Δ, i.e. for high rate. For any given high-rate nonuniform quantizer, consider plotting the MSE and $\mathsf{H}[V]$ on Figure 3.9. The corresponding approximate values of the MSE and $\mathsf{H}[V]$ are then close to the plotted values (with some small difference both in the ordinate and abscissa). These approximate values, however, lie above the approximate values plotted in Figure 3.9 for the scalar quantizer. Thus, in this sense, the performance curve of the MSE versus $\mathsf{H}[V]$ for the approximation to the scalar quantizer either lies below or close to the points for any nonuniform quantizer.

In summary, it has been shown that for large $\mathsf{H}[V]$ (i.e. high-rate quantization), a uniform scalar quantizer approximately minimizes the MSE subject to the entropy constraint. There is little reason to use nonuniform scalar quantizers (except perhaps at low rate). Furthermore the MSE performance at high rate can be easily approximated and depends only on $\mathsf{h}[U]$ and the constraint on $\mathsf{H}[V]$.

3.10.2 Nonuniform 2D quantizers

For completeness, the performance of nonuniform 2D quantizers is now analyzed; the analysis is very similar to that of nonuniform scalar quantizers. Consider an arbitrary set of quantization intervals $\{\mathcal{R}_j\}$. Let $A(\mathcal{R}_j)$ and MSE_j be the area and mean-squared error per dimension, respectively, of $\mathcal{R}_j$, i.e.

$$A(\mathcal{R}_j) = \int_{\mathcal{R}_j} \mathrm{d}\boldsymbol{u}\ ; \qquad \mathrm{MSE}_j = \frac{1}{2}\int_{\mathcal{R}_j} \frac{\|\boldsymbol{u}-\boldsymbol{a}_j\|^2}{A(\mathcal{R}_j)}\,\mathrm{d}\boldsymbol{u},$$

where $\boldsymbol{a}_j$ is the mean of $\mathcal{R}_j$. For each region $\mathcal{R}_j$ and each $\boldsymbol{u} \in \mathcal{R}_j$, let $\overline{f}(\boldsymbol{u}) = \Pr(\mathcal{R}_j)/A(\mathcal{R}_j)$ be the average pdf in $\mathcal{R}_j$. Then

$$p_j = \int_{\mathcal{R}_j} f_{\boldsymbol{U}}(\boldsymbol{u})\mathrm{d}\boldsymbol{u} = \overline{f}(\boldsymbol{u})A(\mathcal{R}_j).$$

The unconditioned mean-squared error is then given by

$$\mathrm{MSE} = \sum_j p_j\ \mathrm{MSE}_j.$$

Let $A(\boldsymbol{u}) = A(\mathcal{R}_j)$ and $\mathrm{MSE}(\boldsymbol{u}) = \mathrm{MSE}_j$ for $\boldsymbol{u} \in A_j$. Then,

$$\mathrm{MSE} = \int f_{\boldsymbol{U}}(\boldsymbol{u})\ \mathrm{MSE}(\boldsymbol{u})\mathrm{d}\boldsymbol{u}. \tag{3.24}$$

Similarly,

$$
\begin{aligned}
\mathsf{H}[V] &= \sum_j -p_j \log p_j \\
&= \int -f_U(\boldsymbol{u}) \log[\overline{f}(\boldsymbol{u}) A(\boldsymbol{u})] \mathrm{d}\boldsymbol{u} \\
&\approx \int -f_U(\boldsymbol{u}) \log[f_U(\boldsymbol{u}) A(\boldsymbol{u})] \mathrm{d}\boldsymbol{u} \qquad (3.25)\\
&= 2\mathsf{h}[U] - \int f_U(\boldsymbol{u}) \log[A(\boldsymbol{u})] \mathrm{d}\boldsymbol{u}. \qquad (3.26)
\end{aligned}
$$

A Lagrange multiplier can again be used to solve for the optimum quantization regions under the high-rate approximation. In particular, from (3.24) and (3.26),

$$
\text{MSE} + \lambda \mathsf{H}[V] \approx \lambda 2\mathsf{h}[U] + \int_{\mathbb{R}^2} f_U(\boldsymbol{u}) \{\text{MSE}(\boldsymbol{u}) - \lambda \log A(\boldsymbol{u})\}\, \mathrm{d}u. \qquad (3.27)
$$

Since each quantization area can be different, the quantization regions need not have geometric shapes whose translates tile the plane. As pointed out earlier, however, the shape that minimizes MSE_c for a given quantization area is a circle. Therefore the MSE can be lowerbounded in the Lagrange multiplier by using this shape. Replacing $\text{MSE}(\boldsymbol{u})$ by $A(\boldsymbol{u})/(4\pi)$ in (3.27) yields

$$
\text{MSE} + \lambda \mathsf{H}[V] \approx 2\lambda \mathsf{h}[U] + \int_{\mathbb{R}^2} f_U(\boldsymbol{u}) \left\{ \frac{A(\boldsymbol{u})}{4\pi} - \lambda \log A(\boldsymbol{u}) \right\} \mathrm{d}u. \qquad (3.28)
$$

Optimizing for each $\boldsymbol{u}$ separately, $A(\boldsymbol{u}) = 4\pi\lambda \log \mathrm{e}$. The optimum is achieved where the same size circle is used for each point $\boldsymbol{u}$ (independent of the probability density). This is unrealizable, but still provides a lowerbound on the MSE for any given $\mathsf{H}[V]$ in the high-rate region. The reduction in MSE over the square region is $\pi/3 = 1.0472$ (0.2 dB). It appears that the uniform quantizer with hexagonal shape is optimal, but this figure of $\pi/3$ provides a simple bound to the possible gain with 2D quantizers. Either way, the improvement by going to two dimensions is small.

The same sort of analysis can be carried out for n-dimensional quantizers. In place of using a circle as a lowerbound, one now uses an n-dimensional sphere. As n increases, the resulting lowerbound to the MSE approaches a gain of $\pi\mathrm{e}/6 = 1.4233$ (1.53 dB) over the scalar quantizer. It is known from a fundamental result in information theory that this gain can be approached arbitrarily closely as $n \to \infty$.

3.11 Exercises

3.1 Let U be an analog rv uniformly distributed between -1 and 1.

(a) Find the 3-bit ($M = 8$) quantizer that minimizes the MSE.
(b) Argue that your quantizer satisfies the necessary conditions for optimality.
(c) Show that the quantizer is unique in the sense that no other 3-bit quantizer satisfies the necessary conditions for optimality.

3.2 Consider a discrete-time, analog source *with memory*, i.e. $U_1, U_2, \ldots$ are dependent rvs. Assume that each U_k is uniformly distributed between 0 and 1, but that $U_{2n} = U_{2n-1}$ for each $n \geq 1$. Assume that $\{U_{2n}\}_{n=1}^{\infty}$ are independent.

(a) Find the 1-bit ($M = 2$) scalar quantizer that minimizes the MSE.
(b) Find the MSE for the quantizer that you have found in part (a).
(c) Find the 1-bit per symbol ($M = 4$) 2D vector quantizer that minimizes the MSE.
(d) Plot the 2D regions and representation points for both your scalar quantizer in part (a) and your vector quantizer in part (c).

3.3 Consider a binary scalar quantizer that partitions the set of reals $\mathbb{R}$ into two subsets $(-\infty, b]$ and (b, ∞) and then represents $(-\infty, b]$ by $a_1 \in \mathbb{R}$ and (b, ∞) by $a_2 \in \mathbb{R}$. This quantizer is used on each letter U_n of a sequence $\ldots, U_{-1}, U_0, U_1, \ldots$ of iid random variables, each having the probability density $f(u)$. Assume throughout this exercise that $f(u)$ is symmetric, i.e. that $f(u) = f(-u)$ for all $u \geq 0$.

(a) Given the representation levels a_1 and $a_2 > a_1$, how should b be chosen to minimize the mean-squared distortion in the quantization? Assume that $f(u) > 0$ for $a_1 \leq u \leq a_2$ and explain why this assumption is relevant.
(b) Given $b \geq 0$, find the values of a_1 and a_2 that minimize the mean-squared distortion. Give both answers in terms of the two functions $Q(x) = \int_x^{\infty} f(u)\mathrm{d}u$ and $y(x) = \int_x^{\infty} uf(u)\mathrm{d}u$.
(c) Show that for $b = 0$, the minimizing values of a_1 and a_2 satisfy $a_1 = -a_2$.
(d) Show that the choice of b, a_1, and a_2 in part (c) satisfies the Lloyd–Max conditions for minimum mean-squared distortion.
(e) Consider the particular symmetric density

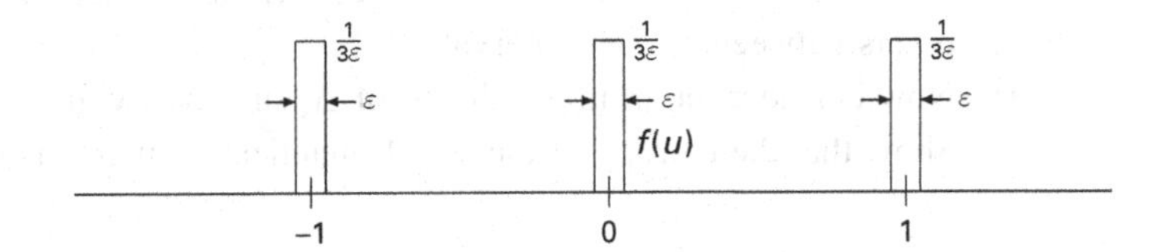

Figure 3.10.

Find all sets of triples $\{b, a_1, a_2\}$ that satisfy the Lloyd–Max conditions and evaluate the MSE for each. You are welcome in your calculation to replace each region of nonzero probability density above with an impulse, i.e. $f(u) = (1/3)[\delta(-1) + \delta(0) + \delta(1)]$, but you should use Figure 3.10 to resolve the ambiguity about regions that occurs when b is -1, 0, or $+1$.

(f) Give the MSE for each of your solutions above (in the limit of $\varepsilon \to 0$). Which of your solutions minimizes the MSE?

3.4 Section 3.4 partly analyzed a minimum-MSE quantizer for a pdf in which $f_U(u) = f_1$ over an interval of size L_1, $f_U(u) = f_2$ over an interval of size L_2, and $f_U(u) = 0$ elsewhere. Let M be the total number of representation points to be used, with M_1

in the first interval and $M_2 = M - M_1$ in the second. Assume (from symmetry) that the quantization intervals are of equal size $\Delta_1 = L_1/M_1$ in interval 1 and of equal size $\Delta_2 = L_2/M_2$ in interval 2. Assume that M is very large, so that we can approximately minimize the MSE over M_1, M_2 without an integer constraint on M_1, M_2 (that is, assume that M_1, M_2 can be arbitrary real numbers).

(a) Show that the MSE is minimized if $\Delta_1 f_1^{1/3} = \Delta_2 f_2^{1/3}$, i.e. the quantization interval sizes are inversely proportional to the cube root of the density. [Hint. Use a Lagrange multiplier to perform the minimization. That is, to minimize a function $\mathrm{MSE}(\Delta_1, \Delta_2)$ subject to a constraint $M = f(\Delta_1, \Delta_2)$, first minimize $\mathrm{MSE}(\Delta_1, \Delta_2) + \lambda f(\Delta_1, \Delta_2)$ without the constraint, and, second, choose λ so that the solution meets the constraint.]

(b) Show that the minimum MSE under the above assumption is given by

$$\mathrm{MSE} = \frac{\left(L_1 f_1^{1/3} + L_2 f_2^{1/3}\right)^3}{12M^2}.$$

(c) Assume that the Lloyd–Max algorithm is started with $0 < M_1 < M$ representation points in the first interval and $M_2 = M - M_1$ points in the second interval. Explain where the Lloyd–Max algorithm converges for this starting point. Assume from here on that the distance between the two intervals is very large.

(d) Redo part (c) under the assumption that the Lloyd–Max algorithm is started with $0 < M_1 \le M - 2$ representation points in the first interval, one point between the two intervals, and the remaining points in the second interval.

(e) Express the exact minimum MSE as a minimum over $M - 1$ possibilities, with one term for each choice of $0 < M_1 < M$. (Assume there are no representation points between the two intervals.)

(f) Now consider an arbitrary choice of Δ_1 and Δ_2 (with no constraint on M). Show that the entropy of the set of quantization points is given by

$$\mathsf{H}(V) = -f_1 L_1 \log(f_1 \Delta_1) - f_2 L_2 \log(f_2 \Delta_2).$$

(g) Show that if the MSE is minimized subject to a constraint on this entropy (ignoring the integer constraint on quantization levels), then $\Delta_1 = \Delta_2$.

3.5 (a) Assume that a continuous-valued rv Z has a probability density that is 0 except over the interval $[-A, +A]$. Show that the differential entropy $\mathsf{h}(Z)$ is upperbounded by $1 + \log_2 A$.

(b) Show that $\mathsf{h}(Z) = 1 + \log_2 A$ if and only if Z is uniformly distributed between $-A$ and $+A$.

3.6 Let $f_U(u) = 1/2 + u$ for $0 < u \le 1$ and $f_U(u) = 0$ elsewhere.

(a) For $\Delta < 1$, consider a quantization region $\mathcal{R} = (x, x + \Delta]$ for $0 < x \le 1 - \Delta$. Find the conditional mean of U conditional on $U \in \mathcal{R}$.

(b) Find the conditional MSE of U conditional on $U \in \mathcal{R}$. Show that, as Δ goes to 0, the difference between the MSE and the approximation $\Delta^2/12$ goes to 0 as Δ^4.

(c) For any given Δ such that $1/\Delta = M$, M a positive integer, let $\{\mathcal{R}_j = ((j-1)\Delta, j\Delta]\}$ be the set of regions for a uniform scalar quantizer with M quantization intervals. Show that the difference between $\mathsf{h}[U] - \log\Delta$ and $\mathsf{H}[V]$ as given in (3.10) is given by

$$\mathsf{h}[U] - \log\Delta - \mathsf{H}[V] = \int_0^1 f_U(u)\log[\overline{f}(u)/f_U(u)]\mathrm{d}u.$$

(d) Show that the difference in (3.6c) is nonnegative. [Hint. Use the inequality $\ln x \le x - 1$.] Note that your argument does not depend on the particular choice of $f_U(u)$.

(e) Show that the difference $\mathsf{h}[U] - \log\Delta - \mathsf{H}[V]$ goes to 0 as Δ^2 as $\Delta \to 0$. [Hint. Use the approximation $\ln x \approx (x-1) - (x-1)^2/2$, which is the second-order Taylor series expansion of $\ln x$ around $x = 1$.]

The major error in the high-rate approximation for small Δ and smooth $f_U(u)$ is due to the slope of $f_U(u)$. Your results here show that this linear term is insignificant for both the approximation of the MSE and for the approximation of $\mathsf{H}[V]$. More work is required to validate the approximation in regions where $f_U(u)$ goes to 0.

3.7 (Example where $\mathsf{h}(U)$ is infinite) Let $f_U(u)$ be given by

$$f_U(u) = \begin{cases} \dfrac{1}{u(\ln u)^2}, & \text{for} \quad u \ge \mathrm{e}; \\ 0, & \text{for} \quad u < \mathrm{e}. \end{cases}$$

(a) Show that $f_U(u)$ is nonnegative and integrates to 1.

(b) Show that $\mathsf{h}(U)$ is infinite.

(c) Show that a uniform scalar quantizer for this source with any separation Δ $(0 < \Delta < \infty)$ has infinite entropy. [Hint. Use the approach in Exercise 3.6, parts (c) and (d).]

3.8 (Divergence and the extremal property of Gaussian entropy) The divergence between two probability densities $f(x)$ and $g(x)$ is defined by

$$D(f\|g) = \int_{-\infty}^{\infty} f(x)\ln\frac{f(x)}{g(x)}\,\mathrm{d}x.$$

(a) Show that $D(f\|g) \ge 0$. [Hint. Use the inequality $\ln y \le y - 1$ for $y \ge 0$ on $-D(f\|g)$.] You may assume that $g(x) > 0$ where $f(x) > 0$.

(b) Let $\int_{-\infty}^{\infty} x^2 f(x)\mathrm{d}x = \sigma^2$ and let $g(x) = \phi(x)$, where $\phi(x) \sim \mathcal{N}(0, \sigma^2)$. Express $D(f\|\phi)$ in terms of the differential entropy (in nats) of an rv with density $f(x)$.

(c) Use parts (a) and (b) to show that the Gaussian rv $\mathcal{N}(0, \sigma^2)$ has the largest differential entropy of any rv with variance σ^2 and that the differential entropy is $(1/2)\ln(2\pi\mathrm{e}\sigma^2)$.

3.9 Consider a discrete source U with a finite alphabet of N real numbers, $r_1 < r_2 < \cdots < r_N$, with the pmf $p_1 > 0, \ldots, p_N > 0$. The set $\{r_1, \ldots, r_N\}$ is to be quantized into a smaller set of $M < N$ representation points, $a_1 < a_2 < \cdots < a_M$.

(a) Let $\mathcal{R}_1, \mathcal{R}_2, \ldots, \mathcal{R}_M$ be a given set of quantization intervals with $\mathcal{R}_1 = (-\infty, b_1], \mathcal{R}_2 = (b_1, b_2], \ldots, \mathcal{R}_M = (b_{M-1}, \infty)$. Assume that at least one source value r_i is in $\mathcal{R}_j$ for each j, $1 \le j \le M$, and give a necessary condition on the representation points $\{a_j\}$ to achieve the minimum MSE.

(b) For a given set of representation points $a_1, \ldots, a_M$, assume that no symbol r_i lies exactly halfway between two neighboring a_i, i.e. that $r_i \neq (a_j + a_{j+1})/2$ for all i, j. For each r_i, find the interval $\mathcal{R}_j$ (and more specifically the representation point a_j) that r_i must be mapped into to minimize the MSE. Note that it is not necessary to place the boundary b_j between $\mathcal{R}_j$ and $\mathcal{R}_{j+1}$ at $b_j = (a_j + a_{j+1})/2$ since there is no probability in the immediate vicinity of $(a_j + a_{j+1})/2$.

(c) For the given representation points $a_1, \ldots, a_M$, assume that $r_i = (a_j + a_{j+1})/2$ for some source symbol r_i and some j. Show that the MSE is the same whether r_i is mapped into a_j or into a_{j+1}.

(d) For the assumption in part (c), show that the set $\{a_j\}$ cannot possibly achieve the minimum MSE. [Hint. Look at the optimal choice of a_j and a_{j+1} for each of the two cases of part (c).]

3.10 Assume an iid discrete-time analog source $U_1, U_2, \ldots$ and consider a scalar quantizer that satisfies the Lloyd–Max conditions. Show that the rectangular 2D quantizer based on this scalar quantizer also satisfies the Lloyd–Max conditions.

3.11 (a) Consider a square 2D quantization region $\mathcal{R}$ defined by $-\Delta/2 \le u_1 \le \Delta/2$ and $-\Delta/2 \le u_2 \le \Delta/2$. Find MSE_c as defined in (3.15) and show that it is proportional to Δ^2.

(b) Repeat part (a) with Δ replaced by $a\Delta$. Show that $\mathrm{MSE}_c/A(\mathcal{R})$ (where $A(\mathcal{R})$ is now the area of the scaled region) is unchanged.

(c) Explain why this invariance to scaling of $\mathrm{MSE}_c/A(\mathcal{R})$ is valid for any 2D region.

4 Source and channel waveforms

4.1 Introduction

This chapter has a dual objective. The first is to understand *analog data compression*, i.e. the compression of sources such as voice for which the output is an arbitrarily varying real- or complex-valued function of time; we denote such functions as *waveforms*. The second is to begin studying the waveforms that are typically transmitted at the input and received at the output of communication channels. The same set of mathematical tools is required for the understanding and representation of both source and channel waveforms; the development of these results is the central topic of this chapter.

These results about waveforms are standard topics in mathematical courses on analysis, real and complex variables, functional analysis, and linear algebra. They are stated here without the precision or generality of a good mathematics text, but with considerably more precision and interpretation than is found in most engineering texts.

4.1.1 Analog sources

The output of many analog sources (voice is the typical example) can be represented as a waveform,[1] $\{u(t) : \mathbb{R} \to \mathbb{R}\}$ or $\{u(t) : \mathbb{R} \to \mathbb{C}\}$. Often, as with voice, we are interested only in real waveforms, but the simple generalization to complex waveforms is essential for Fourier analysis and for baseband modeling of communication channels. Since a real-valued function can be viewed as a special case of a complex-valued function, the results for complex functions are also useful for real functions.

We observed earlier that more complicated analog sources such as video can be viewed as mappings from $\mathbb{R}^n$ to $\mathbb{R}$, e.g. as mappings from horizontal/vertical position and time to real analog values, but for simplicity we consider only waveform sources here.

[1] The notation $\{u(t) : \mathbb{R} \to \mathbb{R}\}$ refers to a function that maps each real number $t \in \mathbb{R}$ into another real number $u(t) \in \mathbb{R}$. Similarly, $\{u(t) : \mathbb{R} \to \mathbb{C}\}$ maps each real number $t \in \mathbb{R}$ into a complex number $u(t) \in \mathbb{C}$. These functions of time, i.e. these waveforms, are usually viewed as dimensionless, thus allowing us to separate physical scale factors in communication problems from the waveform shape.

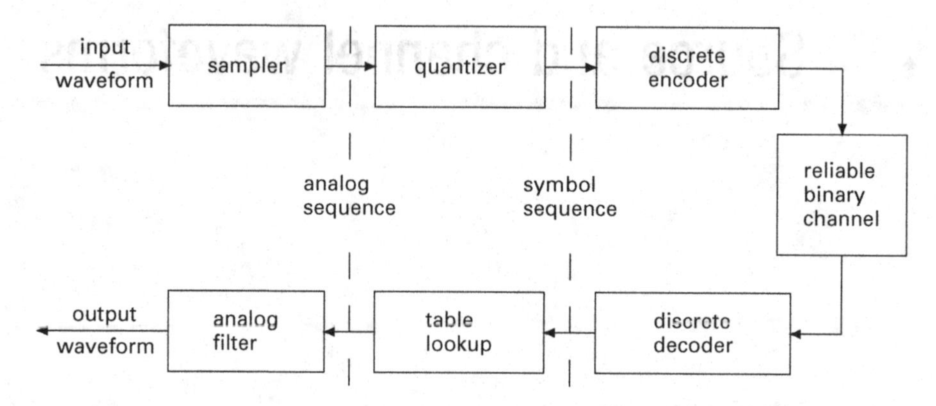

Figure 4.1. Encoding and decoding a waveform source.

We recall in the following why it is desirable to convert analog sources into bits.

- The use of a standard binary interface separates the problem of compressing sources from the problems of channel coding and modulation.
- The outputs from multiple sources can be easily multiplexed together. Multiplexers can work by interleaving bits, 8-bit bytes, or longer packets from different sources.
- When a bit sequence travels serially through multiple links (as in a network), the noisy bit sequence can be cleaned up (regenerated) at each intermediate node, whereas noise tends to accumulate gradually with noisy analog transmission.

A common way of encoding a waveform into a bit sequence is as follows.

(1) Approximate the analog waveform $\{u(t); t \in \mathbb{R}\}$ by its samples[2] $\{u(mT); m \in \mathbb{Z}\}$ at regularly spaced sample times, $\dots, -T, 0, T, 2T, \dots$
(2) Quantize each sample (or n-tuple of samples) into a quantization region.
(3) Encode each quantization region (or block of regions) into a string of bits.

These three layers of encoding are illustrated in Figure 4.1, with the three corresponding layers of decoding.

Example 4.1.1 In standard telephony, the voice is filtered to 4000 Hz (4 kHz) and then sampled[3] at 8000 samples/s. Each sample is then quantized to one of 256 possible levels, represented by 8 bits. Thus the voice signal is represented as a 64 kbps sequence. (Modern digital wireless systems use more sophisticated voice coding schemes that reduce the data rate to about 8 kbps with little loss of voice quality.)

The sampling above may be generalized in a variety of ways for converting waveforms into sequences of real or complex numbers. For example, modern voice

---

[2] $\mathbb{Z}$ denotes the set of integers $-\infty < m < \infty$, so $\{u(mT); m \in \mathbb{Z}\}$ denotes the doubly infinite sequence of samples with $-\infty < m < \infty$.

[3] The sampling theorem, to be discussed in Section 4.6, essentially says that if a waveform is baseband-limited to W Hz, then it can be represented perfectly by 2W samples/s. The highest note on a piano is about 4 kHz, which is considerably higher than most voice frequencies.

compression techniques first segment the voice waveform into 20 ms segments and then use the frequency structure of each segment to generate a vector of numbers. The resulting vector can then be quantized and encoded as previously discussed.

An individual waveform from an analog source should be viewed as a sample waveform from a *random process*. The resulting probabilistic structure on these sample waveforms then determines a probability assignment on the sequences representing these sample waveforms. This random characterization will be studied in Chapter 7; for now, the focus is on ways to map deterministic waveforms to sequences and vice versa. These mappings are crucial both for source coding and channel transmission.

4.1.2 Communication channels

Some examples of communication channels are as follows: a pair of antennas separated by open space; a laser and an optical receiver separated by an optical fiber; a microwave transmitter and receiver separated by a wave guide. For the antenna example, a real waveform at the input in the appropriate frequency band is converted by the input antenna into electromagnetic radiation, part of which is received at the receiving antenna and converted back to a waveform. For many purposes, these physical channels can be viewed as black boxes where the output waveform can be described as a function of the input waveform and noise of various kinds.

Viewing these channels as black boxes is another example of layering. The optical or microwave devices or antennas can be considered as an inner layer around the actual physical channel. This layered view will be adopted here for the most part, since the physics of antennas, optics, and microwaves are largely separable from the digital communication issues developed here. One exception to this is the description of physical channels for wireless communication in Chapter 9. As will be seen, describing a wireless channel as a black box requires some understanding of the underlying physical phenomena.

The function of a channel encoder, i.e. a modulator, is to convert the incoming sequence of binary digits into a waveform in such a way that the noise-corrupted waveform at the receiver can, with high probability, be converted back into the original binary digits. This is typically achieved by first converting the binary sequence into a sequence of analog signals, which are then converted to a waveform. This procession – bit sequence to analog sequence to waveform – is the same procession as performed by a source decoder, and the opposite to that performed by the source encoder. How these functions should be accomplished is very different in the source and channel cases, but both involve converting between waveforms and analog sequences.

The waveforms of interest for channel transmission and reception should be viewed as sample waveforms of random processes (in the same way that source waveforms should be viewed as sample waveforms from a random process). This chapter, however, is concerned only with the relationship between deterministic waveforms and analog sequences; the necessary results about random processes will be postponed until Chapter 7. The reason why so much mathematical precision is necessary here, however, is that these waveforms are a priori unknown. In other words, one cannot use the conventional engineering approach of performing some computation on a function and assuming it is correct if an answer emerges.[4]

4.2 Fourier series

Perhaps the simplest example of an analog sequence that can represent a waveform comes from the Fourier series. The Fourier series is also useful in understanding Fourier transforms and discrete-time Fourier transforms (DTFTs). As will be explained later, our study of these topics will be limited to finite-energy waveforms. Useful models for source and channel waveforms almost invariably fall into the finite-energy class.

The Fourier series represents a waveform, either periodic or time-limited, as a weighted sum of sinusoids. Each weight (coefficient) in the sum is determined by the function, and the function is essentially determined by the sequence of weights. Thus the function and the sequence of weights are essentially equivalent representations.

Our interest here is almost exclusively in time-limited rather than periodic waveforms.[5] Initially the waveforms are assumed to be time-limited to some interval $-T/2 \le t \le T/2$ of an arbitrary duration $T > 0$ around 0. This is then generalized to time-limited waveforms centered at some arbitrary time. Finally, an arbitrary waveform is segmented into equal-length segments each of duration T; each such segment is then represented by a Fourier series. This is closely related to modern voice-compression techniques where voice waveforms are segmented into 20 ms intervals, each of which is separately expanded into a Fourier-like series.

Consider a complex function $\{u(t) : \mathbb{R} \to \mathbb{C}\}$ that is nonzero only for $-T/2 \le t \le T/2$ (i.e. $u(t) = 0$ for $t < -T/2$ and $t > T/2$). Such a function is frequently indicated by $\{u(t) : [-T/2, T/2] \to \mathbb{C}\}$. The *Fourier series* for such a time-limited function is given by[6]

$$u(t) = \begin{cases} \sum_{k=-\infty}^{\infty} \hat{u}_k e^{2\pi i k t/T} & \text{for } -T/2 \le t \le T/2; \\ 0 & \text{elsewhere,} \end{cases} \tag{4.1}$$

where i denotes[7] $\sqrt{-1}$. The Fourier series coefficients $\hat{u}_k$ are, in general, complex (even if $u(t)$ is real), and are given by

$$\hat{u}_k = \frac{1}{T} \int_{-T/2}^{T/2} u(t) e^{-2\pi i k t/T}\, dt, \qquad -\infty < k < \infty. \tag{4.2}$$

[4] This is not to disparage the use of computational (either hand or computer) techniques to get a quick answer without worrying about fine points. Such techniques often provide insight and understanding, and the fine points can be addressed later. For a random process, however, one does not know a priori which sample functions can provide computational insight.

[5] Periodic waveforms are not very interesting for carrying information; after one period, the rest of the waveform carries nothing new.

[6] The conditions and the sense in which (4.1) holds are discussed later.

[7] The use of i for $\sqrt{-1}$ is standard in all scientific fields except electrical engineering. Electrical engineers formerly reserved the symbol i for electrical current and thus often use j to denote $\sqrt{-1}$.

The standard rectangular function,

$$\mathrm{rect}(t) = \begin{cases} 1 & \text{for } -1/2 \le t \le 1/2; \\ 0 & \text{elsewhere,} \end{cases}$$

can be used to simplify (4.1) as follows:

$$u(t) = \sum_{k=-\infty}^{\infty} \hat{u}_k \mathrm{e}^{2\pi i k t/T} \mathrm{rect}\left(\frac{t}{T}\right). \tag{4.3}$$

This expresses $u(t)$ as a linear combination of truncated complex sinusoids,

$$u(t) = \sum_{k \in \mathbb{Z}} \hat{u}_k \theta_k(t), \qquad \text{where} \quad \theta_k(t) = \mathrm{e}^{2\pi i k t/T} \mathrm{rect}\left(\frac{t}{T}\right). \tag{4.4}$$

Assuming that (4.4) holds for some set of coefficients $\{\hat{u}_k;\ k \in \mathbb{Z}\}$, the following simple and instructive argument shows why (4.2) is satisfied for that set of coefficients. Two complex waveforms, $\theta_k(t)$ and $\theta_m(t)$, are defined to be *orthogonal* if $\int_{-\infty}^{\infty} \theta_k(t)\theta_m^*(t)\,\mathrm{d}t = 0$. The truncated complex sinusoids in (4.4) are orthogonal since the interval $[-T/2,\ T/2]$ contains an integral number of cycles of each, i.e., for $k \neq m \in \mathbb{Z}$,

$$\int_{-\infty}^{\infty} \theta_k(t)\theta_m^*(t)\,\mathrm{d}t = \int_{-T/2}^{T/2} \mathrm{e}^{2\pi i (k-m)t/T}\,\mathrm{d}t = 0.$$

Thus, the right side of (4.2) can be evaluated as follows:

$$\begin{aligned} \frac{1}{T}\int_{-T/2}^{T/2} u(t)\mathrm{e}^{-2\pi i k t/T}\mathrm{d}t &= \frac{1}{T}\int_{-\infty}^{\infty} \sum_{m=-\infty}^{\infty} \hat{u}_m \theta_m(t)\theta_k^*(t)\,\mathrm{d}t \\ &= \frac{\hat{u}_k}{T}\int_{-\infty}^{\infty} |\theta_k(t)|^2\,\mathrm{d}t \\ &= \frac{\hat{u}_k}{T}\int_{-T/2}^{T/2} \mathrm{d}t = \hat{u}_k. \end{aligned} \tag{4.5}$$

An expansion such as that of (4.4) is called an *orthogonal expansion*. As shown later, the argument in (4.5) can be used to find the coefficients in any orthogonal expansion. At that point, more care will be taken in exchanging the order of integration and summation above.

Example 4.2.1 This and Example 4.2.2 illustrate why (4.4) need not be valid for all values of t. Let $u(t) = \mathrm{rect}(2t)$ (see Figure 4.2). Consider representing $u(t)$ by a Fourier series over the interval $-1/2 \le t \le 1/2$. As illustrated, the series can be shown to converge to $u(t)$ at all $t \in [-1/2, 1/2]$, except for the discontinuities at $t = \pm 1/4$. At $t = \pm 1/4$, the series converges to the midpoint of the discontinuity and (4.4) is not valid[8] at those points. Section 4.3 will show how to state (4.4) precisely so as to avoid these convergence issues.

[8] Most engineers, including the author, would say "So what? Who cares what the Fourier series converges to at a discontinuity of the waveform?" Unfortunately, this example is only the tip of an iceberg, especially when time-sampling of waveforms and sample waveforms of random processes are considered.

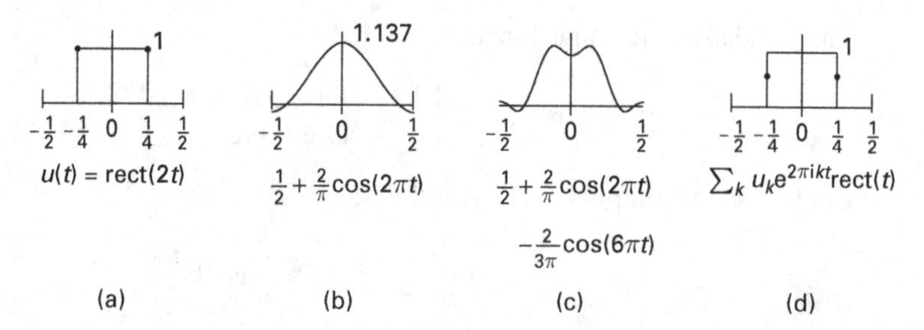

Figure 4.2. The Fourier series (over $[-1/2, 1/2]$) of a rectangular pulse rect$(2t)$, shown in (a). (b) Partial sum with $k = -1, 0, 1$. (c) Partial sum with $-3 \le k \le 3$. Part (d) illustrates that the series converges to $u(t)$ except at the points $t = \pm 1/4$, where it converges to 1/2.

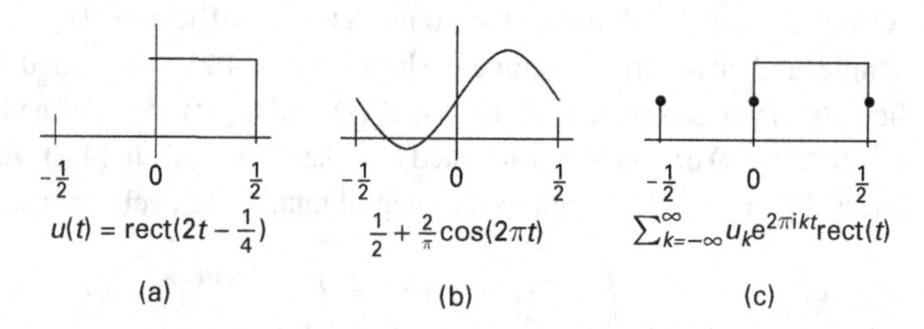

Figure 4.3. The Fourier series over $[-1/2, 1/2]$ of the same rectangular pulse shifted right by 1/4, shown in (a). (b) Partial expansion with $k = -1, 0, 1$. Part (c) depicts that the series converges to $v(t)$ except at the points $t = -1/2, 0$, and 1/2, at each of which it converges to 1/2.

Example 4.2.2 As a variation of the previous example, let $v(t)$ be 1 for $0 \le t \le 1/2$ and 0 elsewhere. Figure 4.3 shows the corresponding Fourier series over the interval $-1/2 \le t \le 1/2$.

A peculiar feature of this example is the isolated discontinuity at $t = -1/2$, where the series converges to 1/2. This happens because the untruncated Fourier series, $\sum_{k=-\infty}^{\infty} \hat{v}_k e^{2\pi ikt}$, is periodic with period 1 and thus must have the same value at both $t = -1/2$ and $t = 1/2$. More generally, if an arbitrary function $\{v(t) : [-T/2, T/2] \to \mathbb{C}\}$ has $v(-T/2) \ne v(T/2)$, then its Fourier series over that interval cannot converge to $v(t)$ at both those points.

4.2.1 Finite-energy waveforms

The *energy* in a real or complex waveform $u(t)$ is defined[9] to be $\int_{-\infty}^{\infty} |u(t)|^2 \, dt$. The energy in source waveforms plays a major role in determining how well the waveforms can be compressed for a given level of distortion. As a preliminary explanation, consider the energy in a time-limited waveform $\{u(t) : [-T/2, T/2] \to \mathbb{R}\}$. This energy

[9] Note that $u^2 = |u|^2$ if u is real, but, for complex u, u^2 can be negative or complex and $|u|^2 = uu^* = [\Re(u)]^2 + [\Im(u)]^2$ is required to correspond to the intuitive notion of energy.

is related to the Fourier series coefficients of $u(t)$ by the following *energy equation* which is derived in Exercise 4.2 by the same argument used in (4.5):

$$\int_{t=-T/2}^{T/2} |u(t)|^2 \, \mathrm{d}t = T \sum_{k=-\infty}^{\infty} |\hat{u}_k|^2. \tag{4.6}$$

Suppose that $u(t)$ is compressed by first generating its Fourier series coefficients, $\{\hat{u}_k;\ k \in \mathbb{Z}\}$, and then compressing those coefficients. Let $\{\hat{v}_k;\ k \in \mathbb{Z}\}$ be this sequence of compressed coefficients. Using a squared distortion measure for the coefficients, the overall distortion is $\sum_k |\hat{u}_k - \hat{v}_k|^2$. Suppose these compressed coefficients are now encoded, sent through a channel, reliably decoded, and converted back to a waveform $v(t) = \sum_k \hat{v}_k \mathrm{e}^{2\pi ikt/T}$ as in Figure 4.1. The difference between the input waveform $u(t)$ and the output $v(t)$ is then $u(t) - v(t)$, which has the Fourier series $\sum_k (\hat{u}_k - \hat{v}_k)\mathrm{e}^{2\pi ikt/T}$. Substituting $u(t) - v(t)$ into (4.6) results in the *difference-energy equation*:

$$\int_{t=-T/2}^{T/2} |u(t) - v(t)|^2 \, \mathrm{d}t = T \sum_{k} |\hat{u}_k - \hat{v}_k|^2. \tag{4.7}$$

Thus the energy in the difference between $u(t)$ and its reconstruction $v(t)$ is simply T times the sum of the squared differences of the quantized coefficients. This means that reducing the squared difference in the quantization of a coefficient leads directly to reducing the energy in the waveform difference. The energy in the waveform difference is a common and reasonable measure of distortion, but the fact that it is directly related to the mean-squared coefficient distortion provides an important added reason for its widespread use.

There must be at least T units of delay involved in finding the Fourier coefficients for $u(t)$ in $[-T/2, T/2]$ and then reconstituting $v(t)$ from the quantized coefficients at the receiver. There is additional processing and propagation delay in the channel. Thus the output waveform must be a delayed approximation to the input. All of this delay is accounted for by timing recovery processes at the receiver. This timing delay is set so that $v(t)$ at the receiver, according to the receiver timing, is the appropriate approximation to $u(t)$ at the transmitter, according to the transmitter timing. Timing recovery and delay are important problems, but they are largely separable from the problems of current interest. Thus, after recognizing that receiver timing is delayed from transmitter timing, delay can be otherwise ignored for now.

Next, visualize the Fourier coefficients $\hat{u}_k$ as sample values of independent random variables and visualize $u(t)$, as given by (4.3), as a sample value of the corresponding random process (this will be explained carefully in Chapter 7). The expected energy in this random process is equal to T times the sum of the mean-squared values of the coefficients. Similarly the expected energy in the difference between $u(t)$ and $v(t)$ is equal to T times the sum of the mean-squared coefficient distortions. It was seen by scaling in Chapter 3 that the the mean-squared quantization error for an analog random variable is proportional to the variance of that random variable. It is thus not surprising that the expected energy in a random waveform will have a similar relation to the mean-squared distortion after compression.

There is an obvious practical problem with compressing a finite-duration waveform by quantizing an *infinite* set of coefficients. One solution is equally obvious: compress only those coefficients with a significant mean-squared value. Since the expected value of $\sum_k |\hat{u}_k|^2$ is finite for finite-energy functions, the mean-squared distortion from ignoring small coefficients can be made as small as desired by choosing a sufficiently large finite set of coefficients. One then simply chooses $\hat{v}_k = 0$ in (4.7) for each ignored value of k.

The above argument will be explained carefully after developing the required tools. For now, there are two important insights. First, the energy in a source waveform is an important parameter in data compression; second, the source waveforms of interest will have finite energy and can be compressed by compressing a finite number of coefficients.

Next consider the waveforms used for channel transmission. The energy used over any finite interval T is limited both by regulatory agencies and by physical constraints on transmitters and antennas. One could consider waveforms of finite power but infinite duration and energy (such as the lowly sinusoid). On one hand, physical waveforms do not last forever (transmitters wear out or become obsolete), but, on the other hand, *models* of physical waveforms can have infinite duration, modeling physical lifetimes that are much longer than any time scale of communication interest. Nonetheless, for reasons that will gradually unfold, the channel waveforms in this text will almost always be restricted to finite energy.

There is another important reason for concentrating on finite-energy waveforms. Not only are they the appropriate models for source and channel waveforms, but they also have remarkably simple and general properties. These properties rely on an additional constraint called *measurability*, which is explained in Section 4.3. These finite-energy measurable functions are called $\mathcal{L}_2$ functions. When time-constrained, they *always* have Fourier series, and without a time constraint, they *always* have Fourier transforms. Perhaps the most important property, however, is that $\mathcal{L}_2$ functions can be treated essentially as conventional vectors (see Chapter 5).

One might question whether a limitation to finite-energy functions is too constraining. For example, a sinusoid is often used to model the carrier in passband communication, and sinusoids have infinite energy because of their infinite duration. As seen later, however, when a finite-energy baseband waveform is modulated by that sinusoid up to passband, the resulting passband waveform has finite energy.

As another example, the unit impulse (the Dirac delta function $\delta(t)$) is a generalized function used to model waveforms of unit area that are nonzero only in a narrow region around $t = 0$, narrow relative to all other intervals of interest. The impulse response of a linear-time-invariant filter is, of course, the response to a unit impulse; this response approximates the response to a physical waveform that is sufficiently narrow and has unit area. The energy in that physical waveform, however, grows wildly as the waveform narrows. A rectangular pulse of width ε and height $1/\varepsilon$, for example, has unit area for all $\varepsilon > 0$, but has energy $1/\varepsilon$, which approaches ∞ as $\varepsilon \to 0$. One could view the energy in a unit impulse as being either undefined or infinite, but in no way could one view it as being finite.

To summarize, there are many useful waveforms outside the finite-energy class. Although they are not physical waveforms, they are useful models of physical waveforms where energy is not important. Energy is such an important aspect of source and channel waveforms, however, that such waveforms can safely be limited to the finite-energy class.

4.3 $\mathcal{L}_2$ functions and Lebesgue integration over $[-T/2, T/2]$

A function $\{u(t) : \mathbb{R} \to \mathbb{C}\}$ is defined to be $\mathcal{L}_2$ if it is Lebesgue measurable and has a finite Lebesgue integral $\int_{-\infty}^{\infty} |u(t)|^2 \, dt$. This section provides a basic and intuitive understanding of what these terms mean. Appendix 4.9 provides proofs of the results, additional examples, and more depth of understanding. Still deeper understanding requires a good mathematics course in real and complex variables. Appendix 4.9 is not required for basic engineering understanding of results in this and subsequent chapters, but it will provide deeper insight.

The basic idea of Lebesgue integration is no more complicated than the more common Riemann integration taught in freshman college courses. Whenever the Riemann integral exists, the Lebesgue integral also exists[10] and has the same value. Thus all the familiar ways of calculating integrals, including tables and numerical procedures, hold without change. The Lebesgue integral is more useful here, partly because it applies to a wider set of functions, but, more importantly, because it greatly simplifies the main results.

This section considers only time-limited functions, $\{u(t) : [-T/2, T/2] \to \mathbb{C}\}$. These are the functions of interest for Fourier series, and the restriction to a finite interval avoids some mathematical details better addressed later.

Figure 4.4 shows intuitively how Lebesgue and Riemann integration differ. Conventional Riemann integration of a nonnegative real-valued function $u(t)$ over an interval $[-T/2, T/2]$ is conceptually performed in Figure 4.4(a) by partitioning $[-T/2, T/2]$ into, say, i_0 intervals each of width T/i_0. The function is then approximated within the

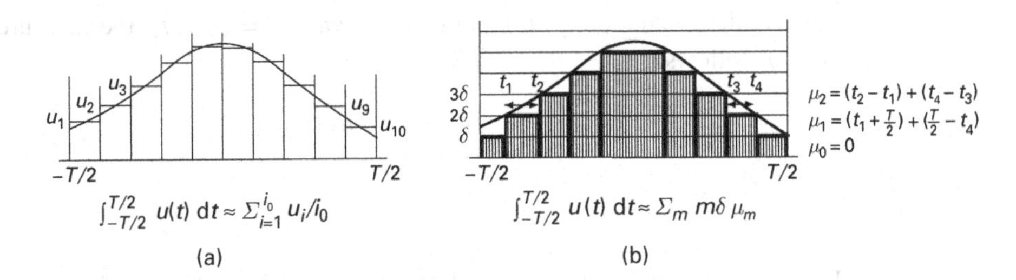

Figure 4.4. Example of (a) Riemann and (b) Lebesgue integration.

[10] There is a slight notional qualification to this which is discussed in the sinc function example of Section 4.5.1.

ith such interval by a single value u_i, such as the midpoint of values in the interval. The integral is then approximated as $\sum_{i=1}^{i_0}(T/i_0)u_i$. If the function is sufficiently smooth, then this approximation has a limit, called the Riemann integral, as $i_0 \to \infty$.

To integrate the same function by Lebesgue integration, the vertical axis is partitioned into intervals each of height δ, as shown in Figure 4.4(b). For the mth such interval,[11] $[m\delta, (m+1)\delta)$, let $\mathcal{E}_m$ be the set of values of t such that $m\delta \leq u(t) < (m+1)\delta$. For example, the set $\mathcal{E}_2$ is illustrated by arrows in Figure 4.4(b) and is given by

$$\mathcal{E}_2 = \{t : 2\delta \leq u(t) < 3\delta\} = [t_1, t_2) \cup (t_3, t_4].$$

As explained below, if $\mathcal{E}_m$ is a finite union of separated[12] intervals, its measure, μ_m is the sum of the widths of those intervals; thus μ_2 in the example above is given by

$$\mu_2 = \mu(\mathcal{E}_2) = (t_2 - t_1) + (t_4 - t_3). \tag{4.8}$$

Similarly, $\mathcal{E}_1 = [-T/2, t_1) \cup (t_4, T/2]$ and $\mu_1 = (t_1 + T/2) + (T/2 - t_4)$.

The Lebesgue integral is approximated as $\sum_m (m\delta)\mu_m$. This approximation is indicated by the vertically shaded area in Figure 4.4(b). The Lebesgue integral is essentially the limit as $\delta \to 0$.

In short, the Riemann approximation to the area under a curve splits the horizontal axis into uniform segments and sums the corresponding rectangular areas. The Lebesgue approximation splits the vertical axis into uniform segments and sums the height times width measure for each segment. In both cases, a limiting operation is required to find the integral, and Section 4.3.3 gives an example where the limit exists in the Lebesgue but not the Riemann case.

4.3.1 Lebesgue measure for a union of intervals

In order to explain Lebesgue integration further, measure must be defined for a more general class of sets.

The *measure* of an interval I from a to b, $a \leq b$, is defined to be $\mu(I) = b - a \geq 0$. For any finite union of, say, ℓ separated intervals, $\mathcal{E} = \bigcup_{j=1}^{\ell} I_j$, the measure $\mu(\mathcal{E})$ is defined as follows:

$$\mu(\mathcal{E}) = \sum_{j=1}^{\ell} \mu(I_j). \tag{4.9}$$

[11] The notation $[a, b)$ denotes the semiclosed interval $a \leq t < b$. Similarly, $(a, b]$ denotes the semiclosed interval $a < t \leq b$, (a, b) the open interval $a < t < b$, and $[a, b]$ the closed interval $a \leq t \leq b$. In the special case where $a = b$, the interval $[a, a]$ consists of the single point a, whereas $[a, a)$, $(a, a]$, and (a, a) are empty.

[12] Two intervals are *separated* if they are both nonempty and there is at least one point between them that lies in neither interval; i.e., $(0, 1)$ and $(1, 2)$ are separated. In contrast, two sets are *disjoint* if they have no points in common. Thus $(0, 1)$ and $[1, 2]$ are disjoint but not separated.

This definition of $\mu(\mathcal{E})$ was used in (4.8) and is necessary for the approximation in Figure 4.4(b) to correspond to the area under the approximating curve. The fact that the measure of an interval does not depend on the inclusion of the endpoints corresponds to the basic notion of area under a curve. Finally, since these separated intervals are all contained in $[-T/2, T/2]$, it is seen that the sum of their widths is at most T, i.e.

$$0 \le \mu(\mathcal{E}) \le T. \tag{4.10}$$

Any finite union of, say, ℓ arbitrary intervals, $\mathcal{E} = \bigcup_{j=1}^{\ell} I_j$, can also be uniquely expressed as a finite union of at most ℓ separated intervals, say $I'_1, \ldots, I'_k$, $k \le \ell$ (see Exercise 4.5), and its measure is then given by

$$\mu(\mathcal{E}) = \sum_{j=1}^{k} \mu(I'_j). \tag{4.11}$$

The union of a countably infinite collection[13] of separated intervals, say $\mathcal{B} = \bigcup_{j=1}^{\infty} I_j$, is also defined to be measurable and has a measure given by

$$\mu(\mathcal{B}) = \lim_{\ell \to \infty} \sum_{j=1}^{\ell} \mu(I_j). \tag{4.12}$$

The summation on the right is bounded between 0 and T for each ℓ. Since $\mu(I_j) \ge 0$, the sum is nondecreasing in ℓ. Thus the limit exists and lies between 0 and T. Also the limit is independent of the ordering of the I_j (see Exercise 4.4).

Example 4.3.1 Let $I_j = (T2^{-2j}, T2^{-2j+1})$ for all integers $j \ge 1$. The jth interval then has measure $\mu(I_j) = 2^{-2j}$. These intervals get smaller and closer to 0 as j increases. They are easily seen to be separated. The union $\mathcal{B} = \bigcup_j I_j$ then has measure $\mu(\mathcal{B}) = \sum_{j=1}^{\infty} T2^{-2j} = T/3$. Visualize replacing the function in Figure 4.4 by one that oscillates faster and faster as $t \to 0$; $\mathcal{B}$ could then represent the set of points on the horizontal axis corresponding to a given vertical slice.

Example 4.3.2 As a variation of Example 4.3.1, suppose $\mathcal{B} = \bigcup_j I_j$, where $I_j = [T2^{-2j}, T2^{-2j}]$ for each j. Then interval I_j consists of the single point $T2^{-2j}$ so $\mu(I_j) = 0$. In this case, $\sum_{j=1}^{\ell} \mu(I_j) = 0$ for each ℓ. The limit of this as $\ell \to \infty$ is also 0, so $\mu(\mathcal{B}) = 0$ in this case. By the same argument, the measure of any countably infinite set of points is 0.

Any countably infinite union of arbitrary (perhaps intersecting) intervals can be uniquely[14] represented as a *countable* (i.e. either a countably infinite or finite) union of separated intervals (see Exercise 4.6); its measure is defined by applying (4.12) to that representation.

---

[13] An elementary discussion of countability is given in Appendix 4.9.1. Readers unfamiliar with ideas such as the countability of the rational numbers are strongly encouraged to read this appendix.

[14] The collection of separated intervals and the limit in (4.12) is unique, but the ordering of the intervals is not.

4.3.2 Measure for more general sets

It might appear that the class of countable unions of intervals is broad enough to represent any set of interest, but it turns out to be too narrow to allow the general kinds of statements that formed our motivation for discussing Lebesgue integration. One vital generalization is to require that the complement $\overline{\mathcal{B}}$ (relative to $[-T/2, T/2]$) of any measurable set $\mathcal{B}$ also be measurable.[15] Since $\mu([-T/2, T/2]) = T$ and every point of $[-T/2, T/2]$ lies in either $\mathcal{B}$ or $\overline{\mathcal{B}}$ but not both, the measure of $\overline{\mathcal{B}}$ should be $T - \mu(\mathcal{B})$. The reason why this property is necessary in order for the Lebesgue integral to correspond to the area under a curve is illustrated in Figure 4.5.

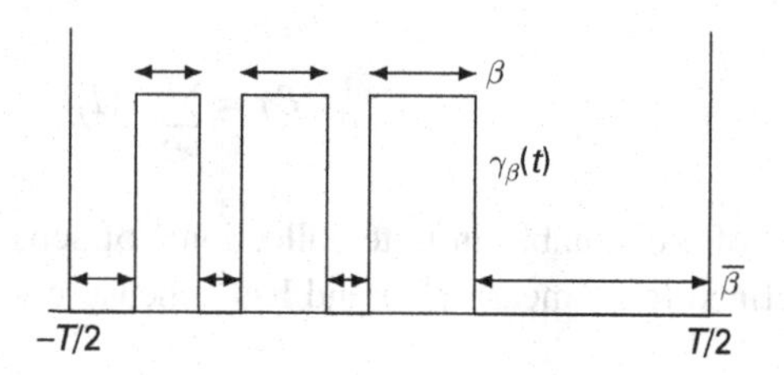

Figure 4.5. Let $f(t)$ have the value 1 on a set $\mathcal{B}$ and the value 0 elsewhere in $[-T/2, T/2]$. Then $\int f(t)\,\mathrm{d}t = \mu(\mathcal{B})$. The complement $\overline{\mathcal{B}}$ of $\mathcal{B}$ is also illustrated, and it is seen that $1 - f(t)$ is 1 on the set $\overline{\mathcal{B}}$ and 0 elsewhere. Thus $\int [1 - f(t)]\,\mathrm{d}t = \mu(\overline{\mathcal{B}})$, which must equal $T - \mu(\mathcal{B})$ for integration to correspond to the area under a curve.

The *subset inequality* is another property that measure should have: this states that if $\mathcal{A}$ and $\mathcal{B}$ are both measurable and $\mathcal{A} \subseteq \mathcal{B}$, then $\mu(\mathcal{A}) \leq \mu(\mathcal{B})$. One can also visualize from Figure 4.5 why this subset inequality is necessary for integration to represent the area under a curve.

Before defining which sets in $[-T/2, T/2]$ are measurable and which are not, a measure-like function called *outer measure* is introduced that exists for *all* sets in $[-T/2, T/2]$. For an arbitrary set $\mathcal{A}$, the set $\mathcal{B}$ is said to *cover* $\mathcal{A}$ if $\mathcal{A} \subseteq \mathcal{B}$ and $\mathcal{B}$ is a countable union of intervals. The outer measure $\mu^{\circ}(\mathcal{A})$ is then essentially the measure of the smallest cover of $\mathcal{A}$. In particular,[16]

$$\mu^{\circ}(\mathcal{A}) = \inf_{\mathcal{B}:\,\mathcal{B}\text{ covers }\mathcal{A}} \mu(\mathcal{B}). \tag{4.13}$$

[15] Appendix 4.9.1 uses the set of rationals in $[-T/2, T/2]$ to illustrate that the complement $\overline{\mathcal{B}}$ of a countable union of intervals $\mathcal{B}$ need not be a countable union of intervals itself. In this case, $\mu(\overline{\mathcal{B}}) = T - \mu(\mathcal{B})$, which is shown to be valid also when $\overline{\mathcal{B}}$ is a countable union of intervals.

[16] The infimum (inf) of a set of real numbers is essentially the minimum of that set. The difference between the minimum and the infimum can be seen in the example of the set of real numbers strictly greater than 1. This set has no minimum, since for each number in the set, there is a smaller number still greater than 1. To avoid this somewhat technical issue, the infimum is defined as the greatest lowerbound of a set. In the example, all numbers less than or equal to 1 are lowerbounds for the set, and 1 is then the greatest lowerbound, i.e. the infimum. Every nonempty set of real numbers has an infinum if one includes $-\infty$ as a choice.

Not surprisingly, the outer measure of a countable union of intervals is equal to its measure as already defined (see Appendix 4.9.3).

Measurable sets and measure over the interval $[-T/2, T/2]$ can now be defined as follows.

Definition 4.3.1 A set $\mathcal{A}$ (over $[-T/2, T/2]$) is *measurable* if $\mu^{\text{o}}(\mathcal{A}) + \mu^{\text{o}}(\overline{\mathcal{A}}) = T$. If $\mathcal{A}$ is measurable, then its *measure*, $\mu(\mathcal{A})$, is the outer measure $\mu^{\text{o}}(\mathcal{A})$.

Intuitively, then, a set is measurable if the set and its complement are sufficiently untangled that each can be covered by countable unions of intervals which have arbitrarily little overlap. The example at the end of Appendix 4.9.4 constructs the simplest nonmeasurable set we are aware of; it should be noted how bizarre it is and how tangled it is with its complement.

The definition of measurability is a "mathematician's definition" in the sense that it is very succinct and elegant, but it does not provide many immediate clues about determining whether a set is measurable and, if so, what its measure is. This is now briefly discussd.

It is shown in Appendix 4.9.3 that countable unions of intervals are measurable according to this definition, and the measure can be found by breaking the set into separated intervals. Also, by definition, the complement of every measurable set is also measurable, so the complements of countable unions of intervals are measurable. Next, if $\mathcal{A} \subseteq \mathcal{A}'$, then any cover of $\mathcal{A}'$ also covers $\mathcal{A}$, so the subset inequality is satisfied. This often makes it possible to find the measure of a set by using a limiting process on a sequence of measurable sets contained in or containing a set of interest. Finally, the following theorem is proven in Appendix 4.9.4.

Theorem 4.3.1 *Let* $\mathcal{A}_1, \mathcal{A}_2, \ldots$ *be any sequence of measurable sets. Then* $\mathcal{S} = \bigcup_{j=1}^{\infty} \mathcal{A}_j$ *and* $\mathcal{D} = \bigcap_{j=1}^{\infty} \mathcal{A}_j$ *are measurable. If* $\mathcal{A}_1, \mathcal{A}_2, \ldots$ *are also disjoint, then* $\mu(\mathcal{S}) = \sum_j \mu(\mathcal{A}_j)$. *If* $\mu^{\text{o}}(\mathcal{A}) = 0$, *then* $\mathcal{A}$ *is measurable and has zero measure.*

This theorem and definition say that the collection of measurable sets is closed under countable unions, countable intersections, and complementation. This partly explains why it is so hard to find nonmeasurable sets and also why their existence can usually be ignored – they simply do not arise in the ordinary process of analysis.

Another consequence concerns sets of zero measure. It was shown earlier that any set containing only countably many points has zero measure, but there are many other sets of zero measure. The Cantor set example in Appendix 4.9.4 illustrates a set of zero measure with uncountably many elements. The theorem implies that a set $\mathcal{A}$ has zero measure if, for any $\varepsilon > 0$, $\mathcal{A}$ has a cover $\mathcal{B}$ such that $\mu(\mathcal{B}) \leq \varepsilon$. The definition of measurability shows that the complement of any set of zero measure has measure T, i.e. $[-T/2, T/2]$ is the cover of smallest measure. It will be seen shortly that, for most purposes, including integration, sets of zero measure can be ignored and sets of measure T can be viewed as the entire interval $[-T/2, T/2]$.

This concludes our study of measurable sets on $[-T/2, T/2]$. The bottom line is that not all sets are measurable, but that nonmeasurable sets arise only from bizarre and artificial constructions and can usually be ignored. The definitions of measure and measurability might appear somewhat arbitrary, but in fact they arise simply through the natural requirement that intervals and countable unions of intervals be measurable with the given measure[17] and that the subset inequality and complement property be satisfied. If we wanted additional sets to be measurable, then at least one of the above properties would have to be sacrificed and integration itself would become bizarre. The major result here, beyond basic familiarity and intuition, is Theorem 4.3.1, which is used repeatedly in the following sections. Appendix 4.9 fills in many important details and proves the results here

4.3.3 Measurable functions and integration over $[-T/2, T/2]$

A function $\{u(t) : [-T/2, T/2] \to \mathbb{R}\}$ is said to be *Lebesgue measurable* (or more briefly *measurable*) if the set of points $\{t : u(t) < \beta\}$ is measurable for each $\beta \in \mathbb{R}$. If $u(t)$ is measurable, then, as shown in Exercise 4.11, the sets $\{t : u(t) \leq \beta\}$, $\{t : u(t) \geq \beta\}$, $\{t : u(t) > \beta\}$, and $\{t : \alpha \leq u(t) < \beta\}$ are measurable for all $\alpha < \beta \in \mathbb{R}$. Thus, if a function is measurable, the measure $\mu_m = \mu(\{t : m\delta \leq u(t) < (m+1)\delta\})$ associated with the mth horizontal slice in Figure 4.4 must exist for each $\delta > 0$ and m.

For the Lebesgue integral to exist, it is also necessary that the Figure 4.4 approximation to the Lebesgue integral has a limit as the vertical interval size δ goes to 0. Initially consider only nonnegative functions, $u(t) \geq 0$ for all t. For each integer $n \geq 1$, define the nth-order approximation to the Lebesgue integral as that arising from partitioning the vertical axis into intervals each of height $\delta_n = 2^{-n}$. Thus a unit increase in n corresponds to halving the vertical interval size as illustrated in Figure 4.6.

Let $\mu_{m,n}$ be the measure of $\{t : m2^{-n} \leq u(t) < (m+1)2^{-n}\}$, i.e. the measure of the set of $t \in [-T/2, T/2]$ for which $u(t)$ is in the mth vertical interval for the nth-order

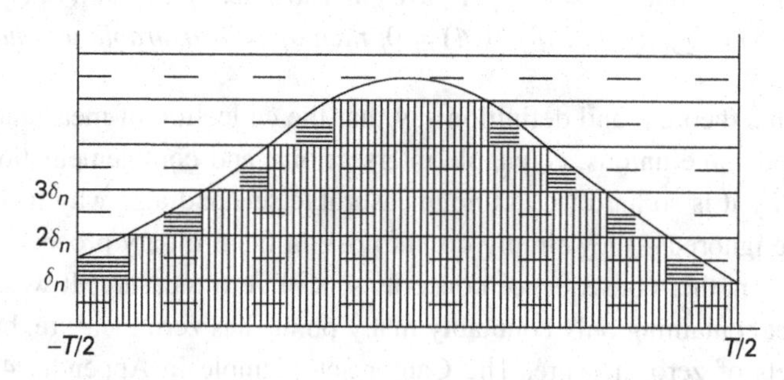

Figure 4.6. Improvement in the approximation to the Lebesgue integral by a unit increase in n is indicated by the horizontal crosshatching.

[17] We have not distinguished between the condition of being measurable and the actual measure assigned a set, which is natural for ordinary integration. The theory can be trivially generalized, however, to random variables restricted to $[-T/2, T/2]$. In this case, the measure of an interval is redefined to be the probability of that interval. Everything else remains the same except that some individual points might have nonzero probability.

approximation. The approximation $\sum_m m2^{-n}\mu_{m,n}$ might be infinite[18] for all n, and in this case the Lebesgue integral is said to be infinite. If the sum is finite for $n = 1$, however, Figure 4.6 shows that the change in going from the approximation of order n to $n+1$ is nonnegative and upperbounded by $T2^{-n-1}$. Thus it is clear that the sequence of approximations has a finite limit which is defined[19] to be the *Lebesgue integral* of $u(t)$. In summary, the Lebesgue integral of an arbitrary measurable nonnegative function $\{u(t) : [-T/2, T/2] \to \mathbb{R}\}$ is finite if any approximation is finite and is then given by

$$\int u(t)\,dt = \lim_{n\to\infty} \sum_{m=0}^{\infty} m2^{-n}\mu_{m,n}, \qquad \text{where} \quad \mu_{m,n} = \mu(t : m2^{-n} \le u(t) < (m+1)2^{-n}). \tag{4.14}$$

Example 4.3.3 Consider a function that has the value 1 for each rational number in $[-T/2, T/2]$ and 0 for all irrational numbers. The set of rationals has zero measure, as shown in Appendix 4.9.1, so that each approximation is zero in Figure 4.6, and thus the Lebesgue integral, as the limit of these approximations, is zero. This is a simple example of a function that has a Lebesgue integral but no Riemann integral.

Next consider two nonnegative measurable functions $u(t)$ and $v(t)$ on $[-T/2, T/2]$ and assume $u(t) = v(t)$ except on a set of zero measure. Then each of the approximations in (4.14) are identical for $u(t)$ and $v(t)$, and thus the two integrals are identical (either both infinite or both the same number). This same property will be seen to carry over for functions that also take on negative values and, more generally, for complex-valued functions. This property says that sets of zero measure can be ignored in integration. This is one of the major simplifications afforded by Lebesgue integration. Two functions that are the same except on a set of zero measure are said to be equal *almost everywhere*, abbreviated a.e. For example, the rectangular pulse and its Fourier series representation illustrated in Figure 4.2 are equal a.e.

For functions taking on both positive and negative values, the function $u(t)$ can be separated into a positive part $u^+(t)$ and a negative part $u^-(t)$. These are defined by

$$u^+(t) = \begin{cases} u(t) & \text{for } t : u(t) \ge 0 \\ 0 & \text{for } t : u(t) < 0 \end{cases}; \qquad u^-(t) = \begin{cases} 0 & \text{for } t : u(t) \ge 0 \\ -u(t) & \text{for } t : u(t) < 0. \end{cases}$$

For all $t \in [-T/2, T/2]$ then,

$$u(t) = u^+(t) - u^-(t). \tag{4.15}$$

[18] For example, this sum is infinite if $u(t) = 1/|t|$ for $-T/2 \le t \le T/2$. The situation here is essentially the same for Riemann and Lebesgue integration.

[19] This limiting operation can be shown to be independent of how the quantization intervals approach 0.

If $u(t)$ is measurable, then $u^+(t)$ and $u^-(t)$ are also.[20] Since these are nonnegative, they can be integrated as before, and each integral exists with either a finite or infinite value. If at most one of these integrals is infinite, the Lebesgue integral of $u(t)$ is defined as

$$\int u(t) = \int u^+(t) - \int u^-(t)\mathrm{d}t. \tag{4.16}$$

If both $\int u^+(t)\,\mathrm{d}t$ and $\int u^-(t)\,\mathrm{d}t$ are infinite, then the integral is undefined.

Finally, a complex function $\{u(t) : [-T/2\, T/2] \to \mathbb{C}\}$ is defined to be *measurable* if the real and imaginary parts of $u(t)$ are measurable. If the integrals of $\Re(u(t))$ and $\Im(u(t))$ are defined, then the Lebesgue integral $\int u(t)\,\mathrm{d}t$ is defined by

$$\int u(t)\mathrm{d}t = \int \Re(u(t))\mathrm{d}t + \mathrm{i}\int \Im(u(t))\mathrm{d}t. \tag{4.17}$$

The integral is undefined otherwise. Note that this implies that any integration property of complex-valued functions $\{u(t) : [-T/2,\, T/2] \to \mathbb{C}\}$ is also shared by real-valued functions $\{u(t) : [-T/2,\, T/2] \to \mathbb{R}\}$.

4.3.4 Measurability of functions defined by other functions

The definitions of measurable functions and Lebesgue integration in Section 4.3.3 were quite simple given the concept of measure. However, functions are often defined in terms of other more elementary functions, so the question arises whether measurability of those elementary functions implies that of the defined function. The bottom-line answer is almost invariably yes. For this reason it is often assumed in the following sections that all functions of interest are measurable. Several results are now given fortifying this bottom-line view.

First, if $\{u(t) : [-T/2,\, T/2] \to \mathbb{R}\}$ is measurable, then $-u(t)$, $|u(t)|$, $u^2(t)$, $\mathrm{e}^{u(t)}$, and $\ln|u(t)|$ are also measurable. These and similar results follow immediately from the definition of measurable functions and are derived in Exercise 4.12.

Next, if $u(t)$ and $v(t)$ are measurable, then $u(t) + v(t)$ and $u(t)v(t)$ are measurable (see Exercise 4.13).

Finally, if $\{u_k(t) : [-T/2,\, T/2] \to \mathbb{R}\}$ is a measurable function for each integer $k \geq 1$, then $\inf_k u_k(t)$ is measurable. This can be seen by noting that $\{t : \inf_k[u_k(t)] \leq \alpha\} = \bigcup_k \{t : u_k(t) \leq \alpha\}$, which is measurable for each α. Using this result, Exercise 4.15 shows that $\lim_k u_k(t)$ is measurable if the limit exists for all $t \in [-T/2,\, T/2]$.

4.3.5 $\mathcal{L}_1$ and $\mathcal{L}_2$ functions over $[-T/2,\, T/2]$

A function $\{u(t) : [-T/2,\, T/2] \to \mathbb{C}\}$ is said to be $\mathcal{L}_1$, or in the class $\mathcal{L}_1$, if $u(t)$ is measurable and the Lebesgue integral of $|u(t)|$ is finite.[21]

---

[20] To see this, note that for $\beta > 0$, $\{t : u^+(t) < \beta\} = \{t : u(t) < \beta\}$. For $\beta \leq 0$, $\{t : u^+(t) < \beta\}$ is the empty set. A similar argument works for $u^-(t)$.

[21] $\mathcal{L}_1$ functions are sometimes called integrable functions.

For the special case of a real function, $\{u(t) : [-T/2, T/2] \to \mathbb{R}\}$, the magnitude $|u(t)|$ can be expressed in terms of the positive and negative parts of $u(t)$ as $|u(t)| = u^+(t) + u^-(t)$. Thus $u(t)$ is $\mathcal{L}_1$ if and only if both $u^+(t)$ and $u^-(t)$ have finite integrals. In other words, $u(t)$ is $\mathcal{L}_1$ if and only if the Lebesgue integral of $u(t)$ is defined and finite.

For a complex function $\{u(t) : [-T/2, T/2] \to \mathbb{C}\}$, it can be seen that $u(t)$ is $\mathcal{L}_1$ if and only if both $\Re[u(t)]$ and $\Im[u(t)]$ are $\mathcal{L}_1$. Thus $u(t)$ is $\mathcal{L}_1$ if and only if $\int u(t)\mathrm{d}t$ is defined and finite.

A function $\{u(t) : [-T/2, T/2] \to \mathbb{R}\}$ or $\{u(t) : [-T/2, T/2] \to \mathbb{C}\}$ is said to be an $\mathcal{L}_2$ function, or a *finite-energy function*, if $u(t)$ is measurable and the Lebesgue integral of $|u(t)|^2$ is finite. All source and channel waveforms discussed in this text will be assumed to be $\mathcal{L}_2$. Although $\mathcal{L}_2$ functions are of primary interest here, the class of $\mathcal{L}_1$ functions is of almost equal importance in understanding Fourier series and Fourier transforms. An important relation between $\mathcal{L}_1$ and $\mathcal{L}_2$ is given in the following simple theorem, illustrated in Figure 4.7.

Theorem 4.3.2 *If* $\{u(t) : [-T/2, T/2] \to \mathbb{C}\}$ *is* $\mathcal{L}_2$*, then it is also* $\mathcal{L}_1$.

Proof: Note that $|u(t)| \le |u(t)|^2$ for all t such that $|u(t)| \ge 1$. Thus $|u(t)| \le |u(t)|^2 + 1$ for all t, so that $\int |u(t)|\,\mathrm{d}t \le \int |u(t)|^2\,\mathrm{d}t + T$. If the function $u(t)$ is $\mathcal{L}_2$, then the right side of this equation is finite, so the function is also $\mathcal{L}_1$. □

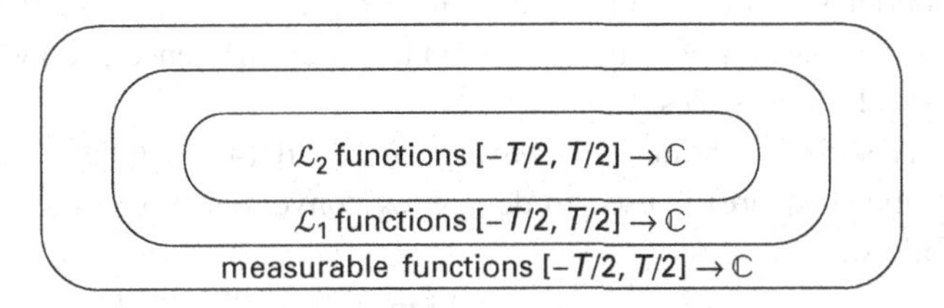

Figure 4.7. Illustration showing that for functions from $[-T/2, T/2]$ to $\mathbb{C}$, the class of $\mathcal{L}_2$ functions is contained in the class of $\mathcal{L}_1$ functions, which in turn is contained in the class of measurable functions. The restriction here to a finite domain such as $[-T/2, T/2]$ is necessary, as seen later.

This completes our basic introduction to measure and Lebesgue integration over the finite interval $[-T/2, T/2]$. The fact that the class of measurable sets is closed under complementation, countable unions, and countable intersections underlies the results about the measurability of functions being preserved over countable limits and sums. These in turn underlie the basic results about Fourier series, Fourier integrals, and orthogonal expansions. Some of those results will be stated without proof, but an understanding of measurability will enable us to understand what those results mean. Finally, ignoring sets of zero measure will simplify almost everything involving integration.

4.4 Fourier series for $\mathcal{L}_2$ waveforms

The most important results about Fourier series for $\mathcal{L}_2$ functions are as follows.

Theorem 4.4.1 (Fourier series) *Let $\{u(t) : [-T/2, T/2] \to \mathbb{C}\}$ be an $\mathcal{L}_2$ function. Then for each $k \in \mathbb{Z}$, the Lebesgue integral*

$$\hat{u}_k = \frac{1}{T}\int_{-T/2}^{T/2} u(t)\mathrm{e}^{-2\pi ikt/T}\,\mathrm{d}t \tag{4.18}$$

exists and satisfies $|\hat{u}_k| \le 1/T\int |u(t)|\,\mathrm{d}t < \infty$. Furthermore,

$$\lim_{\ell\to\infty}\int_{-T/2}^{T/2}\left|u(t) - \sum_{k=-\ell}^{\ell}\hat{u}_k\mathrm{e}^{2\pi ikt/T}\right|^2\mathrm{d}t = 0, \tag{4.19}$$

where the limit is monotonic in ℓ. Also, the energy equation (4.6) is satisfied.

Conversely, if $\{\hat{u}_k; k \in \mathbb{Z}\}$ is a two-sided sequence of complex numbers satisfying $\sum_{k=-\infty}^{\infty}|\hat{u}_k|^2 < \infty$, then an $\mathcal{L}_2$ function $\{u(t) : [-T/2, T/2] \to \mathbb{C}\}$ exists such that (4.6) and (4.19) are satisfied.

The first part of the theorem is simple. Since $u(t)$ is measurable and $\mathrm{e}^{-2\pi ikt/T}$ is measurable for each k, the product $u(t)\mathrm{e}^{-2\pi ikt/T}$ is measurable. Also $|u(t)\mathrm{e}^{-2\pi ikt/T}| = |u(t)|$ so that $u(t)\mathrm{e}^{-2\pi ikt/T}$ is $\mathcal{L}_1$ and the integral exists with the given upperbound (see Exercise 4.17). The rest of the proof is given in, Section 5.3.4.

The integral in (4.19) is the energy in the difference between $u(t)$ and the partial Fourier series using only the terms $-\ell \le k \le \ell$. Thus (4.19) asserts that $u(t)$ can be approximated arbitrarily closely (in terms of difference energy) by finitely many terms in its Fourier series.

A series is defined to *converge in* $\mathcal{L}_2$ if (4.19) holds. The notation l.i.m. (limit in mean-square) is used to denote $\mathcal{L}_2$ convergence, so (4.19) is often abbreviated as follows:

$$u(t) = \text{l.i.m.}\sum_k \hat{u}_k\mathrm{e}^{2\pi ikt/T}\text{rect}\left(\frac{t}{T}\right). \tag{4.20}$$

The notation *does not* indicate that the sum in (4.20) converges pointwise to $u(t)$ at each t; for example, the Fourier series in Figure 4.2 converges to 1/2 rather than 1 at the values $t = \pm 1/4$. In fact, *any* two $\mathcal{L}_2$ functions that are equal a.e. have the same Fourier series coefficients. Thus the best to be hoped for is that $\sum_k \hat{u}_k\mathrm{e}^{2\pi ikt/T}\text{rect}(t/T)$ converges pointwise and yields a "canonical representative" for all the $\mathcal{L}_2$ functions that have the given set of Fourier coefficients, $\{\hat{u}_k; k \in \mathbb{Z}\}$.

Unfortunately, there are some rather bizarre $\mathcal{L}_2$ functions (see the everywhere discontinuous example in Appendix 5.5.1) for which $\sum_k \hat{u}_k\,\mathrm{e}^{2\pi ikt/T}\,\text{rect}(t/T)$ diverges for some values of t.

There is an important theorem due to Carleson (1966), however, stating that if $u(t)$ is $\mathcal{L}_2$, then $\sum_k \hat{u}_k\mathrm{e}^{2\pi ikt/T}\text{rect}(t/T)$ converges a.e. on $[-T/2, T/2]$. Thus for any $\mathcal{L}_2$ function $u(t)$, with Fourier coefficients $\{\hat{u}_k : k \in \mathbb{Z}\}$, there is a well defined function,

$$\tilde{u}(t) = \begin{cases} \sum_{k=-\infty}^{\infty}\hat{u}_k\mathrm{e}^{2\pi ikt/T}\,\text{rect}(t/T) & \text{if the sum converges;} \\ 0 & \text{otherwise.} \end{cases} \tag{4.21}$$

Since the sum above converges a.e., the Fourier coefficients of $\tilde{u}(t)$ given by (4.18) agree with those in (4.21). Thus $\tilde{u}(t)$ can serve as a canonical representative for all the $\mathcal{L}_2$ functions with the same Fourier coefficients $\{\hat{u}_k;\, k \in \mathbb{Z}\}$. From the difference-energy equation (4.7), it follows that the difference between any two $\mathcal{L}_2$ functions with the same Fourier coefficients has zero energy. Two $\mathcal{L}_2$ functions whose difference has zero energy are said to be *$\mathcal{L}_2$-equivalent*; thus all $\mathcal{L}_2$ functions with the same Fourier coefficients are $\mathcal{L}_2$-equivalent. Exercise 4.18 shows that two $\mathcal{L}_2$ functions are $\mathcal{L}_2$-equivalent if and only if they are equal a.e.

In summary, each $\mathcal{L}_2$ function $\{u(t) : [-T/2, T/2] \to \mathbb{C}\}$ belongs to an equivalence class consisting of all $\mathcal{L}_2$ functions with the same set of Fourier coefficients. Each pair of functions in this equivalence class are $\mathcal{L}_2$-equivalent and equal a.e. The canonical representative in (4.21) is determined solely by the Fourier coefficients and is uniquely defined for any given set of Fourier coefficients satisfying $\sum_k |\hat{u}_k|^2 < \infty$; the corresponding equivalence class consists of the $\mathcal{L}_2$ functions that are equal to $\tilde{u}(t)$ a.e.

From an engineering standpoint, the sequence of ever closer approximations in (4.19) is usually more relevant than the notion of an equivalence class of functions with the same Fourier coefficients. In fact, for physical waveforms, there is no physical test that can distinguish waveforms that are $\mathcal{L}_2$-equivalent, since any such physical test requires an energy difference. At the same time, if functions $\{u(t) : [-T/2, T/2] \to \mathbb{C}\}$ are consistently represented by their Fourier coefficients, then equivalence classes can usually be ignored.

For all but the most bizarre $\mathcal{L}_2$ functions, the Fourier series converges everywhere to some function that is $\mathcal{L}_2$-equivalent to the original function, and thus, as with the points $t = \pm 1/4$ in the example of Figure 4.2, it is usually unimportant how one views the function at those isolated points. Occasionally, however, particularly when discussing sampling and vector spaces, the concept of equivalence classes becomes relevant.

4.4.1 The T-spaced truncated sinusoid expansion

There is nothing special about the choice of 0 as the center point of a time-limited function. For a function $\{v(t) : [\Delta - T/2,\, \Delta + T/2] \to \mathbb{C}\}$ centered around some arbitrary time Δ, the *shifted Fourier series* over that interval is given by[22]

$$v(t) = \text{l.i.m.} \sum_k \hat{v}_k \mathrm{e}^{2\pi i k t/T} \operatorname{rect}\left(\frac{t-\Delta}{T}\right), \tag{4.22}$$

where

$$\hat{v}_k = \frac{1}{T} \int_{\Delta - T/2}^{\Delta + T/2} v(t) \mathrm{e}^{-2\pi i k t/T}\, \mathrm{d}t, \qquad -\infty < k < \infty. \tag{4.23}$$

[22] Note that the Fourier relationship between the function $v(t)$ and the sequence $\{v_k\}$ depends implicitly on the interval T and the shift Δ.

To see this, let $u(t) = v(t+\Delta)$. Then $u(0) = v(\Delta)$ and $u(t)$ is centered around 0 and has a Fourier series given by (4.20) and (4.18). Letting $\hat{v}_k = \hat{u}_k e^{-2\pi i k\Delta/T}$ yields (4.22) and (4.23). The results about measure and integration are not changed by this shift in the time axis.

Next, suppose that some given function $u(t)$ is either not time-limited or limited to some very large interval. An important method for source coding is first to break such a function into segments, say of duration T, and then to encode each segment[23] separately. A segment can be encoded by expanding it in a Fourier series and then encoding the Fourier series coefficients.

Most voice-compression algorithms use such an approach, usually breaking the voice waveform into 20 ms segments. Voice-compression algorithms typically use the detailed structure of voice rather than simply encoding the Fourier series coefficients, but the frequency structure of voice is certainly important in this process. Thus understanding the Fourier series approach is a good first step in understanding voice compression.

The implementation of voice compression (as well as most signal processing techniques) usually starts with sampling at a much higher rate than the segment duration above. This sampling is followed by high-rate quantization of the samples, which are then processed digitally. Conceptually, however, it is preferable to work directly with the waveform and with expansions such as the Fourier series. The analog parts of the resulting algorithms can then be implemented by the standard techniques of high-rate sampling and digital signal processing.

Suppose that an $\mathcal{L}_2$ waveform $\{u(t) : \mathbb{R} \to \mathbb{C}\}$ is segmented into segments $u_m(t)$ of duration T. Expressing $u(t)$ as the sum of these segments,[24]

$$u(t) = \text{l.i.m.} \sum_m u_m(t), \qquad \text{where } u_m(t) = u(t)\,\text{rect}\left(\frac{t}{T} - m\right). \tag{4.24}$$

Expanding each segment $u_m(t)$ by the shifted Fourier series of (4.22) and (4.23) we obtain

$$u_m(t) = \text{l.i.m.} \sum_k \hat{u}_{k,m} e^{2\pi i k t/T} \text{rect}\left(\frac{t}{T} - m\right), \tag{4.25}$$

where

$$\begin{aligned} \hat{u}_{k,m} &= \frac{1}{T}\int_{mT-T/2}^{mT+T/2} u_m(t) e^{-2\pi i k t/T}\, dt \\ &= \frac{1}{T}\int_{-\infty}^{\infty} u(t) e^{-2\pi i k t/T} \text{rect}\left(\frac{t}{T} - m\right) dt. \end{aligned} \tag{4.26}$$

[23] Any engineer, experienced or not, when asked to analyze a segment of a waveform, will automatically shift the time axis to be centered at 0. The added complication here simply arises from looking at multiple segments together so as to represent the entire waveform.

[24] This sum double-counts the points at the ends of the segments, but this makes no difference in terms of $\mathcal{L}_2$-convergence. Exercise 4.22 treats the convergence in (4.24) and (4.28) more carefully.

Combining (4.24) and (4.25):

$$u(t) = \text{l.i.m.} \sum_m \sum_k \hat{u}_{k,m} e^{2\pi i k t/T} \operatorname{rect}\left(\frac{t}{T} - m\right).$$

This expands $u(t)$ as a weighted sum[25] of the doubly indexed functions:

$$u(t) = \text{l.i.m.} \sum_m \sum_k \hat{u}_{k,m} \theta_{k,m}(t), \qquad \text{where} \quad \theta_{k,m}(t) = e^{2\pi i k t/T} \operatorname{rect}\left(\frac{t}{T} - m\right). \tag{4.27}$$

The functions $\theta_{k,m}(t)$ are orthogonal, since, for $m \neq m'$, the functions $\theta_{k,m}(t)$ and $\theta_{k',m'}(t)$ do not overlap, and, for $m = m'$ and $k \neq k'$, $\theta_{k,m}(t)$ and $\theta_{k',m}(t)$ are orthogonal as before. These functions, $\{\theta_{k,m}(t); k, m \in \mathbb{Z}\}$, are called the *T-spaced truncated sinusoids* and the expansion in (4.27) is called the *T-spaced truncated sinusoid expansion*.

The coefficients $\hat{u}_{k,m}$ are indexed by $k, m \in \mathbb{Z}$ and thus form a countable set.[26] This permits the conversion of an arbitrary $\mathcal{L}_2$ waveform into a countably infinite sequence of complex numbers, in the sense that the numbers can be found from the waveform, and the waveform can be reconstructed from the sequence, at least up to $\mathcal{L}_2$-equivalence.

The l.i.m. notation in (4.27) denotes $\mathcal{L}_2$-convergence; i.e.,

$$\lim_{n,\ell\to\infty} \int_{-\infty}^{\infty} \left| u(t) - \sum_{m=-n}^{n} \sum_{k=-\ell}^{\ell} \hat{u}_{k,m} \theta_{k,m}(t) \right|^2 dt = 0. \tag{4.28}$$

This shows that any given $u(t)$ can be approximated arbitrarily closely by a finite set of coefficients. In particular, each segment can be approximated by a finite set of coefficients, and a finite set of segments approximates the entire waveform (although the required number of segments and coefficients per segment clearly depend on the particular waveform).

For data compression, a waveform $u(t)$ represented by the coefficients $\{\hat{u}_{k,m}; k, m \in \mathbb{Z}\}$ can be compressed by quantizing each $\hat{u}_{k,m}$ into a representative $\hat{v}_{k,m}$. The energy equation (4.6) and the difference-energy equation (4.7) generalize easily to the T-spaced truncated sinusoid expansion as follows:

$$\int_{-\infty}^{\infty} |u(t)|^2 \, dt = T \sum_{m=-\infty}^{\infty} \sum_{k=-\infty}^{\infty} |\hat{u}_{k,m}|^2, \tag{4.29}$$

$$\int_{-\infty}^{\infty} |u(t) - v(t)|^2 \, dt = T \sum_{k=-\infty}^{\infty} \sum_{m=-\infty}^{\infty} |\hat{u}_{k,m} - \hat{v}_{k,m}|^2. \tag{4.30}$$

[25] Exercise 4.21 shows why (4.27) (and similar later expressions) are independent of the order of the limits.
[26] Example 4.9.2 in Section 4.9.1 explains why the doubly indexed set above is countable.

As in Section 4.2.1, a finite set of coefficients should be chosen for compression and the remaining coefficients should be set to 0. The problem of compression (given this expansion) is then to decide how many coefficients to compress, and how many bits to use for each selected coefficient. This of course requires a probabilistic model for the coefficients; this issue is discussed later.

There is a practical problem with the use of T-spaced truncated sinusoids as an expansion to be used in data compression. The boundaries of the segments usually act like step discontinuities (as in Figure 4.3), and this leads to slow convergence over the Fourier coefficients for each segment. These discontinuities could be removed prior to taking a Fourier series, but the current objective is simply to illustrate one general approach for converting arbitrary $\mathcal{L}_2$ waveforms to sequences of numbers. Before considering other expansions, it is important to look at Fourier transforms.

4.5 Fourier transforms and $\mathcal{L}_2$ waveforms

The T-spaced truncated sinusoid expansion corresponds closely to our physical notion of frequency. For example, musical notes correspond to particular frequencies (and their harmonics), but these notes persist for finite durations and then change to notes at other frequencies. However, the parameter T in the T-spaced expansion is arbitrary, and quantizing frequencies in increments of $1/T$ is awkward.

The Fourier transform avoids the need for segmentation into T-spaced intervals, but also removes the capability of looking at frequencies that change in time. It maps a function of time, $\{u(t) : \mathbb{R} \to \mathbb{C}\}$, into a function of frequency,[27] $\{\hat{u}(f) : \mathbb{R} \to \mathbb{C}\}$. The inverse Fourier transform maps $\hat{u}(f)$ back into $u(t)$, essentially making $\hat{u}(f)$ an alternative representation of $u(t)$.

The Fourier transform and its inverse are defined as follows:

$$\hat{u}(f) = \int_{-\infty}^{\infty} u(t)\mathrm{e}^{-2\pi i f t}\,\mathrm{d}t; \tag{4.31}$$

$$u(t) = \int_{-\infty}^{\infty} \hat{u}(f)\mathrm{e}^{2\pi i f t}\,\mathrm{d}f. \tag{4.32}$$

The time units are seconds and the frequency units hertz (Hz), i.e. cycles per second.

For now we take the conventional engineering viewpoint that any respectable function $u(t)$ has a Fourier transform $\hat{u}(f)$ given by (4.31), and that $u(t)$ can be retrieved from $\hat{u}(f)$ by (4.32). This will shortly be done more carefully for $\mathcal{L}_2$ waveforms.

The following list of equations reviews a few standard Fourier transform relations. In the list, $u(t)$ and $\hat{u}(f)$ denote a Fourier transform pair, written $u(t) \leftrightarrow \hat{u}(f)$, and similarly $v(t) \leftrightarrow \hat{v}(f)$:

[27] The notation $\hat{u}(f)$, rather than the more usual $U(f)$, is used here since capitalization is used to distinguish random variables from sample values. Later, $\{U(t) : \mathbb{R} \to \mathbb{C}\}$ will be used to denote a random process, where, for each t, $U(t)$ is a random variable.

$$au(t) + bv(t) \leftrightarrow a\hat{u}(f) + b\hat{v}(f) \quad \text{linearity;} \tag{4.33}$$

$$u^*(-t) \leftrightarrow \hat{u}^*(f) \quad \text{conjugation;} \tag{4.34}$$

$$\hat{u}(t) \leftrightarrow u(-f) \quad \text{time–frequency duality;} \tag{4.35}$$

$$u(t-\tau) \leftrightarrow \mathrm{e}^{-2\pi \mathrm{i} f\tau}\hat{u}(f) \quad \text{time shift;} \tag{4.36}$$

$$u(t)\,\mathrm{e}^{2\pi \mathrm{i} f_0 t} \leftrightarrow \hat{u}(f-f_0) \quad \text{frequency shift;} \tag{4.37}$$

$$u(t/T) \leftrightarrow T\,\hat{u}(fT) \quad \text{scaling (for } T>0\text{);} \tag{4.38}$$

$$\mathrm{d}u(t)/\mathrm{d}t \leftrightarrow 2\pi \mathrm{i} f\,\hat{u}(f) \quad \text{differentiation;} \tag{4.39}$$

$$\int_{-\infty}^{\infty} u(\tau)v(t-\tau)\mathrm{d}\tau \leftrightarrow \hat{u}(f)\hat{v}(f) \quad \text{convolution;} \tag{4.40}$$

$$\int_{-\infty}^{\infty} u(\tau)v^*(\tau-t)\mathrm{d}\tau \leftrightarrow \hat{u}(f)\hat{v}^*(f) \quad \text{correlation.} \tag{4.41}$$

These relations will be used extensively in what follows. Time–frequency duality is particularly important, since it permits the translation of results about Fourier transforms to inverse Fourier transforms and vice versa.

Exercise 4.23 reviews the convolution relation (4.40). Equation (4.41) results from conjugating $\hat{v}(f)$ in (4.40).

Two useful special cases of any Fourier transform pair are as follows:

$$u(0) = \int_{-\infty}^{\infty} \hat{u}(f)\mathrm{d}f; \tag{4.42}$$

$$\hat{u}(0) = \int_{-\infty}^{\infty} u(t)\mathrm{d}t. \tag{4.43}$$

These are useful in checking multiplicative constants. Also *Parseval's theorem* results from applying (4.42) to (4.41):

$$\int_{-\infty}^{\infty} u(t)v^*(t)\mathrm{d}t = \int_{-\infty}^{\infty} \hat{u}(f)\hat{v}^*(f)\mathrm{d}f. \tag{4.44}$$

As a corollary, replacing $v(t)$ by $u(t)$ in (4.44) results in the *energy equation* for Fourier transforms, namely

$$\int_{-\infty}^{\infty} |u(t)|^2\,\mathrm{d}t = \int_{-\infty}^{\infty} |\hat{u}(f)|^2\,\mathrm{d}f. \tag{4.45}$$

The magnitude squared of the frequency function, $|\hat{u}(f)|^2$, is called the *spectral density* of $u(t)$. It is the energy per unit frequency (for positive and negative frequencies) in the waveform. The energy equation then says that energy can be calculated by integrating over either time or frequency.

As another corollary of (4.44), note that if $u(t)$ and $v(t)$ are orthogonal, then $\hat{u}(f)$ and $\hat{v}(f)$ are orthogonal, i.e.

$$\int_{-\infty}^{\infty} u(t)v^*(t)\,\mathrm{d}t = 0 \quad \text{if and only if} \quad \int_{-\infty}^{\infty} \hat{u}(f)\hat{v}^*(f)\,\mathrm{d}f = 0. \tag{4.46}$$

The following gives a short set of useful and familiar transform pairs:

$$\mathrm{sinc}(t) = \frac{\sin(\pi t)}{\pi t} \leftrightarrow \mathrm{rect}(f) = \begin{cases} 1 & \text{for} \quad |f| \le 1/2; \\ 0 & \text{for} \quad |f| > 1/2, \end{cases} \tag{4.47}$$

$$\mathrm{e}^{-\pi t^2} \leftrightarrow \mathrm{e}^{-\pi f^2}, \tag{4.48}$$

$$\mathrm{e}^{-at};\ t \ge 0 \leftrightarrow \frac{1}{a + 2\pi \mathrm{i} f} \qquad \text{for} \quad a > 0, \tag{4.49}$$

$$\mathrm{e}^{-a|t|} \leftrightarrow \frac{2a}{a^2 + (2\pi \mathrm{i} f)^2} \qquad \text{for} \quad a > 0. \tag{4.50}$$

Equations (4.47)–(4.50), in conjunction with the Fourier relations (4.33)–(4.41), yield a large set of transform pairs. Much more extensive tables of relations are widely available.

4.5.1 Measure and integration over $\mathbb{R}$

A set $\mathcal{A} \subseteq \mathbb{R}$ is defined to be *measurable* if $\mathcal{A} \cap [-T/2, T/2]$ is measurable for all $T > 0$. The definitions of measurability and measure in Section 4.3.2 were given in terms of an overall interval $[-T/2, T/2]$, but Exercise 4.14 verifies that those definitions are in fact independent of T. That is, if $\mathcal{D} \subseteq [-T/2, T/2]$ is measurable relative to $[-T/2, T/2]$, then $\mathcal{D}$ is measurable relative to $[-T_1/2, T_1/2]$, for each $T_1 > T$, and $\mu(\mathcal{D})$ is the same relative to each of those intervals. Thus measure is defined unambiguously for all sets of bounded duration.

For an arbitrary measurable set $\mathcal{A} \in \mathbb{R}$, the measure of $\mathcal{A}$ is defined to be

$$\mu(\mathcal{A}) = \lim_{T\to\infty} \mu(\mathcal{A} \cap [-T/2, T/2]). \tag{4.51}$$

Since $\mathcal{A} \cap [-T/2, T/2]$ is increasing in T, the subset inequality says that $\mu(\mathcal{A} \cap [-T/2, T/2])$ is also increasing, so the limit in (4.51) must exist as either a finite or infinite value. For example, if $\mathcal{A}$ is taken to be $\mathbb{R}$ itself, then $\mu(\mathbb{R} \cap [-T/2, T/2]) = T$ and $\mu(\mathbb{R}) = \infty$. The possibility for measurable sets to have infinite measure is the primary difference between measure over $[-T/2, T/2]$ and $\mathbb{R}$.[28]

Theorem 4.3.1 carries over without change to sets defined over $\mathbb{R}$. Thus the collection of measurable sets over $\mathbb{R}$ is closed under countable unions and intersections. The measure of a measurable set might be infinite in this case, and if a set has finite measure, then its complement (over $\mathbb{R}$) must have infinite measure.

A real function $\{u(t) : \mathbb{R} \to \mathbb{R}\}$ is *measurable* if the set $\{t : u(t) \le \beta\}$ is measurable for each $\beta \in \mathbb{R}$. Equivalently, $\{u(t) : \mathbb{R} \to \mathbb{R}\}$ is measurable if and only if $u(t)\,\mathrm{rect}(t/T)$ is measurable for all $T > 0$. A complex function $\{u(t) : \mathbb{R} \to \mathbb{C}\}$ is measurable if the real and imaginary parts of $u(t)$ are measurable.

[28] In fact, it was the restriction to finite measure that permitted the simple definition of measurability in terms of sets and their complements in Section 4.3.2.

If $\{u(t) : \mathbb{R} \to \mathbb{R}\}$ is measurable and nonnegative, there are two approaches to its Lebesgue integral. The first is to use (4.14) directly and the other is to evaluate first the integral over $[-T/2, T/2]$ and then go to the limit $T \to \infty$. Both approaches give the same result.[29]

For measurable real functions $\{u(t) : \mathbb{R} \to \mathbb{R}\}$ that take on both positive and negative values, the same approach as in the finite duration case is successful. That is, let $u^+(t)$ and $u^-(t)$ be the positive and negative parts of $u(t)$, respectively. If at most one of these has an infinite integral, the integral of $u(t)$ is defined and has the value

$$\int u(t)\mathrm{d}t = \int u^+(t)\mathrm{d}t - \int u^-(t)\mathrm{d}t.$$

Finally, a complex function $\{u(t) : \mathbb{R} \to \mathbb{C}\}$ is defined to be *measurable* if the real and imaginary parts of $u(t)$ are measurable. If the integral of $\Re(u(t))$ and that of $\Im(u(t))$ are defined, then

$$\int u(t)\mathrm{d}t = \int \Re(u(t))\mathrm{d}t + \mathrm{i}\int \Im(u(t))\mathrm{d}t. \tag{4.52}$$

A function $\{u(t) : \mathbb{R} \to \mathbb{C}\}$ is said to be in the class $\mathcal{L}_1$ if $u(t)$ is measurable and the Lebesgue integral of $|u(t)|$ is finite. As with integration over a finite interval, an $\mathcal{L}_1$ function has real and imaginary parts that are both $\mathcal{L}_1$. Also the positive and negative parts of those real and imaginary parts have finite integrals.

Example 4.5.1 The sinc function, $\mathrm{sinc}(t) = \sin(\pi t)/\pi t$ is sketched in Figure 4.8 and provides an interesting example of these definitions. Since $\mathrm{sinc}(t)$ approaches 0 with increasing t only as $1/t$, the Riemann integral of $|\mathrm{sinc}(t)|$ is infinite, and with a little thought it can be seen that the Lebesgue integral is also infinite. Thus $\mathrm{sinc}(t)$ is not an $\mathcal{L}_1$ function. In a similar way, $\mathrm{sinc}^+(t)$ and $\mathrm{sinc}^-(t)$ have infinite integrals, and thus the Lebesgue integral of $\mathrm{sinc}(t)$ over $(-\infty, \infty)$ is undefined.

The Riemann integral in this case is said to be improper, but can still be calculated by integrating from $-A$ to $+A$ and then taking the limit $A \to \infty$. The result of this integration is 1, which is most easily found through the Fourier relationship (4.47) combined with (4.43). Thus, in a sense, the sinc function is an example where the Riemann integral exists but the Lebesgue integral does not. In a deeper sense, however,

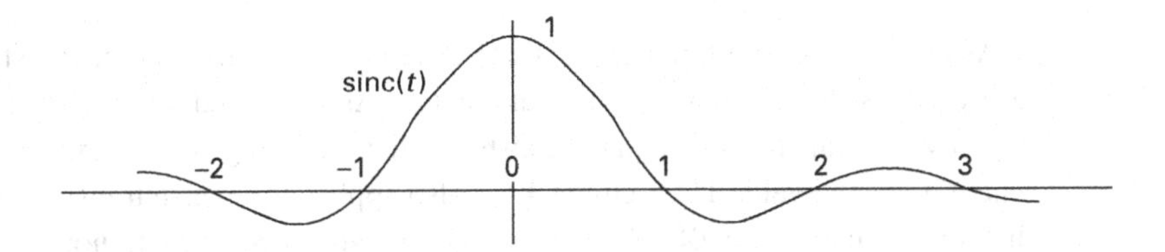

Figure 4.8. The function $\mathrm{sinc}(t)$ goes to 0 as $1/t$ with increasing t.

[29] As explained shortly in the sinc function example, this is not necessarily true for functions taking on positive and negative values.

the issue is simply one of definitions; one can always use Lebesgue integration over $[-A, A]$ and go to the limit $A \to \infty$, getting the same answer as the Riemann integral provides.

A function $\{u(t) : \mathbb{R} \to \mathbb{C}\}$ is said to be in the class $\mathcal{L}_2$ if $u(t)$ is measurable and the Lebesgue integral of $|u(t)|^2$ is finite. All source and channel waveforms will be assumed to be $\mathcal{L}_2$. As pointed out earlier, any $\mathcal{L}_2$ function of finite duration is also $\mathcal{L}_1$. However, $\mathcal{L}_2$ functions of infinite duration need not be $\mathcal{L}_1$; the sinc function is a good example. Since $\text{sinc}(t)$ decays as $1/t$, it is not $\mathcal{L}_1$. However, $|\text{sinc}(t)|^2$ decays as $1/t^2$ as $t \to \infty$, so the integral is finite and $\text{sinc}(t)$ is an $\mathcal{L}_2$ function.

In summary, measure and integration over $\mathbb{R}$ can be treated in essentially the same way as over $[-T/2, T/2]$. The point sets and functions of interest can be truncated to $[-T/2, T/2]$ with a subsequent passage to the limit $T \to \infty$. As will be seen, however, this requires some care with functions that are not $\mathcal{L}_1$.

4.5.2 Fourier transforms of $\mathcal{L}_2$ functions

The Fourier transform does not exist for all functions, and when the Fourier transform does exist, there is not necessarily an inverse Fourier transform. This section first discusses $\mathcal{L}_1$ functions and then $\mathcal{L}_2$ functions. A major result is that $\mathcal{L}_1$ functions always have well defined Fourier transforms, but the inverse transform does not always have very nice properties; $\mathcal{L}_2$ functions also always have Fourier transforms, but only in the sense of $\mathcal{L}_2$-equivalence. Here however, the inverse transform also exists in the sense of $\mathcal{L}_2$-equivalence. We are primarily interested in $\mathcal{L}_2$ functions, but the results about $\mathcal{L}_1$ functions will help in understanding the $\mathcal{L}_2$ transform.

Lemma 4.5.1 *Let $\{u(t) : \mathbb{R} \to \mathbb{C}\}$ be $\mathcal{L}_1$. Then $\hat{u}(f) = \int_{-\infty}^{\infty} u(t)e^{-2\pi i f t}\,dt$ both exists and satisfies $|\hat{u}(f)| \le \int |u(t)|\,dt$ for each $f \in \mathbb{R}$. Furthermore, $\{\hat{u}(f) : \mathbb{R} \to \mathbb{C}\}$ is a continuous function of f.*

Proof Note that $|u(t)e^{-2\pi i f t}| = |u(t)|$ for all real t and f. Thus $u(t)e^{-2\pi i f t}$ is $\mathcal{L}_1$ for each f and the integral exists and satisfies the given bound. This is the same as the argument about Fourier series coefficients in Theorem 4.4.1. The continuity follows from asimple ϵ/δ argument (see Exercise 4.24). □

As an example, the function $u(t) = \text{rect}(t)$ is $\mathcal{L}_1$, and its Fourier transform, defined at each f, is the continuous function $\text{sinc}(f)$. As discussed before, $\text{sinc}(f)$ is not $\mathcal{L}_1$. The inverse transform of $\text{sinc}(f)$ exists at all t, equaling $\text{rect}(t)$ except at $t = \pm 1/2$, where it has the value 1/2. Lemma 4.5.1 also applies to inverse transforms and verifies that $\text{sinc}(f)$ cannot be $\mathcal{L}_1$, since its inverse transform is discontinuous.

Next consider $\mathcal{L}_2$ functions. It will be seen that the pointwise Fourier transform $\int u(t)e^{-2\pi i f t}\,dt$ does not necessarily exist at each f, but that it does exist as an $\mathcal{L}_2$ limit. In exchange for this added complexity, however, the inverse transform exists in exactly the same sense. This result is called Plancherel's theorem and has a nice interpretation in terms of approximations over finite time and frequency intervals.

For any $\mathcal{L}_2$ function $\{u(t) : \mathbb{R} \to \mathbb{C}\}$ and any positive number A, define $\hat{u}_A(f)$ as the Fourier transform of the truncation of $u(t)$ to $[-A, A]$; i.e.,

$$\hat{u}_A(f) = \int_{-A}^{A} u(t)e^{-2\pi i f t}\, dt. \tag{4.53}$$

The function $u(t)\text{rect}(t/2A)$ has finite duration and is thus $\mathcal{L}_1$. It follows that $\hat{u}_A(f)$ is continuous and exists for all f by Lemma 4.5.1. One would normally expect to take the limit in (4.53) as $A \to \infty$ to get the Fourier transform $\hat{u}(f)$, but this limit does not necessarily exist for each f. Plancherel's theorem, however, asserts that this limit exists in the $\mathcal{L}_2$ sense. This theorem is proved in Appendix 5.5.1.

Theorem 4.5.1 (Plancherel, part 1) *For any $\mathcal{L}_2$ function $\{u(t) : \mathbb{R} \to \mathbb{C}\}$, an $\mathcal{L}_2$ function $\{\hat{u}(f) : \mathbb{R} \to \mathbb{C}\}$ exists satisfying both*

$$\lim_{A\to\infty} \int_{-\infty}^{\infty} |\hat{u}(f) - \hat{u}_A(f)|^2\, df = 0 \tag{4.54}$$

and the energy equation, (4.45).

This not only guarantees the existence of a Fourier transform (up to $\mathcal{L}_2$-equivalence), but also guarantees that it is arbitrarily closely approximated (in difference energy) by the continuous Fourier transforms of the truncated versions of $u(t)$. Intuitively what is happening here is that $\mathcal{L}_2$ functions must have an arbitrarily large fraction of their energy within sufficiently large truncated limits; the part of the function outside of these limits cannot significantly affect the $\mathcal{L}_2$-convergence of the Fourier transform.

The inverse transform is treated very similarly. For any $\mathcal{L}_2$ function $\{\hat{u}(f) : \mathbb{R} \to \mathbb{C}\}$ and any B, $0 < B < \infty$, define

$$u_B(t) = \int_{-B}^{B} \hat{u}(f)e^{2\pi i f t}\, df. \tag{4.55}$$

As before, $u_B(t)$ is a continuous $\mathcal{L}_2$ function for all B, $0<B<\infty$. The final part of Plancherel's theorem is then given by Theorem 4.5.2.

Theorem 4.5.2 (Plancherel, part 2) *For any $\mathcal{L}_2$ function $\{u(t) : \mathbb{R} \to \mathbb{C}\}$, let $\{\hat{u}(f) : \mathbb{R} \to \mathbb{C}\}$ be the Fourier transform of Theorem 4.5.1 and let $u_B(t)$ satisfy (4.55). Then*

$$\lim_{B\to\infty} \int_{-\infty}^{\infty} |u(t) - u_B(t)|^2\, dt = 0. \tag{4.56}$$

The interpretation is similar to the first part of the theorem. Specifically the inverse transforms of finite frequency truncations of the transform are continuous and converge to an $\mathcal{L}_2$ limit as $B \to \infty$. It also says that this $\mathcal{L}_2$ limit is equivalent to the original function $u(t)$.

Using the limit in mean-square notation, both parts of the Plancherel theorem can be expressed by stating that every $\mathcal{L}_2$ function $u(t)$ has a Fourier transform $\hat{u}(f)$ satisfying

$$\hat{u}(f) = \underset{A\to\infty}{\text{l.i.m.}} \int_{-A}^{A} u(t)e^{-2\pi i f t}\, dt; \qquad u(t) = \underset{B\to\infty}{\text{l.i.m.}} \int_{-B}^{B} \hat{u}(f)e^{2\pi i f t}\, df;$$

i.e., the inverse Fourier transform of $\hat{u}(f)$ is $\mathcal{L}_2$-equivalent to $u(t)$. The first integral above converges pointwise if $u(t)$ is also $\mathcal{L}_1$, and in this case converges pointwise to a continuous function $\hat{u}(f)$. If $u(t)$ is not $\mathcal{L}_1$, then the first integral need not converge pointwise. The second integral behaves in the analogous way.

It may help in understanding the Plancherel theorem to interpret it in terms of finding Fourier transforms using Riemann integration. Riemann integration over an infinite region is defined as a limit over finite regions. Thus, the Riemann version of the Fourier transform is shorthand for

$$\hat{u}(f) = \lim_{A\to\infty} \int_{-A}^{A} u(t) e^{-2\pi i f t}\, dt = \lim_{A\to\infty} \hat{u}_A(f). \tag{4.57}$$

Thus, the Plancherel theorem can be viewed as replacing the Riemann integral with a Lebesgue integral and replacing the pointwise limit (if it exists) in (4.57) with $\mathcal{L}_2$-convergence. The Fourier transform over the finite limits $-A$ to A is continuous and well behaved, so the major difference comes in using $\mathcal{L}_2$-convergence as $A \to \infty$.

As an example of the Plancherel theorem, let $u(t) = \text{rect}(t)$. Then $\hat{u}_A(f) = \text{sinc}(f)$ for all $A \ge 1/2$, so $\hat{u}(f) = \text{sinc}(f)$. For the inverse transform, $u_B(t) = \int_{-B}^{B} \text{sinc}(f) df$ is messy to compute but can be seen to approach rect(t) as $B \to \infty$ except at $t = \pm 1/2$, where it equals 1/2. At $t = \pm 1/2$, the inverse transform is 1/2, whereas $u(t) = 1$.

As another example, consider the function $u(t)$, where $u(t) = 1$ for rational values of $t \in [0, 1]$ and $u(t) = 0$ otherwise. Since this is 0 a.e, the Fourier transform $\hat{u}(f)$ is 0 for all f and the inverse transform is 0, which is $\mathcal{L}_2$-equivalent to $u(t)$. Finally, Example 5.5.1 in Appendix 5.5.1 illustrates a bizarre $\mathcal{L}_1$ function $g(t)$ that is everywhere discontinuous. Its transform $\hat{g}(f)$ is bounded and continuous by Lemma 4.5.1, but is not $\mathcal{L}_1$. The inverse transform is again discontinuous everywhere in $(0, 1)$ and unbounded over every subinterval. This example makes clear why the inverse transform of a continuous function of frequency might be bizarre, thus reinforcing our focus on $\mathcal{L}_2$ functions rather than a more conventional focus on notions such as continuity.

In what follows, $\mathcal{L}_2$-convergence, as in the Plancherel theorem, will be seen as increasingly friendly and natural. Regarding two functions whose difference has zero energy as being the same (formally, as $\mathcal{L}_2$-equivalent) allows us to avoid many trivialities, such as how to define a discontinuous function at its discontinuities. In this case, engineering commonsense and sophisticated mathematics arrive at the same conclusion.

Finally, it can be shown that all the Fourier transform relations in (4.33)–(4.41) except differentiation hold for all $\mathcal{L}_2$ functions (see Exercises 4.26 and 5.15). The derivative of an $\mathcal{L}_2$ function need not be $\mathcal{L}_2$, and need not have a well defined Fourier transform.

4.6 The DTFT and the sampling theorem

The discrete-time Fourier transform (DTFT) is the time–frequency dual of the Fourier series. It will be shown that the DTFT leads immediately to the sampling theorem.

4.6.1 The discrete-time Fourier transform

Let $\hat{u}(f)$ be an $\mathcal{L}_2$ function of frequency, nonzero only for $-\mathsf{W} \le f \le \mathsf{W}$. The DTFT of $\hat{u}(f)$ over $[-\mathsf{W}, \mathsf{W}]$ is then defined by

$$\hat{u}(f) = \text{l.i.m.} \sum_k u_k e^{-2\pi i k f/2\mathsf{W}} \, \text{rect}\left(\frac{f}{2}\mathsf{W}\right), \tag{4.58}$$

where the DTFT coefficients $\{u_k; k \in \mathbb{Z}\}$ are given by

$$u_k = \frac{1}{2\mathsf{W}} \int_{-\mathsf{W}}^{\mathsf{W}} \hat{u}(f) e^{2\pi i k f/2\mathsf{W}} \, df. \tag{4.59}$$

These are the same as the Fourier series equations, replacing t by f, T by $2\mathsf{W}$, and $e^{2\pi i \cdots}$ by $e^{-2\pi i \cdots}$. Note that $\hat{u}(f)$ has an inverse Fourier transform $u(t)$ which is thus baseband-limited to $[-\mathsf{W}, \mathsf{W}]$. As will be shown shortly, the sampling theorem relates the samples of this baseband waveform to the coefficients in (4.59).

The Fourier series theorem (Theorem 4.4.1) clearly applies to (4.58) and (4.59) with the above notational changes; it is repeated here for convenience.

Theorem 4.6.1 (DTFT) *Let $\{\hat{u}(f) : [-\mathsf{W}, \mathsf{W}] \to \mathbb{C}\}$ be an $\mathcal{L}_2$ function. Then for each $k \in \mathbb{Z}$, the Lebesgue integral (4.59) exists and satisfies $|u_k| \le (1/2\mathsf{W}) \int |\hat{u}(f)| \, df < \infty$. Furthermore,*

$$\lim_{\ell \to \infty} \int_{-\mathsf{W}}^{\mathsf{W}} \left| \hat{u}(f) - \sum_{k=-\ell}^{\ell} u_k e^{-2\pi i k f/2\mathsf{W}} \right|^2 df = 0 \tag{4.60}$$

and

$$\int_{-\mathsf{W}}^{\mathsf{W}} |\hat{u}(f)|^2 \, df = 2\mathsf{W} \sum_{k=-\infty}^{\infty} |u_k|^2. \tag{4.61}$$

Finally, if $\{u_k, k \in \mathbb{Z}\}$ is a sequence of complex numbers satisfying $\sum_k |u_k|^2 < \infty$, then an $\mathcal{L}_2$ function $\{\hat{u}(f) : [-\mathsf{W}, \mathsf{W}] \to \mathbb{C}\}$ exists satisfying (4.60) and (4.61).

As before, (4.58) is shorthand for (4.60). Again, this says that any desired approximation accuracy, in terms of energy, can be achieved by using enough terms in the series.

Both the Fourier series and the DTFT provide a one-to-one transformation (in the sense of $\mathcal{L}_2$-convergence) between a function and a sequence of complex numbers. In the case of the Fourier series, one usually starts with a function $u(t)$ and uses the sequence of coefficients to represent the function (up to $\mathcal{L}_2$-equivalence). In the case of the DTFT, one often starts with the sequence and uses the frequency function to represent the sequence. Since the transformation goes both ways, however, one can view the function and the sequence as equally fundamental.

4.6.2 The sampling theorem

The DTFT is next used to establish the sampling theorem, which in turn will help interpret the DTFT. The DTFT (4.58) expresses $\hat{u}(f)$ as a weighted sum of truncated sinusoids in frequency,

$$\hat{u}(f) = \text{l.i.m.} \sum_k u_k \hat{\phi}_k(f), \quad \text{where} \quad \hat{\phi}_k(f) = e^{-2\pi i k f/2W} \operatorname{rect}\left(\frac{f}{2W}\right). \tag{4.62}$$

Ignoring any questions of convergence for the time being, the inverse Fourier transform of $\hat{u}(f)$ is then given by $u(t) = \sum_k u_k \phi_k(t)$, where $\phi_k(t)$ is the inverse transform of $\hat{\phi}_k(f)$. Since the inverse transform[30] of $\operatorname{rect}(f/2W)$ is $2W \operatorname{sinc}(2Wt)$, the time-shift relation implies that the inverse transform of $\hat{\phi}_k(f)$ is given by

$$\phi_k(t) = 2W \operatorname{sinc}(2Wt - k) \quad \leftrightarrow \quad \hat{\phi}_k(f) = e^{-2\pi i k f/2W} \operatorname{rect}\left(\frac{f}{2W}\right). \tag{4.63}$$

Thus $u(t)$, the inverse transform of $\hat{u}(f)$, is given by

$$u(t) = \sum_{k=-\infty}^{\infty} u_k \phi_k(t) = \sum_{k=-\infty}^{\infty} 2W u_k \operatorname{sinc}(2Wt - k). \tag{4.64}$$

Since the set of truncated sinusoids $\{\hat{\phi}_k; k \in \mathbb{Z}\}$ is orthogonal, the sinc functions $\{\phi_k; k \in \mathbb{Z}\}$ are also orthogonal, from (4.46). Figure 4.9 illustrates ϕ_0 and ϕ_1 for the normalized case where $W = 1/2$.

Note that $\operatorname{sinc}(t)$ equals 1 for $t = 0$ and 0 for all other integer t. Thus if (4.64) is evaluated for $t = k/2W$, the result is that $u(k/2W) = 2W u_k$ for all integer k. Substituting this into (4.64) results in the equation known as the sampling equation:

$$u(t) = \sum_{k=-\infty}^{\infty} u\left(\frac{k}{2W}\right) \operatorname{sinc}(2Wt - k).$$

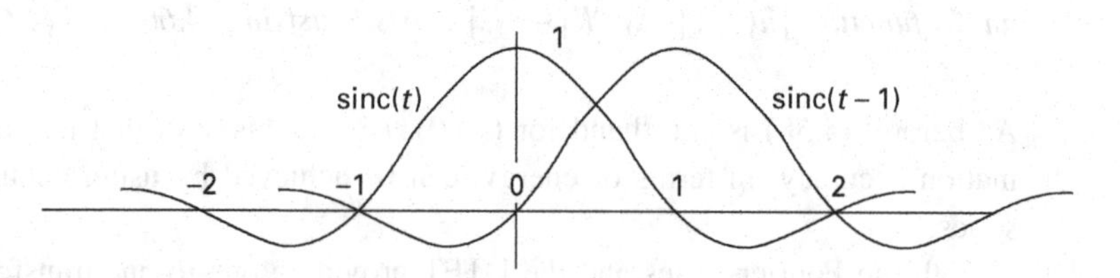

Figure 4.9. Sketch of $\operatorname{sinc}(t) = \sin(\pi t)/\pi t$ and $\operatorname{sinc}(t-1)$. Note that these spaced sinc functions are orthogonal to each other.

[30] This is the time/frequency dual of (4.47). $\hat{u}(f) = \operatorname{rect}(f/2W)$ is both $\mathcal{L}_1$ and $\mathcal{L}_2$; $u(t)$ is continuous and $\mathcal{L}_2$ but not $\mathcal{L}_1$. From the Plancherel theorem, the transform of $u(t)$, in the $\mathcal{L}_2$ sense, is $\hat{u}(f)$.

This says that a baseband-limited function is specified by its samples at intervals $T = 1/2\mathsf{W}$. In terms of this sample interval, the sampling equation is given by

$$u(t) = \sum_{k=-\infty}^{\infty} u(kT)\,\mathrm{sinc}\left(\frac{t}{T} - k\right). \tag{4.65}$$

The following theorem makes this precise. See Appendix 5.5.2 for an insightful proof.

Theorem 4.6.2 (Sampling theorem) *Let $\{u(t) : \mathbb{R} \to \mathbb{C}\}$ be a continuous $\mathcal{L}_2$ function baseband-limited to W. Then (4.65) specifies $u(t)$ in terms of its T-spaced samples with $T = 1/2\mathsf{W}$. The sum in (4.65) converges to $u(t)$ for each $t \in \mathbb{R}$, and $u(t)$ is bounded at each t by $|u(t)| \le \int_{-\mathsf{W}}^{\mathsf{W}} |\hat{u}(f)|\,\mathrm{d}f < \infty$.*

The following example illustrates why $u(t)$ is assumed to be continuous above.

Example 4.6.1 (A discontinuous baseband function) Let $u(t)$ be a continuous $\mathcal{L}_2$ baseband function limited to $|f| \le 1/2$. Let $v(t)$ satisfy $v(t) = u(t)$ for all noninteger t and $v(t) = u(t) + 1$ for all integer t. Then $u(t)$ and $v(t)$ are $\mathcal{L}_2$-equivalent, but their samples at each integer time differ by 1. Their Fourier transforms are the same, say $\hat{u}(f)$, since the differences at isolated points have no effect on the transform. Since $\hat{u}(f)$ is nonzero only in $[-\mathsf{W}, \mathsf{W}]$, it is $\mathcal{L}_1$. According to the time–frequency dual of Lemma 4.5.1, the pointwise inverse Fourier transform of $\hat{u}(f)$ is a continuous function, say $u(t)$. Out of all the $\mathcal{L}_2$-equivalent waveforms that have the transform $\hat{u}(f)$, only $u(t)$ is continuous, and it is that $u(t)$ that satisfies the sampling theorem.

The function $v(t)$ is equal to $u(t)$ except for the isolated discontinuities at each integer point. One could view $u(t)$ as baseband-limited also, but this is clearly not physically meaningful and is not the continuous function of the theorem.

Example 4.6.1 illustrates an ambiguity about the meaning of baseband-limited functions. One reasonable definition is that an $\mathcal{L}_2$ function $v(t)$ is baseband-limited to W if $\hat{u}(f)$ is 0 for $|f| > \mathsf{W}$. Another reasonable definition is that $u(t)$ is baseband-limited to W if $u(t)$ is the pointwise inverse Fourier transform of a function $\hat{u}(f)$ that is 0 for $|f| > \mathsf{W}$. For a given $\hat{u}(f)$, there is a unique $u(t)$ according to the second definition and it is continuous; all the functions that are $\mathcal{L}_2$-equivalent to $u(t)$ are bandlimited by the first definition, and all but $u(t)$ are discontinuous and potentially violate the sampling equation. Clearly the second definition is preferable on both engineering and mathematical grounds.

Definition 4.6.1 An $\mathcal{L}_2$ function is *baseband-limited* to W if it is the pointwise inverse transform of an $\mathcal{L}_2$ function $\hat{u}(f)$ that is 0 for $|f| > \mathsf{W}$. Equivalently, it is baseband-limited to W if it is continuous and its Fourier transform is 0 for $|f| > 0$.

The DTFT can now be further interpreted. Any baseband-limited $\mathcal{L}_2$ function $\{\hat{u}(f) : [-\mathsf{W}, \mathsf{W}] \to \mathbb{C}\}$ has both an inverse Fourier transform $u(t) = \int \hat{u}(f) e^{2\pi i f t}\,\mathrm{d}f$ and a DTFT sequence given by (4.58). The coefficients u_k of the DTFT are the scaled samples, $Tu(kT)$, of $u(t)$, where $T = 1/2\mathsf{W}$. Put in a slightly different way, the DTFT in (4.58) is the Fourier transform of the sampling equation (4.65) with $u(kT) = u_k/T$.[31]

It is somewhat surprising that the sampling theorem holds with pointwise convergence, whereas its transform, the DTFT, holds only in the $\mathcal{L}_2$-equivalence sense. The reason is that the function $\hat{u}(f)$ in the DTFT is $\mathcal{L}_1$ but not necessarily continuous, whereas its inverse transform $u(t)$ is necessarily continuous but not necessarily $\mathcal{L}_1$.

The set of functions $\{\hat{\phi}_k(f);\, k \in \mathbb{Z}\}$ in (4.63) is an orthogonal set, since the interval $[-\mathsf{W}, \mathsf{W}]$ contains an integer number of cycles from each sinusoid. Thus, from (4.46), the set of sinc functions in the sampling equation is also orthogonal. Thus both the DTFT and the sampling theorem expansion are orthogonal expansions. It follows (as will be shown carefully later) that the energy equation,

$$\int_{-\infty}^{\infty} |u(t)|^2 \,\mathrm{d}t = T \sum_{k=-\infty}^{\infty} |u(kT)|^2, \tag{4.66}$$

holds for any continuous $\mathcal{L}_2$ function $u(t)$ baseband-limited to $[-\mathsf{W}, \mathsf{W}]$ with $T = 1/2\mathsf{W}$.

In terms of source coding, the sampling theorem says that any $\mathcal{L}_2$ function $u(t)$ that is baseband-limited to W can be sampled at rate $2\mathsf{W}$ (i.e. at intervals $T = 1/2\mathsf{W}$) and the samples can later be used to reconstruct the function perfectly. This is slightly different from the channel coding situation where a sequence of signal values are mapped into a function from which the signals can later be reconstructed. The sampling theorem shows that any $\mathcal{L}_2$ baseband-limited function can be represented by its samples. The following theorem, proved in Appendix 5.5.2, covers the channel coding variation.

Theorem 4.6.3 (Sampling theorem for transmission) *Let $\{a_k;\, k \in \mathbb{Z}\}$ be an arbitrary sequence of complex numbers satisfying $\sum_k |a_k|^2 < \infty$. Then $\sum_k a_k \,\mathrm{sinc}(2\mathsf{W}t - k)$ converges pointwise to a continuous bounded $\mathcal{L}_2$ function $\{u(t) : \mathbb{R} \to \mathbb{C}\}$ that is baseband-limited to W and satisfies $a_k = u(k/2\mathsf{W})$ for each k.*

4.6.3 Source coding using sampled waveforms

Section 4.1 and Figure 4.1 discuss the sampling of an analog waveform $u(t)$ and quantizing the samples as the first two steps in analog source coding. Section 4.2 discusses an alternative in which successive segments $\{u_m(t)\}$ of the source are each expanded in a Fourier series, and then the Fourier series coefficients are quantized. In this latter case, the received segments $\{v_m(t)\}$ are reconstructed from the quantized coefficients. The energy in $u_m(t) - v_m(t)$ is given in (4.7) as a scaled version of the sum of the squared coefficient differences. This section treats the analogous relationship when quantizing the samples of a baseband-limited waveform.

For a continuous function $u(t)$, baseband-limited to W, the samples $\{u(kT);\, k \in \mathbb{Z}\}$ at intervals $T = 1/2\mathsf{W}$ specify the function. If $u(kT)$ is quantized to $v(kT)$ for each k, and

[31] Note that the DTFT is the time–frequency *dual* of the Fourier series but is the *Fourier transform* of the sampling equation.

$u(t)$ is reconstructed as $v(t) = \sum_k v(kT)\,\text{sinc}(t/T - k)$, then, from (4.66), the mean-squared error (MSE) is given by

$$\int_{-\infty}^{\infty} |u(t) - v(t)|^2\,\mathrm{d}t = T \sum_{k=-\infty}^{\infty} |u(kT) - v(kT)|^2. \tag{4.67}$$

Thus, whatever quantization scheme is used to minimize the MSE between a sequence of samples, that same strategy serves to minimize the MSE between the corresponding waveforms.

The results in Chapter 3 regarding mean-squared distortion for uniform vector quantizers give the distortion at any given bit rate per sample as a linear function of the mean-squared value of the source samples. If any sample has an infinite mean-squared value, then either the quantization rate is infinite or the mean-squared distortion is infinite. This same result then carries over to waveforms. This starts to show why the restriction to $\mathcal{L}_2$ source waveforms is important. It also starts to show why general results about $\mathcal{L}_2$ waveforms are important.

The sampling theorem tells the story for sampling baseband-limited waveforms. However, physical source waveforms are not perfectly limited to some frequency W; rather, their spectra usually drop off rapidly above some nominal frequency W. For example, audio spectra start dropping off well before the nominal cutoff frequency of 4 kHz, but often have small amounts of energy up to 20 kHz. Then the samples at rate $2\mathsf{W}$ do not quite specify the waveform, which leads to an additional source of error, called aliasing. Aliasing will be discussed more fully in Section 4.7.

There is another unfortunate issue with the sampling theorem. The sinc function is nonzero over all noninteger times. Recreating the waveform at the receiver[32] from a set of samples thus requires infinite delay. Practically, of course, sinc functions can be truncated, but the sinc waveform decays to zero as $1/t$, which is impractically slow. Thus the clean result of the sampling theorem is not quite as practical as it first appears.

4.6.4 The sampling theorem for $[\Delta - \mathsf{W}, \Delta + \mathsf{W}]$

Just as the Fourier series generalizes to time intervals centered at some arbitrary time Δ, the DTFT generalizes to frequency intervals centered at some arbitrary frequency Δ.

Consider an $\mathcal{L}_2$ frequency function $\{\hat{v}(f) : [\Delta - \mathsf{W}, \Delta + \mathsf{W}] \to \mathbb{C}\}$. The *shifted DTFT* for $\hat{v}(f)$ is then given by

$$\hat{v}(f) = l.i.m. \sum_k v_k \mathrm{e}^{-2\pi i k f/2\mathsf{W}}\, \text{rect}\left(\frac{f-\Delta}{2\mathsf{W}}\right), \tag{4.68}$$

[32] Recall that the receiver time reference is delayed from that at the source by some constant τ. Thus $v(t)$, the receiver estimate of the source waveform $u(t)$ at source time t, is recreated at source time $t+\tau$. With the sampling equation, even if the sinc function is approximated, τ is impractically large.

where

$$v_k = \frac{1}{2\mathrm{W}} \int_{\Delta-\mathrm{W}}^{\Delta+\mathrm{W}} \hat{v}(f) e^{2\pi i k f/2\mathrm{W}} \, df. \tag{4.69}$$

Equation (4.68) is an orthogonal expansion,

$$\hat{v}(f) = \text{l.i.m.} \sum_k v_k \hat{\theta}_k(f), \quad \text{where} \quad \hat{\theta}_k(f) = e^{-2\pi i k f/2\mathrm{W}} \operatorname{rect}\left(\frac{f-\Delta}{2\mathrm{W}}\right).$$

The inverse Fourier transform of $\hat{\theta}_k(f)$ can be calculated by shifting and scaling as follows:

$$\theta_k(t) = 2\mathrm{W} \operatorname{sinc}(2\mathrm{W}t - k)\, e^{2\pi i \Delta(t-k/2\mathrm{W})} \leftrightarrow \hat{\theta}_k(f) = e^{-2\pi i k f/2\mathrm{W}} \operatorname{rect}\left(\frac{f-\Delta}{2\mathrm{W}}\right). \tag{4.70}$$

Let $v(t)$ be the inverse Fourier transform of $\hat{v}(f)$:

$$v(t) = \sum_k v_k \theta_k(t) = \sum_k 2\mathrm{W} v_k \operatorname{sinc}(2\mathrm{W}t - k) e^{2\pi i \Delta(t-k/2\mathrm{W})}.$$

For $t = k/2\mathrm{W}$, only the kth term is nonzero, and $v(k/2\mathrm{W}) = 2\mathrm{W}v_k$. This generalizes the sampling equation to the frequency band $[\Delta - \mathrm{W}, \Delta + \mathrm{W}]$:

$$v(t) = \sum_k v\left(\frac{k}{2\mathrm{W}}\right) \operatorname{sinc}(2\mathrm{W}t - k) e^{2\pi i \Delta(t-k/2\mathrm{W})}.$$

Defining the sampling interval $T = 1/2\mathrm{W}$ as before, this becomes

$$v(t) = \sum_k v(kT) \operatorname{sinc}\left(\frac{t}{T} - k\right) e^{2\pi i \Delta(t-kT)}. \tag{4.71}$$

Theorems 4.6.2 and 4.6.3 apply to this more general case. That is, with $v(t) = \int_{\Delta-\mathrm{W}}^{\Delta+\mathrm{W}} \hat{v}(f) e^{2\pi i f t} \, df$, the function $v(t)$ is bounded and continuous and the series in (4.71) converges for all t. Similarly, if $\sum_k |v(kT)|^2 < \infty$, there is a unique continuous $\mathcal{L}_2$ function $\{v(t) : [\Delta - \mathrm{W}, \Delta + \mathrm{W}] \to \mathbb{C}\}$, $\mathrm{W} = 1/2T$, with those sample values.

4.7 Aliasing and the sinc-weighted sinusoid expansion

In this section an orthogonal expansion for arbitrary $\mathcal{L}_2$ functions called the *T-spaced sinc-weighted sinusoid expansion* is developed. This expansion is very similar to the T-spaced truncated sinusoid expansion discussed earlier, except that its set of orthogonal waveforms consists of time and frequency shifts of a sinc function rather than a rectangular function. This expansion is then used to discuss the important concept of degrees of freedom. Finally this same expansion is used to develop the concept of aliasing. This will help in understanding sampling for functions that are only approximately frequency-limited.

4.7.1 The T-spaced sinc-weighted sinusoid expansion

Let $u(t) \leftrightarrow \hat{u}(f)$ be an arbitrary $\mathcal{L}_2$ transform pair, and segment $\hat{u}(f)$ into intervals[33] of width 2W. Thus,

$$\hat{u}(f) = \text{l.i.m.} \sum_m \hat{v}_m(f), \qquad \text{where} \quad \hat{v}_m(f) = \hat{u}(f)\,\text{rect}\left(\frac{f}{2\mathsf{W}} - m\right).$$

Note that $\hat{v}_0(f)$ is nonzero only in $[-\mathsf{W}, \mathsf{W}]$ and thus corresponds to an $\mathcal{L}_2$ function $v_0(t)$ baseband-limited to W. More generally, for arbitrary integer m, $\hat{v}_m(f)$ is nonzero only in $[\Delta - \mathsf{W}, \Delta + \mathsf{W}]$ for $\Delta = 2\mathsf{W}m$. From (4.71), the inverse transform with $T = 1/2\mathsf{W}$ satisfies the following:

$$\begin{aligned} v_m(t) &= \sum_k v_m(kT)\,\text{sinc}\left(\frac{t}{T} - k\right) e^{2\pi i (m/T)(t-kT)} \\ &= \sum_k v_m(kT)\,\text{sinc}\left(\frac{t}{T} - k\right) e^{2\pi i m t/T}. \end{aligned} \tag{4.72}$$

Combining all of these frequency segments,

$$u(t) = \text{l.i.m.} \sum_m v_m(t) = \text{l.i.m.} \sum_{m,k} v_m(kT)\,\text{sinc}\left(\frac{t}{T} - k\right) e^{2\pi i m t/T}. \tag{4.73}$$

This converges in $\mathcal{L}_2$, but does not not necessarily converge pointwise because of the infinite summation over m. It expresses an arbitrary $\mathcal{L}_2$ function $u(t)$ in terms of the samples of each frequency slice, $v_m(t)$, of $u(t)$.

This is an orthogonal expansion in the doubly indexed set of functions

$$\{\psi_{m,k}(t) = \text{sinc}\left(\frac{t}{T} - k\right) e^{2\pi i m t/T};\ m, k \in \mathbb{Z}\}. \tag{4.74}$$

These are the time and frequency shifts of the basic function $\psi_{0,0}(t) = \text{sinc}(t/T)$. The time shifts are in multiples of T and the frequency shifts are in multiples of $1/T$. This set of orthogonal functions is called the set of *T-spaced sinc-weighted sinusoids*.

The T-spaced sinc-weighted sinusoids and the T-spaced truncated sinusoids are quite similar. Each function in the first set is a time and frequency translate of $\text{sinc}(t/T)$. Each function in the second set is a time and frequency translate of $\text{rect}(t/T)$. Both sets are made up of functions separated by multiples of T in time and $1/T$ in frequency.

[33] The boundary points between frequency segments can be ignored, as in the case for time segments.

4.7.2 Degrees of freedom

An important rule of thumb used by communication engineers is that the class of real functions that are approximately baseband-limited to W_0 and approximately time-limited to $[-T_0/2, T_0/2]$ have about $2T_0W_0$ real degrees of freedom if $T_0W_0 \gg 1$. This means that any function within that class can be specified approximately by specifying about $2T_0W_0$ real numbers as coefficients in an orthogonal expansion. The same rule is valid for complex functions in terms of complex degrees of freedom.

This somewhat vague statement is difficult to state precisely, since time-limited functions cannot be frequency-limited and vice versa. However, the concept is too important to ignore simply because of lack of precision. Thus several examples are given.

First, consider applying the sampling theorem to real (complex) functions $u(t)$ that are strictly baseband-limited to W_0. Then $u(t)$ is specified by its real (complex) samples at rate $2W_0$. If the samples are nonzero only within the interval $[-T_0/2, T_0/2]$, then there are about $2T_0W_0$ nonzero samples, and these specify $u(t)$ within this class. Here a precise class of functions have been specified, but *functions* that are zero outside of an interval have been replaced with functions whose *samples* are zero outside of the interval.

Second, consider complex functions $u(t)$ that are again strictly baseband-limited to W_0, but now apply the sinc-weighted sinusoid expansion with $W = W_0/(2n+1)$ for some positive integer n. That is, the band $[-W_0, W_0]$ is split into $2n+1$ slices and each slice is expanded in a sampling-theorem expansion. Each slice is specified by samples at rate $2W$, so all slices are specified collectively by samples at an aggregate rate $2W_0$ as before. If the samples are nonzero only within $[-T_0/2, T_0/2]$, then there are about[34] $2T_0W_0$ nonzero complex samples that specify any $u(t)$ in this class.

If the functions in this class are further constrained to be real, then the coefficients for the central frequency slice are real and the negative slices are specified by the positive slices. Thus each real function in this class is specified by about $2T_0W_0$ real numbers.

This class of functions is slightly different for each choice of n, since the detailed interpretation of what "approximately time-limited" means is changing. From a more practical perspective, however, all of these expansions express an approximately baseband-limited waveform by samples at rate $2W_0$. As the overall duration T_0 of the class of waveforms increases, the initial transient due to the samples centered close to $-T_0/2$ and the final transient due to samples centered close to $T_0/2$ should become unimportant relative to the rest of the waveform.

The same conclusion can be reached for functions that are strictly time-limited to $[-T_0/2, T_0/2]$ by using the truncated sinusoid expansion with coefficients outside of $[-W_0, W_0]$ set to 0.

[34] Calculating this number of samples carefully yields $(2n+1)\left(1+\left\lfloor \frac{T_0W_0}{2n+1} \right\rfloor\right)$.

In summary, all the above expansions require roughly $2\mathsf{W}_0 T_0$ numbers for the approximate specification of a waveform essentially limited to time T_0 and frequency W_0 for $T_0\mathsf{W}_0$ large.

It is possible to be more precise about the number of degrees of freedom in a given time and frequency band by looking at the prolate spheroidal waveform expansion (see Appendix 5.5.3). The orthogonal waveforms in this expansion maximize the energy in the given time/frequency region in a certain sense. It is perhaps simpler and better, however, to live with the very approximate nature of the arguments based on the sinc-weighted sinusoid expansion and the truncated sinusoid expansion.

4.7.3 Aliasing – a time-domain approach

Both the truncated sinusoid and the sinc-weighted sinusoid expansions are conceptually useful for understanding waveforms that are approximately time- and bandwidth-limited, but in practice waveforms are usually sampled, perhaps at a rate much higher than twice the nominal bandwidth, before digitally processing the waveforms. Thus it is important to understand the error involved in such sampling.

Suppose an $\mathcal{L}_2$ function $u(t)$ is sampled with T-spaced samples, $\{u(kT);\, k \in \mathbb{Z}\}$. Let $s(t)$ denote the approximation to $u(t)$ that results from the sampling theorem expansion:

$$s(t) = \sum_k u(kT)\,\mathrm{sinc}\left(\frac{t}{T} - k\right). \tag{4.75}$$

If $u(t)$ is baseband-limited to $\mathsf{W} = 1/2T$, then $s(t) = u(t)$, but here it is no longer assumed that $u(t)$ is baseband-limited. The expansion of $u(t)$ into individual frequency slices, repeated below from (4.73), helps in understanding the difference between $u(t)$ and $s(t)$:

$$u(t) = \mathrm{l.i.m.} \sum_{m,k} v_m(kT)\,\mathrm{sinc}\left(\frac{t}{T} - k\right) \mathrm{e}^{2\pi \mathrm{i} mt/T}, \tag{4.76}$$

where

$$v_m(t) = \int \hat{u}(f)\,\mathrm{rect}(fT - m)\mathrm{e}^{2\pi \mathrm{i} ft}\,\mathrm{d}f. \tag{4.77}$$

For an arbitrary $\mathcal{L}_2$ function $u(t)$, the sample points $u(kT)$ might be at points of discontinuity and thus be ill defined. Also (4.75) need not converge, and (4.76) might not converge pointwise. To avoid these problems, $\hat{u}(f)$ will later be restricted beyond simply being $\mathcal{L}_2$. First, however, questions of convergence are disregarded and the relevant equations are derived without questioning when they are correct.

From (4.75), the samples of $s(t)$ are given by $s(kT) = u(kT)$, and combining with (4.76) we obtain

$$s(kT) = u(kT) = \sum_m v_m(kT). \tag{4.78}$$

Thus the samples from different frequency slices are summed together in the samples of $u(t)$. This phenomenon is called *aliasing*. There is no way to tell, from the samples $\{u(kT);\, k \in \mathbb{Z}\}$ alone, how much contribution comes from each frequency slice and thus, as far as the samples are concerned, every frequency band is an "alias" for every other.

Although $u(t)$ and $s(t)$ agree at the sample times, they differ elsewhere (assuming that $u(t)$ is not strictly baseband-limited to $1/2T$). Combining (4.78) and (4.75) we obtain

$$s(t) = \sum_k \sum_m v_m(kT)\,\mathrm{sinc}\left(\frac{t}{T} - k\right). \tag{4.79}$$

The expresssions in (4.79) and (4.76) agree at $m = 0$, so the difference between $u(t)$ and $s(t)$ is given by

$$u(t) - s(t) = \sum_k \sum_{m\neq 0} -v_m(kT)\,\mathrm{sinc}\left(\frac{t}{T} - k\right) + \sum_k \sum_{m\neq 0} v_m(kT) e^{2\pi imt/T} \mathrm{sinc}\left(\frac{t}{T} - k\right).$$

The first term above is $v_0(t) - s(t)$, i.e. the difference in the nominal baseband $[-\mathsf{W}, \mathsf{W}]$. This is the error caused by the aliased terms in $s(t)$. The second term is the energy in the nonbaseband portion of $u(t)$, which is orthogonal to the first error term. Since each term is an orthogonal expansion in the sinc-weighted sinusoids of (4.74), the energy in the error is given by[35]

$$\int \left|u(t) - s(t)\right|^2 \mathrm{d}t = T \sum_k \left| \sum_{m\neq 0} v_m(kT)\right|^2 + T \sum_k \sum_{m\neq 0} \left|v_m(kT)\right|^2. \tag{4.80}$$

Later, when the source waveform $u(t)$ is viewed as a sample function of a random process $U(t)$, it will be seen that under reasonable conditions the expected values of these two error terms are approximately equal. Thus, if $u(t)$ is filtered by an ideal lowpass filter before sampling, then $s(t)$ becomes equal to $v_0(t)$ and only the second error term in (4.80) remains; this reduces the expected MSE roughly by a factor of 2. It is often easier, however, simply to sample a little faster.

4.7.4 Aliasing – a frequency-domain approach

Aliasing can be, and usually is, analyzed from a frequency-domain standpoint. From (4.79), $s(t)$ can be separated into the contribution from each frequency band as follows:

$$s(t) = \sum_m s_m(t), \qquad \text{where} \quad s_m(t) = \sum_k v_m(kT)\,\mathrm{sinc}\left(\frac{t}{T} - k\right). \tag{4.81}$$

Comparing $s_m(t)$ to $v_m(t) = \sum_k v_m(kT)\,\mathrm{sinc}(t/T - k) e^{2\pi imt/T}$, it is seen that

$$v_m(t) = s_m(t) e^{2\pi imt/T}.$$

From the Fourier frequency shift relation, $\hat{v}_m(f) = \hat{s}_m(f - m/T)$, so

$$\hat{s}_m(f) = \hat{v}_m\left(f + \frac{m}{T}\right). \tag{4.82}$$

[35] As shown by example in Exercise 4.38, $s(t)$ need not be $\mathcal{L}_2$ unless the additional restrictions of Theorem 5.5.2 are applied to $\hat{u}(f)$. In these bizarre situations, the first sum in (4.80) is infinite and $s(t)$ is a complete failure as an approximation to $u(t)$.

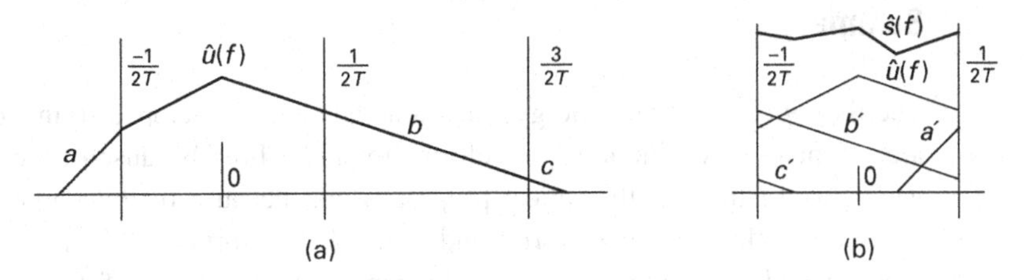

Figure 4.10. The transform $\hat{s}(f)$ of the baseband-sampled approximation $s(t)$ to $u(t)$ is constructed by folding the transform $\hat{u}(f)$ into $[-1/2T, 1/2T]$. For example, using real functions for pictorial clarity, the component a is mapped into a', b into b', and c into c'. These folded components are added to obtain $\hat{s}(f)$. If $\hat{u}(f)$ is complex, then both the real and imaginary parts of $\hat{u}(f)$ must be folded in this way to get the real and imaginary parts, respectively, of $\hat{s}(f)$. The figure further clarifies the two terms on the right of (4.80). The first term is the energy of $\hat{u}(f) - \hat{s}(f)$ caused by the folded components in part (b). The final term is the energy in part (a) outside of $[-T/2, T/2]$.

Finally, since $\hat{v}_m(f) = \hat{u}(f)\,\mathrm{rect}(fT - m)$, one sees that $\hat{v}_m(f + m/T) = \hat{u}(f + m/T)$ $\mathrm{rect}(fT)$. Thus, summing (4.82) over m, we obtain

$$\hat{s}(f) = \sum_m \hat{u}\left(f + \frac{m}{T}\right) \mathrm{rect}(fT). \tag{4.83}$$

Each frequency slice $\hat{v}_m(f)$ is shifted down to baseband in this equation, and then all these shifted frequency slices are summed together, as illustrated in Figure 4.10. This establishes the essence of the following aliasing theorem, which is proved in Appendix 5.5.2.

Theorem 4.7.1 (Aliasing theorem) *Let $\hat{u}(f)$ be $\mathcal{L}_2$, and let $\hat{u}(f)$ satisfy the condition $\lim_{|f|\to\infty} \hat{u}(f)|f|^{1+\varepsilon} = 0$ for some $\varepsilon > 0$. Then $\hat{u}(f)$ is $\mathcal{L}_1$, and the inverse Fourier transform $u(t) = \int \hat{u}(f) e^{2\pi i f t}\,\mathrm{d}f$ converges pointwise to a continuous bounded function. For any given $T > 0$, the sampling approximation $\sum_k u(kT)\,\mathrm{sinc}(t/T - k)$ converges pointwise to a continuous bounded $\mathcal{L}_2$ function $s(t)$. The Fourier transform of $s(t)$ satisfies*

$$\hat{s}(f) = \text{l.i.m.} \sum_m \hat{u}\left(f + \frac{m}{T}\right) \mathrm{rect}(fT). \tag{4.84}$$

The condition that $\lim \hat{u}(f) f^{1+\varepsilon} = 0$ implies that $\hat{u}(f)$ goes to 0 with increasing f at a faster rate than $1/f$. Exercise 4.37 gives an example in which the theorem fails in the absence of this condition.

Without the mathematical convergence details, what the aliasing theorem says is that, corresponding to a Fourier transform pair $u(t) \leftrightarrow \hat{u}(f)$, there is another Fourier transform pair $s(t) \leftrightarrow \hat{s}(f)$; $s(t)$ is a baseband sampling expansion using the T-spaced samples of $u(t)$, and $\hat{s}(f)$ is the result of folding the transform $\hat{u}(f)$ into the band $[-\mathsf{W}, \mathsf{W}]$ with $\mathsf{W} = 1/2T$.

4.8 Summary

The theory of $\mathcal{L}_2$ (finite-energy) functions has been developed in this chapter. These are, in many ways, the ideal waveforms to study, both because of the simplicity and generality of their mathematical properties and because of their appropriateness for modeling both source waveforms and channel waveforms.

For encoding source waveforms, the general approach is as follows:

- expand the waveform into an orthogonal expansion;
- quantize the coefficients in that expansion;
- use discrete source coding on the quantizer output.

The distortion, measured as the energy in the difference between the source waveform and the reconstructed waveform, is proportional to the squared quantization error in the quantized coefficients.

For encoding waveforms to be transmitted over communication channels, the approach is as follows:

- map the incoming sequence of binary digits into a sequence of real or complex symbols;
- use the symbols as coefficients in an orthogonal expansion.

Orthogonal expansions have been discussed in this chapter and will be further discussed in Chapter 5. Chapter 6 will discuss the choice of symbol set, the mapping from binary digits, and the choice of orthogonal expansion.

This chapter showed that every $\mathcal{L}_2$ time-limited waveform has a Fourier series, where each Fourier coefficient is given as a Lebesgue integral and the Fourier series converges in $\mathcal{L}_2$, i.e. as more and more Fourier terms are used in approximating the function, the energy difference between the waveform and the approximation gets smaller and approaches 0 in the limit.

Also, by the Plancherel theorem, every $\mathcal{L}_2$ waveform $u(t)$ (time-limited or not) has a Fourier integral $\hat{u}(f)$. For each truncated approximation, $u_A(t) = u(t)\ \text{rect}(t/2A)$, the Fourier integral $\hat{u}_A(f)$ exists with pointwise convergence and is continuous. The Fourier integral $\hat{u}(f)$ is then the $\mathcal{L}_2$ limit of these approximation waveforms. The inverse transform exists in the same way.

These powerful $\mathcal{L}_2$-convergence results for Fourier series and integrals are not needed for computing the Fourier transforms and series for the conventional waveforms appearing in exercises. They become important both when the waveforms are sample functions of random processes and when one wants to find limits on possible performance. In both of these situations, one is dealing with a large class of potential waveforms, rather than a single waveform, and these general results become important.

The DTFT is the frequency–time dual of the Fourier series, and the sampling theorem is simply the Fourier transform of the DTFT, combined with a little care about convergence.

The T-spaced truncated sinusoid expansion and the T-spaced sinc-weighted sinusoid expansion are two orthogonal expansions of an arbitrary $\mathcal{L}_2$ waveform. The first is formed by segmenting the waveform into T-length segments and expanding each segment in a Fourier series. The second is formed by segmenting the waveform in frequency and sampling each frequency band. The orthogonal waveforms in each are the time–frequency translates of $\text{rect}(t/T)$ for the first case and $\text{sinc}(t/T)$ for the second. Each expansion leads to the notion that waveforms roughly limited to a time interval T_0 and a baseband frequency interval W_0 have approximately $2T_0\mathsf{W}_0$ degrees of freedom when $T_0\mathsf{W}_0$ is large.

Aliasing is the ambiguity in a waveform that is represented by its T-spaced samples. If an $\mathcal{L}_2$ waveform is baseband-limited to $1/2T$, then its samples specify the waveform, but if the waveform has components in other bands, these components are aliased with the baseband components in the samples. The aliasing theorem says that the Fourier transform of the baseband reconstruction from the samples is equal to the original Fourier transform folded into that baseband.

4.9 Appendix: Supplementary material and proofs

The first part of the appendix is an introduction to countable sets. These results are used throughout the chapter, and the material here can serve either as a first exposure or a review. The following three parts of the appendix provide added insight and proofs about the results on measurable sets.

4.9.1 Countable sets

A collection of distinguishable objects is *countably infinite* if the objects can be put into one-to-one correspondence with the positive integers. Stated more intuitively, the collection is countably infinite if the set of elements can be arranged as a sequence $a_1, a_2, \ldots$ A set is countable if it contains either a finite or countably infinite set of elements.

Example 4.9.1 (The set of all integers) The integers can be arranged as the sequence $0, -1, +1, -2, +2, -3, \ldots$, and thus the set is countably infinite. Note that each integer appears once and only once in this sequence, and the one-to-one correspondence is $(0 \leftrightarrow 1)$, $(-1 \leftrightarrow 2)$, $(+1 \leftrightarrow 3)$, $(-2 \leftrightarrow 4)$, etc. There are many other ways to list the integers as a sequence, such as $0, -1, +1, +2, -2, +3, +4, -3, +5, \ldots$, but, for example, listing all the nonnegative integers first followed by all the negative integers is not a valid one-to-one correspondence since there are no positive integers left over for the negative integers to map into.

Example 4.9.2 (The set of 2-tuples of positive integers) Figure 4.11 shows that this set is countably infinite by showing one way to list the elements in a sequence. Note that every 2-tuple is eventually reached in this list. In a weird sense, this means that there are as many positive integers as there are pairs of positive integers, but what is happening is that the integers in the 2-tuple advance much more slowly than the position in the list. For example, it can be verified that (n, n) appears in position $2n(n-1)+1$ of the list.

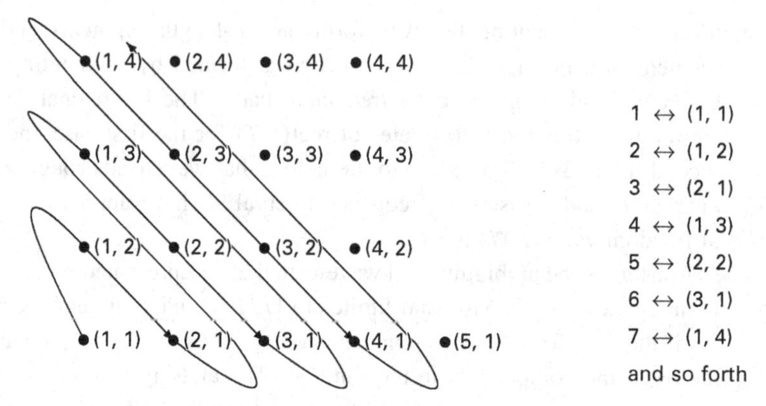

Figure 4.11. One-to-one correspondence between positive integers and 2-tuples of positive integers.

By combining the ideas in the previous two examples, it can be seen that the collection of all integer 2-tuples is countably infinite. With a little more ingenuity, it can be seen that the set of integer n-tuples is countably infinite for all positive integer n. Finally, it is straightforward to verify that any subset of a countable set is also countable. Also a finite union of countable sets is countable, and in fact a countable union of countable sets must be countable.

Example 4.9.3 (The set of rational numbers) Each rational number can be represented by an integer numerator and denominator, and can be uniquely represented by its irreducible numerator and denominator. Thus the rational numbers can be put into one-to-one correspondence with a subset of the collection of 2-tuples of integers, and are thus countable. The rational numbers in the interval $[-T/2, T/2]$ for any given $T > 0$ form a subset of all rational numbers, and therefore are countable also.

As seen in Section 4.3.1, any countable set of numbers $a_1, a_2, \ldots$ can be expressed as a disjoint countable union of zero-measure sets, $[a_1, a_1], [a_2, a_2], \ldots$, so the measure of any countable set is zero. Consider a function that has the value 1 at each rational argument and 0 elsewhere.

The Lebesgue integral of that function is 0. Since rational numbers exist in every positive-sized interval of the real line, no matter how small, the Riemann integral of this function is undefined. This function is not of great practical interest, but provides insight into why Lebesgue integration is so general.

Example 4.9.4 (The set of binary sequences) An example of an *uncountable* set of elements is the set of (unending) sequences of binary digits. It will be shown that this set contains uncountably many elements by assuming the contrary and constructing a contradiction. Thus, suppose we can list all binary sequences, $\boldsymbol{a}_1, \boldsymbol{a}_2, \boldsymbol{a}_3, \ldots$ Each sequence, $\boldsymbol{a}_n$, can be expressed as $\boldsymbol{a}_n = (a_{n,1}, a_{n,2}, \ldots)$, resulting in a doubly infinite array of binary digits. We now construct a new binary sequence $\boldsymbol{b} = b_1, b_2, \ldots$ in the following way. For each integer $n > 0$, choose $b_n \neq a_{n,n}$; since b_n is binary, this specifies b_n for each n and thus specifies $\boldsymbol{b}$. Now $\boldsymbol{b}$ differs from each of the listed sequences in at least one binary digit, so that $\boldsymbol{b}$ is a binary sequence not on the list. This is a contradiction, since by assumption the list contains each binary sequence.

This example clearly extends to ternary sequences and sequences from any alphabet with more than one member.

Example 4.9.5 (The set of real numbers in $[0, 1)$) This is another uncountable set, and the proof is very similar to that of Example 4.9.4. Any real number $r \in [0, 1)$ can be represented as a binary expansion $0.r_1r_2, \dots$ whose elements r_k are chosen to satisfy $r = \sum_{k=1}^{\infty} r_k 2^{-k}$ and where each $r_k \in \{0, 1\}$. For example, $1/2$ can be represented as 0.1, 3/8 as 0.011, etc. This expansion is unique except in the special cases where r can be represented by a finite binary expansion, $r = \sum_{k=1}^{m} r_k$; for example, 1/2 can also be represented as $0.0111\cdots$. By convention, for each such r (other than $r = 0$) choose m as small as possible; thus in the infinite expansion, $r_m = 1$ and $r_k = 0$ for all $k > m$. Each such number can be alternatively represented with $r_m = 0$ and $r_k = 1$ for all $k > m$.

By convention, map each such r into the expansion terminating with an infinite sequence of 0s. The set of binary sequences is then the union of the representations of the reals in $[0, 1)$ and the set of binary sequences terminating in an infinite sequence of 1s. This latter set is countable because it is in one-to-one correspondence with the rational numbers of the form $\sum_{k=1}^{m} r_k 2^{-k}$ with binary r_k and finite m. Thus if the reals were countable, their union with this latter set would be countable, contrary to the known uncountability of the binary sequences.

By scaling the interval [0,1), it can be seen that the set of real numbers in any interval of nonzero size is uncountably infinite. Since the set of rational numbers in such an interval is countable, the irrational numbers must be uncountable (otherwise the union of rational and irrational numbers, i.e. the reals, would be countable).

The set of irrationals in $[-T/2, T/2]$ is the complement of the rationals and thus has measure T. Each pair of distinct irrationals is separated by rational numbers. Thus the irrationals can be represented as a union of intervals only by using an uncountable union[36] of intervals, each containing a single element. The class of uncountable unions of intervals is not very interesting since it includes all subsets of $\mathbb{R}$.

4.9.2 Finite unions of intervals over $[-T/2, T/2]$

Let $\mathcal{M}_f$ be the class of finite unions of intervals, i.e. the class of sets whose elements can each be expressed as $\mathcal{E} = \bigcup_{j=1}^{\ell} I_j$, where $\{I_1, \dots, I_\ell\}$ are intervals and $\ell \geq 1$ is an integer. Exercise 4.5 shows that each such $\mathcal{E} \in \mathcal{M}_f$ can be uniquely expressed as a finite union of $k \leq \ell$ *separated* intervals, say $\mathcal{E} = \bigcup_{j=1}^{k} I'_j$. The measure of $\mathcal{E}$ was defined as $\mu(\mathcal{E}) = \sum_{j=1}^{k} \mu(I'_j)$. Exercise 4.7 shows that $\mu(\mathcal{E}) \leq \sum_{j=1}^{\ell} \mu(I_j)$ for the original

[36] This might be a shock to one's intuition. Each partial union $\bigcup_{j=1}^{k} [a_j, a_j]$ of rationals has a complement which is the union of $k+1$ intervals of nonzero width; each unit increase in k simply causes one interval in the complement to split into two smaller intervals (although maintaining the measure at T). In the limit, however, this becomes an uncountable set of separated points.

intervals making up $\mathcal{E}$ and shows that this holds with equality whenever $I_1, \ldots, I_\ell$ are disjoint.[37]

The class $\mathcal{M}_f$ is closed under the union operation, since if $\mathcal{E}_1$ and $\mathcal{E}_2$ are each finite unions of intervals, then $\mathcal{E}_1 \cup \mathcal{E}_2$ is the union of both sets of intervals. It also follows from this that if $\mathcal{E}_1$ and $\mathcal{E}_2$ are disjoint then

$$\mu(\mathcal{E}_1 \cup \mathcal{E}_2) = \mu(\mathcal{E}_1) + \mu(\mathcal{E}_2). \tag{4.85}$$

The class $\mathcal{M}_f$ is also closed under the intersection operation, since, if $\mathcal{E}_1 = \bigcup_j I_{1,j}$ and $\mathcal{E}_2 = \bigcup_\ell I_{2,\ell}$, then $\mathcal{E}_1 \cap \mathcal{E}_2 = \bigcup_{j,\ell}(I_{1,j} \cap I_{2,\ell})$. Finally, $\mathcal{M}_f$ is closed under complementation. In fact, as illustrated in Figure 4.5, the complement $\overline{\mathcal{E}}$ of a finite union of separated intervals $\mathcal{E}$ is simply the union of separated intervals lying between the intervals of $\mathcal{E}$. Since $\mathcal{E}$ and its complement $\overline{\mathcal{E}}$ are disjoint and fill all of $[-T/2, T/2]$, each $\mathcal{E} \in \mathcal{M}_f$ satisfies the complement property,

$$T = \mu(\mathcal{E}) + \mu(\overline{\mathcal{E}}). \tag{4.86}$$

An important generalization of (4.85) is the following: for any $\mathcal{E}_1, \mathcal{E}_2 \in \mathcal{M}_f$,

$$\mu(\mathcal{E}_1 \cup \mathcal{E}_2) + \mu(\mathcal{E}_1 \cap \mathcal{E}_2) = \mu(\mathcal{E}_1) + \mu(\mathcal{E}_2). \tag{4.87}$$

To see this intuitively, note that each interval in $\mathcal{E}_1 \cap \mathcal{E}_2$ is counted twice on each side of (4.87), whereas each interval in only $\mathcal{E}_1$ or only $\mathcal{E}_2$ is counted once on each side. More formally, $\mathcal{E}_1 \cup \mathcal{E}_2 = \mathcal{E}_1 \cup (\mathcal{E}_2 \cap \overline{\mathcal{E}_1})$. Since this is a disjoint union, (4.85) shows that $\mu(\mathcal{E}_1 \cup \mathcal{E}_2) = \mu(\mathcal{E}_1) + \mu(\mathcal{E}_2 \cap \overline{\mathcal{E}_1})$. Similarly, $\mu(\mathcal{E}_2) = \mu(\mathcal{E}_2 \cap \mathcal{E}_1) + \mu(\mathcal{E}_2 \cap \overline{\mathcal{E}_1})$. Combining these equations results in (4.87).

4.9.3 Countable unions and outer measure over $[-T/2, T/2]$

Let $\mathcal{M}_c$ be the class of countable unions of intervals, i.e. each set $\mathcal{B} \in \mathcal{M}_c$ can be expressed as $\mathcal{B} = \bigcup_j I_j$, where $\{I_1, I_2, \ldots\}$ is either a finite or countably infinite collection of intervals. The class $\mathcal{M}_c$ is closed under both the union operation and the intersection operation by the same argument as used for $\mathcal{M}_f$. Note that $\mathcal{M}_c$ is also closed under countable unions (see Exercise 4.8) but not closed under complements or countable intersections.[38]

Each $\mathcal{B} \in \mathcal{M}_c$ can be uniquely[39] expressed as a countable union of separated intervals, say $\mathcal{B} = \bigcup_j I'_j$, where $\{I'_1, I'_2, \ldots\}$ are separated (see Exercise 4.6). The measure of $\mathcal{B}$ is defined as

---

[37] Recall that intervals such as (0,1], (1,2] are disjoint but not separated. A set $\mathcal{E} \in \mathcal{M}_f$ has many representations as disjoint intervals but only one as separated intervals, which is why the definition refers to separated intervals.

[38] Appendix 4.9.1 shows that the complement of the rationals, i.e. the set of irrationals, does not belong to $\mathcal{M}_c$. The irrationals can also be viewed as the intersection of the complements of the rationals, giving an example where $\mathcal{M}_c$ is not closed under countable intersections.

[39] What is unique here is the *collection* of intervals, not the *particular ordering*; this does not affect the infinite sum in (4.88) (see Exercise 4.4).

$$\mu(\mathcal{B}) = \sum_j \mu(I'_j). \tag{4.88}$$

As shown in Section 4.3.1, the right side of (4.88) always converges to a number between 0 and T. For $B = \bigcup_j I_j$, where $I_1, I_2, \ldots$ are arbitrary intervals, Exercise 4.7 establishes the following union bound:

$$\mu(\mathcal{B}) \leq \sum_j \mu(I_j) \qquad \text{with equality if } I_1, I_2, \ldots \text{ are disjoint.} \tag{4.89}$$

The *outer measure* $\mu^{\circ}(\mathcal{A})$ of an arbitary set $\mathcal{A}$ was defined in (4.13) as

$$\mu^{\circ}(\mathcal{A}) = \inf_{\mathcal{B}\in\mathcal{M}_{\mathrm{c}},\, \mathcal{A}\subseteq\mathcal{B}} \mu(\mathcal{B}). \tag{4.90}$$

Note that $[-T/2,\ T/2]$ is a cover of $\mathcal{A}$ for all $\mathcal{A}$ (recall that only sets in $[-T/2,\ T/2]$ are being considered). Thus $\mu^{\circ}(\mathcal{A})$ must lie between 0 and T for all $\mathcal{A}$. Also, for any two sets $\mathcal{A} \subseteq \mathcal{A}'$, any cover of $\mathcal{A}'$ also covers $\mathcal{A}$. This implies the *subset inequality* for outer measure:

$$\mu^{\circ}(\mathcal{A}) \leq \mu^{\circ}(\mathcal{A}') \qquad \text{for } \mathcal{A} \subseteq \mathcal{A}'. \tag{4.91}$$

The following lemma develops the *union bound* for outer measure called the *union bound*. Its proof illustrates several techniques that will be used frequently.

Lemma 4.9.1 *Let $\mathcal{S} = \bigcup_k \mathcal{A}_k$ be a countable union of arbitrary sets in $[-T/2,\ T/2]$. Then*

$$\mu^{\circ}(\mathcal{S}) \leq \sum_k \mu^{\circ}(\mathcal{A}_k). \tag{4.92}$$

Proof The approach is first to establish an arbitrarily tight cover to each $\mathcal{A}_k$ and then show that the union of these covers is a cover for $\mathcal{S}$. Specifically, let ε be an arbitrarily small positive number. For each $k \geq 1$, the infimum in (4.90) implies that covers exist with measures arbitrarily little greater than that infimum. Thus a cover $\mathcal{B}_k$ to $\mathcal{A}_k$ exists with

$$\mu(\mathcal{B}_k) \leq \varepsilon 2^{-k} + \mu^{\circ}(\mathcal{A}_k).$$

For each k, let $\mathcal{B}_k = \bigcup_j I'_{j,k}$, where $I'_{1,k}, I'_{2,k}, \ldots$ represents $\mathcal{B}_k$ by separated intervals. Then $\mathcal{B} = \bigcup_k \mathcal{B}_k = \bigcup_k \bigcup_j I'_{j,k}$ is a countable union of intervals, so, from (4.89) and Exercise 4.4, we have

$$\mu(\mathcal{B}) \leq \sum_k \sum_j \mu(I'_{j,k}) = \sum_k \mu(\mathcal{B}_k).$$

Since $\mathcal{B}_k$ covers $\mathcal{A}_k$ for each k, it follows that $\mathcal{B}$ covers $\mathcal{S}$. Since $\mu^{\circ}(S)$ is the infimum of its covers,

$$\mu^{\circ}(S) \leq \mu(\mathcal{B}) \leq \sum_k \mu(\mathcal{B}_k) \leq \sum_k \left(\varepsilon 2^{-k} + \mu^{\circ}(\mathcal{A}_k)\right) = \varepsilon + \sum_k \mu^{\circ}(\mathcal{A}_k).$$

Since $\varepsilon > 0$ is arbitrary, (4.92) follows. □

An important special case is the union of any set $\mathcal{A}$ and its complement $\overline{\mathcal{A}}$. Since $[-T/2, T/2] = \mathcal{A} \cup \overline{\mathcal{A}}$,

$$T \leq \mu^{\mathrm{o}}(\mathcal{A}) + \mu^{\mathrm{o}}(\overline{\mathcal{A}}). \tag{4.93}$$

Section 4.9.4 will define measurability and measure for arbitrary sets. Before that, the following theorem shows both that countable unions of intervals are measurable and that their measure, as defined in (4.88), is consistent with the general definition to be given later.

Theorem 4.9.1 *Let $\mathcal{B} = \bigcup_j I_j$, where $\{I_1, I_2, \ldots\}$ is a countable collection of intervals in $[-T/2, T/2]$ (i.e., $\mathcal{B} \in \mathcal{M}_{\mathrm{c}}$). Then*

$$\mu^{\mathrm{o}}(\mathcal{B}) + \mu^{\mathrm{o}}(\overline{\mathcal{B}}) = T \tag{4.94}$$

and

$$\mu^{\mathrm{o}}(\mathcal{B}) = \mu(\mathcal{B}). \tag{4.95}$$

Proof Let $\{I'_j; j \geq 1\}$ be the collection of separated intervals representing $\mathcal{B}$ and let

$$\mathcal{E}^k = \bigcup_{j=1}^{k} I'_j;$$

then

$$\mu(\mathcal{E}^1) \leq \mu(\mathcal{E}^2) \leq \mu(\mathcal{E}^3) \leq \cdots \leq \lim_{k\to\infty} \mu(\mathcal{E}^k) = \mu(\mathcal{B}).$$

For any $\varepsilon > 0$, choose k large enough that

$$\mu(\mathcal{E}^k) \geq \mu(\mathcal{B}) - \varepsilon. \tag{4.96}$$

The idea of the proof is to approximate $\mathcal{B}$ by $\mathcal{E}^k$, which, being in $\mathcal{M}_{\mathrm{f}}$, satisfies $T = \mu(\mathcal{E}^k) + \mu(\overline{\mathcal{E}^k})$. Thus,

$$\mu(\overline{\mathcal{B}}) \leq \mu(\mathcal{E}^k) + \varepsilon = T - \mu(\overline{\mathcal{E}^k}) + \varepsilon \leq T - \mu^{\mathrm{o}}(\overline{\mathcal{B}}) + \varepsilon, \tag{4.97}$$

where the final inequality follows because $\mathcal{E}^k \subseteq \mathcal{B}$, and thus $\overline{\mathcal{B}} \subseteq \overline{\mathcal{E}^k}$ and $\mu^{\mathrm{o}}(\overline{\mathcal{B}}) \leq \mu(\overline{\mathcal{E}^k})$.

Next, since $\mathcal{B} \in \mathcal{M}_{\mathrm{c}}$ and $\mathcal{B} \subseteq \mathcal{B}$, $\mathcal{B}$ is a cover of itself and is a choice in the infimum defining $\mu^{\mathrm{o}}(\mathcal{B})$; thus, $\mu^{\mathrm{o}}(\overline{\mathcal{B}}) \leq \mu(\overline{\mathcal{B}})$. Combining this with (4.97), $\mu^{\mathrm{o}}(\mathcal{B}) + \mu^{\mathrm{o}}(\overline{\mathcal{B}}) \leq T + \varepsilon$. Since $\varepsilon > 0$ is arbitrary, this implies

$$\mu^{\mathrm{o}}(\mathcal{B}) + \mu^{\mathrm{o}}(\overline{\mathcal{B}}) \leq T. \tag{4.98}$$

This combined with (4.93) establishes (4.94). Finally, substituting $T \leq \mu^{\mathrm{o}}(\mathcal{B}) + \mu^{\mathrm{o}}(\overline{\mathcal{B}})$ into (4.97), $\mu(\mathcal{B}) \leq \mu^{\mathrm{o}}(\mathcal{B}) + \varepsilon$. Since $\mu^{\mathrm{o}}(\mathcal{B}) \leq \mu(\mathcal{B})$ and $\varepsilon > 0$ is arbitrary, this establishes (4.95). □

Finally, before proceeding to arbitrary measurable sets, the joint union and intersection property, (4.87), is extended to $\mathcal{M}_{\mathrm{c}}$.

Lemma 4.9.2 *Let $\mathcal{B}_1$ and $\mathcal{B}_2$ be arbitrary sets in $\mathcal{M}_c$. Then*

$$\mu(\mathcal{B}_1 \cup \mathcal{B}_2) + \mu(\mathcal{B}_1 \cap \mathcal{B}_2) = \mu(\mathcal{B}_1) + \mu(\mathcal{B}_2). \tag{4.99}$$

Proof Let $\mathcal{B}_1$ and $\mathcal{B}_2$ be represented, respectively, by separated intervals, $\mathcal{B}_1 = \bigcup_j I_{1,j}$ and $\mathcal{B}_2 = \bigcup_j I_{2,j}$. For $\ell = 1, 2$, let $\mathcal{E}_\ell^k = \bigcup_{j=1}^k I_{\ell,j}$ and $\mathcal{D}_\ell^k = \bigcup_{j=k+1}^\infty I_{\ell,j}$. Thus $\mathcal{B}_\ell = \mathcal{E}_\ell^k \cup \mathcal{D}_\ell^k$ for each integer $k \geq 1$ and $\ell = 1, 2$. The proof is based on using $\mathcal{E}_\ell^k$, which is in $\mathcal{M}_f$ and satisfies the joint union and intersection property, as an approximation to $\mathcal{B}_\ell$. To see how this goes, note that

$$\mathcal{B}_1 \cap \mathcal{B}_2 = (\mathcal{E}_1^k \cup \mathcal{D}_1^k) \cap (\mathcal{E}_2^k \cup \mathcal{D}_2^k) = (\mathcal{E}_1^k \cap \mathcal{E}_2^k) \cup (\mathcal{E}_1^k \cap \mathcal{D}_2^k) \cup (\mathcal{D}_1^k \cap \mathcal{B}_2).$$

For any $\varepsilon > 0$ we can choose k large enough that $\mu(\mathcal{E}_\ell^k) \geq \mu(\mathcal{B}_\ell) - \varepsilon$ and $\mu(\mathcal{D}_\ell^k) \leq \varepsilon$ for $\ell = 1, 2$. Using the subset inequality and the union bound, we then have

$$\begin{aligned}\mu(\mathcal{B}_1 \cap \mathcal{B}_2) &\leq \mu(\mathcal{E}_1^k \cap \mathcal{E}_2^k) + \mu(\mathcal{D}_2^k) + \mu(\mathcal{D}_1^k)\\ &\leq \mu(\mathcal{E}_1^k \cap \mathcal{E}_2^k) + 2\varepsilon.\end{aligned}$$

By a similar but simpler argument,

$$\begin{aligned}\mu(\mathcal{B}_1 \cup \mathcal{B}_2) &\leq \mu(\mathcal{E}_1^k \cup \mathcal{E}_2^k) + \mu(\mathcal{D}_1^k) + \mu(\mathcal{D}_2^k)\\ &\leq \mu(\mathcal{E}_1^k \cup \mathcal{E}_2^k) + 2\varepsilon.\end{aligned}$$

Combining these inequalities and using (4.87) on $\mathcal{E}_1^k \subseteq \mathcal{M}_f$ and $\mathcal{E}_2^k \subseteq \mathcal{M}_f$, we have

$$\begin{aligned}\mu(\mathcal{B}_1 \cap \mathcal{B}_2) + \mu(\mathcal{B}_1 \cup \mathcal{B}_2) &\leq \mu(\mathcal{E}_1^k \cap \mathcal{E}_2^k) + \mu(\mathcal{E}_1^k \cup \mathcal{E}_2^k) + 4\varepsilon\\ &= \mu(\mathcal{E}_1^k) + \mu(\mathcal{E}_2^k) + 4\varepsilon\\ &\leq \mu(\mathcal{B}_1) + \mu(\mathcal{B}_2) + 4\varepsilon,\end{aligned}$$

where we have used the subset inequality in the final inequality.

For a bound in the opposite direction, we start with the subset inequality:

$$\begin{aligned}\mu(\mathcal{B}_1 \cup \mathcal{B}_2) + \mu(\mathcal{B}_1 \cap \mathcal{B}_2) &\geq \mu(\mathcal{E}_1^k \cup \mathcal{E}_2^k) + \mu(\mathcal{E}_1^k \cap \mathcal{E}_2^k)\\ &= \mu(\mathcal{E}_1^k) + \mu(\mathcal{E}_2^k)\\ &\geq \mu(\mathcal{B}_1) + \mu(\mathcal{B}_2) - 2\varepsilon.\end{aligned}$$

Since ε is arbitrary, these two bounds establish (4.99). □

4.9.4 Arbitrary measurable sets over $[-T/2, T/2]$

An arbitrary set $\mathcal{A} \in [-T/2, T/2]$ was defined to be *measurable* if

$$T = \mu^\circ(\mathcal{A}) + \mu^\circ(\overline{\mathcal{A}}). \tag{4.100}$$

The *measure* of a measurable set was defined to be $\mu(\mathcal{A}) = \mu^{\mathrm{o}}(\mathcal{A})$. The class of measurable sets is denoted as $\mathcal{M}$. Theorem 4.9.1 shows that each set $\mathcal{B} \in \mathcal{M}_{\mathrm{c}}$ is measurable, i.e. $\mathcal{B} \in \mathcal{M}$ and thus $\mathcal{M}_{\mathrm{f}} \subseteq \mathcal{M}_{\mathrm{c}} \subseteq \mathcal{M}$. The measure of $\mathcal{B} \in \mathcal{M}_{\mathrm{c}}$ is $\mu(\mathcal{B}) = \sum_j \mu(I_j)$ for any disjoint sequence of intervals, $I_1, I_2, \ldots$, whose union is $\mathcal{B}$.

Although the complements of sets in $\mathcal{M}_{\mathrm{c}}$ are not necessarily in $\mathcal{M}_{\mathrm{c}}$ (as seen from the rational number example), they must be in $\mathcal{M}$; in fact, from (4.100), all sets in $\mathcal{M}$ have complements in $\mathcal{M}$, i.e. $\mathcal{M}$ is closed under complements. We next show that $\mathcal{M}$ is closed under finite, and then countable, unions and intersections. The key to these results is to show first that the joint union and intersection property is valid for outer measure.

Lemma 4.9.3 *For any measurable sets $\mathcal{A}_1$ and $\mathcal{A}_2$,*

$$\mu^{\mathrm{o}}(\mathcal{A}_1 \cup \mathcal{A}_2) + \mu^{\mathrm{o}}(\mathcal{A}_1 \cap \mathcal{A}_2) = \mu^{\mathrm{o}}(\mathcal{A}_1) + \mu^{\mathrm{o}}(\mathcal{A}_2). \tag{4.101}$$

Proof The proof is very similar to that of Lemma 4.9.2, but here we use sets in $\mathcal{M}_{\mathrm{c}}$ to approximate those in $\mathcal{M}$. For any $\varepsilon > 0$, let $\mathcal{B}_1$ and $\mathcal{B}_2$ be covers of $\mathcal{A}_1$ and $\mathcal{A}_2$, respectively, such that $\mu(\mathcal{B}_\ell) \le \mu^{\mathrm{o}}(\mathcal{A}_\ell) + \varepsilon$ for $\ell = 1, 2$. Let $\mathcal{D}_\ell = \mathcal{B}_\ell \cap \overline{\mathcal{A}}_\ell$ for $\ell = 1, 2$. Note that $\mathcal{A}_\ell$ and $\mathcal{D}_\ell$ are disjoint and $\mathcal{B}_\ell = \mathcal{A}_\ell \cup \mathcal{D}_\ell$:

$$\mathcal{B}_1 \cap \mathcal{B}_2 = (\mathcal{A}_1 \cup \mathcal{D}_1) \cap (\mathcal{A}_2 \cup \mathcal{D}_2) = (\mathcal{A}_1 \cap \mathcal{A}_2) \cup (\mathcal{D}_1 \cap \mathcal{A}_2) \cup (\mathcal{B}_1 \cap \mathcal{D}_2).$$

Using the union bound and subset inequality for outer measure on this and the corresponding expansion of $\mathcal{B}_1 \cup \mathcal{B}_2$, we obtain

$$\mu(\mathcal{B}_1 \cap \mathcal{B}_2) \le \mu^{\mathrm{o}}(\mathcal{A}_1 \cap \mathcal{A}_2) + \mu^{\mathrm{o}}(\mathcal{D}_1) + \mu^{\mathrm{o}}(\mathcal{D}_2) \le \mu^{\mathrm{o}}(\mathcal{A}_1 \cap \mathcal{A}_2) + 2\varepsilon,$$
$$\mu(\mathcal{B}_1 \cup \mathcal{B}_2) \le \mu^{\mathrm{o}}(\mathcal{A}_1 \cup \mathcal{A}_2) + \mu^{\mathrm{o}}(\mathcal{D}_1) + \mu^{\mathrm{o}}(\mathcal{D}_2) \le \mu^{\mathrm{o}}(\mathcal{A}_1 \cup \mathcal{A}_2) + 2\varepsilon,$$

where we have also used the fact (see Exercise 4.9) that $\mu^{\mathrm{o}}(\mathcal{D}_\ell) \le \varepsilon$ for $\ell = 1, 2$. Summing these inequalities and rearranging terms, we obtain

$$\begin{aligned} \mu^{\mathrm{o}}(\mathcal{A}_1 \cup \mathcal{A}_2) + \mu^{\mathrm{o}}(\mathcal{A}_1 \cap \mathcal{A}_2) &\ge \mu(\mathcal{B}_1 \cap \mathcal{B}_2) + \mu(\mathcal{B}_1 \cup \mathcal{B}_2) - 4\varepsilon \\ &= \mu(\mathcal{B}_1) + \mu(\mathcal{B}_2) - 4\varepsilon \\ &\ge \mu^{\mathrm{o}}(\mathcal{A}_1) + \mu^{\mathrm{o}}(\mathcal{A}_2) - 4\varepsilon, \end{aligned}$$

where we have used (4.99) and then used $\mathcal{A}_\ell \subseteq \mathcal{B}_\ell$ for $\ell = 1, 2$. Using the subset inequality and (4.99) to bound in the opposite direction,

$$\mu(\mathcal{B}_1) + \mu(\mathcal{B}_2) = \mu(\mathcal{B}_1 \cup \mathcal{B}_2) + \mu(\mathcal{B}_1 \cap \mathcal{B}_2) \ge \mu^{\mathrm{o}}(\mathcal{A}_1 \cup \mathcal{A}_2) + \mu^{\mathrm{o}}(\mathcal{A}_1 \cap \mathcal{A}_2).$$

Rearranging and using $\mu(\mathcal{B}_\ell) \le \mu^{\mathrm{o}}(\mathcal{A}_\ell) + \varepsilon$, we obtain

$$\mu^{\mathrm{o}}(\mathcal{A}_1 \cup \mathcal{A}_2) + \mu^{\mathrm{o}}(\mathcal{A}_1 \cap \mathcal{A}_2) \le \mu^{\mathrm{o}}(\mathcal{A}_1) + \mu^{\mathrm{o}}(\mathcal{A}_2) + 2\varepsilon.$$

Since ε is arbitrary, these bounds establish (4.101). □

Theorem 4.9.2 *Assume* $\mathcal{A}_1, \mathcal{A}_2 \in \mathcal{M}$. *Then* $\mathcal{A}_1 \cup \mathcal{A}_2 \in \mathcal{M}$ *and* $\mathcal{A}_1 \cap \mathcal{A}_2 \in \mathcal{M}$.

Proof Apply (4.101) to $\overline{\mathcal{A}_1}$ and $\overline{\mathcal{A}_2}$, to obtain

$$\mu^{\circ}(\overline{\mathcal{A}_1} \cup \overline{\mathcal{A}_2}) + \mu^{\circ}(\overline{\mathcal{A}_1} \cap \overline{\mathcal{A}_2}) = \mu^{\circ}(\overline{\mathcal{A}_1}) + \mu^{\circ}(\overline{\mathcal{A}_2}).$$

Rewriting $\overline{\mathcal{A}_1} \cup \overline{\mathcal{A}_2}$ as $\overline{\mathcal{A}_1 \cap \mathcal{A}_2}$ and $\overline{\mathcal{A}_1} \cap \overline{\mathcal{A}_2}$ as $\overline{\mathcal{A}_1 \cup \mathcal{A}_2}$ and adding this to (4.101) yields

$$\begin{aligned}\left[\mu^{\circ}(\mathcal{A}_1 \cup \mathcal{A}_2) + \mu^{\circ}(\overline{\mathcal{A}_1 \cup \mathcal{A}_2})\right] + \left[\mu^{\circ}(\mathcal{A}_1 \cap \mathcal{A}_2) + \mu^{\circ}(\overline{\mathcal{A}_1 \cap \mathcal{A}_2})\right] \\ = \mu^{\circ}(\mathcal{A}_1) + \mu^{\circ}(\mathcal{A}_2) + \mu^{\circ}(\overline{\mathcal{A}_1}) + \mu^{\circ}(\overline{\mathcal{A}_2}) = 2T, \end{aligned} \tag{4.102}$$

where we have used (4.100). Each of the bracketed terms above is at least T from (4.93), so each term must be exactly T. Thus $\mathcal{A}_1 \cup \mathcal{A}_2$ and $\mathcal{A}_1 \cap \mathcal{A}_2$ are measurable. □

Since $\mathcal{A}_1 \cup \mathcal{A}_2$ and $\mathcal{A}_1 \cap \mathcal{A}_2$ are measurable if $\mathcal{A}_1$ and $\mathcal{A}_2$ are, the joint union and intersection property holds for measure as well as outer measure for all measurable functions, i.e.

$$\mu(\mathcal{A}_1 \cup \mathcal{A}_2) + \mu(\mathcal{A}_1 \cap \mathcal{A}_2) = \mu(\mathcal{A}_1) + \mu(\mathcal{A}_2). \tag{4.103}$$

If $\mathcal{A}_1$ and $\mathcal{A}_2$ are disjoint, then (4.103) simplifies to the additivity property:

$$\mu(\mathcal{A}_1 \cup \mathcal{A}_2) = \mu(\mathcal{A}_1) + \mu(\mathcal{A}_2). \tag{4.104}$$

Actually, (4.103) shows that (4.104) holds whenever $\mu(\mathcal{A}_1 \cap \mathrm{A}_2) = 0$. That is, $\mathcal{A}_1$ and $\mathcal{A}_2$ need not be disjoint, but need only have an intersection of zero measure. This is another example in which sets of zero measure can be ignored.

The following theorem shows that $\mathcal{M}$ is closed over disjoint countable unions and that $\mathcal{M}$ is countably additive.

Theorem 4.9.3 *Assume that* $\mathcal{A}_j \in \mathcal{M}$ *for each integer* $j \geq 1$ *and that* $\mu(\mathcal{A}_j \cap \mathcal{A}_\ell) = 0$ *for all* $j \neq \ell$. *Let* $\mathcal{A} = \bigcup_j \mathcal{A}_j$. *Then* $\mathcal{A} \in \mathcal{M}$ *and*

$$\mu(\mathcal{A}) = \sum_j \mu(\mathcal{A}_j). \tag{4.105}$$

Proof Let $\mathcal{A}^k = \bigcup_{j=1}^k \mathcal{A}_j$ for each integer $k \geq 1$. Then $\mathcal{A}^{k+1} = \mathcal{A}^k \cup \mathcal{A}_{k+1}$ and, by induction on Theorem 4.9.2, $\mathcal{A}^k \in \mathcal{M}$ for all $k \geq 1$. It also follows that

$$\mu(\mathcal{A}^k) = \sum_{j=1}^{k} \mu(\mathcal{A}_j).$$

The sum on the right is nondecreasing in k and bounded by T, so the limit as $k \to \infty$ exists. Applying the union bound for outer measure to $\mathcal{A}$,

$$\mu^{\circ}(\mathcal{A}) \leq \sum_j \mu^{\circ}(\mathcal{A}_j) = \lim_{k\to\infty} \mu^{\circ}(\mathcal{A}^k) = \lim_{k\to\infty} \mu(\mathcal{A}^k). \tag{4.106}$$

Since $\mathcal{A}^k \subseteq \mathcal{A}$, we see that $\overline{\mathcal{A}} \subseteq \overline{\mathcal{A}^k}$ and $\mu^o(\overline{\mathcal{A}}) \le \mu(\overline{\mathcal{A}^k}) = T - \mu(\mathcal{A}^k)$. Thus

$$\mu^o(\overline{\mathcal{A}}) \le T - \lim_{k\to\infty} \mu(\mathcal{A}^k). \tag{4.107}$$

Adding (4.106) and (4.107) shows that $\mu^o(\mathcal{A}) + \mu^o(\overline{\mathcal{A}}) \le T$. Combining with (4.93), $\mu^o(\mathcal{A}) + \mu^o(\overline{\mathcal{A}}) = T$ and (4.106) and (4.107) are satisfied with equality. Thus $\mathcal{A} \in \mathcal{M}$ and countable additivity, (4.105), is satisfied. □

Next it is shown that $\mathcal{M}$ is closed under arbitrary countable unions and intersections.

Theorem 4.9.4 *Assume that $\mathcal{A}_j \in \mathcal{M}$ for each integer $j \ge 1$. Then $\mathcal{A} = \bigcup_j \mathcal{A}_j$ and $\mathcal{D} = \bigcap_j \mathcal{A}_j$ are both in $\mathcal{M}$.*

Proof Let $\mathcal{A}'_1 = \mathcal{A}_1$ and, for each $k \ge 1$, let $\mathcal{A}^k = \bigcup_{j=1}^k \mathcal{A}_j$ and let $\mathcal{A}'_{k+1} = \mathcal{A}_{k+1} \cap \overline{\mathcal{A}^k}$. By induction, the sets $\mathcal{A}'_1, \mathcal{A}'_2, \ldots$ are disjoint and measurable and $\mathcal{A} = \bigcup_j \mathcal{A}'_j$. Thus, from Theorem 4.9.3, $\mathcal{A}$ is measurable. Next suppose $\mathcal{D} = \cap \mathcal{A}_j$. Then $\overline{\mathcal{D}} = \cup \overline{\mathcal{A}_j}$. Thus, $\overline{\mathcal{D}} \in \mathcal{M}$, so $\mathcal{D} \in \mathcal{M}$ also. □

Proof of Theorem 4.3.1 The first two parts of Theorem 4.3.1 are Theorems 4.9.4 and 4.9.3. The third part, that $\mathcal{A}$ is measurable with zero measure if $\mu^o(\mathcal{A}) = 0$, follows from $T \le \mu^o(\mathcal{A}) + \mu^o(\overline{\mathcal{A}}) = \mu^o(\overline{\mathcal{A}})$ and $\mu^o(\overline{\mathcal{A}}) \le T$, i.e. that $\mu^o(\overline{\mathcal{A}}) = T$. □

Sets of zero measure are quite important in understanding Lebesgue integration, so it is important to know whether there are also uncountable sets of points that have zero measure. The answer is yes; a simple example follows.

Example 4.9.6 (The Cantor set) Express each point in the interval (0,1) by a ternary expansion. Let $\mathcal{B}$ be the set of points in (0,1) for which that expansion contains only 0s and 2s and is also nonterminating. Thus $\mathcal{B}$ excludes the interval $[1/3, 2/3)$, since all these expansions start with 1. Similarly, $\mathcal{B}$ excludes $[1/9, 2/9)$ and $[7/9, 8/9)$, since the second digit is 1 in these expansions. The right endpoint for each of these intervals is also excluded since it has a terminating expansion. Let $\mathcal{B}_n$ be the set of points with no 1 in the first n digits of the ternary expansion. Then $\mu(\mathcal{B}_n) = (2/3)^n$. Since $\mathcal{B}$ is contained in $\mathcal{B}_n$ for each $n \ge 1$, $\mathcal{B}$ is measurable and $\mu(\mathcal{B}) = 0$.

The expansion for each point in $\mathcal{B}$ is a binary sequence (viewing 0 and 2 as the binary digits here). There are uncountably many binary sequences (see Section 4.9.1), and this remains true when the countable number of terminating sequences are removed. Thus we have demonstrated an uncountably infinite set of numbers with zero measure.

Not all point sets are Lebesgue measurable, and an example follows.

Example 4.9.7 (A non-measurable set) Consider the interval [0, 1). We define a collection of equivalence classes where two points in [0, 1) are in the same equivalence class if the difference between them is rational. Thus one equivalence class consists of the rationals in [0,1). Each other equivalence class consists of a countably infinite set of irrationals whose differences are rational. This partitions [0, 1) into an uncountably infinite set of equivalence classes. Now consider a set $\mathcal{A}$ that contains exactly one number chosen from each equivalence class. We will assume that $\mathcal{A}$ is measurable and show that this leads to a contradiction.

For the given set $\mathcal{A}$, let $\mathcal{A}+r$, for r rational in $(0, 1)$, denote the set that results from mapping each $t \in \mathcal{A}$ into either $t+r$ or $t+r-1$, whichever lies in $[0, 1)$. The set $\mathcal{A}+r$ is thus the set $\mathcal{A}$, shifted by r, and then rotated to lie in $[0, 1)$. By looking at outer measures, it is easy to see that $\mathcal{A}+r$ is measurable if $\mathcal{A}$ is and that both then have the same measure. Finally, each $t \in [0, 1)$ lies in exactly one equivalence class, and if τ is the element of $\mathcal{A}$ in that equivalence class, then t lies in $\mathcal{A}+r$, where $r = t - \tau$ or $t - \tau + 1$. In other words, $[0, 1) = \bigcup_r (\mathcal{A}+r)$ and the sets $\mathcal{A}+r$ are disjoint. Assuming that $\mathcal{A}$ is measurable, Theorem 4.9.3 asserts that $1 = \sum_r \mu(\mathcal{A}+r)$. However, the sum on the right is 0 if $\mu(\mathcal{A}) = 0$ and infinite if $\mu(\mathcal{A}) > 0$, establishing the contradiction.

4.10 Exercises

4.1 (Fourier series)

(a) Consider the function $u(t) = \text{rect}(2t)$ of Figure 4.2. Give a general expression for the Fourier series coefficients for the Fourier series over $[-1/2, 1/2]$ and show that the series converges to $1/2$ at each of the endpoints, $-1/4$ and $1/4$. [Hint. You do not need to know anything about convergence here.]

(b) Represent the same function as a Fourier series over the interval $[-1/4, 1/4]$. What does this series converge to at $-1/4$ and $1/4$? Note from this exercise that the Fourier series depends on the interval over which it is taken.

4.2 (Energy equation) Derive (4.6), the energy equation for Fourier series. [Hint. Substitute the Fourier series for $u(t)$ into $\int u(t)u^*(t)\,dt$. Don't worry about convergence or interchange of limits here.]

4.3 (Countability) As shown in Appendix 4.9.1, many subsets of the real numbers, including the integers and the rationals, are countable. Sometimes, however, it is necessary to give up the ordinary numerical ordering in listing the elements of these subsets. This exercise shows that this is sometimes inevitable.

(a) Show that every listing of the integers (such as $0, -1, 1, -2, \ldots$) fails to preserve the numerical ordering of the integers. [Hint. Assume such a numerically ordered listing exists and show that it can have no first element (i.e., no smallest element).]

(b) Show that the rational numbers in the interval $(0, 1)$ cannot be listed in a way that preserves their numerical ordering.

(c) Show that the rationals in $[0,1]$ cannot be listed with a preservation of numerical ordering. (The first element is no problem, but what about the second?)

4.4 (Countable sums) Let $a_1, a_2, \ldots$ be a countable set of nonnegative numbers and assume that $s_a(k) = \sum_{j=1}^{k} a_j \leq A$ for all k and some given $A > 0$.

(a) Show that the limit $\lim_{k\to\infty} s_a(k)$ exists with some value S_a between 0 and A. (Use any level of mathematical care that you feel comfortable with.)

(b) Now let $b_1, b_2, \ldots$ be another ordering of the numbers $a_1, a_2, \ldots$ That is, let $b_1 = a_{j(1)}, b_2 = a_{j(2)}, \ldots, b_\ell = a_{j(\ell)}, \ldots,$ where $j(\ell)$ is a permutation of the positive integers, i.e. a one-to-one function from $\mathbb{Z}^+$ to $\mathbb{Z}^+$. Let $s_b(k) = \sum_{\ell=1}^{k} b_\ell$. Show that $\lim_{k\to\infty} s_b(k) \le S_a$. Note that

$$\sum_{\ell=1}^{k} b_\ell = \sum_{\ell=1}^{k} a_{j(\ell)}.$$

(c) Define $S_b = \lim_{k\to\infty} s_b(k)$ and show that $S_b \ge S_a$. [Hint. Consider the inverse permuation, say $\ell^{-1}(j)$, which for given j' is that ℓ for which $j(\ell) = j'$.] Note that you have shown that a countable sum of nonnegative elements does not depend on the order of summation.

(d) Show that the above result is not necessarily true for a countable sum of numbers that can be positive or negative. [Hint. Consider alternating series.]

4.5 (Finite unions of intervals) Let $\mathcal{E} = \bigcup_{j=1}^{\ell} I_j$ be the union of $\ell \ge 2$ arbitrary nonempty intervals. Let a_j and b_j denote the left and right endpoints, respectively, of I_j; each endpoint can be included or not. Assume the intervals are ordered so that $a_1 \le a_2 \le \cdots \le a_\ell$.

(a) For $\ell = 2$, show that either I_1 and I_2 are separated or that $\mathcal{E}$ is a single interval whose left endpoint is a_1.

(b) For $\ell > 2$ and $2 \le k < \ell$, let $\mathcal{E}^k = \bigcup_{j=1}^{k} I_j$. Give an algorithm for constructing a union of separated intervals for $\mathcal{E}^{k+1}$ given a union of separated intervals for $\mathcal{E}^k$.

(c) Note that using part (b) inductively yields a representation of $\mathcal{E}$ as a union of separated intervals. Show that the left endpoint for each separated interval is drawn from $a_1, \ldots, a_\ell$ and the right endpoint is drawn from $b_1, \ldots, b_\ell$.

(d) Show that this representation is unique, i.e. that $\mathcal{E}$ cannot be represented as the union of any other set of separated intervals. Note that this means that $\mu(\mathcal{E})$ is defined unambiguously in (4.9).

4.6 (Countable unions of intervals) Let $\mathcal{B} = \bigcup_j I_j$ be a countable union of arbitrary (perhaps intersecting) intervals. For each $k \ge 1$, let $\mathcal{B}^k = \bigcup_{j=1}^{k} I_j$, and for each $k \ge j$ let $I_{j,k}$ be the separated interval in $\mathcal{B}^k$ containing I_j (see Exercise 4.5).

(a) For each $k \ge j \ge 1$, show that $I_{j,k} \subseteq I_{j,k+1}$.

(b) Let $\bigcup_{k=j}^{\infty} I_{j,k} = I'_j$. Explain why I'_j is an interval and show that $I'_j \subseteq \mathcal{B}$.

(c) For any i, j, show that either $I'_j = I'_i$ or I'_j and I'_i are separated intervals.

(d) Show that the sequence $\{I'_j;\ 1 \le j < \infty\}$ with repetitions removed is a countable separated-interval representation of $\mathcal{B}$.

(e) Show that the collection $\{I'_j;\ j \ge 1\}$ with repetitions removed is unique; i.e., show that if an arbitrary interval I is contained in $\mathcal{B}$, then it is contained in one of the I'_j. Note, however, that the ordering of the I'_j is not unique.

4.7 (Union bound for intervals) Prove the validity of the union bound for a countable collection of intervals in (4.89). The following steps are suggested.

(a) Show that if $\mathcal{B} = I_1 \bigcup I_2$ for arbitrary intervals I_1, I_2, then $\mu(\mathcal{B}) \leq \mu(I_1) + \mu(I_2)$ with equality if I_1 and I_2 are disjoint. Note: this is true by definition if I_1 and I_2 are separated, so you need only treat the cases where I_1 and I_2 intersect or are disjoint but not separated.

(b) Let $\mathcal{B}^k = \bigcup_{j=1}^{k} I_j$ be represented as the union of, say, m_k separated intervals ($m_k \leq k$), so $\mathcal{B}^k = \bigcup_{j=1}^{m_k} I_j$. Show that $\mu(\mathcal{B}^k \bigcup I_{k+1}) \leq \mu(\mathcal{B}^k) + \mu(I_{k+1})$ with equality if $\mathcal{B}^k$ and I_{k+1} are disjoint.

(c) Use finite induction to show that if $\mathcal{B} = \bigcup_{j=1}^{k} I_j$ is a finite union of arbitrary intervals, then $\mu(\mathcal{B}) \leq \sum_{j=1}^{k} \mu(I_j)$ with equality if the intervals are disjoint.

(d) Extend part (c) to a countably infinite union of intervals.

4.8 For each positive integer n, let $\mathcal{B}_n$ be a countable union of intervals. Show that $\mathcal{B} = \bigcup_{n=1}^{\infty} \mathcal{B}_n$ is also a countable union of intervals. [Hint. Look at Example 4.9.2 in Section 4.9.1.]

4.9 (Measure and covers) Let $\mathcal{A}$ be an arbitrary measurable set in $[-T/2, T/2]$ and let $\mathcal{B}$ be a cover of $\mathcal{A}$. Using only results derived prior to Lemma 4.9.3, show that $\mu^{\circ}(\mathcal{B} \cap \overline{\mathcal{A}}) = \mu(\mathcal{B}) - \mu(\mathcal{A})$. You may use the following steps if you wish.

(a) Show that $\mu^{\circ}(\mathcal{B} \cap \overline{\mathcal{A}}) \geq \mu(\mathcal{B}) - \mu(\mathcal{A})$.

(b) For any $\delta > 0$, let $\mathcal{B}'$ be a cover of $\overline{\mathcal{A}}$ with $\mu(\mathcal{B}') \leq \mu(\overline{\mathcal{A}}) + \delta$. Use Lemma 4.9.2 to show that $\mu(\mathcal{B} \cap \mathcal{B}') = \mu(\mathcal{B}) + \mu(\mathcal{B}') - T$.

(c) Show that $\mu^{\circ}(\mathcal{B} \cap \overline{\mathcal{A}}) \leq \mu(\mathcal{B} \cap \mathcal{B}') \leq \mu(\mathcal{B}) - \mu(\mathcal{A}) + \delta$.

(d) Show that $\mu^{\circ}(\mathcal{B} \cap \overline{\mathcal{A}}) = \mu(\mathcal{B}) - \mu(\mathcal{A})$.

4.10 (Intersection of covers) Let $\mathcal{A}$ be an arbitrary set in $[-T/2, T/2]$.

(a) Show that $\mathcal{A}$ has a sequence of covers, $\mathcal{B}_1, \mathcal{B}_2, \ldots$, such that $\mu^{\circ}(\mathcal{A}) = \mu(\mathcal{D})$, where $\mathcal{D} = \bigcap_n \mathcal{B}_n$.

(b) Show that $\mathcal{A} \subseteq \mathcal{D}$.

(c) Show that if $\mathcal{A}$ is measurable, then $\mu(\mathcal{D} \cap \overline{\mathcal{A}}) = 0$. Note that you have shown that an arbitrary measurable set can be represented as a countable intersection of countable unions of intervals, less a set of zero measure. Argue by example that if $\mathcal{A}$ is not measurable, then $\mu^{\circ}(\mathcal{D} \cap \overline{\mathrm{A}})$ need not be 0.

4.11 (Measurable functions)

(a) For $\{u(t) : [-T/2, T/2] \to R\}$, show that if $\{t : u(t) < \beta\}$ is measurable, then $\{t : u(t) \geq \beta\}$ is measurable.

(b) Show that if $\{t : u(t) < \beta\}$ and $\{t : u(t) < \alpha\}$ are measurable, $\alpha < \beta$, then $\{t : \alpha \leq u(t) < \beta\}$ is measurable.

(c) Show that if $\{t : u(t) < \beta\}$ is measurable for all β, then $\{t : u(t) \leq \beta\}$ is also measurable. [Hint. Express $\{t : u(t) \leq \beta\}$ as a countable intersection of measurable sets.]

(d) Show that if $\{t : u(t) \leq \beta\}$ is measurable for all β, then $\{t : u(t) < \beta\}$ is also measurable, i.e. the definition of measurable function can use either strict or nonstrict inequality.

4.12 (Measurable functions) Assume throughout that $\{u(t) : [-T/2, T/2] \to \mathbb{R}\}$ is measurable.

(a) Show that $-u(t)$ and $|u(t)|$ are measurable.
(b) Assume that $\{g(x) : \mathbb{R} \to \mathbb{R}\}$ is an increasing function (i.e. $x_1 < x_2 \implies g(x_1) < g(x_2)$). Prove that $v(t) = g(u(t))$ is measurable. [Hint. This is a one liner. If the abstraction confuses you, first show that $\exp(u(t))$ is measurable and then prove the more general result.]
(c) Show that $\exp[u(t)]$, $u^2(t)$, and $\ln|u(t)|$ are all measurable.

4.13 (Measurable functions)

(a) Show that if $\{u(t) : [-T/2, T/2] \to \mathbb{R}\}$ and $\{v(t) : [-T/2, T/2] \to \mathbb{R}\}$ are measurable, then $u(t) + v(t)$ is also measurable. [Hint. Use a discrete approximation to the sum and then go to the limit.]
(b) Show that $u(t)v(t)$ is also measurable.

4.14 (Measurable sets) Suppose $\mathcal{A}$ is a subset of $[-T/2, T/2]$ and is measurable over $[-T/2, T/2]$. Show that $\mathcal{A}$ is also measurable, with the same measure, over $[-T'/2, T'/2]$ for any T' satisfying $T' > T$. [Hint. Let $\mu'(\mathcal{A})$ be the outer measure of $\mathcal{A}$ over $[-T'/2, T'/2]$ and show that $\mu'(\mathcal{A}) = \mu^{o}(\mathcal{A})$, where μ^{o} is the outer measure over $[-T/2, T/2]$. Then let $\overline{\mathcal{A}}'$ be the complement of $\mathcal{A}$ over $[-T'/2, T'/2]$ and show that $\mu'(\overline{\mathcal{A}}') = \mu^{o}(\overline{\mathcal{A}}) + T' - T$.]

4.15 (Measurable limits)

(a) Assume that $\{u_n(t) : [-T/2, T/2] \to \mathbb{R}\}$ is measurable for each $n \geq 1$. Show that $\liminf_n u_n(t)$ is measurable ($\liminf_n u_n(t)$ means $\lim_m v_m(t)$, where $v_m(t) = \inf_{n=m}^{\infty} u_n(t)$ and infinite values are allowed).
(b) Show that $\lim_n u_n(t)$ exists for a given t if and only if $\liminf_n u_n(t) = \limsup_n u_n(t)$.
(c) Show that the set of t for which $\lim_n u_n(t)$ exists is measurable. Show that a function $u(t)$ that is $\lim_n u_n(t)$ when the limit exists and is 0 otherwise is measurable.

4.16 (Lebesgue integration) For each integer $n \geq 1$, define $u_n(t) = 2^n \,\mathrm{rect}(2^n t - 1)$. Sketch the first few of these waveforms. Show that $\lim_{n\to\infty} u_n(t) = 0$ for all t. Show that $\int \lim_n u_n(t)\,dt \neq \lim_n \int u_n(t)dt$.

4.17 ($\mathcal{L}_1$ integrals)

(a) Assume that $\{u(t) : [-T/2, T/2] \to \mathbb{R}\}$ is $\mathcal{L}_1$. Show that

$$\left|\int u(t)\,dt\right| = \left|\int u^{+}(t)dt - \int u^{-}(t)dt\right| \leq \int |u(t)|\,dt.$$

(b) Assume that $\{u(t) : [-T/2, T/2] \to \mathbb{C}\}$ is $\mathcal{L}_1$. Show that

$$\left|\int u(t)dt\right| \leq \int |u(t)|\,dt.$$

[Hint. Choose α such that $\alpha \int u(t)\,dt$ is real and nonnegative and $|\alpha| = 1$. Use part (a) on $\alpha u(t)$.]

4.18 ($\mathcal{L}_2$-equivalence) Assume that $\{u(t) : [-T/2, T/2] \to \mathbb{C}\}$ and $\{v(t) : [-T/2, T/2] \to \mathbb{C}\}$ are $\mathcal{L}_2$ functions.

(a) Show that if $u(t)$ and $v(t)$ are equal a.e., then they are $\mathcal{L}_2$-equivalent.
(b) Show that if $u(t)$ and $v(t)$ are $\mathcal{L}_2$-equivalent, then for any $\varepsilon > 0$ the set $\{t : |u(t) - v(t)|^2 \geq \varepsilon\}$ has zero measure.
(c) Using (b), show that $\mu\{t : |u(t) - v(t)| > 0\} = 0$, i.e. that $u(t) = v(t)$ a.e.

4.19 (Orthogonal expansions) Assume that $\{u(t) : \mathbb{R} \to \mathbb{C}\}$ is $\mathcal{L}_2$. Let $\{\theta_k(t);\ 1 \leq k < \infty\}$ be a set of orthogonal waveforms and assume that $u(t)$ has the following orthogonal expansion:

$$u(t) = \sum_{k=1}^{\infty} u_k \theta_k(t).$$

Assume the set of orthogonal waveforms satisfy

$$\int_{-\infty}^{\infty} \theta_k(t)\theta_j^*(t)\mathrm{d}t = \begin{cases} 0 & \text{for} \quad k \neq j; \\ A_j & \text{for} \quad k = j, \end{cases}$$

where $\{A_j;\ j \in \mathbb{Z}^+\}$ is an arbitrary set of positive numbers. Do not concern yourself with convergence issues in this exercise.

(a) Show that each u_k can be expressed in terms of $\int_{-\infty}^{\infty} u(t)\theta_k^*(t)\mathrm{d}t$ and A_k.
(b) Find the energy $\int_{-\infty}^{\infty} |u(t)|^2\, \mathrm{d}t$ in terms of $\{u_k\}$ and $\{A_k\}$.
(c) Suppose that $v(t) = \sum_k v_k \theta_k(t)$, where $v(t)$ also has finite energy. Express $\int_{-\infty}^{\infty} u(t)v^*(t)\,\mathrm{d}t$ as a function of $\{u_k, v_k, A_k;\ k \in \mathbb{Z}\}$.

4.20 (Fourier series)

(a) Verify that (4.22) and (4.23) follow from (4.20) and (4.18) using the transformation $u(t) = v(t + \Delta)$.
(b) Consider the Fourier series in periodic form, $w(t) = \sum_k \hat{w}_k \mathrm{e}^{2\pi i k t/T}$, where $\hat{w}_k = (1/T)\int_{-T/2}^{T/2} w(t)\mathrm{e}^{-2\pi i k t/T}\,\mathrm{d}t$. Show that for any real Δ, $(1/T)\int_{-T/2+\Delta}^{T/2+\Delta} w(t)\mathrm{e}^{-2\pi i k t/T}\,\mathrm{d}t$ is also equal to $\hat{w}_k$, providing an alternative derivation of (4.22) and (4.23).

4.21 Equation (4.27) claims that

$$\lim_{n\to\infty,\ell\to\infty} \int \left| u(t) - \sum_{m=-n}^{n} \sum_{k=-\ell}^{\ell} \hat{u}_{k,m}\theta_{k,m}(t) \right|^2 \mathrm{d}t = 0.$$

(a) Show that the integral above is nonincreasing in both ℓ and n.
(b) Show that the limit is independent of how n and ℓ approach ∞. [Hint. See Exercise 4.4.]
(c) More generally, show that the limit is the same if the pair (k, m), $k \in \mathbb{Z}$, $m \in \mathbb{Z}$, is ordered in an arbitrary way and the limit above is replaced by a limit on the partial sums according to that ordering.

4.22 (Truncated sinusoids)

(a) Verify (4.24) for $\mathcal{L}_2$ waveforms, i.e. show that

$$\lim_{n\to\infty} \int \left| u(t) - \sum_{m=-n}^{n} u_m(t) \right|^2 \mathrm{d}t = 0.$$

(b) Break the integral in (4.28) into separate integrals for $|t| > (n+1/2)T$ and $|t| \le (n+1/2)T$. Show that the first integral goes to 0 with increasing n.
(c) For given n, show that the second integral above goes to 0 with increasing ℓ.

4.23 (Convolution) The left side of (4.40) is a function of t. Express the Fourier transform of this as a double integral over t and τ. For each t, make the substitution $r = t - \tau$ and integrate over r. Then integrate over τ to get the right side of (4.40). Do not concern yourself with convergence issues here.

4.24 (Continuity of $\mathcal{L}_1$ transform) Assume that $\{u(t) : \mathbb{R} \to \mathbb{C}\}$ is $\mathcal{L}_1$ and let $\hat{u}(f)$ be its Fourier transform. Let ε be any given positive number.

(a) Show that for sufficiently large T, $\int_{|t|>T} |u(t)\mathrm{e}^{-2\pi \mathrm{i} ft} - u(t)\mathrm{e}^{-2\pi \mathrm{i}(f-\delta)t}|\,\mathrm{d}t < \varepsilon/2$ for all f and all $\delta > 0$.
(b) For the ε and T selected above, show that $\int_{|t|\le T} |u(t)\mathrm{e}^{-2\pi \mathrm{i} ft} - u(t)\mathrm{e}^{-2\pi \mathrm{i}(f-\delta)t}|\,\mathrm{d}t < \varepsilon/2$ for all f and sufficiently small $\delta > 0$. This shows that $\hat{u}(f)$ is continuous.

4.25 (Plancherel) The purpose of this exercise is to get some understanding of the Plancherel theorem. Assume that $u(t)$ is $\mathcal{L}_2$ and has a Fourier transform $\hat{u}(f)$.

(a) Show that $\hat{u}(f) - \hat{u}_A(f)$ is the Fourier transform of the function $x_A(t)$ that is 0 from $-A$ to A and equal to $u(t)$ elsewhere.
(b) Argue that since $\int_{-\infty}^{\infty} |u(t)|^2\,\mathrm{d}t$ is finite, the integral $\int_{-\infty}^{\infty} |x_A(t)|^2\,\mathrm{d}t$ must go to 0 as $A \to \infty$. Use whatever level of mathematical care and common sense that you feel comfortable with.
(c) Using the energy equation (4.45), argue that

$$\lim_{A\to\infty} \int_{-\infty}^{\infty} |\hat{u}(f) - \hat{u}_A(f)|^2\,\mathrm{d}t = 0.$$

Note: this is only the easy part of the Plancherel theorem. The difficult part is to show the existence of $\hat{u}(f)$. The limit as $A \to \infty$ of the integral $\int_{-A}^{A} u(t)\mathrm{e}^{-2\pi \mathrm{i} ft}\,\mathrm{d}t$ need not exist for all f, and the point of the Plancherel theorem is to forget about this limit for individual f and focus instead on the energy in the difference between the hypothesized $\hat{u}(f)$ and the approximations.

4.26 (Fourier transform for $\mathcal{L}_2$) Assume that $\{u(t) : \mathbb{R} \to \mathbb{C}\}$ and $\{v(t) : \mathbb{R} \to \mathbb{C}\}$ are $\mathcal{L}_2$ and that a and b are complex numbers. Show that $au(t) + bv(t)$ is $\mathcal{L}_2$. For $T > 0$, show that $u(t-T)$ and $u(t/T)$ are $\mathcal{L}_2$ functions.

4.27 (Relation of Fourier series to Fourier integral) Assume that $\{u(t) : [-T/2, T/2] \to \mathbb{C}\}$ is $\mathcal{L}_2$. Without being very careful about the mathematics, the Fourier series expansion of $\{u(t)\}$ is given by

$$u(t) = \lim_{\ell\to\infty} u^{(\ell)}(t), \quad \text{where} \quad u^{(\ell)}(t) = \sum_{k=-\ell}^{\ell} \hat{u}_k e^{2\pi ikt/T} \operatorname{rect}\left(\frac{t}{T}\right);$$

$$\hat{u}_k = \frac{1}{T}\int_{-T/2}^{T/2} u(t)e^{-2\pi ikt/T}\, dt.$$

(a) Does the above limit hold for all $t \in [-T/2, T/2]$? If not, what can you say about the type of convergence?

(b) Does the Fourier transform $\hat{u}(f) = \int_{-T/2}^{T/2} u(t)e^{-2\pi ift}\, dt$ exist for all f? Explain.

(c) The Fourier transform of the finite sum $u^{(\ell)}(t)$ is $\hat{u}^{(\ell)}(f) = \sum_{k=-\ell}^{\ell} \hat{u}_k T \operatorname{sinc}(fT - k)$. In the limit $\ell \to \infty$, $\hat{u}(f) = \lim_{\ell\to\infty} \hat{u}^{(\ell)}(f)$, so

$$\hat{u}(f) = \lim_{\ell\to\infty} \sum_{k=-\ell}^{\ell} \hat{u}_k T \operatorname{sinc}(fT - k).$$

Give a brief explanation why this equation must hold with equality for all $f \in \mathbb{R}$. Also show that $\{\hat{u}(f) : f \in \mathbb{R}\}$ is completely specified by its values, $\{\hat{u}(k/T) : k \in \mathbb{Z}\}$ at multiples of $1/T$.

4.28 (Sampling) One often approximates the value of an integral by a discrete sum, i.e.

$$\int_{-\infty}^{\infty} g(t)dt \approx \delta \sum_k g(k\delta).$$

(a) Show that if $u(t)$ is a real finite-energy function, lowpass-limited to W Hz, then the above approximation is exact for $g(t) = u^2(t)$ if $\delta \le 1/2\mathsf{W}$; i.e., show that

$$\int_{-\infty}^{\infty} u^2(t)dt = \delta \sum_k u^2(k\delta).$$

(b) Show that if $g(t)$ is a real finite-energy function, lowpass-limited to W Hz, then for $\delta \le 1/2\mathsf{W}$,

$$\int_{-\infty}^{\infty} g(t)dt = \delta \sum_k g(k\delta).$$

(c) Show that if $\delta > 1/2\mathsf{W}$, then there exists no such relation in general.

4.29 (Degrees of freedom) This exercise explores how much of the energy of a baseband-limited function $\{u(t) : [-1/2, 1/2] \to \mathbb{R}\}$ can reside outside the region where the sampling coefficients are nonzero. Let $T = 1/2\mathsf{W} = 1$, and let n be a positive even integer. Let $u_k = (-1)^k$ for $-n \le k \le n$ and $u_k = 0$ for $|k| > n$. Show that $|u(n + 1/2)|$ increases without bound as the endpoint n is increased. Show that $|u(n+m+1/2)| > |u(n-m-1/2)|$ for all integers m, $0 \le m < n$. In other words, shifting the sample points by 1/2 leads to most of the sample point energy being outside the interval $[-n, n]$.

4.30 (Sampling theorem for $[\Delta - \mathsf{W}, \Delta + \mathsf{W})]$)

(a) Verify the Fourier transform pair in (4.70). [Hint. Use the scaling and shifting rules on $\text{rect}(f) \leftrightarrow \text{sinc}(t)$.]

(b) Show that the functions making up that expansion are orthogonal. [Hint. Show that the corresponding Fourier transforms are orthogonal.]

(c) Show that the functions in (4.74) are orthogonal.

4.31 (Amplitude-limited functions) Sometimes it is important to generate baseband waveforms with bounded amplitude. This problem explores pulse shapes that can accomplish this

(a) Find the Fourier transform of $g(t) = \text{sinc}^2(\mathsf{W}t)$. Show that $g(t)$ is bandlimited to $f \leq \mathsf{W}$ and sketch both $g(t)$ and $\hat{g}(f)$. [Hint. Recall that multiplication in the time domain corresponds to convolution in the frequency domain.]

(b) Let $u(t)$ be a continuous real $\mathcal{L}_2$ function baseband-limited to $f \leq \mathsf{W}$ (i.e. a function such that $u(t) = \sum_k u(kT)\,\text{sinc}(t/T - k)$, where $T = 1/2\mathsf{W}$. Let $v(t) = u(t) * g(t)$. Express $v(t)$ in terms of the samples $\{u(kT); k \in \mathbb{Z}\}$ of $u(t)$ and the shifts $\{g(t - kT); k \in \mathbb{Z}\}$ of $g(t)$. [Hint. Use your sketches in part (a) to evaluate $g(t) * \text{sinc}(t/T)$.]

(c) Show that if the T-spaced samples of $u(t)$ are nonnegative, then $v(t) \geq 0$ for all t.

(d) Explain why $\sum_k \text{sinc}(t/T - k) = 1$ for all t.

(e) Using (d), show that $\sum_k g(t - kT) = c$ for all t and find the constant c. [Hint. Use the hint in (b) again.]

(f) Now assume that $u(t)$, as defined in part (b), also satisfies $u(kT) \leq 1$ for all $k \in \mathbb{Z}$. Show that $v(t) \leq 2$ for all t.

(g) Allow $u(t)$ to be complex now, with $|u(kT)| \leq 1$. Show that $|v(t)| \leq 2$ for all t.

4.32 (Orthogonal sets) The function $\text{rect}(t/T)$ has the very special property that it, plus its time and frequency shifts, by kT and j/T, respectively, form an orthogonal set. The function $\text{sinc}(t/T)$ has this same property. We explore other functions that are generalizations of $\text{rect}(t/T)$ and which, as you will show in parts (a)–(d), have this same interesting property. For simplicity, choose $T = 1$.

These functions take only the values 0 and 1 and are allowed to be nonzero only over $[-1, 1]$ rather than $[-1/2, 1/2]$ as with $\text{rect}(t)$. Explicitly, the functions considered here satisfy the following constraints:

$$p(t) = p^2(t) \qquad \text{for all } t \quad \text{(0/1 property)}; \tag{4.108}$$

$$p(t) = 0 \qquad \text{for } |t| > 1; \tag{4.109}$$

$$p(t) = p(-t) \qquad \text{for all } t \quad \text{(symmetry)}; \tag{4.110}$$

$$p(t) = 1 - p(t-1) \qquad \text{for } 0 \leq t < 1/2. \tag{4.111}$$

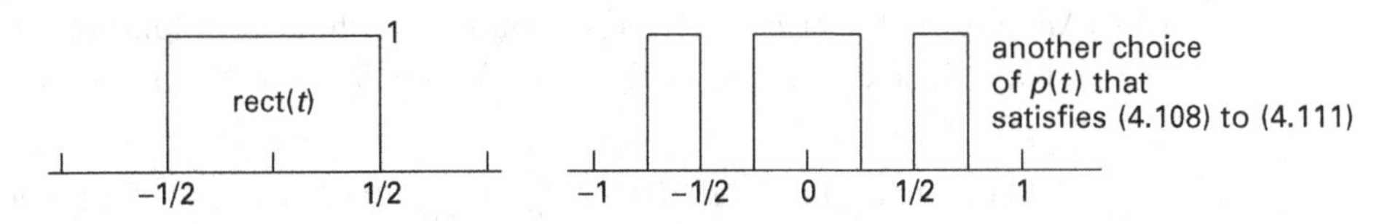

Figure 4.12. Two functions that each satisfy (4.108)–(4.111)

Note: because of property (4.110), condition (4.111) also holds for $1/2 < t \leq 1$. Note also that $p(t)$ at the single points $t = \pm 1/2$ does not affect any orthogonality properties, so you are free to ignore these points in your arguments. Figure 4.12 illustrates two examples of functions satisfying (4.108)–(4.111).

(a) Show that $p(t)$ is orthogonal to $p(t-1)$. [Hint. Evaluate $p(t)p(t-1)$ for each $t \in [0, 1]$ other than $t = 1/2$.]
(b) Show that $p(t)$ is orthogonal to $p(t-k)$ for all integer $k \neq 0$.
(c) Show that $p(t)$ is orthogonal to $p(t-k)e^{i2\pi mt}$ for integer $m \neq 0$ and $k \neq 0$.
(d) Show that $p(t)$ is orthogonal to $p(t)e^{2\pi imt}$ for integer $m \neq 0$. [Hint. Evaluate $p(t)e^{-2\pi imt} + p(t-1)e^{-2\pi im(t-1)}$.]
(e) Let $h(t) = \hat{p}(t)$ where $\hat{p}(f)$ is the Fourier transform of $p(t)$. If $p(t)$ satisfies properties (4.108)–(4.111), does it follow that $h(t)$ has the property that it is orthogonal to $h(t-k)e^{2\pi imt}$ whenever either the integer k or m is nonzero?

Note: almost no calculation is required in this problem.

4.33 (Limits) Construct an example of a sequence of $\mathcal{L}_2$ functions $v^{(m)}(t)$, $m \in \mathbb{Z}$, $m > 0$, such that $\lim_{m\to\infty} v^{(m)}(t) = 0$ for all t but for which $\text{l.i.m.}_{m\to\infty} v^{(m)}(t)$ does not exist. In other words show that pointwise convergence does not imply $\mathcal{L}_2$-convergence. [Hint. Consider time shifts.]

4.34 (Aliasing) Find an example where $\hat{u}(f)$ is 0 for $|f| > 3\mathrm{W}$ and nonzero for $\mathrm{W} < |f| < 3\mathrm{W}$, but where, $s(kT) = v_0(kT)$ for all $k \in \mathbb{Z}$. Here $v_0(kT)$ is defined in (4.77) and $T = 1/2\mathrm{W}$. [Hint. Note that it is equivalent to achieve equality between $\hat{s}(f)$ and $\hat{u}(f)$ for $|f| \leq \mathrm{W}$. Look at Figure 4.10.]

4.35 (Aliasing) The following exercise is designed to illustrate the sampling of an approximately baseband waveform. To avoid messy computation, we look at a waveform baseband-limited to 3/2 which is sampled at rate 1 (i.e. sampled at only 1/3 the rate that it should be sampled at). In particular, let $u(t) = \text{sinc}(3t)$.

(a) Sketch $\hat{u}(f)$. Sketch the function $\hat{v}_m(f) = \text{rect}(f - m)$ for each integer m such that $v_m(f) \neq 0$. Note that $\hat{u}(f) = \sum_m \hat{v}_m(f)$.
(b) Sketch the inverse transforms $v_m(t)$ (real and imaginary parts if complex).
(c) Verify directly from the equations that $u(t) = \sum v_m(t)$. [Hint. This is easier if you express the sine part of the sinc function as a sum of complex exponentials.]
(d) Verify the sinc-weighted sinusoid expansion, (4.73). (There are only three nonzero terms in the expansion.)
(e) For the approximation $s(t) = u(0)\,\text{sinc}(t)$, find the energy in the difference between $u(t)$ and $s(t)$ and interpret the terms.

4.36 (Aliasing) Let $u(t)$ be the inverse Fourier transform of a function $\hat{u}(f)$ which is both $\mathcal{L}_1$ and $\mathcal{L}_2$. Let $v_m(t) = \int \hat{u}(f)\, \text{rect}(fT - m) e^{2\pi i f t}\, df$ and let $v^{(n)}(t) = \sum_{-n}^{n} v_m(t)$.

(a) Show that $|u(t) - v^{(n)}(t)| \leq \int_{|f| \geq (2n+1)/T} |\hat{u}(f)|\, df$ and thus that $u(t) = \lim_{n\to\infty} v^{(n)}(t)$ for all t.

(b) Show that the sinc-weighted sinusoid expansion of (4.76) then converges pointwise for all t. [Hint. For any t and any $\varepsilon > 0$, choose n so that $|u(t) - v^n(t)| \leq \varepsilon/2$. Then for each m, $|m| \leq n$, expand $v_m(t)$ in a sampling expansion using enough terms to keep the error less than $\varepsilon/4n+2$.]

4.37 (Aliasing)

(a) Show that $\hat{s}(f)$ in (4.83) is $\mathcal{L}_1$ if $\hat{u}(f)$ is.

(b) Let $\hat{u}(f) = \sum_{k \neq 0} \text{rect}[k^2(f - k)]$. Show that $\hat{u}(f)$ is $\mathcal{L}_1$ and $\mathcal{L}_2$. Let $T = 1$ for $\hat{s}(f)$ and show that $\hat{s}(f)$ is not $\mathcal{L}_2$. [Hint. Sketch $\hat{u}(f)$ and $\hat{s}(f)$.]

(c) Show that $\hat{u}(f)$ does not satisfy $\lim_{|f|\to\infty} \hat{u}(f)|f|^{1+\varepsilon} = 0$.

4.38 (Aliasing) Let $u(t) = \sum_{k \neq 0} \text{rect}[k^2(t - k)]$ and show that $u(t)$ is $\mathcal{L}_2$. Find $s(t) = \sum_k u(k)\, \text{sinc}(t - k)$ and show that it is neither $\mathcal{L}_1$ nor $\mathcal{L}_2$. Find $\sum_k u^2(k)$ and explain why the sampling theorem energy equation (4.66) does not apply here.

5 Vector spaces and signal space

In Chapter 4, we showed that any $\mathcal{L}_2$ function $u(t)$ can be expanded in various orthogonal expansions, using such sets of orthogonal functions as the T-spaced truncated sinusoids or the sinc-weighted sinusoids. Thus $u(t)$ may be specified (up to $\mathcal{L}_2$-equivalence) by a countably infinite sequence such as $\{u_{k,m};\ -\infty < k, m < \infty\}$ of coefficients in such an expansion.

In engineering, n-tuples of numbers are often referred to as *vectors*, and the use of vector notation is very helpful in manipulating these n-tuples. The collection of n-tuples of real numbers is called $\mathbb{R}^n$ and that of complex numbers $\mathbb{C}^n$. It turns out that the most important properties of these n-tuples also apply to countably infinite sequences of real or complex numbers. It should not be surprising, after the results of the previous chapters, that these properties also apply to $\mathcal{L}_2$ waveforms.

A vector space is essentially a collection of objects (such as the collection of real n-tuples) along with a set of rules for manipulating those objects. There is a set of axioms describing precisely how these objects and rules work. Any properties that follow from those axioms must then apply to any vector space, i.e. any set of objects satisfying those axioms. These axioms are satisfied by $\mathbb{R}^n$ and $\mathbb{C}^n$, and we will soon see that they are also satisfied by the class of countable sequences and the class of $\mathcal{L}_2$ waveforms.

Fortunately, it is just as easy to develop the general properties of vector spaces from these axioms as it is to develop specific properties for the special case of $\mathbb{R}^n$ or $\mathbb{C}^n$ (although we will constantly use $\mathbb{R}^n$ and $\mathbb{C}^n$ as examples). Also, we can use the example of $\mathbb{R}^n$ (and particularly $\mathbb{R}^2$) to develop geometric insight about general vector spaces.

The collection of $\mathcal{L}_2$ functions, viewed as a vector space, will be called *signal space*. The signal-space viewpoint has been one of the foundations of modern digital communication theory since its popularization in the classic text of Wozencraft and Jacobs (1965).

The signal-space viewpoint has the following merits.

- Many insights about waveforms (signals) and signal sets do not depend on time and frequency (as does the development up until now), but depend only on vector relationships.
- Orthogonal expansions are best viewed in vector space terms.
- Questions of limits and approximation are often easily treated in vector space terms. It is for this reason that many of the results in Chapter 4 are proved here.

5.1 Axioms and basic properties of vector spaces

A *vector space* $\mathcal{V}$ is a set of elements $\boldsymbol{v} \in \mathcal{V}$, called *vectors*, along with a set of rules for operating on both these vectors and a set of ancillary elements $\alpha \in \mathbb{F}$, called *scalars*. For the treatment here, the set $\mathbb{F}$ of scalars will be either the real field $\mathbb{R}$ (which is the set of real numbers along with their familiar rules of addition and multiplication) or the complex field $\mathbb{C}$ (which is the set of complex numbers with their addition and multiplication rules).[1] A vector space with real scalars is called a *real vector space*, and one with complex scalars is called a *complex vector space*.

The most familiar example of a real vector space is $\mathbb{R}^n$. Here the vectors are n-tuples of real numbers.[2] Note that $\mathbb{R}^2$ is represented geometrically by a plane, and the vectors in $\mathbb{R}^2$ are represented by points in the plane. Similarly, $\mathbb{R}^3$ is represented geometrically by three-dimensional Euclidean space.

The most familiar example of a complex vector space is $\mathbb{C}^n$, the set of n-tuples of complex numbers.

The axioms of a vector space $\mathcal{V}$ are listed below; they apply to arbitrary vector spaces, and in particular to the real and complex vector spaces of interest here.

(1) **Addition** For each $\boldsymbol{v} \in \mathcal{V}$ and $\boldsymbol{u} \in \mathcal{V}$, there is a unique vector $\boldsymbol{v}+\boldsymbol{u} \in \mathcal{V}$ called the sum of $\boldsymbol{v}$ and $\boldsymbol{u}$, satisfying

(a) commutativity: $\boldsymbol{v}+\boldsymbol{u} = \boldsymbol{u}+\boldsymbol{v}$;
(b) associativity: $\boldsymbol{v}+(\boldsymbol{u}+\boldsymbol{w}) = (\boldsymbol{v}+\boldsymbol{u})+\boldsymbol{w}$ for each $\boldsymbol{v}, \boldsymbol{u}, \boldsymbol{w} \in \mathcal{V}$;
(c) zero: there is a unique element $\boldsymbol{0} \in \mathcal{V}$ satisfying $\boldsymbol{v}+\boldsymbol{0} = \boldsymbol{v}$ for all $\boldsymbol{v} \in \mathcal{V}$;
(d) negation: for each $\boldsymbol{v} \in \mathcal{V}$, there is a unique $-\boldsymbol{v} \in \mathcal{V}$ such that $\boldsymbol{v}+(-\boldsymbol{v}) = \boldsymbol{0}$.

(2) **Scalar multiplication** For each scalar[3] α and each $\boldsymbol{v} \in \mathcal{V}$, there is a unique vector $\alpha\boldsymbol{v} \in \mathcal{V}$ called the scalar product of α and $\boldsymbol{v}$ satisfying

(a) scalar associativity: $\alpha(\beta\boldsymbol{v}) = (\alpha\beta)\boldsymbol{v}$ for all scalars α, β, and all $\boldsymbol{v} \in \mathcal{V}$;
(b) unit multiplication: for the unit scalar 1, $1\boldsymbol{v} = \boldsymbol{v}$ for all $\boldsymbol{v} \in \mathcal{V}$.

(3) **Distributive laws:**

(a) for all scalars α and all $\boldsymbol{v}, \boldsymbol{u} \in \mathcal{V}$, $\alpha(\boldsymbol{v}+\boldsymbol{u}) = \alpha\boldsymbol{v}+\alpha\boldsymbol{u}$;
(b) for all scalars α, β and all $\boldsymbol{v} \in \mathcal{V}$, $(\alpha+\beta)\boldsymbol{v} = \alpha\boldsymbol{v}+\beta\boldsymbol{v}$.

Example 5.1.1 For $\mathbb{R}^n$, a vector $\boldsymbol{v}$ is an n-tuple $(v_1, \ldots, v_n)$ of real numbers. Addition is defined by $\boldsymbol{v}+\boldsymbol{u} = (v_1+u_1, \ldots, v_n+u_n)$. The zero vector is defined

[1] It is not necessary here to understand the general notion of a field, although Chapter 8 will also briefly discuss another field, $\mathbb{F}_2$, consisting of binary elements with mod 2 addition.

[2] Some authors prefer to define $\mathbb{R}^n$ as the class of real vector spaces of dimension n, but almost everyone visualizes $\mathbb{R}^n$ as the space of n-tuples. More importantly, the space of n-tuples will be constantly used as an example and $\mathbb{R}^n$ is a convenient name for it.

[3] Addition, subtraction, multiplication, and division between scalars is performed according to the familiar rules of $\mathbb{R}$ or $\mathbb{C}$ and will not be restated here. Neither $\mathbb{R}$ nor $\mathbb{C}$ includes ∞.

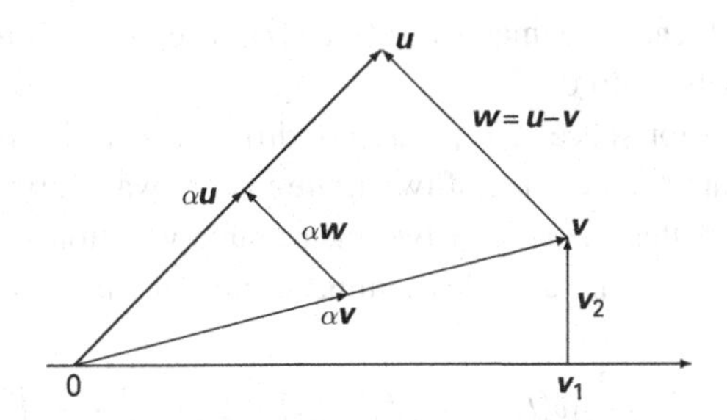

Figure 5.1. Geometric interpretation of $\mathbb{R}^2$. The vector $\boldsymbol{v} = (v_1, v_2)$ is represented as a point in the Euclidean plane with abscissa v_1 and ordinate v_2. It can also be viewed as the directed line from $\mathbf{0}$ to the point $\boldsymbol{v}$. Sometimes, as in the case of $\boldsymbol{w} = \boldsymbol{u} - \boldsymbol{v}$, a vector is viewed as a directed line from some nonzero point ($\boldsymbol{v}$ in this case) to another point $\boldsymbol{u}$. Note that the scalar multiple $\alpha\boldsymbol{u}$ lies on the straight line from 0 to $\boldsymbol{u}$. The scaled triangles also illustrate the distributive law. This geometric interpretation also suggests the concepts of length and angle, which are not included in the axioms. This is discussed more fully later.

by $\mathbf{0} = (0, \ldots, 0)$. The scalars α are the real numbers, and $\alpha\boldsymbol{v}$ is defined to be $(\alpha v_1, \ldots, \alpha v_n)$. This is illustrated geometrically in Figure 5.1 for $\mathbb{R}^2$.

Example 5.1.2 The vector space $\mathbb{C}^n$ is similar to $\mathbb{R}^n$ except that $\boldsymbol{v}$ is an n-tuple of complex numbers and the scalars are complex. Note that $\mathbb{C}^2$ cannot be easily illustrated geometrically, since a vector in $\mathbb{C}^2$ is specified by four real numbers. The reader should verify the axioms for both $\mathbb{R}^n$ and $\mathbb{C}^n$.

Example 5.1.3 There is a trivial vector space whose only element is the zero vector $\mathbf{0}$. For both real and complex scalars, $\alpha\mathbf{0} = \mathbf{0}$. The vector spaces of interest here are nontrivial spaces, i.e. spaces with more than one element, and this will usually be assumed without further mention.

Because of the commutative and associative axioms, we see that a finite sum $\sum_j \alpha_j \boldsymbol{v}_j$, where each α_j is a scalar and $\boldsymbol{v}_j$ a vector, is unambiguously defined without the need for parentheses. This sum is called a *linear combination* of the vectors $\boldsymbol{v}_1, \boldsymbol{v}_2, \ldots$

We next show that the set of finite-energy complex waveforms can be viewed as a complex vector space.[4] When we view a waveform $v(t)$ as a vector, we denote it by $\boldsymbol{v}$. There are two reasons for this: first, it reminds us that we are viewing the waveform as a vector; second, $v(t)$ sometimes denotes a function and sometimes denotes the value of that function at a particular argument t. Denoting the function as $\boldsymbol{v}$ avoids this ambiguity.

The vector sum $\boldsymbol{v} + \boldsymbol{u}$ is defined in the obvious way as the waveform for which each t is mapped into $v(t) + u(t)$; the scalar product $\alpha\boldsymbol{v}$ is defined as the waveform

[4] There is a small but important technical difference between the vector space being defined here and what we will later define to be the vector space $\mathcal{L}_2$. This difference centers on the notion of $\mathcal{L}_2$-equivalence, and will be discussed later.

for which each t is mapped into $\alpha v(t)$. The vector $\mathbf{0}$ is defined as the waveform that maps each t into 0.

The vector space axioms are not difficult to verify for this space of waveforms. To show that the sum $\boldsymbol{v}+\boldsymbol{u}$ of two finite-energy waveforms $\boldsymbol{v}$ and $\boldsymbol{u}$ also has finite energy, recall first that the sum of two measurable waveforms is also measurable. Next recall that if v and u are complex numbers, then $|v+u|^2 \leq 2|v|^2+2|u|^2$. Thus,

$$\int_{-\infty}^{\infty} |v(t)+u(t)|^2\, dt \leq \int_{-\infty}^{\infty} 2|v(t)|^2\, dt + \int_{-\infty}^{\infty} 2|u(t)|^2\, dt < \infty. \tag{5.1}$$

Similarly, if $\boldsymbol{v}$ has finite energy, then $\alpha\boldsymbol{v}$ has $|\alpha|^2$ times the energy of $\boldsymbol{v}$, which is also finite. The other axioms can be verified by inspection.

The above argument has shown that the set of finite-energy complex waveforms forms a complex vector space with the given definitions of complex addition and scalar multiplication. Similarly, the set of finite-energy real waveforms forms a real vector space with real addition and scalar multiplication.

5.1.1 Finite-dimensional vector spaces

A set of vectors $\boldsymbol{v}_1, \ldots, \boldsymbol{v}_n \in \mathcal{V}$ *spans* $\mathcal{V}$ (and is called a *spanning set* of $\mathcal{V}$) if every vector $\boldsymbol{v} \in \mathcal{V}$ is a linear combination of $\boldsymbol{v}_1, \ldots, \boldsymbol{v}_n$. For $\mathbb{R}^n$, let $\boldsymbol{e}_1 = (1, 0, 0, \ldots, 0)$, $\boldsymbol{e}_2 = (0, 1, 0, \ldots, 0), \ldots, \boldsymbol{e}_n = (0, \ldots 0, 1)$ be the n *unit vectors* of $\mathbb{R}^n$. The unit vectors span $\mathbb{R}^n$ since every vector $\boldsymbol{v} \in \mathbb{R}^n$ can be expressed as a linear combination of the unit vectors, i.e.

$$\boldsymbol{v} = (\alpha_1, \ldots, \alpha_n) = \sum_{j=1}^{n} \alpha_j \boldsymbol{e}_j.$$

A vector space $\mathcal{V}$ is *finite-dimensional* if there exists a finite set of vectors $\boldsymbol{u}_1, \ldots, \boldsymbol{u}_n$ that span $\mathcal{V}$. Thus $\mathbb{R}^n$ is finite-dimensional since it is spanned by $\boldsymbol{e}_1, \ldots, \boldsymbol{e}_n$. Similarly, $\mathbb{C}^n$ is finite-dimensional, and is spanned by the same unit vectors, $\boldsymbol{e}_1, \ldots, \boldsymbol{e}_n$, now viewed as vectors in $\mathbb{C}^n$. If $\mathcal{V}$ is not finite-dimensional, then it is *infinite-dimensional.* We will soon see that $\mathcal{L}_2$ is infinite-dimensional.

A set of vectors $\boldsymbol{v}_1, \ldots, \boldsymbol{v}_n \in \mathcal{V}$ is *linearly dependent* if $\sum_{j=1}^n \alpha_j \boldsymbol{v}_j = 0$ for some set of scalars not all equal to 0. This implies that each vector $\boldsymbol{v}_k$ for which $\alpha_k \neq 0$ is a linear combination of the others, i.e.

$$\boldsymbol{v}_k = \sum_{j \neq k} \frac{-\alpha_j}{\alpha_k} \boldsymbol{v}_j.$$

A set of vectors $\boldsymbol{v}_1, \ldots, \boldsymbol{v}_n \in \mathcal{V}$ is linearly independent if it is not linearly dependent, i.e. if $\sum_{j=1}^n \alpha_j \boldsymbol{v}_j = 0$ implies that each α_j is 0. For brevity we often omit the word "linear" when we refer to independence or dependence.

It can be seen that the unit vectors $\boldsymbol{e}_1, \ldots, \boldsymbol{e}_n$ are linearly independent as elements of $\mathbb{R}^n$. Similarly, they are linearly independent as elements of $\mathbb{C}^n$.

A set of vectors $\boldsymbol{v}_1, \ldots, \boldsymbol{v}_n \in \mathcal{V}$ is defined to be a *basis* for $\mathcal{V}$ if the set is linearly independent and spans $\mathcal{V}$. Thus the unit vectors $\boldsymbol{e}_1, \ldots, \boldsymbol{e}_n$ form a basis for both $\mathbb{R}^n$ and $\mathbb{C}^n$, in the first case viewing them as vectors in $\mathbb{R}^n$, and in the second as vectors in $\mathbb{C}^n$.

The following theorem is both important and simple; see Exercise 5.1 or any linear algebra text for a proof.

Theorem 5.1.1 (Basis for finite-dimensional vector space) *Let $\mathcal{V}$ be a non-trivial finite-dimensional vector space. Then the following hold.*[5]

- *If $\boldsymbol{v}_1, \ldots, \boldsymbol{v}_m$ span $\mathcal{V}$ but are linearly dependent, then a subset of $\boldsymbol{v}_1, \ldots, \boldsymbol{v}_m$ forms a basis for $\mathcal{V}$ with $n < m$ vectors.*
- *If $\boldsymbol{v}_1, \ldots, \boldsymbol{v}_m$ are linearly independent but do not span $\mathcal{V}$, then there exists a basis for $\mathcal{V}$ with $n > m$ vectors that includes $\boldsymbol{v}_1, \ldots, \boldsymbol{v}_m$.*
- *Every basis of $\mathcal{V}$ contains the same number of vectors.*

The *dimension* of a finite-dimensional vector space may thus be defined as the number of vectors in any basis. The theorem implicitly provides two conceptual algorithms for finding a basis. First, start with any linearly independent set (such as a single nonzero vector) and successively add independent vectors until a spanning set is reached. Second, start with any spanning set and successively eliminate dependent vectors until a linearly independent set is reached.

Given any basis $\{\boldsymbol{v}_1, \ldots, \boldsymbol{v}_n\}$ for a finite-dimensional vector space $\mathcal{V}$, any vector $\boldsymbol{v} \in \mathcal{V}$ can be expressed as follows:

$$\boldsymbol{v} = \sum_{j=1}^{n} \alpha_j \boldsymbol{v}_j, \tag{5.2}$$

where $\alpha_1, \ldots, \alpha_n$ are unique scalars.

In terms of the given basis, each $\boldsymbol{v} \in \mathcal{V}$ can be uniquely represented by the n-tuple of coefficients $(\alpha_1, \ldots, \alpha_n)$ in (5.2). Thus any n-dimensional vector space $\mathcal{V}$ over $\mathbb{R}$ or $\mathbb{C}$ may be viewed (relative to a given basis) as a version[6] of $\mathbb{R}^n$ or $\mathbb{C}^n$. This leads to the elementary vector/matrix approach to linear algebra. What is gained by the axiomatic ("coordinate-free") approach is the ability to think about vectors without first specifying a basis. The value of this will be clear after subspaces are defined and infinite-dimensional vector spaces such as $\mathcal{L}_2$ are viewed in terms of various finite-dimensional subspaces.

[5] The trivial vector space whose only element is $\mathbf{0}$ is conventionally called a zero-dimensional space and could be viewed as having the empty set as a basis.

[6] More precisely, $\mathcal{V}$ and $\mathbb{R}^n$ ($\mathbb{C}^n$) are isomorphic in the sense that that there is a one-to-one correspondence between vectors in $\mathcal{V}$ and n-tuples in $\mathbb{R}^n$ ($\mathbb{C}^n$) that preserves the vector space operations. In plain English, solvable problems concerning vectors in $\mathcal{V}$ can always be solved by first translating to n-tuples in a basis and then working in $\mathbb{R}^n$ or $\mathbb{C}^n$.

5.2 Inner product spaces

The vector space axioms listed in Section 5.1 contain no inherent notion of length or angle, although such geometric properties are clearly present in Figure 5.1 and in our intuitive view of $\mathbb{R}^n$ or $\mathbb{C}^n$. The missing ingredient is that of an *inner product*.

An *inner product* on a complex vector space $\mathcal{V}$ is a complex-valued function of two vectors, $\boldsymbol{v}, \boldsymbol{u} \in \mathcal{V}$, denoted by $\langle \boldsymbol{v}, \boldsymbol{u} \rangle$, that satisfies the following axioms.

(a) Hermitian symmetry: $\langle \boldsymbol{v}, \boldsymbol{u} \rangle = \langle \boldsymbol{u}, \boldsymbol{v} \rangle^*$;
(b) Hermitian bilinearity: $\langle \alpha\boldsymbol{v} + \beta\boldsymbol{u}, \boldsymbol{w} \rangle = \alpha\langle \boldsymbol{v}, \boldsymbol{w} \rangle + \beta\langle \boldsymbol{u}, \boldsymbol{w} \rangle$ (and consequently $\langle \boldsymbol{v}, \alpha\boldsymbol{u} + \beta\boldsymbol{w} \rangle = \alpha^*\langle \boldsymbol{v}, \boldsymbol{u} \rangle + \beta^*\langle \boldsymbol{v}, \boldsymbol{w} \rangle$);
(c) strict positivity: $\langle \boldsymbol{v}, \boldsymbol{v} \rangle \geq 0$, with equality if and only if $\boldsymbol{v} = \boldsymbol{0}$.

A vector space with an inner product satisfying these axioms is called an *inner product space*.

The same definition applies to a real vector space, but the inner product is always real and the complex conjugates can be omitted.

The *norm* or *length* $\|\boldsymbol{v}\|$ of a vector $\boldsymbol{v}$ in an inner product space is defined as

$$\|\boldsymbol{v}\| = \sqrt{\langle \boldsymbol{v}, \boldsymbol{v} \rangle}.$$

Two vectors $\boldsymbol{v}$ and $\boldsymbol{u}$ are defined to be *orthogonal* if $\langle \mathbf{v}, \boldsymbol{u} \rangle = 0$. Thus we see that the important geometric notions of length and orthogonality are both defined in terms of the inner product.

5.2.1 The inner product spaces $\mathbb{R}^n$ and $\mathbb{C}^n$

For the vector space $\mathbb{R}^n$ of real n-tuples, the inner product of vectors $\boldsymbol{v} = (v_1, \ldots, v_n)$ and $\boldsymbol{u} = (u_1, \ldots, u_n)$ is usually defined (and is defined here) as

$$\langle \boldsymbol{v}, \boldsymbol{u} \rangle = \sum_{j=1}^{n} v_j u_j.$$

You should verify that this definition satisfies the inner product axioms given in Section 5.1.

The length $\|\boldsymbol{v}\|$ of a vector $\boldsymbol{v}$ is then given by $\sqrt{\sum_j v_j^2}$, which agrees with Euclidean geometry. Recall that the formula for the cosine between two arbitrary nonzero vectors in $\mathbb{R}^2$ is given by

$$\cos(\angle(\boldsymbol{v}, \boldsymbol{u})) = \frac{v_1 u_1 + v_2 u_2}{\sqrt{v_1^2 + v_2^2}\sqrt{u_1^2 + u_1^2}} = \frac{\langle \boldsymbol{v}, \boldsymbol{u} \rangle}{\|\boldsymbol{v}\|\,\|\boldsymbol{u}\|}, \tag{5.3}$$

where the final equality expresses this in terms of the inner product. Thus the inner product determines the angle between vectors in $\mathbb{R}^2$. This same inner product formula will soon be seen to be valid in any real vector space, and the derivation is much simpler in the coordinate-free environment of general vector spaces than in the unit-vector context of $\mathbb{R}^2$.

For the vector space $\mathbb{C}^n$ of complex n-tuples, the inner product is defined as

$$\langle \boldsymbol{v}, \boldsymbol{u} \rangle = \sum_{j=1}^{n} v_j u_j^*. \tag{5.4}$$

The norm, or length, of $\boldsymbol{v}$ is then given by

$$\sqrt{\sum_j |v_j|^2} = \sqrt{\sum_j [\Re(v_j)^2 + \Im(v_j)^2]}.$$

Thus, as far as length is concerned, a complex n-tuple $\boldsymbol{u}$ can be regarded as the real $2n$-vector formed from the real and imaginary parts of $\boldsymbol{u}$. Warning: although a complex n-tuple can be viewed as a real $2n$-tuple for some purposes, such as length, many other operations on complex n-tuples are very different from those operations on the corresponding real $2n$-tuple. For example, scalar multiplication and inner products in $\mathbb{C}^n$ are very different from those operations in $\mathbb{R}^{2n}$.

5.2.2 One-dimensional projections

An important problem in constructing orthogonal expansions is that of breaking a vector $\boldsymbol{v}$ into two components relative to another vector $\boldsymbol{u} \neq \boldsymbol{0}$ in the same inner product space. One component, $\boldsymbol{v}_{\perp \boldsymbol{u}}$, is to be orthogonal (i.e. perpendicular) to $\boldsymbol{u}$ and the other, $\boldsymbol{v}_{|\boldsymbol{u}}$, is to be collinear with $\boldsymbol{u}$ (two vectors $\boldsymbol{v}_{|\boldsymbol{u}}$ and $\boldsymbol{u}$ are collinear if $\boldsymbol{v}_{|\boldsymbol{u}} = \alpha \boldsymbol{u}$ for some scalar α). Figure 5.2 illustrates this decomposition for vectors in $\mathbb{R}^2$. We can view this geometrically as dropping a perpendicular from $\boldsymbol{v}$ to $\boldsymbol{u}$. From the geometry of Figure 5.2, $\|\boldsymbol{v}_{|\boldsymbol{u}}\| = \|\boldsymbol{v}\| \cos(\angle(\boldsymbol{v}, \boldsymbol{u}))$. Using (5.3), $\|\boldsymbol{v}_{|\boldsymbol{u}}\| = \langle \boldsymbol{v}, \boldsymbol{u} \rangle / \|\boldsymbol{u}\|$. Since $\boldsymbol{v}_{|\boldsymbol{u}}$ is also collinear with $\boldsymbol{u}$, it can be seen that

$$\boldsymbol{v}_{|\boldsymbol{u}} = \frac{\langle \boldsymbol{v}, \boldsymbol{u} \rangle}{\|\boldsymbol{u}\|^2} \boldsymbol{u}. \tag{5.5}$$

The vector $\boldsymbol{v}_{|\boldsymbol{u}}$ is called the *projection* of $\boldsymbol{v}$ onto $\boldsymbol{u}$.

Rather surprisingly, (5.5) is valid for any inner product space. The general proof that follows is also simpler than the derivation of (5.3) and (5.5) using plane geometry.

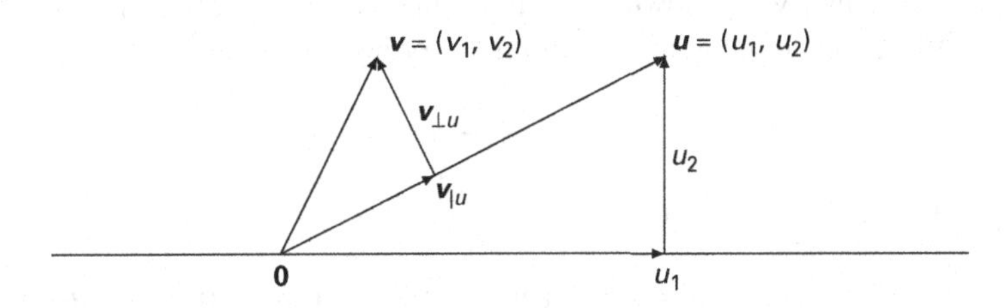

Figure 5.2. Two vectors, $\boldsymbol{v} = (v_1, v_2)$ and $\boldsymbol{u} = (u_1, u_2)$, in $\mathbb{R}^2$. Note that $\|\boldsymbol{u}\|^2 = \langle \boldsymbol{u}, \boldsymbol{u} \rangle = u_1^2 + u_2^2$ is the squared length of $\boldsymbol{u}$. The vector $\boldsymbol{v}$ is also expressed as $\boldsymbol{v} = \boldsymbol{v}_{|\boldsymbol{u}} + \boldsymbol{v}_{\perp \boldsymbol{u}}$, where $\boldsymbol{v}_{|\boldsymbol{u}}$ is collinear with $\boldsymbol{u}$ and $\boldsymbol{v}_{\perp \boldsymbol{u}}$ is perpendicular to $\boldsymbol{u}$.

Theorem 5.2.1 (One-dimensional projection theorem) *Let v and u be arbitrary vectors with $u \neq 0$ in a real or complex inner product space. Then there is a unique scalar α for which $\langle v - \alpha u, u\rangle = 0$, namely $\alpha = \langle v, u\rangle / \|u\|^2$.*

Remark The theorem states that $v - \alpha u$ is perpendicular to u if and only if $\alpha = \langle v, u\rangle / \|u\|^2$. Using that value of α, $v - \alpha u$ is called the perpendicular to u and is denoted as $v_{\perp u}$; similarly, αu is called the projection of v onto u and is denoted as $u_{|u}$. Finally, $v = v_{\perp u} + v_{|u}$, so v has been split into a perpendicular part and a collinear part.

Proof Calculating $\langle v - \alpha u, u\rangle$ for an arbitrary scalar α, the conditions can be found under which this inner product is zero:

$$\langle v - \alpha u, u\rangle = \langle v, u\rangle - \alpha\langle u, u\rangle = \langle v, u\rangle - \alpha\|u\|^2,$$

which is equal to zero if and only if $\alpha = \langle v, u\rangle / \|u\|^2$. □

The reason why $\|u\|^2$ is in the denominator of the projection formula can be understood by rewriting (5.5) as follows:

$$v_{|u} = \left\langle v, \frac{u}{\|u\|}\right\rangle \frac{u}{\|u\|}.$$

In words, the projection of v onto u is the same as the projection of v onto the normalized version, $u/\|u\|$, of u. More generally, the value of $v_{|u}$ is invariant to scale changes in u, i.e.

$$v_{|\beta u} = \frac{\langle v, \beta u\rangle}{\|\beta u\|^2}\beta u = \frac{\langle v, u\rangle}{\|u\|^2} u = v_{|u}. \tag{5.6}$$

This is clearly consistent with the geometric picture in Figure 5.2 for $\mathbb{R}^2$, but it is also valid for arbitrary inner product spaces where such figures cannot be drawn.

In $\mathbb{R}^2$, the cosine formula can be rewritten as

$$\cos(\angle(u, v)) = \left\langle \frac{u}{\|u\|} \frac{v}{\|v\|}\right\rangle. \tag{5.7}$$

That is, the cosine of $\angle(u, v)$ is the inner product of the normalized versions of u and v.

Another well known result in $\mathbb{R}^2$ that carries over to any inner product space is the Pythagorean theorem: if v and u are orthogonal, then

$$\|v + u\|^2 = \|v\|^2 + \|u\|^2. \tag{5.8}$$

To see this, note that

$$\langle v + u, v + u\rangle = \langle v, v\rangle + \langle v, u\rangle + \langle u, v\rangle + \langle u, u\rangle.$$

The cross terms disappear by orthogonality, yielding (5.8).

Theorem 5.2.1 has an important corollary, called the *Schwarz inequality*.

Corollary 5.2.1 (Schwarz inequality) *Let $\boldsymbol{v}$ and $\boldsymbol{u}$ be vectors in a real or complex inner product space. Then*

$$|\langle \boldsymbol{v}, \boldsymbol{u} \rangle| \le \|\boldsymbol{v}\| \, \|\boldsymbol{u}\|. \tag{5.9}$$

Proof Assume $\boldsymbol{u} \neq \boldsymbol{0}$ since (5.9) is obvious otherwise. Since $\boldsymbol{v}_{|\boldsymbol{u}}$ and $\boldsymbol{v}_{\perp \boldsymbol{u}}$ are orthogonal, (5.8) shows that

$$\|\boldsymbol{v}\|^2 = \|\boldsymbol{v}_{|\boldsymbol{u}}\|^2 + \|\boldsymbol{v}_{\perp \boldsymbol{u}}\|^2.$$

Since $\|\boldsymbol{v}_{\perp \boldsymbol{u}}\|^2$ is nonnegative, we have

$$\|\boldsymbol{v}\|^2 \ge \|\boldsymbol{v}_{|\boldsymbol{u}}\|^2 = \left| \frac{\langle \boldsymbol{v}, \boldsymbol{u} \rangle}{\|\boldsymbol{u}\|^2} \right|^2 \|\boldsymbol{u}\|^2 = \frac{|\langle \boldsymbol{v}, \boldsymbol{u} \rangle|^2}{\|\boldsymbol{u}\|^2},$$

which is equivalent to (5.9). □

For $\boldsymbol{v}$ and $\boldsymbol{u}$ both nonzero, the Schwarz inequality may be rewritten in the following form:

$$\left| \left\langle \frac{\boldsymbol{v}}{\|\boldsymbol{v}\|}, \frac{\boldsymbol{u}}{\|\boldsymbol{u}\|} \right\rangle \right| \le 1.$$

In $\mathbb{R}^2$, the Schwarz inequality is thus equivalent to the familiar fact that the cosine function is upperbounded by 1.

As shown in Exercise 5.6, the triangle inequality is a simple consequence of the Schwarz inequality:

$$\|\boldsymbol{v} + \boldsymbol{u}\| \le \|\boldsymbol{v}\| + \|\boldsymbol{u}\|. \tag{5.10}$$

5.2.3 The inner product space of $\mathcal{L}_2$ functions

Consider again the set of complex finite-energy waveforms. We attempt to define the inner product of two vectors $\boldsymbol{v}$ and $\boldsymbol{u}$ in this set as follows:

$$\langle \boldsymbol{v}, \boldsymbol{u} \rangle = \int_{-\infty}^{\infty} v(t) u^*(t) \mathrm{d}t. \tag{5.11}$$

It is shown in Exercise 5.8 that $\langle \boldsymbol{v}, \boldsymbol{u} \rangle$ is always finite. The Schwarz inequality cannot be used to prove this, since we have not yet shown that $\mathcal{L}_2$ satisfies the axioms of an inner product space. However, the first two inner product axioms follow immediately from the existence and finiteness of the inner product, i.e. the integral in (5.11). This existence and finiteness is a vital and useful property of $\mathcal{L}_2$.

The final inner product axiom is that $\langle \boldsymbol{v}, \boldsymbol{v} \rangle \ge 0$, with equality if and only if $\boldsymbol{v} = \boldsymbol{0}$. This axiom does not hold for finite-energy waveforms, because, as we have already seen, if a function $v(t)$ is zero almost everywhere (a.e.), then its energy is 0, even though the function is not the zero function. This is a nit-picking issue at some level, but axioms cannot be ignored simply because they are inconvenient.

The resolution of this problem is to define *equality* in an $\mathcal{L}_2$ inner product space as $\mathcal{L}_2$-equivalence between $\mathcal{L}_2$ functions. What this means is that a *vector* in an $\mathcal{L}_2$ inner product space is an *equivalence class* of $\mathcal{L}_2$ functions that are equal a.e. For example, the zero equivalence class is the class of zero-energy functions, since each is $\mathcal{L}_2$-equivalent to the all-zero function. With this modification, the inner product axioms all hold. We then have the following definition.

Definition 5.2.1 An $\mathcal{L}_2$ inner product space is an inner product space whose vectors are $\mathcal{L}_2$-equivalence classes in the set of $\mathcal{L}_2$ functions. The inner product in this vector space is given by (5.11).

Viewing a vector as an equivalence class of $\mathcal{L}_2$ functions seems very abstract and strange at first. From an engineering perspective, however, the notion that all zero-energy functions are the same is more natural than the notion that two functions that differ in only a few isolated points should be regarded as different.

From a more practical viewpoint, it will be seen later that $\mathcal{L}_2$ functions (in this equivalence class sense) can be represented by the coefficients in any orthogonal expansion whose elements span the $\mathcal{L}_2$ space. Two ordinary functions have the same coefficients in such an orthogonal expansion if and only if they are $\mathcal{L}_2$-equivalent. Thus each element $\boldsymbol{u}$ of the $\mathcal{L}_2$ inner product space is in one-to-one correspondence to a finite-energy sequence $\{u_k;\ k \in \mathbb{Z}\}$ of coefficients in an orthogonal expansion. Thus we can now avoid the awkwardness of having many $\mathcal{L}_2$-equivalent ordinary functions map into a single sequence of coefficients and having no very good way of going back from sequence to function. Once again, engineering common sense and sophisticated mathematics agree.

From now on, we will simply view $\mathcal{L}_2$ as an inner product space, referring to the notion of $\mathcal{L}_2$-equivalence only when necessary. With this understanding, we can use all the machinery of inner product spaces, including projections and the Schwarz inequality.

5.2.4 Subspaces of inner product spaces

A *subspace* $\mathcal{S}$ of a vector space $\mathcal{V}$ is a subset of the vectors in $\mathcal{V}$ which forms a vector space in its own right (over the same set of scalars as used by $\mathcal{V}$). An equivalent definition is that for all $\boldsymbol{v}$ and $\boldsymbol{u} \in \mathcal{S}$, the linear combination $\alpha\boldsymbol{v} + \beta\boldsymbol{u}$ is in $\mathcal{S}$ for all scalars α and β. If $\mathcal{V}$ is an inner product space, then it can be seen that $\mathcal{S}$ is also an inner product space using the same inner product definition as $\mathcal{V}$.

Example 5.2.1 (Subspaces of $\mathbb{R}^3$) Consider the real inner product space $\mathbb{R}^3$, namely the inner product space of real 3-tuples $\boldsymbol{v} = (v_1, v_2, v_3)$. Geometrically, we regard this as a space in which there are three orthogonal coordinate directions, defined by the three unit vectors $\boldsymbol{e}_1, \boldsymbol{e}_2, \boldsymbol{e}_3$. The 3-tuple v_1, v_2, v_3 then specifies the length of $\boldsymbol{v}$ in each of those directions, so that $\boldsymbol{v} = v_1\boldsymbol{e}_1 + v_2\boldsymbol{e}_2 + v_3\boldsymbol{e}_3$.

Let $\boldsymbol{u} = (1, 0, 1)$ and $\boldsymbol{w} = (0, 1, 1)$ be two fixed vectors, and consider the subspace of $\mathbb{R}^3$ composed of all linear combinations, $\boldsymbol{v} = \alpha\boldsymbol{u} + \beta\boldsymbol{w}$, of $\boldsymbol{u}$ and $\boldsymbol{w}$. Geometrically, this subspace is a plane going through the points $\mathbf{0}, \boldsymbol{u}$, and $\boldsymbol{w}$. In this plane, as in the original vector space, $\boldsymbol{u}$ and $\boldsymbol{w}$ each have length $\sqrt{2}$ and $\langle \boldsymbol{u}, \boldsymbol{w} \rangle = 1$.

Since neither $\boldsymbol{u}$ nor $\boldsymbol{w}$ is a scalar multiple of the other, they are linearly independent. They span $\mathcal{S}$ by definition, so $\mathcal{S}$ is a 2D subspace with a basis $\{\boldsymbol{u}, \boldsymbol{w}\}$.

The projection of $\boldsymbol{u}$ onto $\boldsymbol{w}$ is $\boldsymbol{u}_{|\boldsymbol{w}} = (0, 1/2, 1/2)$, and the perpendicular is $\boldsymbol{u}_{\perp \boldsymbol{w}} = (1, -1/2, 1/2)$. These vectors form an orthogonal basis for $\mathcal{S}$. Using these vectors as an orthogonal basis, we can view $\mathcal{S}$, pictorially and geometrically, in just the same way as we view vectors in R^2.

Example 5.2.2 (General 2D subspace) Let $\mathcal{V}$ be an arbitrary real or complex inner product space that contains two noncollinear vectors, say $\boldsymbol{u}$ and $\boldsymbol{w}$. Then the set $\mathcal{S}$ of linear combinations of $\boldsymbol{u}$ and $\boldsymbol{w}$ is a 2D subspace of $\mathcal{V}$ with basis $\{\boldsymbol{u}, \boldsymbol{w}\}$. Again, $\boldsymbol{u}_{|\boldsymbol{w}}$ and $\boldsymbol{u}_{\perp \boldsymbol{w}}$ form an orthogonal basis of $\mathcal{S}$. We will soon see that this procedure for generating subspaces and orthogonal bases from two vectors in an arbitrary inner product space can be generalized to orthogonal bases for subspaces of arbitrary dimension.

Example 5.2.3 ($\mathbb{R}^2$ is a subset but not a subspace of $\mathbb{C}^2$) Consider the complex vector space $\mathbb{C}^2$. The set of real 2-tuples is a subset of $\mathbb{C}^2$, but this subset is not closed under multiplication by scalars in $\mathbb{C}$. For example, the real 2-tuple $\boldsymbol{u} = (1, 2)$ is an element of $\mathbb{C}^2$, but the scalar product $\mathrm{i}\boldsymbol{u}$ is the vector (i, 2i), which is not a real 2-tuple. More generally, the notion of linear combination (which is at the heart of both the use and theory of vector spaces) depends on what the scalars are.

We cannot avoid dealing with both complex and real $\mathcal{L}_2$ waveforms without enormously complicating the subject (as a simple example, consider using the sine and cosine forms of the Fourier transform and series). We also cannot avoid inner product spaces without great complication. Finally, we cannot avoid going back and forth between complex and real $\mathcal{L}_2$ waveforms. The price of this is frequent confusion between real and complex scalars. The reader is advised to use considerable caution with linear combinations and to be very clear about whether real or complex scalars are involved.

5.3 Orthonormal bases and the projection theorem

In an inner product space, a set of vectors $\boldsymbol{\phi}_1, \boldsymbol{\phi}_2, \ldots$ is *orthonormal* if

$$\langle \boldsymbol{\phi}_j, \boldsymbol{\phi}_k \rangle = \begin{cases} 0 & \text{for } j \neq k; \\ 1 & \text{for } j = k. \end{cases} \tag{5.12}$$

In other words, an orthonormal set is a set of nonzero orthogonal vectors where each vector is *normalized* to unit length. It can be seen that if a set of vectors $\boldsymbol{u}_1, \boldsymbol{u}_2, \ldots$ is orthogonal, then the set

$$\boldsymbol{\phi}_j = \frac{1}{\|\boldsymbol{u}_j\|} \boldsymbol{u}_j$$

is orthonormal. Note that if two nonzero vectors are orthogonal, then any scaling (including normalization) of each vector maintains orthogonality.

If a vector $\boldsymbol{v}$ is projected onto a normalized vector $\boldsymbol{\phi}$, then the 1D projection theorem states that the projection is given by the simple formula

$$\boldsymbol{v}_{|\boldsymbol{\phi}} = \langle \boldsymbol{v}, \boldsymbol{\phi} \rangle \boldsymbol{\phi}. \tag{5.13}$$

Furthermore, the theorem asserts that $\boldsymbol{v}_{\perp\boldsymbol{\phi}} = \boldsymbol{v} - \boldsymbol{v}_{|\boldsymbol{\phi}}$ is orthogonal to $\boldsymbol{\phi}$. We now generalize the projection theorem to the projection of a vector $\boldsymbol{v} \in \mathcal{V}$ onto any finite-dimensional subspace $\mathcal{S}$ of $\mathcal{V}$.

5.3.1 Finite-dimensional projections

If $\mathcal{S}$ is a subspace of an inner product space $\mathcal{V}$, and $\boldsymbol{v} \in \mathcal{V}$, then a *projection of* $\boldsymbol{v}$ *onto* $\mathcal{S}$ is defined to be a vector $\boldsymbol{v}_{|\mathcal{S}} \in \mathcal{S}$ such that $\boldsymbol{v} - \boldsymbol{v}_{|\mathcal{S}}$ is orthogonal to all vectors in $\mathcal{S}$. The theorem to follow shows that $\boldsymbol{v}_{|\mathcal{S}}$ always exists and has the unique value given in the theorem.

The earlier definition of projection is a special case in which $\mathcal{S}$ is taken to be the 1D subspace spanned by a vector $\boldsymbol{u}$ (the orthonormal basis is then $\boldsymbol{\phi} = \boldsymbol{u}/\|\boldsymbol{u}\|$).

Theorem 5.3.1 (Projection theorem) *Let $\mathcal{S}$ be an n-dimensional subspace of an inner product space $\mathcal{V}$, and assume that $\{\boldsymbol{\phi}_1, \boldsymbol{\phi}_2, \ldots, \boldsymbol{\phi}_n\}$ is an orthonormal basis for $\mathcal{S}$. Then for any $\boldsymbol{v} \in \mathcal{V}$, there is a unique vector $\boldsymbol{v}_{|\mathcal{S}} \in \mathcal{S}$ such that $\langle \boldsymbol{v} - \boldsymbol{v}_{|\mathcal{S}}, \boldsymbol{s} \rangle = 0$ for all $\boldsymbol{s} \in \mathcal{S}$. Furthermore, $\boldsymbol{v}_{|\mathcal{S}}$ is given by*

$$\boldsymbol{v}_{|\mathcal{S}} = \sum_{j=1}^{n} \langle \boldsymbol{v}, \boldsymbol{\phi}_j \rangle \boldsymbol{\phi}_j. \tag{5.14}$$

Remark Note that the theorem assumes that $\mathcal{S}$ has a set of orthonormal vectors as a basis. It will be shown later that any nontrivial finite-dimensional inner product space has such an orthonormal basis, so that the assumption does not restrict the generality of the theorem.

Proof Let $\boldsymbol{w} = \sum_{j=1}^{n} \alpha_j \boldsymbol{\phi}_j$ be an arbitrary vector in $\mathcal{S}$. First consider the conditions on $\boldsymbol{w}$ under which $\boldsymbol{v} - \boldsymbol{w}$ is orthogonal to all vectors $\boldsymbol{s} \in \mathcal{S}$. It can be seen that $\boldsymbol{v} - \boldsymbol{w}$ is orthogonal to all $\boldsymbol{s} \in \mathcal{S}$ if and only if

$$\langle \boldsymbol{v} - \boldsymbol{w}, \boldsymbol{\phi}_j \rangle = 0, \quad \text{for all } j,\ 1 \le j \le n,$$

or equivalently if and only if

$$\langle \boldsymbol{v}, \boldsymbol{\phi}_j \rangle = \langle \boldsymbol{w}, \boldsymbol{\phi}_j \rangle, \quad \text{for all } j,\ 1 \le j \le n. \tag{5.15}$$

Since $\boldsymbol{w} = \sum_{\ell=1}^{n} \alpha_\ell \boldsymbol{\phi}_\ell$,

$$\langle \boldsymbol{w}, \boldsymbol{\phi}_j \rangle = \sum_{\ell=1}^{n} \alpha_\ell \langle \boldsymbol{\phi}_\ell, \boldsymbol{\phi}_j \rangle = \alpha_j, \quad \text{for all } j,\ 1 \le j \le n. \tag{5.16}$$

Combining this with (5.15), $v-w$ is orthogonal to all $s \in \mathcal{S}$ if and only if $\alpha_j = \langle v, \phi_j \rangle$ for each j, i.e. if and only if $w = \sum_j \langle v, \phi_j \rangle \phi_j$. Thus $v_{|\mathcal{S}}$ as given in (5.14) is the unique vector $w \in \mathcal{S}$ for which $v - v_{|\mathcal{S}}$ is orthogonal to all $s \in \mathcal{S}$. □

The vector $v - v_{|\mathcal{S}}$ is denoted as $v_{\perp\mathcal{S}}$, the *perpendicular from* v *to* $\mathcal{S}$. Since $v_{|\mathcal{S}} \in \mathcal{S}$, we see that $v_{|\mathcal{S}}$ and $v_{\perp\mathcal{S}}$ are orthogonal. The theorem then asserts that v can be uniquely split into two orthogonal components, $v = v_{|\mathcal{S}} + v_{\perp\mathcal{S}}$, where the projection $v_{|\mathcal{S}}$ is in $\mathcal{S}$ and the perpendicular $v_{\perp\mathcal{S}}$ is orthogonal to all vectors $s \in \mathcal{S}$.

5.3.2 Corollaries of the projection theorem

There are three important corollaries of the projection theorem that involve the norm of the projection. First, for any scalars $\alpha_1, \ldots, \alpha_n$, the squared norm of $w = \sum_j \alpha_j \phi_j$ is given by

$$\|w\|^2 = \left\langle w \sum_{j=1}^{n} \alpha_j \phi_j \right\rangle = \sum_{j=1}^{n} \alpha_j^* \langle w, \phi_j \rangle = \sum_{j=1}^{n} |\alpha_j|^2,$$

where (5.16) has been used in the last step. For the projection $v_{|\mathcal{S}}$, $\alpha_j = \langle v, \phi_j \rangle$, so

$$\|v_{|\mathcal{S}}\|^2 = \sum_{j=1}^{n} |\langle v, \phi_j \rangle|^2. \tag{5.17}$$

Also, since $v = v_{|\mathcal{S}} + v_{\perp\mathcal{S}}$ and $v_{|\mathcal{S}}$ is orthogonal to $v_{\perp\mathcal{S}}$, it follows from the Pythagorean theorem (5.8) that

$$\|v\|^2 = \|v_{|\mathcal{S}}\|^2 + \|v_{\perp\mathcal{S}}\|^2. \tag{5.18}$$

Since $\|v_{\perp\mathcal{S}}\|^2 \geq 0$, the following corollary has been proven.

Corollary 5.3.1 (Norm bound)

$$0 \leq \|v_{|\mathcal{S}}\|^2 \leq \|v\|^2, \tag{5.19}$$

with equality on the right if and only if $v \in \mathcal{S}$, *and equality on the left if and only if* v *is orthogonal to all vectors in* $\mathcal{S}$.

Substituting (5.17) into (5.19), we get *Bessel's inequality*, which is the key to understanding the convergence of orthonormal expansions.

Corollary 5.3.2 (Bessel's inequality) *Let* $\mathcal{S} \subseteq \mathcal{V}$ *be the subspace spanned by the set of orthonormal vectors* $\{\phi_1, \ldots, \phi_n\}$. *For any* $v \in \mathcal{V}$,

$$0 \leq \sum_{j=1}^{n} |\langle v, \phi_j \rangle|^2 \leq \|v\|^2,$$

with equality on the right if and only if $v \in \mathcal{S}$, *and equality on the left if and only if* v *is orthogonal to all vectors in* $\mathcal{S}$.

Another useful characterization of the projection $\boldsymbol{v}_{|\mathcal{S}}$ is that it is the vector in $\mathcal{S}$ that is closest to $\boldsymbol{v}$. In other words, using some $\boldsymbol{s} \in \mathcal{S}$ as an approximation to $\boldsymbol{v}$, the squared error is $\|\boldsymbol{v}-\boldsymbol{s}\|^2$. The following corollary says that $\boldsymbol{v}_{|\mathcal{S}}$ is the choice for $\boldsymbol{s}$ that yields the minimum squared error (MSE).

Corollary 5.3.3 (MSE property) *The projection $\boldsymbol{v}_{|\mathcal{S}}$ is the unique closest vector in $\mathcal{S}$ to $\boldsymbol{v}$; i.e., for all $\boldsymbol{s} \in \mathcal{S}$,*

$$\|\boldsymbol{v}-\boldsymbol{v}_{|\mathcal{S}}\|^2 \leq \|\boldsymbol{v}-\boldsymbol{s}\|^2,$$

with equality if and only if $\boldsymbol{s} = \boldsymbol{v}_{|\mathcal{S}}$.

Proof Decomposing $\boldsymbol{v}$ into $\boldsymbol{v}_{|\mathcal{S}} + \boldsymbol{v}_{\perp\mathcal{S}}$, we have $\boldsymbol{v}-\boldsymbol{s} = (\boldsymbol{v}_{|\mathcal{S}}-\boldsymbol{s}) + \boldsymbol{v}_{\perp\mathcal{S}}$. Since $\boldsymbol{v}_{|\mathcal{S}}$ and $\boldsymbol{s}$ are in $\mathcal{S}$, $\boldsymbol{v}_{|\mathcal{S}}-\boldsymbol{s}$ is also in $\mathcal{S}$, so, by Pythagoras,

$$\|\boldsymbol{v}-\boldsymbol{s}\|^2 = \|\boldsymbol{v}_{|\mathcal{S}}-\boldsymbol{s}\|^2 + \|\boldsymbol{v}_{\perp\mathcal{S}}\|^2 \geq \|\boldsymbol{v}_{\perp\mathcal{S}}\|^2,$$

with equality if and only if $\|\boldsymbol{v}_{|\mathcal{S}}-\boldsymbol{s}\|^2 = 0$, i.e. if and only if $\boldsymbol{s} = \boldsymbol{v}_{|\mathcal{S}}$. Since $\boldsymbol{v}_{\perp\mathcal{S}} = \boldsymbol{v} - \boldsymbol{v}_{|\mathcal{S}}$, this completes the proof. □

5.3.3 Gram–Schmidt orthonormalization

Theorem 5.3.1, the projection theorem, assumed an orthonormal basis $\{\boldsymbol{\phi}_1, \ldots, \boldsymbol{\phi}_n\}$ for any given n-dimensional subspace $\mathcal{S}$ of $\mathcal{V}$. The use of orthonormal bases simplifies almost everything concerning inner product spaces, and for infinite-dimensional expansions orthonormal bases are even more useful.

This section presents the Gram–Schmidt procedure, which, starting from an arbitrary basis $\{\boldsymbol{s}_1, \ldots, \boldsymbol{s}_n\}$ for an n-dimensional inner product subspace $\mathcal{S}$, generates an orthonormal basis for $\mathcal{S}$. The procedure is useful in finding orthonormal bases, but is even more useful theoretically, since it shows that such bases always exist. In particular, since every n-dimensional subspace contains an orthonormal basis, the projection theorem holds for each such subspace.

The procedure is almost obvious in view of the previous subsections. First an orthonormal basis, $\boldsymbol{\phi}_1 = \boldsymbol{s}_1/\|\boldsymbol{s}_1\|$, is found for the 1D subspace $\mathcal{S}_1$ spanned by $\boldsymbol{s}_1$. Projecting $\boldsymbol{s}_2$ onto this 1D subspace, a second orthonormal vector can be found. Iterating, a complete orthonormal basis can be constructed.

In more detail, let $(\boldsymbol{s}_2)_{|\mathcal{S}_1}$ be the projection of $\boldsymbol{s}_2$ onto $\mathcal{S}_1$. Since $\boldsymbol{s}_2$ and $\boldsymbol{s}_1$ are linearly independent, $(\boldsymbol{s}_2)_{\perp\mathcal{S}_1} = \boldsymbol{s}_2 - (\boldsymbol{s}_2)_{|\mathcal{S}_1}$ is nonzero. It is orthogonal to $\boldsymbol{\phi}_1$ since $\boldsymbol{\phi}_1 \in \mathcal{S}_1$. It is normalized as $\boldsymbol{\phi}_2 = (\boldsymbol{s}_2)_{\perp\mathcal{S}_1}/\|(\boldsymbol{s}_2)_{\perp\mathcal{S}_1}\|$. Then $\boldsymbol{\phi}_1$ and $\boldsymbol{\phi}_2$ span the space $\mathcal{S}_2$ spanned by $\boldsymbol{s}_1$ and $\boldsymbol{s}_2$.

Now, using induction, suppose that an orthonormal basis $\{\boldsymbol{\phi}_1, \ldots, \boldsymbol{\phi}_k\}$ has been constructed for the subspace $\mathcal{S}_k$ spanned by $\{\boldsymbol{s}_1, \ldots, \boldsymbol{s}_k\}$. The result of projecting $\boldsymbol{s}_{k+1}$ onto $\mathcal{S}_k$ is $(\boldsymbol{s}_{k+1})_{|\mathcal{S}_k} = \sum_{j=1}^{k} \langle \boldsymbol{s}_{k+1}, \boldsymbol{\phi}_j\rangle \boldsymbol{\phi}_j$. The perpendicular, $(\boldsymbol{s}_{k+1})_{\perp\mathcal{S}_k} = \boldsymbol{s}_{k+1} - (\boldsymbol{s}_{k+1})_{|\mathcal{S}_k}$, is given by

$$(\boldsymbol{s}_{k+1})_{\perp\mathcal{S}_k} = \boldsymbol{s}_{k+1} - \sum_{j=1}^{k} \langle \boldsymbol{s}_{k+1}, \boldsymbol{\phi}_j\rangle \boldsymbol{\phi}_j. \tag{5.20}$$

This is nonzero since $\boldsymbol{s}_{k+1}$ is not in $\mathcal{S}_k$ and thus not a linear combination of $\boldsymbol{\phi}_1, \ldots, \boldsymbol{\phi}_k$. Normalizing, we obtain

$$\boldsymbol{\phi}_{k+1} = \frac{(\boldsymbol{s}_{k+1})_{\perp \mathcal{S}_k}}{\|(\boldsymbol{s}_{k+1})_{\perp \mathcal{S}_k}\|}. \tag{5.21}$$

From (5.20) and (5.21), $\boldsymbol{s}_{k+1}$ is a linear combination of $\boldsymbol{\phi}_1, \ldots, \boldsymbol{\phi}_{k+1}$ and $\boldsymbol{s}_1, \ldots, \boldsymbol{s}_k$ are linear combinations of $\boldsymbol{\phi}_1, \ldots, \boldsymbol{\phi}_k$, so $\boldsymbol{\phi}_1, \ldots, \boldsymbol{\phi}_{k+1}$ is an orthonormal basis for the space $\mathcal{S}_{k+1}$ spanned by $\boldsymbol{s}_1, \ldots, \boldsymbol{s}_{k+1}$.

In summary, given any n-dimensional subspace $\mathcal{S}$ with a basis $\{\boldsymbol{s}_1, \ldots, \boldsymbol{s}_n\}$, the Gram–Schmidt orthonormalization procedure produces an orthonormal basis $\{\boldsymbol{\phi}_1, \ldots, \boldsymbol{\phi}_n\}$ for $\mathcal{S}$.

Note that if a set of vectors is not necessarily independent, then the procedure will automatically find any vector $\boldsymbol{s}_j$ that is a linear combination of previous vectors via the projection theorem. It can then simply discard such a vector and proceed. Consequently, it will still find an orthonormal basis, possibly of reduced size, for the space spanned by the original vector set.

5.3.4 Orthonormal expansions in $\mathcal{L}_2$

The background has now been developed to understand countable orthonormal expansions in $\mathcal{L}_2$. We have already looked at a number of *orthogonal* expansions, such as those used in the sampling theorem, the Fourier series, and the T-spaced truncated or sinc-weighted sinusoids. Turning these into *orthonormal* expansions involves only minor scaling changes.

The Fourier series will be used both to illustrate these changes and as an example of a general orthonormal expansion. The vector space view will then allow us to understand the Fourier series at a deeper level.

Define $\theta_k(t) = e^{2\pi ikt/T}\text{rect}(t/T)$ for $k \in \mathbb{Z}$. The set $\{\theta_k(t);\ k \in \mathbb{Z}\}$ of functions is orthogonal with $\|\boldsymbol{\theta}_k\|^2 = T$. The corresponding orthonormal expansion is obtained by scaling each $\boldsymbol{\theta}_k$ by $\sqrt{1/T}$; i.e.

$$\phi_k(t) = \sqrt{\frac{1}{T}} e^{2\pi ikt/T}\ \text{rect}\left(\frac{t}{T}\right). \tag{5.22}$$

The Fourier series of an $\mathcal{L}_2$ function $\{v(t) : [-T/2, T/2] \to \mathbb{C}\}$ then becomes $\sum_k \alpha_k \phi_k(t)$, where $\alpha_k = \int v(t)\phi_k^*(t)\,dt = \langle \boldsymbol{v}, \boldsymbol{\phi}_k \rangle$. For any integer $n > 0$, let $\mathcal{S}_n$ be the $(2n+1)$-dimensional subspace spanned by the vectors $\{\boldsymbol{\phi}_k, -n \le k \le n\}$. From the projection theorem, the projection $\boldsymbol{v}_{|\mathcal{S}_n}$ of $\boldsymbol{v}$ onto $\mathcal{S}_n$ is given by

$$\boldsymbol{v}_{|\mathcal{S}_n} = \sum_{k=-n}^{n} \langle \boldsymbol{v}, \boldsymbol{\phi}_k \rangle \boldsymbol{\phi}_k.$$

That is, the projection $\boldsymbol{v}_{|\mathcal{S}_n}$ is simply the approximation to $\boldsymbol{v}$ resulting from truncating the expansion to $-n \le k \le n$. The error in the approximation, $\boldsymbol{v}_{\perp \mathcal{S}_n} = \boldsymbol{v} - \boldsymbol{v}_{|\mathcal{S}_n}$, is orthogonal to all vectors in $\mathcal{S}_n$, and, from the MSE property, $\boldsymbol{v}_{|\mathcal{S}_n}$ is the closest point in

$\mathcal{S}_n$ to $\boldsymbol{v}$. As n increases, the subspace $\mathcal{S}_n$ becomes larger and $\boldsymbol{v}_{|\mathcal{S}_n}$ gets closer to $\boldsymbol{v}$ (i.e. $\|\boldsymbol{v} - \boldsymbol{v}_{|\mathcal{S}_n}\|$ is nonincreasing).

As the preceding analysis applies equally well to any orthonormal sequence of functions, the general case can now be considered. The main result of interest is the following infinite-dimensional generalization of the projection theorem.

Theorem 5.3.2 (Infinite-dimensional projection) *Let $\{\boldsymbol{\phi}_m, 1 \le m < \infty\}$ be a sequence of orthonormal vectors in $\mathcal{L}_2$, and let $\mathbf{v}$ be an arbitrary $\mathcal{L}_2$ vector. Then there exists a unique*[7] *$\mathcal{L}_2$ vector $\boldsymbol{u}$ such that $\mathbf{v} - \boldsymbol{u}$ is orthogonal to each $\boldsymbol{\phi}_m$ and*

$$\lim_{n\to\infty} \|\boldsymbol{u} - \sum_{m=1}^{n} \alpha_m \boldsymbol{\phi}_m\| = 0, \qquad \text{where} \quad \alpha_m = \langle \boldsymbol{v}, \boldsymbol{\phi}_m \rangle; \tag{5.23}$$

$$\|\boldsymbol{u}\|^2 = \sum |\alpha_m|^2. \tag{5.24}$$

Conversely, for any complex sequence $\{\alpha_m;\ 1 \le m \le \infty\}$ such that $\sum_k |\alpha_k|^2 < \infty$, an $\mathcal{L}_2$ function $\boldsymbol{u}$ exists satisfying (5.23) and (5.24).

Remark This theorem says that the orthonormal expansion $\sum_m \alpha_m \boldsymbol{\phi}_m$ converges in the $\mathcal{L}_2$ sense to an $\mathcal{L}_2$ function $\boldsymbol{u}$, which we later interpret as the projection of $\boldsymbol{v}$ onto the infinite-dimensional subspace $\mathcal{S}$ spanned by $\{\boldsymbol{\phi}_m, 1 \le m < \infty\}$. For example, in the Fourier series case, the orthonormal functions span the subspace of $\mathcal{L}_2$ functions time-limited to $[-T/2, T/2]$, and $\mathbf{u}$ is then $v(t)\,\text{rect}(t/T)$. The difference $v(t) - v(t)\,\text{rect}(t/T)$ is then $\mathcal{L}_2$-equivalent to $\mathbf{0}$ over $[-T/2, T/2]$, and thus orthogonal to each $\boldsymbol{\phi}_m$.

Proof Let $\mathcal{S}_n$ be the subspace spanned by $\{\boldsymbol{\phi}_1, \ldots, \boldsymbol{\phi}_n\}$. From the finite-dimensional projection theorem, the projection of $\boldsymbol{v}$ onto $\mathcal{S}_n$ is then $\boldsymbol{v}_{|\mathcal{S}_n} = \sum_{k=1}^{n} \alpha_k \boldsymbol{\phi}_k$. From (5.17),

$$\|\boldsymbol{v}_{|\mathcal{S}_n}\|^2 = \sum_{k=1}^{n} |\alpha_k|^2, \qquad \text{where} \quad \alpha_k = \langle \boldsymbol{v}, \boldsymbol{\phi}_k \rangle. \tag{5.25}$$

This quantity is nondecreasing with n, and from Bessel's inequality it is upperbounded by $\|\boldsymbol{v}\|^2$, which is finite since $\boldsymbol{v}$ is $\mathcal{L}_2$. It follows that, for any n and any $m > n$,

$$\|\boldsymbol{v}_{|\mathcal{S}_m} - \boldsymbol{v}_{|\mathcal{S}_n}\|^2 = \sum_{n<|k|\le m} |\alpha_k|^2 \le \sum_{|k|>n} |\alpha_k|^2 \overset{n\to\infty}{\longrightarrow} 0. \tag{5.26}$$

This says that the projections $\{\boldsymbol{v}_{|\mathcal{S}_n};\ n \in \mathbb{Z}^+\}$ approach each other as $n \to \infty$ in terms of their energy difference.

A sequence whose terms approach each other is called a *Cauchy sequence*. The Riesz–Fischer theorem[8] is a central theorem of analysis stating that any Cauchy

[7] Recall that the vectors in the $\mathcal{L}_2$ class of functions are equivalence classes, so this uniqueness specifies only the equivalence class and not an individual function within that class.

[8] See any text on real and complex analysis, such as Rudin 1966.

sequence of $\mathcal{L}_2$ waveforms has an $\mathcal{L}_2$ limit. Taking $\boldsymbol{u}$ to be this $\mathcal{L}_2$ limit, i.e. $\boldsymbol{u} = \text{l.i.m.}_{n\to\infty} \boldsymbol{v}_{|\mathcal{S}_n}$, we obtain (5.23) and (5.24).[9]

Essentially the same use of the Riesz-Fischer theorem establishes (5.23) and (5.24), starting with the sequence $\alpha_1, \alpha_2, \ldots$ □

Let $\mathcal{S}$ be the space of functions (or, more precisely, of equivalence classes) that can be represented as $\text{l.i.m.} \sum_k \alpha_k \phi_k(t)$ over all sequences $\alpha_1, \alpha_2, \ldots$ such that $\sum_k |\alpha_k|^2 < \infty$. It can be seen that this is an inner product space. It is the *space spanned* by the orthonormal sequence $\{\boldsymbol{\phi}_k; k \in \mathbb{Z}\}$.

The following proof of the Fourier series theorem illustrates the use of the infinite-dimensional projection theorem and infinite-dimensional spanning sets.

Proof of Theorem 4.4.1 Let $\{v(t) : [-T/2, T/2]] \to \mathbb{C}\}$ be an arbitrary $\mathcal{L}_2$ function over $[-T/2, T/2]$. We have already seen that $v(t)$ is $\mathcal{L}_1$, that $\hat{v}_k = (1/T)\int v(t)\mathrm{e}^{-2\pi ikt/T}\,\mathrm{d}t$ exists, and that $|\hat{v}_k| \le \int |v(t)|\,\mathrm{d}t$ for all $k \in \mathbb{Z}$. From Theorem 5.3.2, there is an $\mathcal{L}_2$ function $u(t) = \text{l.i.m.} \sum_k \hat{v}_k \mathrm{e}^{2\pi ikt/T}\,\mathrm{rect}(t/T)$ such that $v(t) - u(t)$ is orthogonal to $\theta_k(t) = \mathrm{e}^{2\pi ikt/T}\,\mathrm{rect}(t/T)$ for each $k \in \mathbb{Z}$.

We now need an additional basic fact:[10] the above set of orthogonal functions $\{\theta_k(t) = \mathrm{e}^{2\pi ikt/T}\,\mathrm{rect}(t/T); k \in \mathbb{Z}\}$ spans the space of $\mathcal{L}_2$ functions over $[-T/2, T/2]$, i.e. there is no function of positive energy over $[-T/2, T/2]$ that is orthogonal to each $\theta_k(t)$. Using this fact, $v(t) - u(t)$ has zero energy and is equal to 0 a.e. Thus $v(t) = \text{l.i.m.} \sum_k \hat{v}_k \mathrm{e}^{2\pi ikt/T}\,\mathrm{rect}(t/T)$. The energy equation then follows from (5.24). The final part of the theorem follows from the final part of Theorem 5.3.2. □

As seen by the above proof, the infinite-dimensional projection theorem can provide simple and intuitive proofs and interpretations of limiting arguments and the approximations suggested by those limits. Appendix 5.5 uses this theorem to prove both parts of the Plancherel theorem, the sampling theorem, and the aliasing theorem.

Another, more pragmatic, use of the theorem lies in providing a uniform way to treat all orthonormal expansions. As in the above Fourier series proof, though, the theorem does not necessarily provide a simple characterization of the space spanned by the orthonormal set. Fortunately, however, knowing that the truncated sinusoids span $[-T/2, T/2]$ shows us, by duality, that the T-spaced sinc functions span the space of baseband-limited $\mathcal{L}_2$ functions. Similarly, both the T-spaced truncated sinusoids and the sinc-weighted sinusoids span all of $\mathcal{L}_2$.

5.4 Summary

We have developed the theory of $\mathcal{L}_2$ waveforms, viewed as vectors in the inner product space known as signal space. The most important consequence of this viewpoint is

[9] An inner product space in which all Cauchy sequences have limits is said to be *complete*, and is called a *Hilbert space*. Thus the Riesz–Fischer theorem states that $\mathcal{L}_2$ is a Hilbert space.

[10] Again, see any basic text on real and complex analysis.

that *all* orthonormal expansions in $\mathcal{L}_2$ may be viewed in a common framework. The Fourier series is simply one example.

Another important consequence is that, as additional terms are added to a partial orthonormal expansion of an $\mathcal{L}_2$ waveform, the partial expansion changes by increasingly small amounts, approaching a limit in $\mathcal{L}_2$. A major reason for restricting attention to finite-energy waveforms (in addition to physical reality) is that as their energy is used up in different degrees of freedom (i.e. expansion coefficients), there is less energy available for other degrees of freedom, so that some sort of convergence must result. The $\mathcal{L}_2$ limit above simply makes this intuition precise.

Another consequence is the realization that if $\mathcal{L}_2$ functions are represented by orthonormal expansions, or approximated by partial orthonormal expansions, then there is no further need to deal with sophisticated mathematical issues such as $\mathcal{L}_2$-equivalence. Of course, how the truncated expansions converge may be tricky mathematically, but the truncated expansions themselves are very simple and friendly.

5.5 Appendix: Supplementary material and proofs

The first part of this appendix uses the inner product results of this chapter to prove the theorems about Fourier transforms in Chapter 4. The second part uses inner products to prove the theorems in Chapter 4 about sampling and aliasing. The final part discusses prolate spheroidal waveforms; these provide additional insight about the degrees of freedom in a time/bandwidth region.

5.5.1 The Plancherel theorem

Proof of Theorem 4.5.1 (Plancherel 1) The idea of the proof is to expand the waveform $\boldsymbol{u}$ into an orthonormal expansion for which the partial sums have known Fourier transforms; the $\mathcal{L}_2$ limit of these transforms is then identified as the $\mathcal{L}_2$ transform $\hat{\boldsymbol{u}}$ of $\boldsymbol{u}$.

First expand an arbitrary $\mathcal{L}_2$ function $u(t)$ in the T-spaced truncated sinusoid expansion, using $T = 1$. This expansion spans $\mathcal{L}_2$, and the orthogonal functions $e^{2\pi ikt}\,\text{rect}(t-m)$ are orthonormal since $T = 1$. Thus the infinite-dimensional projection, as specified by Theorem 5.3.2, is given by[11]

$$u(t) = \underset{n\to\infty}{\text{l.i.m.}}\, u^{(n)}(t), \qquad \text{where} \quad u^{(n)}(t) = \sum_{m=-n}^{n}\sum_{k=-n}^{n} \hat{u}_{k,m}\theta_{k,m}(t);$$

$$\theta_{k,m}(t) = e^{2\pi ikt}\,\text{rect}(t-m) \quad \text{and} \quad \hat{u}_{k,m} = \int u(t)\theta^*_{k,m}(t)\text{d}t.$$

[11] Note that $\{\theta_{k,m};\, k, m \in \mathbb{Z}\}$ is a countable set of orthonormal vectors, and they have been arranged in an order so that, for all $n \in \mathbb{Z}^+$, all terms with $|k| \le n$ and $|m| \le n$ come before all other terms.

Since $u^{(n)}(t)$ is time-limited, it is $\mathcal{L}_1$, and thus has a continuous Fourier transform which is defined pointwise by

$$\hat{u}^{(n)}(f) = \sum_{m=-n}^{n} \sum_{k=-n}^{n} \hat{u}_{k,m} \psi_{k,m}(f), \tag{5.27}$$

where $\psi_{k,m}(f) = e^{2\pi i f m} \operatorname{sinc}(f-k)$ is the k, m term of the T-spaced sinc-weighted orthonormal set with $T = 1$. By the final part of Theorem 5.3.2, the sequence of vectors $\hat{\boldsymbol{u}}^{(n)}$ converges to an $\mathcal{L}_2$ vector $\hat{\boldsymbol{u}}$ (equivalence class of functions) denoted as the Fourier transform of $u(t)$ and satisfying

$$\lim_{n\to\infty} \|\hat{\boldsymbol{u}} - \hat{\boldsymbol{u}}^{(n)}\| = 0. \tag{5.28}$$

This must now be related to the functions $u_A(t)$ and $\hat{u}_A(f)$ in the theorem. First, for each integer $\ell > n$, define

$$\hat{u}^{(n,\ell)}(f) = \sum_{m=-n}^{n} \sum_{k=-\ell}^{\ell} \hat{u}_{k,m} \psi_{k,m}(f). \tag{5.29}$$

Since this is a more complete partial expansion than $\hat{u}^{(n)}(f)$,

$$\|\hat{\boldsymbol{u}} - \hat{\boldsymbol{u}}^{(n)}\| \geq \|\hat{\boldsymbol{u}} - \hat{\boldsymbol{u}}^{(n,\ell)}\|.$$

In the limit $\ell \to \infty$, $\hat{\boldsymbol{u}}^{(n,\ell)}$ is the Fourier transform $\hat{u}_A(f)$ of $u_A(t)$ for $A = n + 1/2$. Combining this with (5.28), we obtain

$$\lim_{n\to\infty} \|\hat{\boldsymbol{u}} - \hat{\boldsymbol{u}}_{n+1/2}\| = 0. \tag{5.30}$$

Finally, taking the limit of the finite-dimensional energy equation,

$$\|\boldsymbol{u}^{(n)}\|^2 = \sum_{k=-n}^{n} \sum_{m=-n}^{n} |\hat{u}_{k,m}|^2 = \|\hat{\boldsymbol{u}}^{(n)}\|^2,$$

we obtain the $\mathcal{L}_2$ energy equation, $\|\boldsymbol{u}\|^2 = \|\hat{\boldsymbol{u}}\|^2$. This also shows that $\|\hat{\boldsymbol{u}} - \hat{\boldsymbol{u}}_A\|$ is monotonic in A so that (5.30) can be replaced by

$$\lim_{A\to\infty} \|\hat{\boldsymbol{u}} - \hat{\boldsymbol{u}}_{n+1/2}\| = 0.$$

□

Proof of Theorem 4.5.2 (Plancherel 2) By time/frequency duality with Theorem 4.5.1, we see that $\text{l.i.m.}_{B\to\infty} u_B(t)$ exists; we call this limit $\mathcal{F}^{-1}(\hat{u}(f))$. The only remaining thing to prove is that this inverse transform is $\mathcal{L}_2$-equivalent to the original $u(t)$. Note first that the Fourier transform of $\theta_{0,0}(t) = \operatorname{rect}(t)$ is $\operatorname{sinc}(f)$ and that the inverse transform, defined as above, is $\mathcal{L}_2$-equivalent to $\operatorname{rect}(t)$. By time and frequency shifts, we see that $u^{(n)}(t)$ is the inverse transform, defined as above, of $\hat{u}^{(n)}(f)$. It follows that $\lim_{n\to\infty} \|\mathcal{F}^{-1}(\hat{\boldsymbol{u}}) - \boldsymbol{u}^{(n)}\| = 0$, so we see that $\|\mathcal{F}^{-1}(\hat{\boldsymbol{u}}) - \boldsymbol{u}\| = 0$. □

As an example of the Plancherel theorem, let $h(t)$ be defined as 1 on the rationals in (0, 1) and as 0 elsewhere. Then $\boldsymbol{h}$ is both $\mathcal{L}_1$ and $\mathcal{L}_2$, and has a Fourier transform $\hat{h}(f) = 0$ which is continuous, $\mathcal{L}_1$, and $\mathcal{L}_2$. The inverse transform is also 0 and equal to $h(t)$ a.e.

The above function $h(t)$ is in some sense trivial, since it is $\mathcal{L}_2$-equivalent to the zero function. The next example to be discussed is $\mathcal{L}_2$, nonzero only on the interval $(0, 1)$, and thus also $\mathcal{L}_1$. This function is discontinuous and unbounded over every open interval within $(0, 1)$, and yet has a continuous Fourier transform. This example will illustrate how bizarre functions can have nice Fourier transforms and vice versa. It will also be used later to illustrate some properties of $\mathcal{L}_2$ functions.

Example 5.5.1 (a bizarre $\mathcal{L}_2$ and $\mathcal{L}_1$ function) List the rationals in (0,1) in order of increasing denominator, i.e. as $a_1 = 1/2$, $a_2 = 1/3$, $a_3 = 2/3$, $a_4 = 1/4$, $a_5 = 3/4$, $a_6 = 1/5, \ldots$ Define

$$g_n(t) = \begin{cases} 1 & \text{for } \; a_n \le t < a_n + 2^{-n-1}; \\ 0 & \text{elsewhere} \end{cases}$$

and

$$g(t) = \sum_{n=1}^{\infty} g_n(t).$$

Thus $g(t)$ is a sum of rectangular functions, one for each rational number, with the width of the function going to 0 rapidly with the index of the rational number (see Figure 5.3). The integral of $g(t)$ can be calculated as

$$\int_0^1 g(t)\mathrm{d}t = \sum_{n=1}^{\infty} \int g_n(t)\mathrm{d}t = \sum_{n=1}^{\infty} 2^{-n-1} = \frac{1}{2}.$$

Thus $g(t)$ is an $\mathcal{L}_1$ function, as illustrated in Figure 5.3.

Consider the interval $[2/3, 2/3 + 1/8)$ corresponding to the rectangle g_3 in Figure 5.3. Since the rationals are dense over the real line, there is a rational, say a_j, in the interior of this interval, and thus a new interval starting at a_j over which g_1, g_3, and g_j all have value 1; thus $g(t) \ge 3$ within this new interval. Moreover, this same

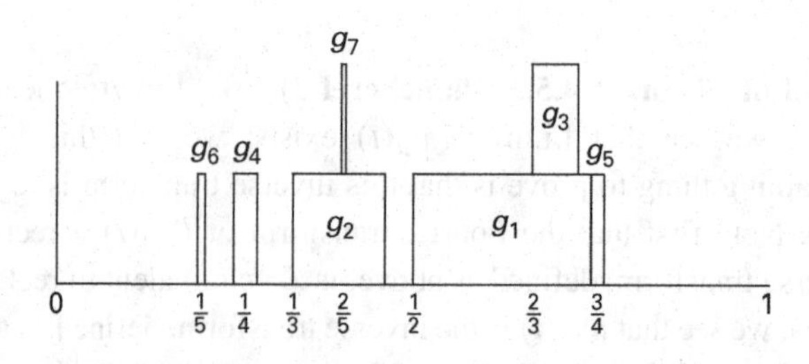

Figure 5.3. First seven terms of $\sum_i g_i(t)$.

argument can be repeated within this new interval, which again contains a rational, say $a_{j'}$. Thus there is an interval starting at $a_{j'}$ where g_1, g_3, g_j, and $g_{j'}$ are 1 and thus $g(t) \geq 4$.

Iterating this argument, we see that $[2/3, 2/3 + 1/8)$ contains subintervals within which $g(t)$ takes on arbitrarily large values. In fact, by taking the limit $a_1, a_3, a_j, a_{j'}, \ldots$, we find a limit point a for which $g(a) = \infty$. Moreover, we can apply the same argument to any open interval within $(0, 1)$ to show that $g(t)$ takes on infinite values within that interval.[12] More explicitly, for every $\varepsilon > 0$ and every $t \in (0, 1)$, there is a t' such that $|t - t'| < \varepsilon$ and $g(t') = \infty$. This means that $g(t)$ is discontinuous and unbounded in each region of $(0, 1)$.

The function $g(t)$ is also in $\mathcal{L}_2$ as seen in the following:

$$\int_0^1 g^2(t)\,\mathrm{d}t = \sum_{n,m} \int g_n(t) g_m(t) \mathrm{d}t \tag{5.31}$$

$$= \sum_n \int g_n^2(t)\,\mathrm{d}t + 2 \sum_n \sum_{m=n+1}^{\infty} \int g_n(t) g_m(t) \mathrm{d}t \tag{5.32}$$

$$\leq \frac{1}{2} + 2 \sum_n \sum_{m=n+1}^{\infty} \int g_m(t) \mathrm{d}t = \frac{3}{2}, \tag{5.33}$$

where in (5.33) we have used the fact that $g_n^2(t) = g_n(t)$ in the first term and $g_n(t) \leq 1$ in the second term.

In conclusion, $g(t)$ is both $\mathcal{L}_1$ and $\mathcal{L}_2$, but is discontinuous everywhere and takes on infinite values at points in every interval. The transform $\hat{g}(f)$ is continuous and $\mathcal{L}_2$ but not $\mathcal{L}_1$. The inverse transform, $g_B(t)$, of $\hat{g}(f)\,\mathrm{rect}(f/2B)$ is continuous, and converges in $\mathcal{L}_2$ to $g(t)$ as $B \to \infty$. For $B = 2^k$, the function $g_B(t)$ is roughly approximated by $g_1(t) + \cdots + g_k(t)$, all somewhat rounded at the edges.

This is a nice example of a continuous function $\hat{g}(f)$ which has a bizarre inverse Fourier transform. Note that $g(t)$ and the function $h(t)$ that is 1 on the rationals in (0,1) and 0 elsewhere are both discontinuous everywhere in (0,1). However, the function $h(t)$ is 0 a.e., and thus is weird only in an artificial sense. For most purposes, it is the same as the zero function. The function $g(t)$ is weird in a more fundamental sense. It cannot be made respectable by changing it on a countable set of points.

One should not conclude from this example that intuition cannot be trusted, or that it is necessary to take a few graduate math courses before feeling comfortable with functions. One can conclude, however, that the simplicity of the results about Fourier transforms and orthonormal expansions for $\mathcal{L}_2$ functions is truly extraordinary in view of the bizarre functions included in the $\mathcal{L}_2$ class.

[12] The careful reader will observe that $g(t)$ is not really a function $\mathbb{R} \to \mathbb{R}$, but rather a function from $\mathbb{R}$ to the extended set of real values including ∞. The set of t on which $g(t) = \infty$ has zero measure and this can be ignored in Lebesgue integration. Do not confuse a function that takes on an infinite value at some isolated point with a unit impulse at that point. The first integrates to 0 around the singularity, whereas the second is a generalized function that, by definition, integrates to 1.

In summary, Plancherel's theorem has taught us two things. First, Fourier transforms and inverse transforms exist for *all* $\mathcal{L}_2$ functions. Second, finite-interval and finite-bandwidth approximations become arbitrarily good (in the sense of $\mathcal{L}_2$ convergence) as the interval or the bandwidth becomes large.

5.5.2 The sampling and aliasing theorems

This section contains proofs of the sampling and aliasing theorems. The proofs are important and not available elsewhere in this form. However, they involve some careful mathematical analysis that might be beyond the interest and/or background of many students.

Proof of Theorem 4.6.2 (Sampling theorem) Let $\hat{u}(f)$ be an $\mathcal{L}_2$ function that is zero outside of $[-\mathsf{W}, \mathsf{W}]$. From Theorem 4.3.2, $\hat{u}(f)$ is $\mathcal{L}_1$, so, by Lemma 4.5.1,

$$u(t) = \int_{-\mathsf{W}}^{\mathsf{W}} \hat{u}(f) e^{2\pi i f t} \, df \tag{5.34}$$

holds at each $t \in \mathbb{R}$. We want to show that the sampling theorem expansion also holds at each t. By the DTFT theorem,

$$\hat{u}(f) = \operatorname*{l.i.m.}_{\ell\to\infty} \hat{u}^{(\ell)}(f), \quad \text{where} \quad \hat{u}^{(\ell)}(f) = \sum_{k=-\ell}^{\ell} u_k \hat{\phi}_k(f), \tag{5.35}$$

and where $\hat{\phi}_k(f) = e^{-2\pi i k f/2\mathsf{W}} \operatorname{rect}(f/2\mathsf{W})$ and

$$u_k = \frac{1}{2\mathsf{W}} \int_{-\mathsf{W}}^{\mathsf{W}} \hat{u}(f) e^{2\pi i k f/2\mathsf{W}} \, df. \tag{5.36}$$

Comparing (5.34) and (5.36), we see as before that $2\mathsf{W}u_k = u(k/2\mathsf{W})$. The functions $\hat{\phi}_k(f)$ are in $\mathcal{L}_1$, so the finite sum $\hat{u}^{(\ell)}(f)$ is also in $\mathcal{L}_1$. Thus the inverse Fourier transform,

$$u^{(\ell)}(t) = \int \hat{u}^{(\ell)}(f) df = \sum_{k=-\ell}^{\ell} u\left(\frac{k}{2\mathsf{W}}\right) \operatorname{sinc}(2\mathsf{W}t - k),$$

is defined pointwise at each t. For each $t \in \mathbb{R}$, the difference $u(t) - u^{(\ell)}(t)$ is then given by

$$u(t) - u^{(\ell)}(t) = \int_{-\mathsf{W}}^{\mathsf{W}} [\hat{u}(f) - \hat{u}^{(\ell)}(f)] e^{2\pi i f t} \, df.$$

This integral can be viewed as the inner product of $\hat{u}(f) - \hat{u}^{(\ell)}(f)$ and $e^{-2\pi i f t} \operatorname{rect}(f/2\mathsf{W})$, so, by the Schwarz inequality, we have

$$|u(t) - u^{(\ell)}(t)| \le \sqrt{2\mathsf{W}} \|\hat{\boldsymbol{u}} - \hat{\boldsymbol{u}}^{(\ell)}\|.$$

From the $\mathcal{L}_2$-convergence of the DTFT, the right side approaches 0 as $\ell \to \infty$, so the left side also approaches 0 for each t, establishing pointwise convergence. □

Proof of Theorem 4.6.3 [Sampling theorem for transmission] For a given W, assume that the sequence $\{u(k/2\mathrm{W});\, k \in \mathbb{Z}\}$ satisfies $\sum_k |u(k/2\mathrm{W})|^2 < \infty$. Define $u_k = (1/2\mathrm{W})u(k/2\mathrm{W})$ for each $k \in \mathbb{Z}$. By the DTFT theorem, there is a frequency function $\hat{u}(f)$, nonzero only over $[-\mathrm{W}, \mathrm{W}]$, that satisfies (4.60) and (4.61). By the sampling theorem, the inverse transform $u(t)$ of $\hat{u}(f)$ has the desired properties. □

Proof of Theorem 4.7.1 [Aliasing theorem] We start by separating $\hat{u}(f)$ into frequency slices $\{\hat{v}_m(f);\, m \in \mathbb{Z}\}$:

$$\hat{u}(f) = \sum_m \hat{v}_m(f), \quad \text{where} \quad \hat{v}_m(f) = \hat{u}(f)\,\mathrm{rect}^{\dagger}(fT - m). \tag{5.37}$$

The function $\mathrm{rect}^{\dagger}(f)$ is defined to equal 1 for $-1/2 < f \le 1/2$ and 0 elsewhere. It is $\mathcal{L}_2$-equivalent to $\mathrm{rect}(f)$, but gives us pointwise equality in (5.37). For each positive integer n, define $\hat{v}^{(n)}(f)$ as follows:

$$\hat{v}^{(n)}(f) = \sum_{m=-n}^{n} \hat{v}_m(f) = \begin{cases} \hat{u}(f) & \text{for } \frac{2n-1}{2T} < f \le \frac{2n+1}{2T}; \\ 0 & \text{elsewhere.} \end{cases} \tag{5.38}$$

It is shown in Exercise 5.16 that the given conditions on $\hat{u}(f)$ imply that $\hat{u}(f)$ is in $\mathcal{L}_1$. In conjunction with (5.38), this implies that

$$\lim_{n\to\infty} \int_{-\infty}^{\infty} |\hat{u}(f) - \hat{v}^{(n)}(f)|\, \mathrm{d}f = 0.$$

Since $\hat{u}(f) - \hat{v}^{(n)}(f)$ is in $\mathcal{L}_1$, the inverse transform at each t satisfies

$$\begin{aligned} \left|u(t) - v^{(n)}(t)\right| &= \left|\int_{-\infty}^{\infty} [\hat{u}(f) - \hat{v}^{(n)}(f)] \mathrm{e}^{2\pi i f t}\, \mathrm{d}f\right| \\ &\le \int_{-\infty}^{\infty} \left|\hat{u}(f) - \hat{v}^{(n)}(f)\right| \mathrm{d}f = \int_{|f| \ge (2n+1)/2T} |\hat{u}(f)|\, \mathrm{d}f. \end{aligned}$$

Since $\hat{u}(f)$ is in $\mathcal{L}_1$, the final integral above approaches 0 with increasing n. Thus, for each t, we have

$$u(t) = \lim_{n\to\infty} v^{(n)}(t). \tag{5.39}$$

Next define $\hat{s}_m(f)$ as the frequency slice $\hat{v}_m(f)$ shifted down to baseband, i.e.

$$\hat{s}_m(f) = \hat{v}_m\left(f - \frac{m}{T}\right) = \hat{u}\left(f - \frac{m}{T}\right) \mathrm{rect}^{\dagger}(fT). \tag{5.40}$$

Applying the sampling theorem to $v_m(t)$, we obtain

$$v_m(t) = \sum_k v_m(kT)\,\mathrm{sinc}\left(\frac{t}{T} - k\right) \mathrm{e}^{2\pi i m t/T}. \tag{5.41}$$

Applying the frequency shift relation to (5.40), we see that $s_m(t) = v_m(t)\mathrm{e}^{-2\pi i f t}$, and thus

$$s_m(t) = \sum_k v_m(kT)\,\mathrm{sinc}\left(\frac{t}{T} - k\right). \tag{5.42}$$

Now define $\hat{s}^{(n)}(f) = \sum_{m=-n}^{n} \hat{s}_m(f)$. From (5.40), we see that $\hat{s}^{(n)}(f)$ is the aliased version of $\hat{v}^{(n)}(f)$, as illustrated in Figure 4.10. The inverse transform is then given by

$$s^{(n)}(t) = \sum_{k=-\infty}^{\infty} \sum_{m=-n}^{n} v_m(kT) \operatorname{sinc}\left(\frac{t}{T} - k\right). \tag{5.43}$$

We have interchanged the order of summation, which is valid since the sum over m is finite. Finally, define $\hat{s}(f)$ to be the "folded" version of $\hat{u}(f)$ summing over all m, i.e.

$$\hat{s}(f) = \underset{n\to\infty}{\text{l.i.m.}}\, \hat{s}^{(n)}(f). \tag{5.44}$$

Exercise 5.16 shows that this limit converges in the $\mathcal{L}_2$ sense to an $\mathcal{L}_2$ function $\hat{s}(f)$. Exercise 4.38 provides an example where $\hat{s}(f)$ is not in $\mathcal{L}_2$ if the condition $\lim_{|f|\to\infty} \hat{u}(f)|f|^{1+\varepsilon} = 0$ is not satisfied.

Since $\hat{s}(f)$ is in $\mathcal{L}_2$ and is 0 outside $[-1/2T, 1/2T]$, the sampling theorem shows that the inverse transform $s(t)$ satisfies

$$s(t) = \sum_k s(kT) \operatorname{sinc}\left(\frac{t}{T} - k\right). \tag{5.45}$$

Combining this with (5.43), we obtain

$$s(t) - s^{(n)}(t) = \sum_k \left[s(kT) - \sum_{m=-n}^{n} v_m(kT) \right] \operatorname{sinc}\left(\frac{t}{T} - k\right). \tag{5.46}$$

From (5.44), we see that $\lim_{n\to\infty} \|s - s^{(n)}\| = 0$, and thus

$$\lim_{n\to\infty} \sum_k |s(kT) - v^{(n)}(kT)|^2 = 0.$$

This implies that $s(kT) = \lim_{n\to\infty} v^{(n)}(kT)$ for each integer k. From (5.39), we also have $u(kT) = \lim_{n\to\infty} v^{(n)}(kT)$, and thus $s(kT) = u(kT)$ for each $k \in \mathbb{Z}$:

$$s(t) = \sum_k u(kT) \operatorname{sinc}\left(\frac{t}{T} - k\right). \tag{5.47}$$

This shows that (5.44) implies (5.47). Since $s(t)$ is in $\mathcal{L}_2$, it follows that $\sum_k |u(kT)|^2 < \infty$. Conversely, (5.47) defines a unique $\mathcal{L}_2$ function, and thus its Fourier transform must be $\mathcal{L}_2$-equivalent to $\hat{s}(f)$, as defined in (5.44). □

5.5.3 Prolate spheroidal waveforms

The prolate spheroidal waveforms (see Slepian and Pollak (1961)) are a set of orthonormal functions that provide a more precise way to view the degree-of-freedom arguments of Section 4.7.2. For each choice of baseband bandwidth W and time interval $[-T/2, T/2]$, these functions form an orthonormal set $\{\boldsymbol{\phi}_0(t), \boldsymbol{\phi}_1(t), \dots\}$ of real $\mathcal{L}_2$ functions time-limited to $[-T/2, T/2]$. In a sense to be described, these functions have the maximum possible energy in the frequency band $(-\mathsf{W}, \mathsf{W})$ subject to their constraint to $[-T/2, T/2]$.

To be more precise, for each $n \geq 0$ let $\hat{\phi}_n(f)$ be the Fourier transform of $\phi_n(t)$, and define

$$\hat{\theta}_n(f) = \begin{cases} \hat{\phi}_n(f) & \text{for } -\mathsf{W} < t < \mathsf{W}; \\ 0 & \text{elsewhere.} \end{cases} \tag{5.48}$$

That is, $\theta_n(t)$ is $\phi_n(t)$ truncated in frequency to $(-\mathsf{W}, \mathsf{W})$; equivalently, $\theta_n(t)$ may be viewed as the result of passing $\phi_n(t)$ through an ideal lowpass filter.

The function $\phi_0(t)$ is chosen to be the normalized function $\phi_0(t) : (-T/2, T/2) \to \mathbb{R}$ that maximizes the energy in $\theta_0(t)$. We will not show how to solve this optimization problem. However, $\phi_0(t)$ turns out to resemble $\sqrt{1/T}\,\mathrm{rect}(t/T)$, except that it is rounded at the edges to reduce the out-of-band energy.

Similarly, for each $n > 0$, the function $\phi_n(t)$ is chosen to be the normalized function $\{\phi_n(t) : (-T/2, T/2) \to \mathbb{R}\}$ that is orthonormal to $\phi_m(t)$ for each $m < n$ and, subject to this constraint, maximizes the energy in $\theta_n(t)$.

Finally, define $\lambda_n = \|\boldsymbol{\theta}_n\|^2$. It can be shown that $1 > \lambda_0 > \lambda_1 > \cdots$. We interpret λ_n as the fraction of energy in $\boldsymbol{\phi}_n$ that is baseband-limited to $(-\mathsf{W}, \mathsf{W})$. The number of degrees of freedom in $(-T/2, T/2)$, $(-\mathsf{W}, \mathsf{W})$, is then reasonably defined as the largest n for which λ_n is close to 1.

The values λ_n depend on the product $T\mathsf{W}$, so they can be denoted by $\lambda_n(T\mathsf{W})$. The main result about prolate spheroidal wave functions, which we do not prove, is that, for any $\varepsilon > 0$,

$$\lim_{T\mathsf{W}\to\infty} \lambda_n(T\mathsf{W}) = \begin{cases} 1 & \text{for } n < 2T\mathsf{W}(1-\varepsilon); \\ 0 & \text{for } n > 2T\mathsf{W}(1+\varepsilon). \end{cases}$$

This says that when $T\mathsf{W}$ is large, there are close to $2T\mathsf{W}$ orthonormal functions for which most of the energy in the time-limited function is also frequency-limited, but there are not significantly more orthonormal functions with this property.

The prolate spheroidal wave functions $\phi_n(t)$ have many other remarkable properties, of which we list a few:

- for each n, $\phi_n(t)$ is continuous and has n zero crossings;
- $\phi_n(t)$ is even for n even and odd for n odd;
- $\theta_n(t)$ is an orthogonal set of functions;
- in the interval $(-T/2, T/2)$, $\theta_n(t) = \lambda_n \phi_n(t)$.

5.6 Exercises

5.1 (Basis) Prove Theorem 5.1.1 by first suggesting an algorithm that establishes the first item and then an algorithm to establish the second item.

5.2 Show that the $\mathbf{0}$ vector can be part of a spanning set but cannot be part of a linearly independent set.

5.3 (Basis) Prove that if a set of n vectors *uniquely* spans a vector space $\mathcal{V}$, in the sense that each $\boldsymbol{v} \in \mathcal{V}$ has a unique representation as a linear combination of the n vectors, then those n vectors are linearly independent and $\mathcal{V}$ is an n-dimensional space.

5.4 ($\mathbb{R}^2$)

(a) Show that the vector space $\mathbb{R}^2$ with vectors $\{\boldsymbol{v} = (v_1, v_2)\}$ and inner product $\langle \boldsymbol{v}, \boldsymbol{u} \rangle = v_1 u_1 + v_2 u_2$ satisfies the axioms of an inner product space.

(b) Show that, in the Euclidean plane, the length of $\boldsymbol{v}$ (i.e. the distance from $\mathbf{0}$ to $\boldsymbol{v}$) is $\|\boldsymbol{v}\|$.

(c) Show that the distance from $\boldsymbol{v}$ to $\boldsymbol{u}$ is $\|\boldsymbol{v} - \boldsymbol{u}\|$.

(d) Show that $\cos(\angle(\boldsymbol{v}, \boldsymbol{u})) = \langle \boldsymbol{v}, \boldsymbol{u} \rangle / \|\boldsymbol{v}\| \, \|\boldsymbol{u}\|$; assume that $\|\boldsymbol{u}\| > 0$ and $\|\boldsymbol{v}\| > 0$.

(e) Suppose that the definition of the inner product is now changed to $\langle \boldsymbol{v}, \boldsymbol{u} \rangle = v_1 u_1 + 2 v_2 u_2$. Does this still satisfy the axioms of an inner product space? Do the length formula and the angle formula still correspond to the usual Euclidean length and angle?

5.5 Consider $\mathbb{C}^n$ and define $\langle \boldsymbol{v}, \boldsymbol{u} \rangle$ as $\sum_{j=1}^{n} c_j v_j u_j^*$, where $c_1, \ldots, c_n$ are complex numbers. For each of the following cases, determine whether $\mathbb{C}^n$ must be an inner product space and explain why or why not.

(a) The c_j are all equal to the same positive real number.

(b) The c_j are all positive real numbers.

(c) The c_j are all nonnegative real numbers.

(d) The c_j are all equal to the same nonzero complex number.

(e) The c_j are all nonzero complex numbers.

5.6 (Triangle inequality) Prove the triangle inequality, (5.10). [Hint. Expand $\|\boldsymbol{v} + \boldsymbol{u}\|^2$ into four terms and use the Schwarz inequality on each of the two cross terms.]

5.7 Let $\boldsymbol{u}$ and $\boldsymbol{v}$ be orthonormal vectors in $\mathbb{C}^n$ and let $\boldsymbol{w} = w_u \boldsymbol{u} + w_v \boldsymbol{v}$ and $\boldsymbol{x} = x_u \boldsymbol{u} + x_v \boldsymbol{v}$ be two vectors in the subspace spanned by $\boldsymbol{u}$ and $\boldsymbol{v}$.

(a) Viewing $\boldsymbol{w}$ and $\boldsymbol{x}$ as vectors in the subspace $\mathbb{C}^2$, find $\langle \boldsymbol{w}, \boldsymbol{x} \rangle$.

(b) Now view $\boldsymbol{w}$ and $\boldsymbol{x}$ as vectors in $\mathbb{C}^n$, e.g. $\boldsymbol{w} = (w_1, \ldots, w_n)$, where $w_j = w_u u_j + w_v v_j$ for $1 \le j \le n$. Calculate $\langle \boldsymbol{w}, \boldsymbol{x} \rangle$ this way and show that the answer agrees with that in part (a).

5.8 ($\mathcal{L}_2$ inner product) Consider the vector space of $\mathcal{L}_2$ functions $\{u(t) : \mathbb{R} \to \mathbb{C}\}$. Let $\boldsymbol{v}$ and $\boldsymbol{u}$ be two vectors in this space represented as $v(t)$ and $u(t)$. Let the inner product be defined by

$$\langle \boldsymbol{v}, \boldsymbol{u} \rangle = \int_{-\infty}^{\infty} v(t) u^*(t) \mathrm{d}t.$$

(a) Assume that $u(t) = \sum_{k,m} \hat{u}_{k,m} \theta_{k,m}(t)$, where $\{\theta_{k,m}(t)\}$ is an orthogonal set of functions each of energy T. Assume that $v(t)$ can be expanded similarly. Show that

$$\langle \boldsymbol{u}, \boldsymbol{v} \rangle = T \sum_{k,m} \hat{u}_{k,m} \hat{v}_{k,m}^*.$$

(b) Show that $\langle \boldsymbol{u}, \boldsymbol{v} \rangle$ is finite. Do not use the Schwarz inequality, because the purpose of this exercise is to show that $\mathcal{L}_2$ is an inner product space, and the Schwarz inequality is based on the assumption of an inner product space.

Use the result in (a) along with the properties of complex numbers (you can use the Schwarz inequality for the 1D vector space $\mathbb{C}^1$ if you choose).

(c) Why is this result necessary in showing that $\mathcal{L}_2$ is an inner product space?

5.9 ($\mathcal{L}_2$ inner product) Given two waveforms $\boldsymbol{u}_1, \boldsymbol{u}_2 \in \mathcal{L}_2$, let $\mathcal{V}$ be the set of all waveforms $\boldsymbol{v}$ that are equidistant from $\boldsymbol{u}_1$ and $\boldsymbol{u}_2$. Thus

$$\mathcal{V} = \{\boldsymbol{v} : \|\boldsymbol{v} - \boldsymbol{u}_1\| = \|\boldsymbol{v} - \boldsymbol{u}_2\|\}.$$

(a) Is $\mathcal{V}$ a vector subspace of $\mathcal{L}_2$?

(b) Show that

$$\mathcal{V} = \left\{\boldsymbol{v} : \Re(\langle \boldsymbol{v}, \boldsymbol{u}_2 - \boldsymbol{u}_1 \rangle) = \frac{\|\boldsymbol{u}_2\|^2 - \|\boldsymbol{u}_1\|^2}{2}\right\}.$$

(c) Show that $(\boldsymbol{u}_1 + \boldsymbol{u}_2)/2 \in \mathcal{V}$.

(d) Give a geometric interpretation for $\mathcal{V}$.

5.10 (Sampling) For any $\mathcal{L}_2$ function $\boldsymbol{u}$ bandlimited to $[-\mathrm{W}, \mathrm{W}]$ and any t, let $a_k = u(k/2\mathrm{W})$ and let $b_k = \mathrm{sinc}(2\mathrm{W}t - k)$. Show that $\sum_k |a_k|^2 < \infty$ and $\sum_k |b_k|^2 < \infty$. Use this to show that $\sum_k |a_k b_k| < \infty$. Use this to show that the sum in the sampling equation (4.65) converges for each t.

5.11 (Projection) Consider the following set of functions $\{u_m(t)\}$ for integer $m \geq 0$:

$$u_0(t) = \begin{cases} 1 & 0 \leq t < 1, \\ 0 & \text{otherwise}; \end{cases}$$

$$\vdots$$

$$u_m(t) = \begin{cases} 1 & 0 \leq t < 2^{-m}, \\ 0 & \text{otherwise}; \end{cases}$$

$$\vdots$$

Consider these functions as vectors $\boldsymbol{u}_0, \boldsymbol{u}_1, \ldots$ over real $\mathcal{L}_2$ vector space. Note that $\boldsymbol{u}_0$ is normalized; we denote it as $\boldsymbol{\phi}_0 = \boldsymbol{u}_0$.

(a) Find the projection $(\boldsymbol{u}_1)_{|\boldsymbol{\phi}_0}$ of $\boldsymbol{u}_1$ onto $\boldsymbol{\phi}_0$, find the perpendicular $(\boldsymbol{u}_1)_{\perp\boldsymbol{\phi}_0}$, and find the normalized form $\boldsymbol{\phi}_1$ of $(\boldsymbol{u}_1)_{\perp\boldsymbol{\phi}_0}$. Sketch each of these as functions of t.

(b) Express $u_1(t - 1/2)$ as a linear combination of $\boldsymbol{\phi}_0$ and $\boldsymbol{\phi}_1$. Express (in words) the subspace of real $\mathcal{L}_2$ spanned by $u_1(t)$ and $u_1(t - 1/2)$. What is the subspace $\mathcal{S}_1$ of real $\mathcal{L}_2$ spanned by $\boldsymbol{\phi}_0$ and $\boldsymbol{\phi}_1$?

(c) Find the projection $(\boldsymbol{u}_2)_{|\mathcal{S}_1}$ of $\boldsymbol{u}_2$ onto $\mathcal{S}_1$, find the perpendicular $(\boldsymbol{u}_2)_{\perp\mathcal{S}_1}$, and find the normalized form of $(\boldsymbol{u}_2)_{\perp\mathcal{S}_1}$. Denote this normalized form as $\boldsymbol{\phi}_{2,0}$; it will be clear shortly why a double subscript is used here. Sketch $\boldsymbol{\phi}_{2,0}$ as a function of t.

(d) Find the projection of $u_2(t - 1/2)$ onto $\mathcal{S}_1$ and find the perpendicular $u_2(t - 1/2)_{\perp\mathcal{S}_1}$. Denote the normalized form of this perpendicular by $\boldsymbol{\phi}_{2,1}$. Sketch $\boldsymbol{\phi}_{2,1}$ as a function of t and explain why $\langle \boldsymbol{\phi}_{2,0}, \boldsymbol{\phi}_{2,1} \rangle = 0$.

(e) Express $u_2(t-1/4)$ and $u_2(t-3/4)$ as linear combinations of $\{\boldsymbol{\phi}_0, \boldsymbol{\phi}_1, \boldsymbol{\phi}_{2,0}, \boldsymbol{\phi}_{2,1}\}$. Let $\mathcal{S}_2$ be the subspace of real $\mathcal{L}_2$ spanned by $\boldsymbol{\phi}_0, \boldsymbol{\phi}_1, \boldsymbol{\phi}_{2,0}, \boldsymbol{\phi}_{2,1}$ and describe this subspace in words.

(f) Find the projection $(\boldsymbol{u}_3)_{|\mathcal{S}_2}$ of $\boldsymbol{u}_3$ onto $\mathcal{S}_2$, find the perpendicular $(\boldsymbol{u}_2)_{\perp\mathcal{S}_1}$, and find its normalized form, $\boldsymbol{\phi}_{3,0}$. Sketch $\boldsymbol{\phi}_{3,0}$ as a function of t.

(g) For $j=1,2,3$, find $u_3(t-j/4)_{\perp\mathcal{S}_2}$ and find its normalized form $\boldsymbol{\phi}_{3,j}$. Describe the subspace $\mathcal{S}_3$ spanned by $\boldsymbol{\phi}_0, \boldsymbol{\phi}_1, \boldsymbol{\phi}_{2,0}, \boldsymbol{\phi}_{2,1}, \boldsymbol{\phi}_{3,0}, \ldots, \boldsymbol{\phi}_{3,3}$.

(h) Consider iterating this process to form $\mathcal{S}_4, \mathcal{S}_5, \ldots$ What is the dimension of $\mathcal{S}_m$? Describe this subspace. Describe the projection of an arbitrary real $\mathcal{L}_2$ function constrained to the interval [0,1) onto $\mathcal{S}_m$.

5.12 (Orthogonal subspaces) For any subspace $\mathcal{S}$ of an inner product space $\mathcal{V}$, define $\mathcal{S}^\perp$ as the set of vectors $\boldsymbol{v} \in \mathcal{V}$ that are orthogonal to all $\boldsymbol{w} \in \mathcal{S}$.

(a) Show that $\mathcal{S}^\perp$ is a subspace of $\mathcal{V}$.

(b) Assuming that $\mathcal{S}$ is finite-dimensional, show that any $\boldsymbol{u} \in \mathcal{V}$ can be uniquely decomposed into $\boldsymbol{u} = \boldsymbol{u}_{|\mathcal{S}} + \boldsymbol{u}_{\perp\mathcal{S}}$, where $\boldsymbol{u}_{|\mathcal{S}} \in \mathcal{S}$ and $\boldsymbol{u}_{\perp\mathcal{S}} \in \mathcal{S}^\perp$.

(c) Assuming that $\mathcal{V}$ is finite-dimensional, show that $\mathcal{V}$ has an orthonormal basis where some of the basis vectors form a basis for $\boldsymbol{S}$ and the remaining basis vectors form a basis for $\mathcal{S}^\perp$.

5.13 (Orthonormal expansion) Expand the function $\text{sinc}(3t/2)$ as an orthonormal expansion in the set of functions $\{\text{sinc}(t-n)\,;\ -\infty < n < \infty\}$.

5.14 (Bizarre function)

(a) Show that the pulses $g_n(t)$ in Example 5.5.1 of Appendix 5.5.1 overlap each other either completely or not at all.

(b) Modify each pulse $g_n(t)$ to $h_n(t)$ as follows: let $h_n(t) = g_n(t)$ if $\sum_{i=1}^{n-1} g_i(t)$ is even and let $h_n(t) = -g_n(t)$ if $\sum_{i=1}^{n-1} g_i(t)$ is odd. Show that $\sum_{i=1}^{n} h_i(t)$ is bounded between 0 and 1 for each $t \in (0,1)$ and each $n \geq 1$.

(c) Show that there are a countably infinite number of points t at which $\sum_n h_n(t)$ does not converge.

5.15 (Parseval) Prove Parseval's relation, (4.44), for $\mathcal{L}_2$ functions. Use the same argument as used to establish the energy equation in the proof of Plancherel's theorem.

5.16 (Aliasing theorem) Assume that $\hat{u}(f)$ is $\mathcal{L}_2$ and $\lim_{|f|\to\infty} \hat{u}(f)|f|^{1+\varepsilon} = 0$ for some $\varepsilon > 0$.

(a) Show that for large enough $A > 0$, $|\hat{u}(f)| \leq |f|^{-1-\varepsilon}$ for $|f| > A$.

(b) Show that $\hat{u}(f)$ is $\mathcal{L}_1$. [Hint. For the A above, split the integral $\int |\hat{u}(f)|\,df$ into one integral for $|f| > A$ and another for $|f| \leq A$.]

(c) Show that, for $T = 1$, $\hat{s}(f)$ as defined in (5.44), satisfies

$$|\hat{s}(f)| \leq \sqrt{(2A+1)\sum_{|m|\leq A} |\hat{u}(f+m)|^2} + \sum_{m\geq A} m^{-1-\varepsilon}.$$

(d) Show that $\hat{s}(f)$ is $\mathcal{L}_2$ for $T = 1$. Use scaling to show that $\hat{s}(f)$ is $\mathcal{L}_2$ for any $T > 0$.

6 Channels, modulation, and demodulation

6.1 Introduction

Digital modulation (or channel encoding) is the process of converting an input sequence of bits into a waveform suitable for transmission over a communication channel. Demodulation (channel decoding) is the corresponding process at the receiver of converting the received waveform into a (perhaps noisy) replica of the input bit sequence. Chapter 1 discussed the reasons for using a bit sequence as the interface between an arbitrary source and an arbitrary channel, and Chapters 2 and 3 discussed how to encode the source output into a bit sequence.

Chapters 4 and 5 developed the signal-space view of waveforms. As explained in those chapters, the source and channel waveforms of interest can be represented as real or complex[1] $\mathcal{L}_2$ vectors. Any such vector can be viewed as a conventional function of time, $x(t)$. Given an orthonormal basis $\{\phi_1(t), \phi_2(t), \dots\}$ of $\mathcal{L}_2$, any such $x(t)$ can be represented as

$$x(t) = \sum_j x_j \phi_j(t). \tag{6.1}$$

Each x_j in (6.1) can be uniquely calculated from $x(t)$, and the above series converges in $\mathcal{L}_2$ to $x(t)$. Moreover, starting from any sequence satisfying $\sum_j |x_j|^2 < \infty$, there is an $\mathcal{L}_2$ function $x(t)$ satisfying (6.1) with $\mathcal{L}_2$-convergence. This provides a simple and generic way of going back and forth between functions of time and sequences of numbers. The basic parts of a modulator will then turn out to be a procedure for mapping a sequence of binary digits into a sequence of real or complex numbers, followed by the above approach for mapping a sequence of numbers into a waveform.

[1] As explained later, the actual transmitted waveforms are real. However, they are usually bandpass real waveforms that are conveniently represented as complex baseband waveforms.

In most cases of modulation, the set of waveforms $\phi_1(t), \phi_2(t), \ldots$ in (6.1) will be chosen not as a basis for $\mathcal{L}_2$ but as a basis for some subspace[2] of $\mathcal{L}_2$ such as the set of functions that are baseband-limited to some frequency W_b or passband-limited to some range of frequencies. In some cases, it will also be desirable to use a sequence of waveforms that are not orthonormal.

We can view the mapping from bits to numerical signals and the conversion of signals to a waveform as separate layers. The demodulator then maps the received waveform to a sequence of received signals, which is then mapped to a bit sequence, hopefully equal to the input bit sequence. A major objective in designing the modulator and demodulator is to maximize the rate at which bits enter the encoder, subject to the need to retrieve the original bit stream with a suitably small error rate. Usually this must be done subject to constraints on the transmitted power and bandwidth. In practice there are also constraints on delay, complexity, compatibility with standards, etc., but these need not be a major focus here.

Example 6.1.1 As a particularly simple example, suppose a sequence of binary symbols enters the encoder at T-spaced instants of time. These symbols can be mapped into real numbers using the mapping $0 \to +1$ and $1 \to -1$. The resulting sequence $u_1, u_2, \ldots$ of real numbers is then mapped into a transmitted waveform given by

$$u(t) = \sum_k u_k \operatorname{sinc}\left(\frac{t}{T} - k\right). \tag{6.2}$$

This is baseband-limited to $\mathsf{W}_\mathrm{b} = 1/2T$. At the receiver, in the absence of noise, attenuation, and other imperfections, the received waveform is $u(t)$. This can be sampled at times $T, 2T, \ldots$ to retrieve $u_1, u_2, \ldots,$ which can be decoded into the original binary symbols.

The above example contains rudimentary forms of the two layers discussed above. The first is the mapping of binary symbols into numerical signals[3] and the second is the conversion of the sequence of signals into a waveform. In general, the set of T-spaced sinc functions in (6.2) can be replaced by any other set of orthogonal functions (or even nonorthogonal functions). Also, the mapping $0 \to +1$, $1 \to -1$ can be generalized by segmenting the binary stream into b-tuples of binary symbols, which can then be mapped into n-tuples of real or complex numbers. The set of 2^b possible n-tuples resulting from this mapping is called a *signal constellation*.

[2] Equivalently, $\phi_1(t), \phi_2(t), \ldots$ can be chosen as a basis of $\mathcal{L}_2$, but the set of indices for which x_j is allowed to be nonzero can be restricted.

[3] The word *signal* is often used in the communication literature to refer to symbols, vectors, waveforms, or almost anything else. Here we use it only to refer to real or complex numbers (or n-tuples of numbers) in situations where the numerical properties are important. For example, in (6.2) the *signals* (numerical values) $u_1, u_2, \ldots$ determine the real-valued waveform $u(t)$, whereas the binary input *symbols* could be 'Alice' and 'Bob' as easily as 0 and 1.

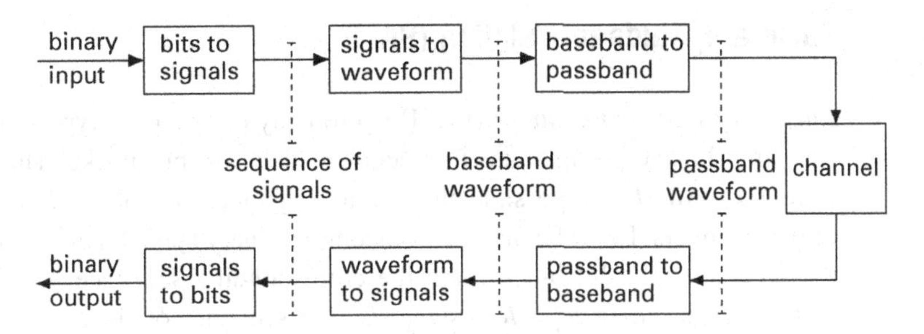

Figure 6.1. Layers of a modulator (channel encoder) and demodulator (channel decoder).

Modulators usually include a third layer, which maps a baseband-encoded waveform, such as $u(t)$ in (6.2), into a passband waveform $x(t) = \Re\{u(t)e^{2\pi i f_c t}\}$ centered on a given carrier frequency f_c. At the decoder, this passband waveform is mapped back to baseband before the other components of decoding are performed. This frequency conversion operation at encoder and decoder is often referred to as modulation and demodulation, but it is more common today to use the word modulation for the entire process of mapping bits to waveforms. Figure 6.1 illustrates these three layers.

We have illustrated the channel as a one-way device going from source to destination. Usually, however, communication goes both ways, so that a physical location can send data to another location and also receive data from that remote location. A physical device that both encodes data going out over a channel and also decodes oppositely directed data coming in from the channel is called a *modem* (for modulator/demodulator). As described in Chapter 1, feedback on the reverse channel can be used to request retransmissions on the forward channel, but in practice this is usually done as part of an automatic retransmission request (ARQ) strategy in the data link control layer. Combining coding with more sophisticated feedback strategies than ARQ has always been an active area of communication and information-theoretic research, but it will not be discussed here for the following reasons:

- it is important to understand communication in a single direction before addressing the complexities of two directions;
- feedback does not increase channel capacity for typical channels (see Shannon (1956));
- simple error detection and retransmission is best viewed as a topic in data networks.

There is an interesting analogy between analog source coding and digital modulation. With analog source coding, an analog waveform is first mapped into a sequence of real or complex numbers (e.g. the coefficients in an orthogonal expansion). This sequence of signals is then quantized into a sequence of symbols from a discrete alphabet, and finally the symbols are encoded into a binary sequence. With modulation, a sequence of bits is encoded into a sequence of signals from a signal constellation. The elements of this constellation are real or complex points in one or several dimensions. This sequence of signal points is then mapped into a waveform by the inverse of the process for converting waveforms into sequences.

6.2 Pulse amplitude modulation (PAM)

Pulse amplitude modulation[4] (PAM) is probably the simplest type of modulation. The incoming binary symbols are first segmented into b-bit blocks. There is a mapping from the set of $M = 2^b$ possible blocks into a signal constellation $\mathcal{A} = \{a_1, a_2, \ldots, a_M\}$ of real numbers. Let R be the rate of incoming binary symbols in bits per second. Then the sequence of b-bit blocks, and the corresponding sequence $u_1, u_2, \ldots$ of M-ary signals, has a rate of $R_s = R/b$ signals/s. The sequence of signals is then mapped into a waveform $u(t)$ by the use of time shifts of a basic pulse waveform $p(t)$, i.e.

$$u(t) = \sum_k u_k p(t - kT), \tag{6.3}$$

where $T = 1/R_s$ is the interval between successive signals. The special case where $b = 1$ is called *binary* PAM and the case $b > 1$ is called *multilevel* PAM. Example 6.1.1 is an example of binary PAM where the basic pulse shape $p(t)$ is a sinc function. Comparing (6.1) with (6.3), we see that PAM is a special case of digital modulation in which the underlying set of functions $\phi_1(t), \phi_2(t), \ldots$ is replaced by functions that are T-spaced time shifts of a basic function $p(t)$.

In what follows, signal constellations (i.e. the outer layer in Figure 6.1) are discussed in Sections 6.2.1 and 6.2.2. Pulse waveforms (i.e. the middle layer in Figure 6.1) are then discussed in Sections 6.2.3 and 6.2.4. In most cases,[5] the pulse waveform $p(t)$ is a baseband waveform, and the resulting modulated waveform $u(t)$ is then modulated up to some passband (i.e. the inner layer in Figure 6.1). Section 6.4 discusses modulation from baseband to passband and back.

6.2.1 Signal constellations

A *standard* M-PAM signal constellation $\mathcal{A}$ (see Figure 6.2) consists of $M = 2^b$ d-spaced real numbers located symmetrically about the origin; i.e.,

$$\mathcal{A} = \left\{ \frac{-d(M-1)}{2}, \ldots, \frac{-d}{2}, \frac{d}{2}, \ldots, \frac{d(M-1)}{2} \right\}.$$

In other words, the signal points are the same as the representation points of a symmetric M-point uniform scalar quantizer.

If the incoming bits are independent equiprobable random symbols (which is a good approximation with effective source coding), then each signal u_k is a sample value of a random variable U_k that is equiprobable over the constellation (alphabet) $\mathcal{A}$. Also the

[4] This terminology comes from analog amplitude modulation, where a baseband waveform is modulated up to some passband for communication. For digital communication, the more interesting problem is turning a bit stream into a waveform at baseband.

[5] Ultra-wide-band modulation (UWB) is an interesting modulation technique where the transmitted waveform is essentially a baseband PAM system over a "baseband" of multiple gigahertz. This is discussed briefly in Chapter 9.

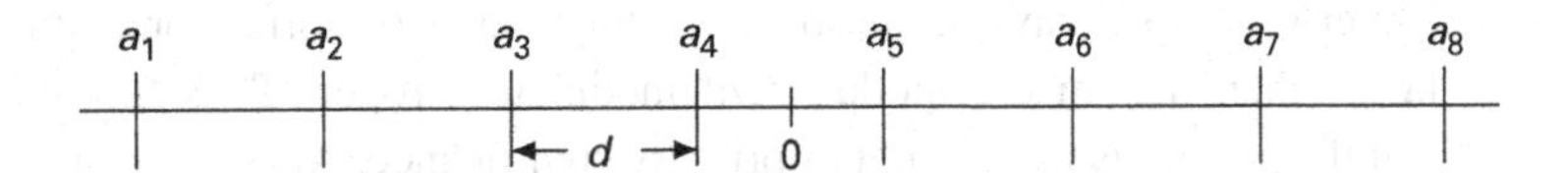

Figure 6.2. An 8-PAM signal set.

sequence $U_1, U_2, \ldots$ is independent and identically distributed (iid). As derived in Exercise 6.1, the mean squared signal value, or "energy per signal," $E_s = \mathsf{E}[U_k^2]$, is then given by

$$E_s = \frac{d^2(M^2-1)}{12} = \frac{d^2(2^{2b}-1)}{12}. \tag{6.4}$$

For example, for $M = 2, 4$, and 8, we have $E_s = d^2/4$, $5d^2/4$, and $21d^2/4$, respectively.

For $b > 2$, $2^{2b} - 1$ is approximately 2^{2b}, so we see that each unit increase in b increases E_s by a factor of 4. Thus, increasing the rate R by increasing b requires impractically large energy for large b.

Before explaining why standard M-PAM is a good choice for PAM and what factors affect the choice of constellation size M and distance d, a brief introduction to channel imperfections is required.

6.2.2 Channel imperfections: a preliminary view

Physical waveform channels are always subject to propagation delay, attenuation, and noise. Many wireline channels can be reasonably modeled using only these degradations, whereas wireless channels are subject to other degradations discussed in Chapter 9. This section provides a preliminary look at delay, then attenuation, and finally noise.

The time reference at a communication receiver is conventionally delayed relative to that at the transmitter. If a waveform $u(t)$ is transmitted, the received waveform (in the absence of other distortion) is $u(t - \tau)$, where τ is the delay due to propagation and filtering. The receiver clock (as a result of tracking the transmitter's timing) is ideally delayed by τ, so that the received waveform, according to the receiver clock, is $u(t)$. With this convention, the channel can be modeled as having no delay, and all equations are greatly simplified. This explains why communication engineers often model filters in the modulator and demodulator as being noncausal, since responses before time 0 can be added to the difference between the two clocks. *Estimating* the above fixed delay at the receiver is a significant problem called timing recovery, but is largely separable from the problem of recovering the transmitted data.

The magnitude of delay in a communication system is often important. It is one of the parameters included in the *quality of service* of a communication system. Delay is important for voice communication and often critically important when the communication is in the feedback loop of a real-time control system. In addition to the fixed delay in time reference between modulator and demodulator, there is also delay in source encoding and decoding. Coding for error correction adds additional delay, which might or might not be counted as part of the modulator/demodulator delay.

Either way, the delays in the source coding and error-correction coding are often much larger than that in the modulator/demodulator proper. Thus this latter delay can be significant, but is usually not of primary significance. Also, as channel speeds increase, the filtering delays in the modulator/demodulator become even less significant.

Amplitudes are usually measured on a different scale at transmitter and receiver. The actual power attenuation suffered in transmission is a product of amplifier gain, antenna coupling losses, antenna directional gain, propagation losses, etc. The process of finding all these gains and losses (and perhaps changing them) is called "the link budget." Such gains and losses are invariably calculated in decibels (dB). Recall that the number of decibels corresponding to a power gain α is defined to be $10\log_{10}\alpha$. The use of a logarithmic measure of gain allows the various components of gain to be added rather than multiplied.

The link budget in a communication system is largely separable from other issues, so the amplitude scale at the transmitter is usually normalized to that at the receiver.

By treating attenuation and delay as issues largely separable from modulation, we obtain a model of the channel in which a baseband waveform $u(t)$ is converted to passband and transmitted. At the receiver, after conversion back to baseband, a waveform $v(t) = u(t) + z(t)$ is received, where $z(t)$ is noise. This noise is a fundamental limitation to communication and arises from a variety of causes, including thermal effects and unwanted radiation impinging on the receiver. Chapter 7 is largely devoted to understanding noise waveforms by modeling them as sample values of random processes. Chapter 8 then explains how best to decode signals in the presence of noise. These issues are briefly summarized here to see how they affect the choice of signal constellation.

For reasons to be described shortly, the basic pulse waveform $p(t)$ used in PAM often has the property that it is orthonormal to all its shifts by multiples of T. In this case, the transmitted waveform $u(t) = \sum_k u_k p(t-k/T)$ is an orthonormal expansion, and, in the absence of noise, the transmitted signals $u_1, u_2, \ldots$ can be retrieved from the baseband waveform $u(t)$ by the inner product operation

$$u_k = \int u(t)p(t-kT)\mathrm{d}t.$$

In the presence of noise, this same operation can be performed, yielding

$$v_k = \int v(t)p(t-kT)\mathrm{d}t = u_k + z_k, \tag{6.5}$$

where $z_k = \int z(t)p(t-kT)\mathrm{d}t$ is the projection of $z(t)$ onto the shifted pulse $p(t-kT)$.

The most common (and often the most appropriate) model for noise on channels is called the additive white Gaussian noise model. As shown in Chapters 7 and 8, the above coefficients $\{z_k;\ k \in \mathbb{Z}\}$ in this model are the sample values of zero-mean, iid Gaussian random variables $\{Z_k;\ k \in \mathbb{Z}\}$. This is true no matter how the orthonormal functions $\{p(t-kT);\ k \in \mathbb{Z}\}$ are chosen, and these noise random variables $\{Z_k;\ k \in \mathbb{Z}\}$ are also independent of the signal random variables $\{U_k;\ k \in \mathbb{Z}\}$. Chapter 8 also shows that the operation in (6.5) is the appropriate operation to go from waveform to signal sequence in the layered demodulator of Figure 6.1.

Now consider the effect of the noise on the choice of M and d in a PAM modulator. Since the transmitted signal reappears at the receiver with a zero-mean Gaussian random variable added to it, any attempt to retrieve U_k from V_k directly with reasonably small probability of error[6] will require d to exceed several standard deviations of the noise. Thus the noise determines how large d must be, and this, combined with the power constraint, determines M.

The relation between error probability and signal-point spacing also helps explain why multi-level PAM systems almost invariably use a standard M-PAM signal set. Because the Gaussian density drops off so fast with increasing distance, the error probability due to confusion of nearest neighbors drops off equally fast. Thus error probability is dominated by the points in the constellation that are closest together. If the signal points are constrained to have some minimum distance d between points, it can be seen that the minimum energy E_s for a given number of points M is achieved by the standard M-PAM set.[7]

To be more specific about the relationship between M, d, and the variance σ^2 of the noise Z_k, suppose that d is selected to be $\alpha\sigma$, where α is chosen to make the detection sufficiently reliable. Then with $M = 2^b$, where b is the number of bits encoded into each PAM signal, (6.4) becomes

$$E_s = \frac{\alpha^2\sigma^2(2^{2b}-1)}{12}; \qquad b = \frac{1}{2}\log\left(1+\frac{12E_s}{\alpha^2\sigma^2}\right). \tag{6.6}$$

This expression looks strikingly similar to Shannon's capacity formula for additive white Gaussian noise, which says that, for the appropriate PAM bandwidth, the capacity per signal is $C = (1/2)\log(1+E_s/\sigma^2)$. The important difference is that in (6.6) α must be increased, thus decreasing b, in order to decrease error probability. Shannon's result, on the other hand, says that error probability can be made arbitrarily small for any number of bits per signal less than C. Both equations, however, show the same basic form of relationship between bits per signal and the signal-to-noise ratio E_s/σ^2. Both equations also say that if there is no noise ($\sigma^2 = 0$), then the the number of transmitted bits per signal can be infinitely large (i.e. the distance d between signal points can be made infinitesimally small). Thus both equations suggest that noise is a fundamental limitation on communication.

6.2.3 Choice of the modulation pulse

As defined in (6.3), the baseband transmitted waveform, $u(t) = \sum_k u_k p(t-kT)$, for a PAM modulator is determined by the signal constellation $\mathcal{A}$, the signal interval T, and the real $\mathcal{L}_2$ modulation pulse $p(t)$.

[6] If error-correction coding is used with PAM, then d can be smaller, but for any given error-correction code, d still depends on the standard deviation of Z_k.

[7] On the other hand, if we choose a set of M signal points to minimize E_s for a given error probability, then the standard M-PAM signal set is not quite optimal (see Exercise 6.3).

It may be helpful to visualize $p(t)$ as the impulse response of a linear time-invariant filter. Then $u(t)$ is the response of that filter to a sequence of T-spaced impulses $\sum_k u_k \delta(t-kT)$. The problem of choosing $p(t)$ for a given T turns out to be largely separable from that of choosing $\mathcal{A}$. The choice of $p(t)$ is also the more challenging and interesting problem.

The following objectives contribute to the choice of $p(t)$.

- $p(t)$ must be 0 for $t < -\tau$ for some finite τ. To see this, assume that the kth input signal to the modulator arrives at time $Tk - \tau$. The contribution of u_k to the transmitted waveform $u(t)$ cannot start until $kT - \tau$, which implies $p(t) = 0$ for $t < -\tau$ as stated. This rules out $\operatorname{sinc}(t/T)$ as a choice for $p(t)$ (although $\operatorname{sinc}(t/T)$ could be truncated at $t = -\tau$ to satisfy the condition).
- In most situations, $\hat{p}(f)$ should be essentially baseband-limited to some bandwidth B_b slightly larger than $\mathsf{W}_b = 1/2T$. We will see shortly that it cannot be baseband-limited to less than $\mathsf{W}_b = 1/2T$, which is called the nominal, or Nyquist, bandwidth. There is usually an upper limit on B_b because of regulatory constraints at bandpass or to allow for other transmission channels in neighboring bands. If this limit were much larger than $\mathsf{W}_b = 1/2T$, then T could be increased, increasing the rate of transmission.
- The retrieval of the sequence $\{u_k;\, k \in \mathbb{Z}\}$ from the noisy received waveform should be simple and relatively reliable. In the absence of noise, $\{u_k;\, k \in \mathbb{Z}\}$ should be uniquely specified by the received waveform.

The first condition above makes it somewhat tricky to satisfy the second condition. In particular, the Paley–Wiener theorem (Paley and Wiener, 1934) states that a necessary and sufficient condition for a nonzero $\mathcal{L}_2$ function $p(t)$ to be zero for all $t < 0$ is that its Fourier transform satisfies

$$\int_{-\infty}^{\infty} \frac{|\ln|\hat{p}(f)||}{1+f^2}\, \mathrm{d}f < \infty. \tag{6.7}$$

Combining this with the shift condition for Fourier transforms, it says that any $\mathcal{L}_2$ function that is 0 for all $t < -\tau$ for any finite delay τ must also satisfy (6.7). This is a particularly strong statement of the fact that functions cannot be both time- and frequency-limited. One consequence of (6.7) is that if $p(t) = 0$ for $t < -\tau$, then $\hat{p}(f)$ must be nonzero except on a set of measure 0. Another consequence is that $\hat{p}(f)$ must go to 0 with increasing f more slowly than exponentially.

The Paley–Wiener condition turns out to be useless as a tool for choosing $p(t)$. First, it distinguishes whether the delay τ is finite or infinite, but gives no indication of its value when finite. Second, if an $\mathcal{L}_2$ function $p(t)$ is chosen with no concern for (6.7), it can then be truncated to be 0 for $t < -\tau$. The resulting $\mathcal{L}_2$ error caused by truncation can be made arbitrarily small by choosing τ to be sufficiently large. The tradeoff between truncation error and delay is clearly improved by choosing $p(t)$ to approach 0 rapidly as $t \to -\infty$.

In summary, we will replace the first objective above with the objective of choosing $p(t)$ to approach 0 rapidly as $t \to -\infty$. The resulting $p(t)$ will then be truncated to satisfy the original objective. Thus $p(t) \leftrightarrow \hat{p}(f)$ will be an approximation to the transmit pulse in what follows. This also means that $\hat{p}(f)$ can be strictly bandlimited to a frequency slightly larger than $1/2T$.

We now turn to the third objective, particularly that of easily retrieving the sequence $u_1, u_2, \ldots$ from $u(t)$ in the absence of noise. This problem was first analyzed in 1928 in a classic paper by Harry Nyquist (Nyquist, 1928). Before looking at Nyquist's results, however, we must consider the demodulator.

6.2.4 PAM demodulation

For the time being, ignore the channel noise. Assume that the time reference and the amplitude scaling at the receiver have been selected so that the received baseband waveform is the same as the transmitted baseband waveform $u(t)$. This also assumes that no noise has been introduced by the channel.

The problem at the demodulator is then to retrieve the transmitted signals $u_1, u_2, \ldots$ from the received waveform $u(t) = \sum_k u_k p(t-kT)$. The middle layer of a PAM demodulator is defined by a signal interval T (the same as at the modulator) and a real $\mathcal{L}_2$ waveform $q(t)$. The demodulator first filters the received waveform using a filter with impulse response $q(t)$. It then samples the output at T-spaced sample times. That is, the received filtered waveform is given by

$$r(t) = \int_{-\infty}^{\infty} u(\tau)q(t-\tau)\,d\tau, \tag{6.8}$$

and the received samples are $r(T), r(2T), \ldots$

Our objective is to choose $p(t)$ and $q(t)$ so that $r(kT) = u_k$ for each k. If this objective is met for all choices of $u_1, u_2, \ldots$, then the PAM system involving $p(t)$ and $q(t)$ is said to have *no intersymbol interference*. Otherwise, intersymbol interference is said to exist. The reader should verify that $p(t) = q(t) = (1/\sqrt{T})\text{sinc}(t/T)$ is one solution.

This problem of choosing filters to avoid intersymbol interference appears at first to be somewhat artificial. First, the form of the receiver is restricted to be a filter followed by a sampler. Exercise 6.4 shows that if the detection of each signal is restricted to a linear operation on the received waveform, then there is no real loss of generality in further restricting the operation to be a filter followed by a T-spaced sampler. This does not explain the restriction to linear operations, however.

The second artificiality is neglecting the noise, thus neglecting the fundamental limitation on the bit rate. The reason for posing this artificial problem is, first, that avoiding intersymbol interference is significant in choosing $p(t)$, and, second, that there is a simple and elegant solution to this problem. This solution also provides part of the solution when noise is brought into the picture.

Recall that $u(t) = \sum_k u_k p(t-kT)$; thus, from (6.8),

$$r(t) = \int_{-\infty}^{\infty} \sum_k u_k p(\tau - kT) q(t-\tau)\,d\tau. \tag{6.9}$$

Let $g(t)$ be the convolution $g(t) = p(t) * q(t) = \int p(\tau)q(t-\tau)\mathrm{d}\tau$ and assume[8] that $g(t)$ is $\mathcal{L}_2$. We can then simplify (6.9) as follows:

$$r(t) = \sum_k u_k g(t - kT). \tag{6.10}$$

This should not be surprising. The filters $p(t)$ and $q(t)$ are in cascade with each other. Thus $r(t)$ does not depend on which part of the filtering is done in one and which in the other; it is only the convolution $g(t)$ that determines $r(t)$. Later, when channel noise is added, the individual choice of $p(t)$ and $q(t)$ will become important.

There is no intersymbol interference if $r(kT) = u_k$ for each integer k, and from (6.10) this is satisfied if $g(0) = 1$ and $g(kT) = 0$ for each nonzero integer k. Waveforms with this property are said to be *ideal Nyquist* or, more precisely, *ideal Nyquist with interval T*.

Even though the clock at the receiver is delayed by some finite amount relative to that at the transmitter, and each signal u_k can be generated at the transmitter at some finite time before kT, $g(t)$ must still have the property that $g(t) = 0$ for $t < -\tau$ for some finite τ. As before with the transmit pulse $p(t)$, this finite delay constraint will be replaced with the objective that $g(t)$ should approach 0 rapidly as $|t| \to \infty$. Thus the function $\operatorname{sinc}(t/T)$ is ideal Nyquist with interval T, but is unsuitable because of the slow approach to 0 as $|t| \to \infty$.

As another simple example, the function $\operatorname{rect}(t/T)$ is ideal Nyquist with interval T and can be generated with finite delay, but is not remotely close to being baseband-limited.

In summary, we want to find functions $g(t)$ that are ideal Nyquist but are approximately baseband-limited and approximately time-limited. The Nyquist criterion, discussed in Section 6.3, provides a useful frequency characterization of functions that are ideal Nyquist. This characterization will then be used to study ideal Nyquist functions that are approximately baseband-limited and approximately time-limited.

6.3 The Nyquist criterion

The ideal Nyquist property is determined solely by the T-spaced samples of the waveform $g(t)$. This suggests that the results about aliasing should be relevant. Let $s(t)$ be the baseband-limited waveform generated by the samples of $g(t)$, i.e.

$$s(t) = \sum_k g(kT) \operatorname{sinc}\left(\frac{t}{T} - k\right). \tag{6.11}$$

If $g(t)$ is ideal Nyquist, then all the above terms except $k = 0$ disappear and $s(t) = \operatorname{sinc}(t/T)$. Conversely, if $s(t) = \operatorname{sinc}(t/T)$, then $g(t)$ must be ideal Nyquist. Thus $g(t)$

[8] By looking at the frequency domain, it is not difficult to construct a $g(t)$ of infinite energy from $\mathcal{L}_2$ functions $p(t)$ and $q(t)$. When we study noise, however, we find that there is no point in constructing such a $g(t)$, so we ignore the possibility.

is ideal Nyquist if and only if $s(t) = \operatorname{sinc}(t/T)$. Fourier transforming this, $g(t)$ is ideal Nyqist if and only if

$$\hat{s}(f) = T\,\operatorname{rect}(fT). \tag{6.12}$$

From the aliasing theorem,

$$\hat{s}(f) = \text{l.i.m.} \sum_m \hat{g}\left(f + \frac{m}{T}\right) \operatorname{rect}(fT). \tag{6.13}$$

The result of combining (6.12) and (6.13) is the Nyquist criterion.

Theorem 6.3.1 (Nyquist criterion) *Let $\hat{g}(f)$ be $\mathcal{L}_2$ and satisfy the condition $\lim_{|f|\to\infty} \hat{g}(f)|f|^{1+\varepsilon} = 0$ for some $\varepsilon > 0$. Then the inverse transform, $g(t)$, of $\hat{g}(f)$ is ideal Nyquist with interval T if and only if $\hat{g}(f)$ satisfies the "Nyquist criterion" for T, defined as*[9]

$$\text{l.i.m.} \sum_m \hat{g}(f + m/T)\operatorname{rect}(fT) = T\,\operatorname{rect}(fT). \tag{6.14}$$

Proof From the aliasing theorem, the baseband approximation $s(t)$ in (6.11) converges pointwise and is $\mathcal{L}_2$. Similarly, the Fourier transform $\hat{s}(f)$ satisfies (6.13). If $g(t)$ is ideal Nyquist, then $s(t) = \operatorname{sinc}(t/T)$. This implies that $\hat{s}(f)$ is $\mathcal{L}_2$-equivalent to $T\,\operatorname{rect}(fT)$, which in turn implies (6.14). Conversely, satisfaction of the Nyquist criterion (6.14) implies that $\hat{s}(f) = T\,\operatorname{rect}(fT)$. This implies $s(t) = \operatorname{sinc}(t/T)$, implying that $g(t)$ is ideal Nyquist. □

There are many choices for $\hat{g}(f)$ that satisfy (6.14), but the ones of major interest are those that are approximately both bandlimited and time-limited. We look specifically at cases where $\hat{g}(f)$ is strictly bandlimited, which, as we have seen, means that $g(t)$ is not strictly time-limited. Before these filters can be used, of course, they must be truncated to be strictly time-limited. It is strange to look for strictly bandlimited and approximately time-limited functions when it is the opposite that is required, but the reason is that the frequency constraint is the more important. The time constraint is usually more flexible and can be imposed as an approximation.

6.3.1 Band-edge symmetry

The *nominal* or *Nyquist bandwidth* associated with a PAM pulse $g(t)$ with signal interval T is defined to be $\mathsf{W}_\mathrm{b} = 1/(2T)$. The actual baseband bandwidth[10] B_b is defined as the smallest number B_b such that $\hat{g}(f) = 0$ for $|f| > B_\mathrm{b}$. Note that if $\hat{g}(f) = 0$

[9] It can be seen that $\sum_m \hat{g}(f + m/T)$ is periodic and thus the $\operatorname{rect}(fT)$ could be essentially omitted from both sides of (6.14). Doing this, however, would make the limit in the mean meaningless and would also complicate the intuitive understanding of the theorem.

[10] It might be better to call this the design bandwidth, since after the truncation necessary for finite delay, the resulting frequency function is nonzero a.e. However, if the delay is large enough, the energy outside of B_b is negligible. On the other hand, Exercise 6.9 shows that these approximations must be handled with great care.

for $|f| > W_b$, then the left side of (6.14) is zero except for $m = 0$, so $\hat{g}(f) = T\,\mathrm{rect}(fT)$. This means that $B_b \geq W_b$, with equality if and only if $g(t) = \mathrm{sinc}(t/T)$.

As discussed above, if W_b is much smaller than B_b, then W_b can be increased, thus increasing the rate R_s at which signals can be transmitted. Thus $g(t)$ should be chosen in such a way that B_b exceeds W_b by a relatively small amount. In particular, we now focus on the case where $W_b \leq B_b < 2W_b$.

The assumption $B_b < 2W_b$ means that $\hat{g}(f) = 0$ for $|f| \geq 2W_b$. Thus for $0 \leq f \leq W_b$, $\hat{g}(f + 2mW_b)$ can be nonzero only for $m = 0$ and $m = -1$. Thus the Nyquist criterion (6.14) in this positive frequency interval becomes

$$\hat{g}(f) + \hat{g}(f - 2W_b) = T \qquad \text{for } 0 \leq f \leq W_b. \tag{6.15}$$

Since $p(t)$ and $q(t)$ are real, $g(t)$ is also real, so $\hat{g}(f - 2W_b) = \hat{g}^*(2W_b - f)$. Substituting this in (6.15) and letting $\Delta = f - W_b$, (6.15) becomes

$$T - \hat{g}(W_b + \Delta) = \hat{g}^*(W_b - \Delta). \tag{6.16}$$

This is sketched and interpreted in Figure 6.3. The figure assumes the typical situation in which $\hat{g}(f)$ is real. In the general case, the figure illustrates the real part of $\hat{g}(f)$ and the imaginary part satisfies $\Im\{\hat{g}(W_b + \Delta)\} = \Im\{\hat{g}(W_b - \Delta)\}$.

Figure 6.3 makes it particularly clear that B_b must satisfy $B_b \geq W_b$ to avoid intersymbol interference. We then see that the choice of $\hat{g}(f)$ involves a tradeoff between making $\hat{g}(f)$ smooth, so as to avoid a slow time decay in $g(t)$, and reducing the excess of B_b over the Nyquist bandwidth W_b. This excess is expressed as a *rolloff factor*,[11] defined to be $(B_b/W_b) - 1$, usually expressed as a percentage. Thus $\hat{g}(f)$ in the figure has about a 30% rolloff.

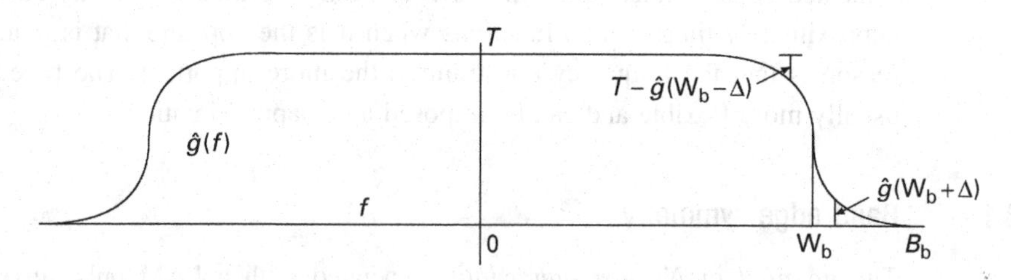

Figure 6.3. Band-edge symmetry illustrated for real $\hat{g}(f)$. For each Δ, $0 \leq \Delta \leq W_b$, $\hat{g}(W_b + \Delta) = T - \hat{g}(W_b - \Delta)$. The portion of the curve for $f \geq W_b$, rotated by 180° around the point $(W_b, T/2)$, is equal to the portion of the curve for $f \leq W_b$.

[11] The requirement for a small rolloff actually arises from a requirement on the transmitted pulse $p(t)$, i.e. on the actual bandwidth of the transmitted channel waveform, rather than on the cascade $g(t) = p(t) * q(t)$. The tacit assumption here is that $\hat{p}(f) = 0$ when $\hat{g}(f) = 0$. One reason for this is that it is silly to transmit energy in a part of the spectrum that is going to be completely filtered out at the receiver. We see later that $\hat{p}(f)$ and $\hat{q}(f)$ are usually chosen to have the same magnitude, ensuring that $\hat{p}(f)$ and $\hat{g}(f)$ have the same rolloff.

PAM filters in practice often have *raised cosine* transforms. The raised cosine frequency function, for any given rolloff α between 0 and 1, is defined by

$$\hat{g}_\alpha(f) = \begin{cases} T, & 0 \le |f| \le (1-\alpha)/2T; \\ T\cos^2\left[\frac{\pi T}{2\alpha}(|f| - \frac{1-\alpha}{2T})\right], & (1-\alpha)/2T) \le |f| \le (1+\alpha)/2T); \\ 0, & |f| \ge (1+\alpha)/2T). \end{cases} \tag{6.17}$$

The inverse transform of $\hat{g}_\alpha(f)$ can be shown to be (see Exercise 6.8)

$$g_\alpha(t) = \text{sinc}\left(\frac{t}{T}\right)\frac{\cos(\pi\alpha t/T)}{1-4\alpha^2t^2/T^2}, \tag{6.18}$$

which decays asymptotically as $1/t^3$, compared to $1/t$ for $\text{sinc}(t/T)$. In particular, for a rolloff $\alpha = 1$, $\hat{g}_\alpha(f)$ is nonzero from $-2\mathsf{W}_\mathsf{b} = -1/T$ to $2\mathsf{W}_\mathsf{b} = 1/T$ and $g_\alpha(t)$ has most of its energy between $-T$ and T. Rolloffs as sharp as 5–10% are used in current practice. The resulting $g_\alpha(t)$ goes to 0 with increasing $|t|$ much faster than $\text{sinc}(t/T)$, but the ratio of $g_\alpha(t)$ to $\text{sinc}(t/T)$ is a function of $\alpha t/T$ and reaches its first zero at $t = 1.5T/\alpha$. In other words, the required filtering delay is proportional to $1/\alpha$.

The motivation for the raised cosine shape is that $\hat{g}(f)$ should be smooth in order for $g(t)$ to decay quickly in time, but $\hat{g}(f)$ must decrease from T at $\mathsf{W}_\mathsf{b}(1-\alpha)$ to 0 at $\mathsf{W}_\mathsf{b}(1+\alpha)$. As seen in Figure 6.3, the raised cosine function simply rounds off the step discontinuity in $\text{rect}(f/2\mathsf{W}_\mathsf{b})$ in such a way as to maintain the Nyquist criterion while making $\hat{g}(f)$ continuous with a continuous derivative, thus guaranteeing that $g(t)$ decays asymptotically with $1/t^3$.

6.3.2 Choosing $\{p(t-kT);\ k \in \mathbb{Z}\}$ as an orthonormal set

In Section 6.3.1, the choice of $\hat{g}(f)$ was described as a compromise between rolloff and smoothness, subject to band-edge symmetry. As illustrated in Figure 6.3, it is not a serious additional constraint to restrict $\hat{g}(f)$ to be real and nonnegative. (Why let $\hat{g}(f)$ go negative or imaginary in making a smooth transition from T to 0?) After choosing $\hat{g}(f) \ge 0$, however, there is still the question of how to choose the transmit filter $p(t)$ and the receive filter $q(t)$ subject to $\hat{p}(f)\hat{q}(f) = \hat{g}(f)$. When studying white Gaussian noise later, we will find that $\hat{q}(f)$ should be chosen to equal $\hat{p}^*(f)$. Thus,[12]

$$|\hat{p}(f)| = |\hat{q}(f)| = \sqrt{\hat{g}(f)}. \tag{6.19}$$

The phase of $\hat{p}(f)$ can be chosen in an arbitrary way, but this determines the phase of $\hat{q}(f) = \hat{p}^*(f)$. The requirement that $\hat{p}(f)\hat{q}(f) = \hat{g}(f) \ge 0$ means that $\hat{q}(f) = \hat{p}^*(f)$. In addition, if $p(t)$ is real then $\hat{p}(-f) = \hat{p}^*(f)$, which determines the phase for negative f in terms of an arbitrary phase for $f > 0$. It is convenient here, however, to be slightly

[12] A function $p(t)$ satisfying (6.19) is often called square root of Nyquist, although it is the magnitude of the transform that is the square root of the transform of an ideal Nyquist pulse.

more general and allow $p(t)$ to be complex. We will prove the following important theorem.

Theorem 6.3.2 (Orthonormal shifts) *Let $p(t)$ be an $\mathcal{L}_2$ function such that $\hat{g}(f) = |\hat{p}(f)|^2$ satisfies the Nyquist criterion for T. Then $\{p(t-kT);\ k \in \mathbb{Z}\}$ is a set of orthonormal functions. Conversely, if $\{p(t-kT);\ k \in \mathbb{Z}\}$ is a set of orthonormal functions, then $|\hat{p}(f)|^2$ satisfies the Nyquist criterion.*

Proof Let $q(t) = p^*(-t)$. Then $g(t) = p(t) * q(t)$, so that

$$g(kT) = \int_{-\infty}^{\infty} p(\tau)q(kT-\tau)\mathrm{d}\tau = \int_{-\infty}^{\infty} p(\tau)p^*(\tau - kT)\mathrm{d}\tau. \tag{6.20}$$

If $\hat{g}(f)$ satisfies the Nyquist criterion, then $g(t)$ is ideal Nyquist and (6.20) has the value 0 for each integer $k \neq 0$ and has the value 1 for $k = 0$. By shifting the variable of integration by jT for any integer j in (6.20), we see also that $\int p(\tau - jT)p^*(\tau - (k+j)T)\mathrm{d}\tau = 0$ for $k \neq 0$ and 1 for $k = 0$. Thus $\{p(t-kT);\ k \in \mathbb{Z}\}$ is an orthonormal set. Conversely, assume that $\{p(t-kT);\ k \in \mathbb{Z}\}$ is an orthonormal set. Then (6.20) has the value 0 for integer $k \neq 0$ and 1 for $k = 0$. Thus $g(t)$ is ideal Nyquist and $\hat{g}(f)$ satisfies the Nyquist criterion. □

Given this orthonormal shift property for $p(t)$, the PAM transmitted waveform $u(t) = \sum_k u_k p(t-kT)$ is simply an orthonormal expansion. Retrieving the coefficient u_k then corresponds to projecting $u(t)$ onto the 1D subspace spanned by $p(t-kT)$. Note that this projection is accomplished by filtering $u(t)$ by $q(t)$ and then sampling at time kT. The filter $q(t)$ is called the *matched filter* to $p(t)$. These filters will be discussed later when noise is introduced into the picture.

Note that we have restricted the pulse $p(t)$ to have unit energy. There is no loss of generality here, since the input signals $\{u_k\}$ can be scaled arbitrarily, and there is no point in having an arbitrary scale factor in both places.

For $|\hat{p}(f)|^2 = \hat{g}(f)$, the actual bandwidth of $\hat{p}(f)$, $\hat{q}(f)$, and $\hat{g}(f)$ are the same, say B_{b}. Thus if $B_{\mathrm{b}} < \infty$, we see that $p(t)$ and $q(t)$ can be realized only with infinite delay, which means that both must be truncated. Since $q(t) = p^*(-t)$, they must be truncated for both positive and negative t. We assume that they are truncated at such a large value of delay that the truncation error is negligible. Note that the delay generated by both the transmitter and receiver filter (i.e. from the time that $u_k p(t-kT)$ starts to be formed at the transmitter to the time when u_k is sampled at the receiver) is twice the duration of $p(t)$.

6.3.3 Relation between PAM and analog source coding

The main emphasis in PAM modulation has been that of converting a sequence of T-spaced signals into a waveform. Similarly, the first part of analog source coding is often to convert a waveform into a T-spaced sequence of samples. The major difference is that, with PAM modulation, we have control over the PAM pulse $p(t)$

and thus some control over the class of waveforms. With source coding, we are stuck with whatever class of waveforms describes the source of interest.

For both systems, the nominal bandwidth is $\mathsf{W}_b = 1/2T$, and B_b can be defined as the actual baseband bandwidth of the waveforms. In the case of source coding, $B_b \leq \mathsf{W}_b$ is a necessary condition for the sampling approximation $\sum_k u(kT)\,\mathrm{sinc}(t/T-k)$ to recreate perfectly the waveform $u(t)$. The aliasing theorem and the T-spaced sinc-weighted sinusoid expansion were used to analyze the squared error if $B_b > \mathsf{W}_b$.

For PAM, on the other hand, the necessary condition for the PAM demodulator to recreate the initial PAM sequence is $B_b \geq \mathsf{W}_b$. With $B_b > \mathsf{W}_b$, aliasing can be used to advantage, creating an aggregate pulse $g(t)$ that is ideal Nyquist. There is considerable choice in such a pulse, and it is chosen by using contributions from both $f < \mathsf{W}_b$ and $f > \mathsf{W}_b$. Finally we saw that the transmission pulse $p(t)$ for PAM can be chosen so that its T-spaced shifts form an orthonormal set. The sinc functions have this property; however, many other waveforms with slightly greater bandwidth have the same property, but decay much faster with t.

6.4 Modulation: baseband to passband and back

The discussion of PAM in Sections 6.2 and 6.3 focussed on converting a T-spaced sequence of real signals into a real waveform of bandwidth B_b slightly larger than the Nyquist bandwidth $\mathsf{W}_b = 1/2T$. This section focuses on converting that baseband waveform into a passband waveform appropriate for the physical medium, regulatory constraints, and avoiding other transmission bands.

6.4.1 Double-sideband amplitude modulation

The objective of modulating a baseband PAM waveform $u(t)$ to some high-frequency passband around some carrier f_c is simply to shift $u(t)$ up in frequency to $u(t)e^{2\pi i f_c t}$. Thus if $\hat{u}(f)$ is zero except for $-B_b \leq f \leq B_b$, then the shifted version would be zero except for $f_c - B_b \leq f \leq f_c + B_b$. This does not quite work since it results in a complex waveform, whereas only real waveforms can actually be transmitted. Thus $u(t)$ is also multiplied by the complex conjugate of $e^{2\pi i f_c t}$, i.e. $e^{-2\pi i f_c t}$, resulting in the following passband waveform:

$$x(t) = u(t)[e^{2\pi i f_c t} + e^{-2\pi i f_c t}] = 2u(t)\cos(2\pi f_c t), \tag{6.21}$$

$$\hat{x}(f) = \hat{u}(f - f_c) + \hat{u}(f + f_c). \tag{6.22}$$

As illustrated in Figure 6.4, $u(t)$ is both translated up in frequency by f_c and also translated down by f_c. Since $x(t)$ must be real, $\hat{x}(f) = \hat{x}^*(-f)$, and the negative frequencies cannot be avoided. Note that the entire set of frequencies in $[-B_b, B_b]$ is both translated up to $[-B_b + f_c, B_b + f_c]$ and down to $[-B_b - f_c, B_b - f_c]$. Thus (assuming $f_c > B_b$) the range of nonzero frequencies occupied by $x(t)$ is twice as large as that occupied by $u(t)$.

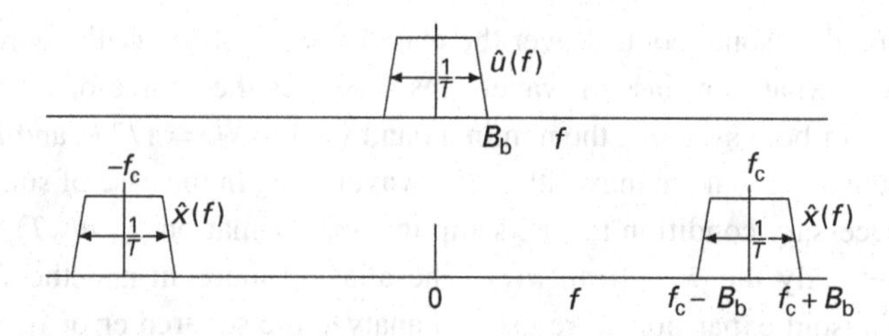

Figure 6.4. Frequency-domain representation of a baseband waveform $u(t)$ shifted up to a passband around the carrier f_c. Note that the baseband bandwidth B_b of $u(t)$ has been doubled to the passband bandwidth $B = 2B_b$ of $x(t)$.

In the communication field, the *bandwidth* of a system is universally defined as the range of *positive* frequencies used in transmission. Since transmitted waveforms are real, the negative frequency part of those waveforms is determined by the positive part and is not counted. This is consistent with our earlier baseband usage, where B_b is the bandwidth of the baseband waveform $u(t)$ in Figure 6.4, and with our new usage for passband waveforms, where $B = 2B_b$ is the bandwidth of $\hat{x}(f)$.

The passband modulation scheme described by (6.21) is called *double-sideband amplitude modulation*. The terminology comes not from the negative frequency band around $-f_c$ and the positive band around f_c, but rather from viewing $[f_c - B_b,\ f_c + B_b]$ as two sidebands, the upper, $[f_c,\ f_c + B_b]$, coming from the positive frequency components of $u(t)$ and the lower, $[f_c - B_b,\ f_c]$ from its negative components. Since $u(t)$ is real, these two bands are redundant and either could be reconstructed from the other.

Double-sideband modulation is quite wasteful of bandwidth since half of the band is redundant. Redundancy is often useful for added protection against noise, but such redundancy is usually better achieved through digital coding.

The simplest and most widely employed solution for using this wasted bandwidth[13] is *quadrature amplitude modulation* (QAM), which is described in Section 6.5. PAM at passband is appropriately viewed as a special case of QAM, and thus the demodulation of PAM from passband to baseband is discussed at the same time as the demodulation of QAM.

6.5 Quadrature amplitude modulation (QAM)

QAM is very similar to PAM except that with QAM the baseband waveform $u(t)$ is chosen to be complex. The complex QAM waveform $u(t)$ is then shifted up to passband

[13] An alternative approach is single-sideband modulation. Here either the positive or negative sideband of a double-sideband waveform is filtered out, thus reducing the transmitted bandwidth by a factor of 2. This used to be quite popular for analog communication, but is harder to implement for digital communication than QAM.

as $u(t)e^{2\pi i f_c t}$. This waveform is complex and is converted into a real waveform for transmission by adding its complex conjugate. The resulting real passband waveform is then given by

$$x(t) = u(t)e^{2\pi i f_c t} + u^*(t)e^{-2\pi i f_c t}. \tag{6.23}$$

Note that the passband waveform for PAM in (6.21) is a special case of this in which $u(t)$ is real. The passband waveform $x(t)$ in (6.23) can also be written in the following equivalent ways:

$$x(t) = 2\Re\{u(t)e^{2\pi i f_c t}\} \tag{6.24}$$

$$= 2\Re\{u(t)\}\cos(2\pi f_c t) - 2\Im\{u(t)\}\sin(2\pi f_c t). \tag{6.25}$$

The factor of 2 in (6.24) and (6.25) is an arbitrary scale factor. Some authors leave it out (thus requiring a factor of 1/2 in (6.23)) and others replace it by $\sqrt{2}$ (requiring a factor of $1/\sqrt{2}$ in (6.23)). This scale factor (however chosen) causes additional confusion when we look at the energy in the waveforms. With the scaling here, $\|\boldsymbol{x}\|^2 = 2\|\boldsymbol{u}\|^2$. Using the scale factor $\sqrt{2}$ solves this problem, but introduces many other problems, not least of which is an extraordinary number of $\sqrt{2}$s in equations. At one level, scaling is a trivial matter, but although the literature is inconsistent, we have tried to be consistent here. One intuitive advantage of the convention here, as illustrated in Figure 6.4, is that the positive frequency part of $x(t)$ is simply $u(t)$ shifted up by f_c.

The remainder of this section provides a more detailed explanation of QAM, and thus also of a number of issues about PAM. A QAM modulator (see Figure 6.5) has the same three layers as a PAM modulator, i.e. first mapping a sequence of bits to a sequence of complex signals, then mapping the complex sequence to a complex baseband waveform, and finally mapping the complex baseband waveform to a real passband waveform.

The demodulator, not surprisingly, performs the inverse of these operations in reverse order, first mapping the received bandpass waveform into a baseband waveform, then recovering the sequence of signals, and finally recovering the binary digits. Each of these layers is discussed in turn.

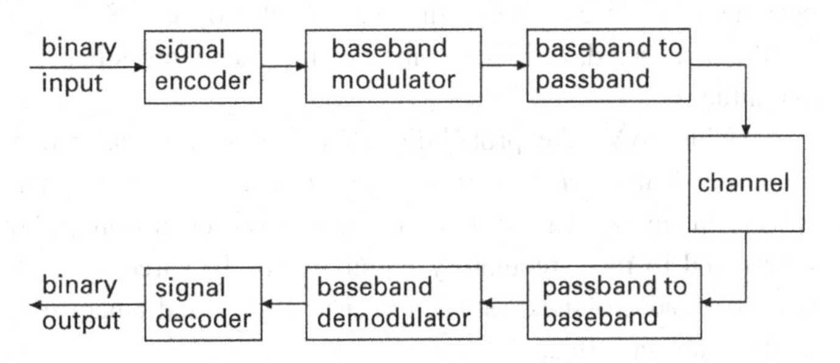

Figure 6.5. QAM modulator and demodulator.

6.5.1 QAM signal set

The input bit sequence arrives at a rate of R bps and is converted, b bits at a time, into a sequence of complex signals u_k chosen from a *signal set* (alphabet, constellation) $\mathcal{A}$ of size $M = |\mathcal{A}| = 2^b$. The *signal rate* is thus $R_s = R/b$ signals/s, and the *signal interval* is $T = 1/R_s = b/R$.

In the case of QAM, the transmitted signals u_k are complex numbers $u_k \in \mathbb{C}$, rather than real numbers. Alternatively, we may think of each signal as a real 2-tuple in $\mathbb{R}^2$.

A *standard* $(M' \times M')$*-QAM signal set*, where $M = (M')^2$ is the Cartesian product of two M'-PAM sets; i.e.,

$$\mathcal{A} = \{(a' + \mathrm{i}a'') \mid a' \in \mathcal{A}', a'' \in \mathcal{A}'\},$$

where

$$\mathcal{A}' = \{-d(M'-1)/2, \ldots, -d/2, d/2, \ldots, d(M'-1)/2\}.$$

The signal set $\mathcal{A}$ thus consists of a square array of $M = (M')^2 = 2^b$ signal points located symmetrically about the origin, as illustrated for $M = 16$:

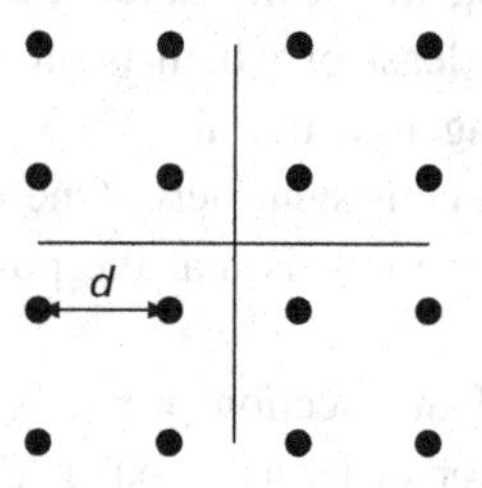

The minimum distance between the 2D points is denoted by d. The average energy per 2D signal, which is denoted by E_s, is simply twice the average energy per dimension:

$$E_s = \frac{d^2[(M')^2 - 1]}{6} = \frac{d^2[M-1]}{6}.$$

In the case of QAM, there are clearly many ways to arrange the signal points other than on a square grid as above. For example, in an M-PSK (phase-shift keyed) signal set, the signal points consist of M equally spaced points on a circle centered on the origin. Thus 4-PSK = 4-QAM. For large M it can be seen that the signal points become very close to each other on a circle so that PSK is rarely used for large M. On the other hand, PSK has some practical advantages because of the uniform signal magnitudes.

As with PAM, the probability of decoding error is primarily a function of the minimum distance d. Not surprisingly, E_s is linear in the signal power of the passband waveform. In wireless systems the signal power is limited both to conserve battery power and to meet regulatory requirements. In wired systems, the power is limited both to avoid crosstalk between adjacent wires and adjacent frequencies, and also to avoid nonlinear effects.

For all of these reasons, it is desirable to choose signal constellations that approximately minimize E_s for a given d and M. One simple result here is that a hexagonal grid of signal points achieves smaller E_s than a square grid for very large M and fixed minimum distance. Unfortunately, finding the optimal signal set to minimize E_s for practical values of M is a messy and ugly problem, and the minima have few interesting properties or symmetries (a possible exception is discussed in Exercise 6.3).

The standard $(M' \times M')$-QAM signal set is almost universally used in practice and will be assumed in what follows.

6.5.2 QAM baseband modulation and demodulation

A QAM baseband modulator is determined by the signal interval T and a complex $\mathcal{L}_2$ waveform $p(t)$. The discrete-time sequence $\{u_k\}$ of complex signal points modulates the amplitudes of a sequence of time shifts $\{p(t-kT)\}$ of the basic pulse $p(t)$ to create a complex transmitted signal $u(t)$ as follows:

$$u(t) = \sum_{k\in\mathbb{Z}} u_k p(t-kT). \tag{6.26}$$

As in the PAM case, we could choose $p(t)$ to be $\text{sinc}(t/T)$, but, for the same reasons as before, $p(t)$ should decay with increasing $|t|$ faster than the sinc function. This means that $\hat{p}(f)$ should be a continuous function that goes to zero rapidly but not instantaneously as f increases beyond $1/2T$. As with PAM, we define $\mathsf{W}_b = 1/2T$ to be the nominal baseband bandwidth of the QAM modulator and B_b to be the actual design bandwidth.

Assume for the moment that the process of conversion to passband, channel transmission, and conversion back to baseband, is ideal, recreating the baseband modulator output $u(t)$ at the input to the baseband demodulator. The baseband demodulator is determined by the interval T (the same as at the modulator) and an $\mathcal{L}_2$ waveform $q(t)$. The demodulator filters $u(t)$ by $q(t)$ and samples the output at T-spaced sample times. Denoting the filtered output by

$$r(t) = \int_{-\infty}^{\infty} u(\tau)q(t-\tau)\mathrm{d}\tau,$$

we see that the received samples are $r(T), r(2T), \ldots$ Note that this is the same as the PAM demodulator except that real signals have been replaced by complex signals. As before, the output $r(t)$ can be represented as

$$r(t) = \sum_k u_k g(t-kT),$$

where $g(t)$ is the convolution of $p(t)$ and $q(t)$. As before, $r(kT) = u_k$ if $g(t)$ is ideal Nyquist, namely if $g(0) = 1$ and $g(kT) = 0$ for all nonzero integer k.

The proof of the Nyquist criterion, Theorem 6.3.1, is valid whether or not $g(t)$ is real. For the reasons explained earlier, however, $\hat{g}(f)$ is usually real and symmetric (as with the raised cosine functions), and this implies that $g(t)$ is also real and symmetric.

Finally, as discussed with PAM, $\hat{p}(f)$ is usually chosen to satisfy $|\hat{p}(f)| = \sqrt{\hat{g}(f)}$. Choosing $\hat{p}(f)$ in this way does not specify the phase of $\hat{p}(f)$, and thus $\hat{p}(f)$ might be real or complex. However $\hat{p}(f)$ is chosen, subject to $|\hat{g}(f)|^2$ satisfying the Nyquist criterion, the set of time shifts $\{p(t-kT)\}$ form an orthonormal set of functions. With this choice also, the baseband bandwidth of $u(t)$, $p(t)$, and $g(t)$ are all the same. Each has a nominal baseband bandwidth given by $1/2T$ and each has an actual baseband bandwidth that exceeds $1/2T$ by some small rolloff factor. As with PAM, $p(t)$ and $q(t)$ must be truncated in time to allow finite delay. The resulting filters are then not quite bandlimited, but this is viewed as a negligible implementation error.

In summary, QAM baseband modulation is virtually the same as PAM baseband modulation. The signal set for QAM is of course complex, and the modulating pulse $p(t)$ can be complex, but the Nyquist results about avoiding intersymbol interference are unchanged.

6.5.3 QAM: baseband to passband and back

Next we discuss modulating the complex QAM baseband waveform $u(t)$ to the passband waveform $x(t)$. Alternative expressions for $x(t)$ are given by (6.23), (6.24), and (6.25), and the frequency representation is illustrated in Figure 6.4.

As with PAM, $u(t)$ has a nominal baseband bandwidth $\mathsf{W}_b = 1/2T$. The actual baseband bandwidth B_b exceeds W_b by some small rolloff factor. The corresponding passband waveform $x(t)$ has a nominal passband bandwidth $\mathsf{W} = 2\mathsf{W}_b = 1/T$ and an actual passband bandwidth $B = 2B_b$. We will assume in everything to follow that $B/2 < f_c$. Recall that $u(t)$ and $x(t)$ are idealized approximations of the true baseband and transmitted waveforms. These true baseband and transmitted waveforms must have finite delay and thus infinite bandwidth, but it is assumed that the delay is large enough that the approximation error is negligible. The assumption[14] $B/2 < f_c$ implies that $u(t)e^{2\pi i f_c t}$ is constrained to positive frequencies and $u(t)e^{-2\pi i f_c t}$ to negative frequencies. Thus the Fourier transform $\hat{u}(f-f_c)$ does not overlap with $\hat{u}(f+f_c)$.

As with PAM, the modulation from baseband to passband is viewed as a two-step process. First $u(t)$ is translated up in frequency by an amount f_c, resulting in a complex passband waveform $x^+(t) = u(t)e^{2\pi i f_c t}$. Next $x^+(t)$ is converted to the real passband waveform $x(t) = [x^+(t)]^* + x^+(t)$.

Assume for now that $x(t)$ is transmitted to the receiver with no noise and no delay. In principle, the received $x(t)$ can be modulated back down to baseband by the reverse of the two steps used in going from baseband to passband. That is, $x(t)$ must first be converted back to the complex positive passband waveform $x^+(t)$, and then $x^+(t)$ must be shifted down in frequency by f_c.

[14] Exercise 6.11 shows that when this assumption is violated, $u(t)$ cannot be perfectly retrieved from $x(t)$, even in the absence of noise. The negligible frequency components of the truncated version of $u(t)$ outside of $B/2$ are assumed to cause negligible error in demodulation.

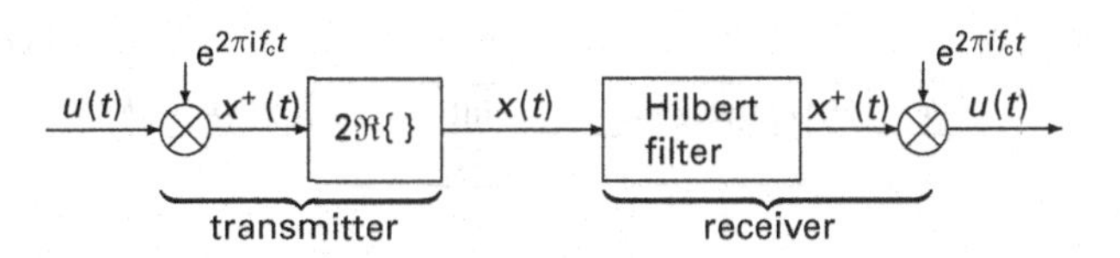

Figure 6.6. Baseband to passband and back.

Mathematically, $x^+(t)$ can be retrieved from $x(t)$ simply by filtering $x(t)$ by a complex filter $h(t)$ such that $\hat{h}(f) = 0$ for $f < 0$ and $\hat{h}(f) = 1$ for $f > 0$. This filter is called a *Hilbert filter*. Note that $h(t)$ is not an $\mathcal{L}_2$ function, but it can be converted to $\mathcal{L}_2$ by making $\hat{h}(f)$ have the value 0 except in the positive passband $[-B/2+f_c, B/2+f_c]$ where it has the value 1. We can then easily retrieve $u(t)$ from $x^+(t)$ simply by a frequency shift. Figure 6.6 illustrates the sequence of operations from $u(t)$ to $x(t)$ and back again.

6.5.4 Implementation of QAM

From an implementation standpoint, the baseband waveform $u(t)$ is usually implemented as two real waveforms, $\Re\{u(t)\}$ and $\Im\{u(t)\}$. These are then modulated up to passband using multiplication by in-phase and out-of-phase carriers as in (6.25), i.e.

$$x(t) = 2\Re\{u(t)\}\cos(2\pi f_c t) - 2\Im\{u(t)\}\sin(2\pi f_c t).$$

There are many other possible implementations, however, such as starting with $u(t)$ given as magnitude and phase. The positive frequency expression $x^+(t) = u(t)e^{2\pi i f_c t}$ is a complex multiplication of complex waveforms which requires four real multiplications rather than the two above used to form $x(t)$ directly. Thus, going from $u(t)$ to $x^+(t)$ to $x(t)$ provides insight but not ease of implementation.

The baseband waveforms $\Re\{u(t)\}$ and $\Im\{u(t)\}$ are easier to generate and visualize if the modulating pulse $p(t)$ is also real. From the discussion of the Nyquist criterion, this is not a fundamental limitation, and there are few reasons for desiring a complex $p(t)$. For real $p(t)$,

$$\Re\{u(t)\} = \sum_k \Re\{u_k\}p(t-kT),$$

$$\Im\{u(t)\} = \sum_k \Im\{u_k\}p(t-kT).$$

Letting $u'_k = \Re\{u_k\}$ and $u''_k = \Im\{u_k\}$, the transmitted passband waveform becomes

$$x(t) = 2\cos(2\pi f_c t)\left(\sum_k u'_k p(t-kT)\right) - 2\sin(2\pi f_c t)\left(\sum_k u''_k p(t-kT)\right). \quad (6.27)$$

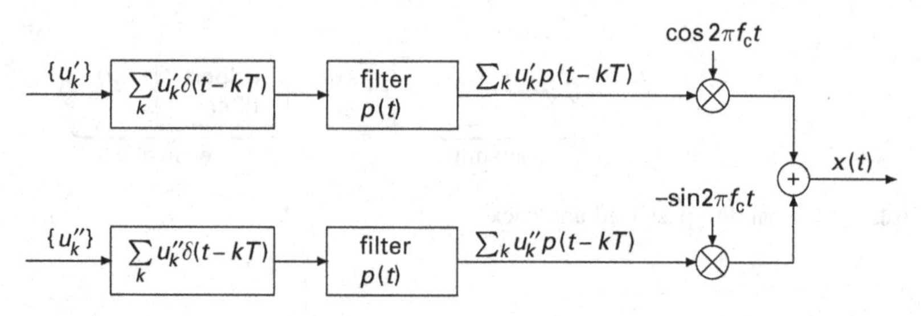

Figure 6.7. DSB-QC modulation.

If the QAM signal set is a standard QAM set, then $\sum_k u'_k p(t-kT)$ and $\sum_k u''_k p(t-kT)$ are parallel baseband PAM systems. They are modulated to passband using "double-sideband" modulation by "quadrature carriers" $\cos(2\pi f_c t)$ and $-\sin(2\pi f_c t)$. These are then summed (with the usual factor of 2), as shown in Figure 6.7. This realization of QAM is called *double-sideband quadrature-carrier* (DSB-QC) modulation.[15]

We have seen that $u(t)$ can be recovered from $x(t)$ by a Hilbert filter followed by shifting down in frequency. A more easily implemented but equivalent procedure starts by multiplying $x(t)$ both by $\cos(2\pi f_c t)$ and by $-\sin(2\pi f_c t)$. Using the trigonometric identities $2\cos^2(\alpha) = 1 + \cos(2\alpha)$, $2\sin(\alpha)\cos(\alpha) = \sin(2\alpha)$, and $2\sin^2(\alpha) = 1 - \cos(2\alpha)$, these terms can be written as follows:

$$x(t)\cos(2\pi f_c t) = \Re\{u(t)\} + \Re\{u(t)\}\cos(4\pi f_c t) + \Im\{u(t)\}\sin(4\pi f_c t), \quad (6.28)$$

$$-x(t)\sin(2\pi f_c t) = \Im\{u(t)\} - \Re\{u(t)\}\sin(4\pi f_c t) + \Im\{u(t)\}\cos(4\pi f_c t). \quad (6.29)$$

To interpret this, note that multiplying by $\cos(2\pi f_c t) = 1/2\mathrm{e}^{2\pi \mathrm{i} f_c t} + 1/2\mathrm{e}^{-2\pi \mathrm{i} f_c t}$ both shifts $x(t)$ up[16] and down in frequency by f_c. Thus the positive frequency part of $x(t)$ gives rise to a baseband term and a term around $2f_c$, and the negative frequency part gives rise to a baseband term and a term at $-2f_c$. Filtering out the double-frequency terms then yields $\Re\{u(t)\}$. The interpretation of the sine multiplication is similar.

As another interpretation, recall that $x(t)$ is real and consists of one band of frequencies around f_c and another around $-f_c$. Note also that (6.28) and (6.29) are the real and imaginary parts of $x(t)\mathrm{e}^{-2\pi \mathrm{i} f_c t}$, which shifts the positive frequency part of $x(t)$ down to baseband and shifts the negative frequency part down to a band around $-2f_c$. In the Hilbert filter approach, the lower band is filtered out before the frequency shift, and in the approach here it is filtered out after the frequency shift. Clearly the two are equivalent.

[15] The terminology comes from analog modulation in which two real analog waveforms are modulated, respectively, onto cosine and sine carriers. For analog modulation, it is customary to transmit an additional component of carrier from which timing and phase can be recovered. As we see shortly, no such additional carrier is necessary here.

[16] This shift up in frequency is a little confusing, since $x(t)\mathrm{e}^{-2\pi \mathrm{i} f_c t} = x(t)\cos(2\pi f_c t) - \mathrm{i}x(t)\sin(2\pi f_c t)$ is only a shift down in frequency. What is happening is that $x(t)\cos(2\pi f_c t)$ is the real part of $x(t)\mathrm{e}^{-2\pi \mathrm{i} f_c t}$ and thus needs positive frequency terms to balance the negative frequency terms.

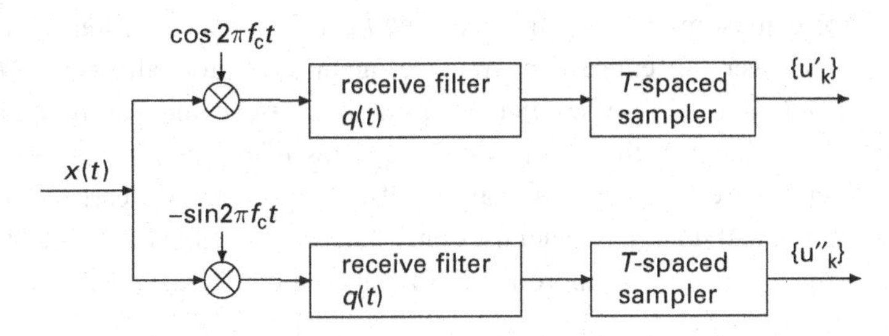

Figure 6.8. DSB-QC demodulation.

It has been assumed throughout that f_c is greater than the baseband bandwidth of $u(t)$. If this is not true, then, as shown in Exercise 6.11, $u(t)$ cannot be retrieved from $x(t)$ by any approach.

Now assume that the baseband modulation filter $p(t)$ is real and a standard QAM signal set is used. Then $\Re\{u(t)\} = \sum u'_k p(t-kT)$ and $\Im\{u(t)\} = \sum u''_k p(t-kT)$ are parallel baseband PAM modulations. Assume also that a receiver filter $q(t)$ is chosen so that $\hat{g}(f) = \hat{p}(f)\hat{q}(f)$ satisfies the Nyquist criterion and all the filters have the common bandwidth $B_b < f_c$. Then, from (6.28), if $x(t)\cos(2\pi f_c t)$ is filtered by $q(t)$, it can be seen that $q(t)$ will filter out the component around $2f_c$. The output from the remaining component $\Re\{u(t)\}$ can then be sampled to retrieve the real signal sequence $u'_1, u'_2, \ldots$ This, plus the corresponding analysis of $-x(t)\sin(2\pi f_c t)$, is illustrated in the DSB-QC receiver in Figure 6.8. Note that the use of the filter $q(t)$ eliminates the need for either filtering out the double-frequency terms or using a Hilbert filter.

The above description of demodulation ignores the noise. As explained in Section 6.3.2, however, if $p(t)$ is chosen so that $\{p(t-kT);\ k \in \mathbb{Z}\}$ is an orthonormal set (i.e. so that $|\hat{p}(f)|^2$ satisfies the Nyquist criterion), then the receiver filter should satisfy $q(t) = p(-t)$. It will be shown later that in the presence of white Gaussian noise, this is the optimal thing to do (in a sense to be described later).

6.6 Signal space and degrees of freedom

Using PAM, real signals can be generated at T-spaced intervals and transmitted in a baseband bandwidth arbitrarily little more than $\mathsf{W}_b = 1/2T$. Thus, over an asymptotically long interval T_0, and in a baseband bandwidth asymptotically close to W_b, $2\mathsf{W}_b T_0$ real signals can be transmitted using PAM.

Using QAM, complex signals can be generated at T-spaced intervals and transmitted in a passband bandwidth arbitrarily little more than $\mathsf{W} = 1/T$. Thus, over an asymptotically long interval T_0, and in a passband bandwidth asymptotically close to W, $\mathsf{W}T_0$ complex signals, and thus $2\mathsf{W}T_0$ real signals, can be transmitted using QAM.

The above description describes PAM at baseband and QAM at passband. To achieve a better comparison of the two, consider an overall large baseband bandwidth W_0 broken into m passbands each of bandwidth W_0/m. Using QAM in each band, we can asymptotically transmit $2W_0T_0$ real signals in a long interval T_0. With PAM used over the entire band W_0, we again asymptotically send $2W_0T_0$ real signals in a duration T_0. We see that, in principle, QAM and baseband PAM with the same overall bandwidth are equivalent in terms of the number of degrees of freedom that can be used to transmit real signals. As pointed out earlier, however, PAM when modulated up to passband uses only half the available degrees of freedom. Also, QAM offers considerably more flexibility since it can be used over an arbitrary selection of frequency bands.

Recall that when we were looking at T-spaced truncated sinusoids and T-spaced sinc-weighted sinusoids, we argued that the class of real waveforms occupying a time interval $(-T_0/2, T_0/2)$ and a frequency interval $(-W_0, W_0)$ has about $2T_0W_0$ degrees of freedom for large W_0, T_0. What we see now is that baseband PAM and passband QAM each employ about $2T_0W_0$ degrees of freedom. In other words, these simple techniques essentially use all the degrees of freedom available in the given bands.

The use of Nyquist theory here has added to our understanding of waveforms that are "essentially" time-and frequency-limited. That is, we can start with a family of functions that are bandlimited within a rolloff factor and then look at asymptotically small rolloffs. The discussion of noise in Chapters 7 and 8 will provide a still better understanding of degrees of freedom subject to essential time and frequency limits.

6.6.1 Distance and orthogonality

Previous sections have shown how to modulate a complex QAM baseband waveform $u(t)$ up to a real passband waveform $x(t)$ and how to retrieve $u(t)$ from $x(t)$ at the receiver. They have also discussed signal constellations that minimize energy for given minimum distance. Finally, the use of a modulation waveform $p(t)$ with orthonormal shifts has connected the energy difference between two baseband signal waveforms, say $u(t) = \sum u_k p(t-kT)$ and $v(t) = \sum_k v_k p(t-kt)$, and the energy difference in the signal points by

$$\|\boldsymbol{u} - \boldsymbol{v}\|^2 = \sum_k |u_k - v_k|^2.$$

Now consider this energy difference at passband. The energy $\|\boldsymbol{x}\|^2$ in the passband waveform $x(t)$ is twice that in the corresponding baseband waveform $u(t)$. Next suppose that $x(t)$ and $y(t)$ are the passband waveforms arising from the baseband waveforms $u(t)$ and $v(t)$, respectively. Then

$$x(t) - y(t) = 2\Re\{u(t)e^{2\pi i f_c t}\} - 2\Re\{v(t)e^{2\pi i f_c t}\} = 2\Re\{[u(t) - v(t)]e^{2\pi i f_c t}\}.$$

Thus $x(t) - y(t)$ is the passband waveform corresponding to $u(t) - v(t)$, so

$$\|x(t) - y(t)\|^2 = 2\|u(t) - v(t)\|^2.$$

This says that, for QAM and PAM, distances between waveforms are preserved (aside from the scale factor of 2 in energy or $\sqrt{2}$ in distance) in going from baseband to passband. Thus distances are preserved in going from signals to baseband waveforms to passband waveforms and back. We will see later that the error probability caused by noise is essentially determined by the distances between the set of passband source waveforms. This error probability is then simply related to the choice of signal constellation and the discrete coding that precedes the mapping of data into signals.

This preservation of distance through the modulation to passband and back is a crucial aspect of the signal-space viewpoint of digital communication. It provides a practical focus to viewing waveforms at baseband and passband as elements of related $\mathcal{L}_2$ inner product spaces.

There is unfortunately a mathematical problem in this very nice story. The set of baseband waveforms forms a complex inner product space, whereas the set of passband waveforms constitutes a real inner product space. The transformation $x(t) = \Re\{u(t)e^{2\pi i f_c t}\}$ is not linear, since, for example, $iu(t)$ does not map into $ix(t)$ for $u(t) \neq 0$. In fact, the notion of a linear transformation does not make much sense, since the transformation goes from complex $\mathcal{L}_2$ to real $\mathcal{L}_2$ and the scalars are different in the two spaces.

Example 6.6.1 As an important example, suppose the QAM modulation pulse is a real waveform $p(t)$ with orthonormal T-spaced shifts. The set of complex baseband waveforms spanned by the orthonormal set $\{p(t-kT);\ k \in \mathbb{Z}\}$ has the form $\sum_k u_k p(t-kT)$, where each u_k is complex. As in (6.27), this is transformed at passband to

$$\sum_k u_k p(t-kT) \to \sum_k 2\Re\{u_k\}p(t-kT)\cos(2\pi f_c t) - 2\sum_k \Im\{u_k\}p(t-kT)\sin(2\pi f_c t).$$

Each baseband function $p(t-kT)$ is modulated to the passband waveform $2p(t-kT)\cos(2\pi f_c t)$. The set of functions $\{p(t-kT)\cos(2\pi f_c t);\ k \in \mathbb{Z}\}$ is not enough to span the space of modulated waveforms, however. It is necessary to add the additional set $\{p(t-kT)\sin(2\pi f_c t);\ k \in \mathbb{Z}\}$. As shown in Exercise 6.15, this combined set of waveforms is an orthogonal set, each with energy 2.

Another way to look at this example is to observe that modulating the baseband function $u(t)$ into the positive passband function $x^+(t) = u(t)e^{2\pi i f_c t}$ is somewhat easier to understand in that the orthonormal set $\{p(t-kT);\ k \in \mathbb{Z}\}$ is modulated to the orthonormal set $\{p(t-kT)e^{2\pi i f_c t};\ k \in \mathbb{Z}\}$, which can be seen to span the space of complex positive frequency passband source waveforms. The additional set of orthonormal waveforms $\{p(t-kT)e^{-2\pi i f_c t};\ k \in \mathbb{Z}\}$ is then needed to span the real passband source waveforms. We then see that the sine/cosine series is simply another way to express this. In the sine/cosine formulation all the coefficients in the series are real, whereas in the complex exponential formulation there is a real and complex coefficient for each term, but they are pairwise-dependent. It will be easier to understand the effects of noise in the sine/cosine formulation.

In the above example, we have seen that each orthonormal function at baseband gives rise to two real orthonormal functions at passband. It can be seen from a degrees-of-freedom argument that this is inevitable no matter what set of orthonormal functions are used at baseband. For a nominal passband bandwidth W, there are 2W real degrees of freedom per second in the baseband complex source waveform, which means there are two real degrees of freedom for each orthonormal baseband waveform. At passband, we have the same 2W degrees of freedom per second, but with a real orthonormal expansion, there is only one real degree of freedom for each orthonormal waveform. Thus there must be two passband real orthonormal waveforms for each baseband complex orthonormal waveform.

The sine/cosine expansion above generalizes in a nice way to an arbitrary set of complex orthonormal baseband functions. Each complex function in this baseband set generates two real functions in an orthogonal passband set. This is expressed precisely in the following theorem, which is proven in Exercise 6.16.

Theorem 6.6.1 *Let $\{\theta_k(t) : k \in \mathbb{Z}\}$ be an orthonormal set limited to the frequency band $[-B/2, B/2]$. Let f_c be greater than $B/2$, and for each $k \in \mathbb{Z}$ let*

$$\psi_{k,1}(t) = \Re\left\{2\theta_k(t)\mathrm{e}^{2\pi \mathrm{i} f_c t}\right\},$$
$$\psi_{k,2}(t) = \Im\left\{-2\theta_k(t)\mathrm{e}^{2\pi \mathrm{i} f_c t}\right\}.$$

The set $\{\psi_{k,j};\, k \in \mathbb{Z}, j \in \{1,2\}\}$ is an orthogonal set of functions, each with energy 2. Furthermore, if $u(t) = \sum_k u_k\theta_k(t)$, then the corresponding passband function $x(t) = 2\Re\{u(t)\mathrm{e}^{2\pi \mathrm{i} f_c t}\}$ is given by

$$x(t) = \sum_k \Re\{u_k\}\,\psi_{k,1}(t) + \Im\{u_k\}\,\psi_{k,2}(t).$$

This provides a very general way to map any orthonormal set at baseband into a related orthonormal set at passband, with two real orthonormal functions at passband corresponding to each orthonormal function at baseband. It is not limited to any particular type of modulation, and thus will allow us to make general statements about signal space at baseband and passband.

6.7 Carrier and phase recovery in QAM systems

Consider a QAM receiver and visualize the passband-to-baseband conversion as multiplying the positive frequency passband by the complex sinusoid $\mathrm{e}^{-2\pi \mathrm{i} f_c t}$. If the receiver has a phase error $\phi(t)$ in its estimate of the phase of the transmitted carrier, then it will instead multiply the incoming waveform by $\mathrm{e}^{-2\pi \mathrm{i} f_c t + \mathrm{i}\phi(t)}$. We assume in this analysis that the time reference at the receiver is perfectly known, so that the sampling of the filtered output is carried out at the correct time. Thus the assumption is that the oscillator at the receiver is not quite in phase with the oscillator at the transmitter. Note that the carrier frequency is usually orders of magnitude higher than the baseband bandwidth, and thus a small error in timing is significant in terms of carrier phase but not in terms of sampling. The carrier phase error will rotate the correct complex baseband signal $u(t)$ by $\phi(t)$; i.e. the actual received baseband signal $r(t)$ will be

$$r(t) = \mathrm{e}^{\mathrm{i}\phi(t)}u(t).$$

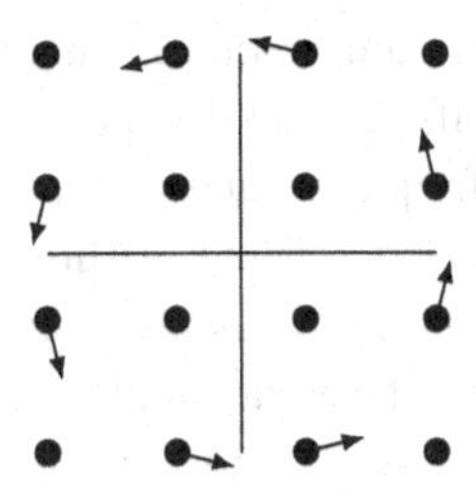

Figure 6.9. Rotation of constellation points by phase error.

If $\phi(t)$ is slowly time-varying relative to the response $q(t)$ of the receiver filter, then the samples $\{r(kT)\}$ of the filter output will be given by

$$r(kT) \approx e^{i\phi(kT)} u_k,$$

as illustrated in Figure 6.9. The phase error $\phi(t)$ is said to come through *coherently*. This phase coherence makes carrier recovery easy in QAM systems.

As can be seen from Figure 6.9, if the phase error is small enough, and the set of points in the constellation are well enough separated, then the phase error can be simply corrected by moving to the closest signal point and adjusting the phase of the demodulating carrier accordingly.

There are two complicating factors here. The first is that we have not taken noise into account yet. When the received signal $y(t)$ is $x(t) + n(t)$, then the output of the T-spaced sampler is not the original signals $\{u_k\}$, but, rather, a noise-corrupted version of them. The second problem is that if a large phase error ever occurs, it cannot be corrected. For example, in Figure 6.9, if $\phi(t) = \pi/2$, then, even in the absence of noise, the received samples always line up with signals from the constellation (but of course not the transmitted signals).

6.7.1 Tracking phase in the presence of noise

The problem of *deciding on* or *detecting* the signals $\{u_k\}$ from the received samples $\{r(kT)\}$ in the presence of noise is a major topic of Chapter 8. Here, however, we have the added complication of both detecting the transmitted signals and tracking and eliminating the phase error.

Fortunately, the problem of decision making and that of phase tracking are largely separable. The oscillators used to generate the modulating and demodulating carriers are relatively stable and have phases which change quite slowly relative to each other. Thus the phase error with any kind of reasonable tracking will be quite small, and thus the data signals can be detected from the received samples almost as if the phase error were zero. The difference between the received sample and the detected data signal will still be nonzero, mostly due to noise but partly due to phase error. However, the noise has zero mean (as we understand later) and thus tends to average out over many sample times. Thus the general approach is to make decisions on the data signals as if the phase error were zero, and then to make slow changes to the phase based on averaging over many sample times. This approach is called *decision-directed carrier recovery*. Note that if we track the phase as phase errors occur, we are also tracking the carrier, in both frequency and phase.

In a decision-directed scheme, assume that the received sample $r(kT)$ is used to make a decision d_k on the transmitted signal point u_k. Also assume that $d_k = u_k$ with very high probability. The apparent phase error for the kth sample is then the difference between the phase of $r(kT)$ and the phase of d_k. Any method for feeding back the apparent phase error to the generator of the sinusoid $\mathrm{e}^{-2\pi \mathrm{i} f_c t + \mathrm{i}\phi(t)}$ in such a way as to reduce the apparent phase error slowly will tend to produce a robust carrier-recovery system.

In one popular method, the feedback signal is taken as the imaginary part of $r(kT)d_k^*$. If the phase angle from d_k to $r(kT)$ is ϕ_k, then

$$r(kT)d_k^* = |r(kT)||d_k|\mathrm{e}^{\mathrm{i}\phi_k},$$

so the imaginary part is $|r(kT)||d_k| \sin \phi_k \approx |r(kT)||d_k|\phi_k$, when ϕ_k is small. Decision-directed carrier recovery based on such a feedback signal can be extremely robust even in the presence of substantial distortion and large initial phase errors. With a second-order phase-locked carrier-recovery loop, it turns out that the carrier frequency f_c can be recovered as well.

6.7.2 Large phase errors

A problem with decision-directed carrier recovery, as with many other approaches, is that the recovered phase may settle into any value for which the received eye pattern (i.e. the pattern of a long string of received samples as viewed on a scope) "looks OK." With $(M \times M)$-QAM signal sets, as in Figure 6.9, the signal set has four-fold symmetry, and phase errors of 90°, 180°, or 270° are not detectable. Simple differential coding methods that transmit the "phase" (quadrantal) part of the signal information as a change of phase from the previous signal rather than as an absolute phase can easily overcome this problem. Another approach is to resynchronize the system frequently by sending some known pattern of signals. This latter approach is frequently used in wireless systems, where fading sometimes causes a loss of phase synchronization.

6.8 Summary of modulation and demodulation

This chapter has used the signal space developed in Chapters 4 and 5 to study the mapping of binary input sequences at a modulator into the waveforms to be transmitted over the channel. Figure 6.1 summarized this process, mapping bits to signals, then signals to baseband waveforms, and then baseband waveforms to passband waveforms. The demodulator goes through the inverse process, going from passband waveforms to baseband waveforms, to signals, to bits. This breaks the modulation process into three layers that can be studied more or less independently.

The development used PAM and QAM throughout, both as widely used systems and as convenient ways to bring out the principles that can be applied more widely.

The mapping from binary digits to signals segments the incoming binary sequence into b-tuples of bits and then maps the set of $M = 2^b$ n-tuples into a constellation of M signal points in $\mathbb{R}^m$ or $\mathbb{C}^m$ for some convenient m. Since the m components of these signal points are going to be used as coefficients in an orthogonal expansion to generate the waveforms, the objectives are to choose a signal constellation with small average energy but with a large distance between each pair of points. PAM is an example where the signal space is $\mathbb{R}^1$, and QAM is an example where the signal space is $\mathbb{C}^1$. For both of these, the standard mapping is the same as the representation points of a uniform quantizer. These are not quite optimal in terms of minimizing the average energy for a given minimum point spacing, but they are almost universally used because of the near optimality and the simplicity.

The mapping of signals into baseband waveforms for PAM chooses a fixed waveform $p(t)$ and modulates the sequence of signals $u_1, u_2, \dots$ into the baseband waveform $\sum_j u_j p(t - jT)$. One of the objectives in choosing $p(t)$ is to be able to retrieve the sequence $u_1, u_2, \dots$ from the received waveform. This involves an output filter $q(t)$ which is sampled each T seconds to retrieve $u_1, u_2, \dots$ The Nyquist criterion was derived, specifying the properties that the product $\hat{g}(f) = \hat{p}(f)\hat{q}(f)$ must satisfy to avoid intersymbol interference. The objective in choosing $\hat{g}(f)$ is a trade-off between the closeness of $\hat{g}(f)$ to $T\,\mathrm{rect}(fT)$ and the time duration of $g(t)$, subject to satisfying the Nyquist criterion. The raised cosine functions are widely used as a good compromise between these dual objectives. For a given real $\hat{g}(f)$, the choice of $\hat{p}(f)$ usually satisfies $\hat{g}(f) = |\hat{p}(f)|^2$, and in this case $\{p(t - kT); k \in \mathbb{Z}\}$ is a set of orthonormal functions.

Most of the remainder of the chapter discussed modulation from baseband to passband. This is an elementary topic in manipulating Fourier transforms, and need not be summarized here.

6.9 Exercises

6.1 (PAM) Consider standard M-PAM and assume that the signals are used with equal probability. Show that the average energy per signal $E_s = \overline{U_k^2}$ is equal to the average energy $\overline{U^2} = d^2M^2/12$ of a uniform continuous distribution over the interval $[-dM/2, dM/2]$, minus the average energy $\overline{(U - U_k)^2} = d^2/12$ of a uniform continuous distribution over the interval $[-d/2, d/2]$:

$$E_s = \frac{d^2(M^2 - 1)}{12}.$$

This establishes (6.4). Verify the formula for $M = 4$ and $M = 8$.

6.2 (PAM) A discrete memoryless source emits binary equiprobable symbols at a rate of 1000 symbols/s. The symbols from a 1 s interval are grouped into pairs and sent over a bandlimited channel using a standard 4-PAM signal set. The modulation uses a signal interval 0.002 and a pulse $p(t) = \mathrm{sinc}(t/T)$.

(a) Suppose that a sample sequence $u_1, \ldots, u_{500}$ of transmitted signals includes 115 appearances of $3d/2$, 130 appearances of $d/2$, 120 appearances of $-d/2$, and 135 appearances of $-3d/2$. Find the energy in the corresponding transmitted waveform $u(t) = \sum_{k=1}^{500} u_k \,\mathrm{sinc}(t/T - k)$ as a function of d.

(b) What is the bandwidth of the waveform $u(t)$ in part (a)?

(c) Find $\mathsf{E}\left[\int U^2(t)\mathrm{d}t\right]$, where $U(t)$ is the random waveform given by $\sum_{k=1}^{500} U_k \,\mathrm{sinc}(t/T - k)$.

(d) Now suppose that the binary source is not memoryless, but is instead generated by a Markov chain, where

$$\Pr(X_i = 1 | X_{i-1} = 1) = \Pr(X_i = 0 | X_{i-1} = 0) = 0.9.$$

Assume the Markov chain starts in steady state with $\Pr(X_1 = 1) = 1/2$. Using the mapping $(00 \to a_1), (01 \to a_2), (10 \to a_3), (11 \to a_4)$, find $\mathsf{E}[U_k^2]$ for $1 \le k \le 500$.

(e) Find $\mathsf{E}\left[\int U^2(t)\,\mathrm{d}t\right]$ for this new source.

(f) For the above Markov chain, explain how the above mapping could be changed to reduce the expected energy without changing the separation between signal points.

6.3 (a) Assume that the received signal in a 4-PAM system is $V_k = U_k + Z_k$, where U_k is the transmitted 4-PAM signal at time k. Let Z_k be independent of U_k and Gaussian with density $f_Z(z) = \sqrt{1/2\pi}\,\exp(-z^2/2)$. Assume that the receiver chooses the signal $\tilde{U}_k$ closest to V_k. (It is shown in Chapter 8 that this detection rule minimizes P_e for equiprobable signals.) Find the probability P_e (in terms of Gaussian integrals) that $U_k \neq \tilde{U}_k$.

(b) Evaluate the partial derivitive of P_e with respect to the third signal point a_3 (i.e. the positive inner signal point) at the point where a_3 is equal to its value $d/2$ in standard 4-PAM and all other signal points are kept at their 4-PAM values. [Hint. This does not require any calculation.]

(c) Evaluate the partial derivitive of the signal energy E_s with respect to a_3.

(d) Argue from this that the signal constellation with minimum-error probability for four equiprobable signal points is not 4-PAM, but rather a constellation, where the distance between the inner points is smaller than the distance from inner point to outer point on either side. (This is quite surprising intuitively to the author.)

6.4 (Nyquist) Suppose that the PAM modulated baseband waveform $u(t) = \sum_{k=-\infty}^{\infty} u_k p(t - kT)$ is received. That is, $u(t)$ is known, T is known, and $p(t)$ is known. We want to determine the signals $\{u_k\}$ from $u(t)$. Assume only linear operations can be used. That is, we wish to find some waveform $d_k(t)$ for each integer k such that $\int_{-\infty}^{\infty} u(t)d_k(t)\mathrm{d}t = u_k$.

(a) What properites must be satisfied by $d_k(t)$ such that the above equation is satisfied no matter what values are taken by the other signals, $\ldots, u_{k-2}, u_{k-1}, u_{k+1}, u_{k+2}, \ldots$? These properties should take the form of constraints on the inner products $\langle p(t-kT), d_j(t)\rangle$. Do not worry about convergence, interchange of limits, etc.

(b) Suppose you find a function $d_0(t)$ that satisfies these constraints for $k = 0$. Show that, for each k, a function $d_k(t)$ satisfying these constraints can be found simply in terms of $d_0(t)$.

(c) What is the relationship between $d_0(t)$ and a function $q(t)$ that avoids intersymbol interference in the approach taken in Section 6.3 (i.e. a function $q(t)$ such that $p(t) * q(t)$ is ideal Nyquist)?

You have shown that the filter/sample approach in Section 6.3 is no less general than the arbitrary linear operation approach here. Note that, in the absence of noise and with a known signal constellation, it might be possible to retrieve the signals from the waveform using nonlinear operations even in the presence of intersymbol interference.

6.5 (Nyquist) Let $v(t)$ be a continuous $\mathcal{L}_2$ waveform with $v(0) = 1$ and define $g(t) = v(t)\,\mathrm{sinc}(t/T)$.

(a) Show that $g(t)$ is ideal Nyquist with interval T.

(b) Find $\hat{g}(f)$ as a function of $\hat{v}(f)$.

(c) Give a direct demonstration that $\hat{g}(f)$ satisfies the Nyquist criterion.

(d) If $v(t)$ is baseband-limited to B_b, what is $g(t)$ baseband-limited to?

Note: the usual form of the Nyquist criterion helps in choosing waveforms that avoid intersymbol interference with prescribed rolloff properties in frequency. The approach above show how to avoid intersymbol interference with prescribed attenuation in time and in frequency.

6.6 (Nyquist) Consider a PAM baseband system in which the modulator is defined by a signal interval T and a waveform $p(t)$, the channel is defined by a filter $h(t)$, and the receiver is defined by a filter $q(t)$ which is sampled at T-spaced intervals. The received waveform, after the receiver filter $q(t)$, is then given by $r(t) = \sum_k u_k g(t-kT)$, where $g(t) = p(t) * h(t) * q(t)$.

(a) What property must $g(t)$ have so that $r(kT) = u_k$ for all k and for all choices of input $\{u_k\}$? What is the Nyquist criterion for $\hat{g}(f)$?

(b) Now assume that $T = 1/2$ and that $p(t), h(t), q(t)$ and all their Fourier transforms are restricted to be real. Assume further that $\hat{p}(f)$ and $\hat{h}(f)$ are specified by Figure 6.10, i.e. by

$$\hat{p}(f) = \begin{cases} 1 & |f| \le 0.5; \\ 1.5 - t & 0.5 < |f| \le 1.5; \\ 0 & |f| > 1.5; \end{cases} \qquad \hat{h}(f) = \begin{cases} 1 & |f| \le 0.75; \\ 0 & 0.75 < |f| \le 1; \\ 1 & 1 < |f| \le 1.25; \\ 0 & |f| > 1.25. \end{cases}$$

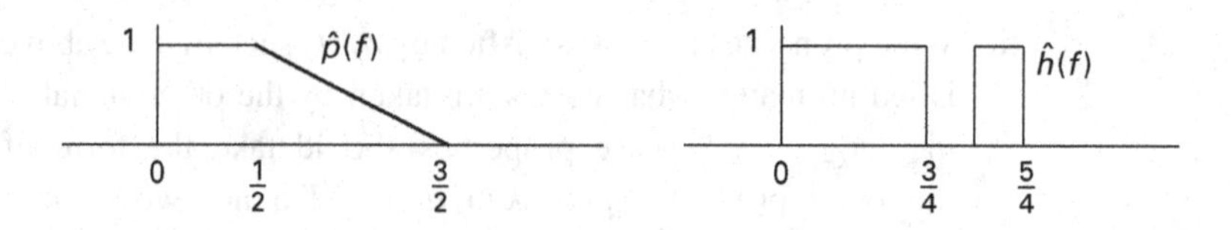

Figure 6.10.

Is it possible to choose a receiver filter transform $\hat{q}(f)$ so that there is no intersymbol interference? If so, give such a $\hat{q}(f)$ and indicate the regions in which your solution is nonunique.

(c) Redo part (b) with the modification that now $\hat{h}(f) = 1$ for $|f| \leq 0.75$ and $\hat{h}(f) = 0$ for $|f| > 0.75$.

(d) Explain the conditions on $\hat{p}(f)\hat{h}(f)$ under which intersymbol interference can be avoided by proper choice of $\hat{q}(f)$. (You may assume, as above, that $\hat{p}(f)$, $\hat{h}(f)$, $p(t)$, and $h(t)$ are all real.)

6.7 (Nyquist) Recall that the rect(t/T) function has the very special property that it, plus its time and frequency shifts by kT and j/T, respectively, form an orthogonal set of functions. The function sinc(t/T) has this same property. This problem is about some other functions that are generalizations of rect(t/T) and which, as you will show in parts (a)–(d), have this same interesting property. For simplicity, choose $T = 1$.

These functions take only the values 0 and 1 and are allowed to be nonzero only over $[-1, 1]$ rather than $[-1/2, 1/2]$ as with rect(t). Explicitly, the functions considered here satisfy the following constraints:

$$p(t) = p^2(t) \qquad \text{for all } t \quad \text{(0/1 property);} \tag{6.30}$$

$$p(t) = 0 \qquad \text{for } |t| > 1; \tag{6.31}$$

$$p(t) = p(-t) \qquad \text{for all } t \quad \text{(symmetry);} \tag{6.32}$$

$$p(t) = 1 - p(t-1) \qquad \text{for } 0 \leq t < 1/2. \tag{6.33}$$

Two examples of functions $P(t)$ satisfying (6.30)–(6.33) are illustrated in Figure 6.11. Note: because of property (6.32), condition (6.33) also holds for $1/2 < t \leq 1$. Note also that $p(t)$ at the single points $t = \pm 1/2$ does not affect any orthogonality properties, so you are free to ignore these points in your arguments.

(a) Show that $p(t)$ is orthogonal to $p(t-1)$. [Hint. Evaluate $p(t)p(t-1)$ for each $t \in [0, 1]$ other than $t = 1/2$.]

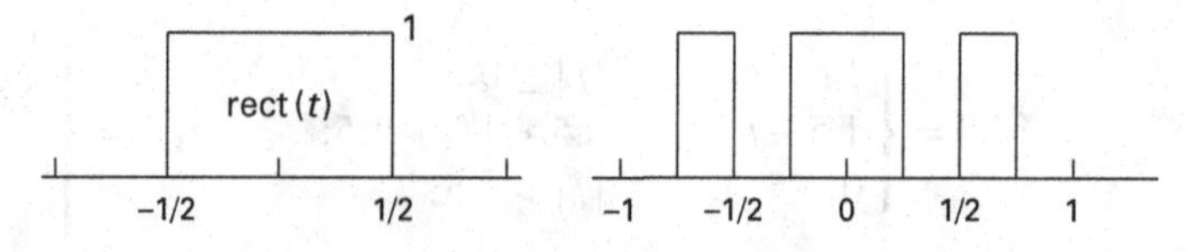

Figure 6.11. Two simple functions $p(t)$ that satisfy (6.30)–(6.33).

(b) Show that $p(t)$ is orthogonal to $p(t-k)$ for all integer $k \neq 0$.
(c) Show that $p(t)$ is orthogonal to $p(t-k)e^{2\pi imt}$ for integer $m \neq 0$ and $k \neq 0$.
(d) Show that $p(t)$ is orthogonal to $p(t)e^{2\pi imt}$ for integer $m \neq 0$. [Hint. Evaluate $p(t)e^{-2\pi imt} + p(t-1)e^{-2\pi im(t-1)}$.]
(e) Let $h(t) = \hat{p}(t)$, where $\hat{p}(f)$ is the Fourier transform of $p(t)$. If $p(t)$ satisfies properties (6.30) to (6.33), does it follow that $h(t)$ has the property that it is orthogonal to $h(t-k)e^{2\pi imt}$ whenever either the integer k or m is nonzero?

Note: almost no calculation is required in this exercise.

6.8 (Nyquist)

(a) For the special case $\alpha = 1$, $T = 1$, verify the formula in (6.18) for $g_1(t)$ given $\hat{g}_1(f)$ in (6.17). [Hint. As an intermediate step, verify that $g_1(t) = \text{sinc}(2t) + (1/2)\,\text{sinc}(2t+1) + (1/2)\,\text{sinc}(2t-1)$.] Sketch $g_1(t)$, in particular showing its value at $mT/2$ for each $m \geq 0$.
(b) For the general case $0 < \alpha < 1$, $T = 1$, show that $\hat{g}_\alpha(f)$ is the convolution of $\text{rect}(f)$ with a half cycle of $\beta\cos(\pi f/\alpha)$ and find β.
(c) Verify (6.18) for $0 < \alpha < 1$, $T = 1$, and then verify for arbitrary $T > 0$.

6.9 (Approximate Nyquist) This exercise shows that approximations to the Nyquist criterion must be treated with great care. Define $\hat{g}_k(f)$, for integer $k \geq 0$ as in Figure 6.12 for $k = 2$. For arbitrary k, there are k small pulses on each side of the main pulse, each of height $1/k$.

(a) Show that $\hat{g}_k(f)$ satisfies the Nyquist criterion for $T = 1$ and for each $k \geq 1$.
(b) Show that $\text{l.i.m.}_{k\to\infty}\, \hat{g}_k(f)$ is simply the central pulse above. That is, this $\mathcal{L}_2$ limit satisfies the Nyquist criterion for $T = 1/2$. To put it another way, $\hat{g}_k(f)$, for large k, satisfies the Nyquist criterion for $T = 1$ using "approximately" the bandwidth $1/4$ rather than the necessary bandwidth $1/2$. The problem is that the $\mathcal{L}_2$ notion of approximation (done carefully here as a limit in the mean of a sequence of approximations) is not always appropriate, and it is often inappropriate with sampling issues.

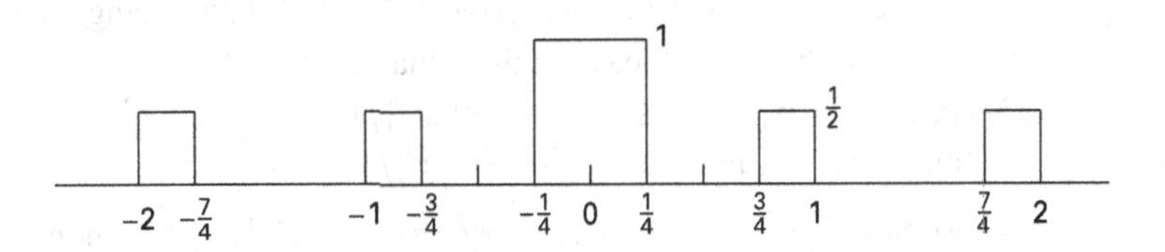

Figure 6.12.

6.10 (Nyquist)

(a) Assume that $\hat{p}(f) = \hat{q}^*(f)$ and $\hat{g}(f) = \hat{p}(f)\hat{q}(f)$. Show that if $p(t)$ is real, then $\hat{g}(f) = \hat{g}(-f)$ for all f.
(b) Under the same assumptions, find an example where $p(t)$ is not real but $\hat{g}(f) \neq \hat{g}(-f)$ and $\hat{g}(f)$ satisifes the Nyquist criterion. [Hint. Show that

$\hat{g}(f) = 1$ for $0 \leq f \leq 1$ and $\hat{g}(f) = 0$ elsewhere satisfies the Nyquist criterion for $T = 1$ and find the corresponding $p(t)$.]

6.11 (Passband)

(a) Let $u_k(t) = \exp(2\pi i f_k t)$ for $k = 1, 2$ and let $x_k(t) = 2\Re\{u_k(t)\exp(2\pi i f_c t)\}$. Assume $f_1 > -f_c$ and find the $f_2 \neq f_1$ such that $x_1(t) = x_2(t)$.

(b) Explain that what you have done is to show that, without the assumption that the bandwidth of $u(t)$ is less than f_c, it is impossible to always retrieve $u(t)$ from $x(t)$, even in the absence of noise.

(c) Let $y(t)$ be a real $\mathcal{L}_2$ function. Show that the result in part (a) remains valid if $u_k(t) = y(t)\exp(2\pi i f_k t)$ (i.e. show that the result in part (a) is valid with a restriction to $\mathcal{L}_2$ functions).

(d) Show that if $u(t)$ is restricted to be real, then $u(t)$ can be retrieved a.e. from $x(t) = 2\Re\{u(t)\exp(2\pi i f_c t)\}$. [Hint. Express $x(t)$ in terms of $\cos(2\pi f_c t)$.]

(e) Show that if the bandwidth of $u(t)$ exceeds f_c, then neither Figure 6.6 nor Figure 6.8 work correctly, even when $u(t)$ is real.

6.12 (QAM)

(a) Let $\theta_1(t)$ and $\theta_2(t)$ be orthonormal complex waveforms. Let $\phi_j(t) = \theta_j(t)e^{2\pi i f_c t}$ for $j = 1, 2$. Show that $\phi_1(t)$ and $\phi_2(t)$ are orthonormal for any f_c.

(b) Suppose that $\theta_2(t) = \theta_1(t-T)$. Show that $\phi_2(t) = \phi_1(t-T)$ if f_c is an integer multiple of $1/T$.

6.13 (QAM)

(a) Assume $B/2 < f_c$. Let $u(t)$ be a real function and let $v(t)$ be an imaginary function, both baseband-limited to $B/2$. Show that the corresponding passband functions, $\Re\{u(t)e^{2\pi i f_c t}\}$ and $\Re\{v(t)e^{2\pi i f_c t}\}$, are orthogonal.

(b) Give an example where the functions in part (a) are not orthogonal if $B/2 > f_c$.

6.14 (a) Derive (6.28) and (6.29) using trigonometric identities.

(b) View the left side of (6.28) and (6.29) as the real and imaginary part, respectively, of $x(t)e^{-2\pi i f_c t}$. Rederive (6.28) and (6.29) using complex exponentials. (Note how much easier this is than part (a).)

6.15 (Passband expansions) Assume that $\{p(t-kT) : k \in \mathbb{Z}\}$ is a set of orthonormal functions. Assume that $\hat{p}(f) = 0$ for $|f| \geq f_c$).

(a) Show that $\{\sqrt{2}p(t-kT)\cos(2\pi f_c t);\ k \in \mathbb{Z}\}$ is an orthonormal set.

(b) Show that $\{\sqrt{2}p(t-kT)\sin(2\pi f_c t);\ k \in \mathbb{Z}\}$ is an orthonormal set and that each function in it is orthonormal to the cosine set in part (a).

6.16 (Passband expansions) Prove Theorem 6.6.1. [Hint. First show that the set of functions $\{\hat{\psi}_{k,1}(f)\}$ and $\{\hat{\psi}_{k,2}(f)\}$ are orthogonal with energy 2 by comparing the integral over negative frequencies with that over positive frequencies.] Indicate explicitly why you need $f_c > B/2$.

6.17 (Phase and envelope modulation) This exercise shows that any real passband waveform can be viewed as a combination of phase and amplitude modulation. Let $x(t)$ be an $\mathcal{L}_2$ real passband waveform of bandwidth B around a carrier frequency $f_c > B/2$. Let $x^+(t)$ be the positive frequency part of $x(t)$ and let $u(t) = x^+(t)\exp\{-2\pi i f_c t\}$.

(a) Express $x(t)$ in terms of $\Re\{u(t)\}$, $\Im\{u(t)\}$, $\cos[2\pi f_c t]$, and $\sin[2\pi f_c t]$.

(b) Define $\phi(t)$ implicitly by $e^{i\phi(t)} = u(t)/|u(t)|$. Show that $x(t)$ can be expressed as $x(t) = 2|u(t)|\cos[2\pi f_c t + \phi(t)]$. Draw a sketch illustrating that $2|u(t)|$ is a baseband waveform upperbounding $x(t)$ and touching $x(t)$ roughly once per cycle. Either by sketch or words illustrate that $\phi(t)$ is a phase modulation on the carrier.

(c) Define the *envelope* of a passband waveform $x(t)$ as twice the magnitude of its positive frequency part, i.e. as $2|x^+(t)|$. Without changing the waveform $x(t)$ (or $x^+(t)$) from that before, change the carrier frequency from f_c to some other frequency f_c'. Thus $u'(t) = x^+(t)\exp\{-2\pi i f_c' t\}$. Show that $|x^+(t)| = |u(t)| = |u'(t)|$. Note that you have shown that the envelope does not depend on the assumed carrier frequency, but has the interpretation of part (b).

(d) Show the relationship of the phase $\phi'(t)$ for the carrier f_c' to that for the carrier f_c.

(e) Let $p(t) = |x(t)|^2$ be the power in $x(t)$. Show that if $p(t)$ is lowpass filtered to bandwidth B, the result is $2|u(t)|^2$. Interpret this filtering as a short term average over $|x(t)|^2$ to interpret why the envelope squared is twice the short term average power (and thus why the envelope is $\sqrt{2}$ times the short term root-mean-squared amplitude).

6.18 (Carrierless amplitude-phase modulation (CAP)) We have seen how to modulate a baseband QAM waveform up to passband and then demodulate it by shifting down to baseband, followed by filtering and sampling. This exercise explores the interesting concept of eliminating the baseband operations by modulating and demodulating directly at passband. This approach is used in one of the North American standards for asymmetrical digital subscriber loop (ADSL).

(a) Let $\{u_k\}$ be a complex data sequence and let $u(t) = \sum_k u_k p(t - kT)$ be the corresponding modulated output. Let $\hat{p}(f)$ be equal to $\sqrt{T}$ over $f \in [3/2T, 5/2T]$ and be equal to 0 elsewhere. At the receiver, $u(t)$ is filtered using $p(t)$ and the output $y(t)$ is then T-space sampled at time instants kT. Show that $y(kT) = u_k$ for all $k \in \mathbb{Z}$. Don't worry about the fact that the transmitted waveform $u(t)$ is complex.

(b) Now suppose that $\hat{p}(f) = \sqrt{T}\,\text{rect}(T(f - f_c))$ for some arbitrary f_c rather than $f_c = 2/T$ as in part (a). For what values of f_c does the scheme still work?

(c) Suppose that $\Re\{u(t)\}$ is now sent over a communication channel. Suppose that the received waveform is filtered by a Hilbert filter before going through the demodulation procedure above. Does the scheme still work?

7 Random processes and noise

7.1 Introduction

Chapter 6 discussed modulation and demodulation, but replaced any detailed discussion of the noise by the assumption that a minimal separation is required between each pair of signal points. This chapter develops the underlying principles needed to understand noise, and Chapter 8 shows how to use these principles in detecting signals in the presence of noise.

Noise is usually the fundamental limitation for communication over physical channels. This can be seen intuitively by accepting for the moment that different possible transmitted waveforms must have a difference of some minimum energy to overcome the noise. This difference reflects back to a required distance between signal points, which, along with a transmitted power constraint, limits the number of bits per signal that can be transmitted.

The transmission rate in bits per second is then limited by the product of the number of bits per signal times the number of signals per second, i.e. the number of degrees of freedom per second that signals can occupy. This intuitive view is substantially correct, but must be understood at a deeper level, which will come from a probabilistic model of the noise.

This chapter and the next will adopt the assumption that the channel output waveform has the form $y(t) = x(t) + z(t)$, where $x(t)$ is the channel input and $z(t)$ is the noise. The channel input $x(t)$ depends on the random choice of binary source digits, and thus $x(t)$ has to be viewed as a particular selection out of an ensemble of possible channel inputs. Similarly, $z(t)$ is a particular selection out of an ensemble of possible noise waveforms.

The assumption that $y(t) = x(t) + z(t)$ implies that the channel attenuation is known and removed by scaling the received signal and noise. It also implies that the input is not filtered or distorted by the channel. Finally it implies that the delay and carrier phase between input and output are known and removed at the receiver.

The noise should be modeled probabilistically. This is partly because the noise is a priori unknown, but can be expected to behave in statistically predictable ways. It is also because encoders and decoders are designed to operate successfully on a variety of different channels, all of which are subject to different noise waveforms. The noise is usually modeled as zero mean, since a mean can be trivially removed.

Modeling the waveforms $x(t)$ and $z(t)$ probabilistically will take considerable care. If $x(t)$ and $z(t)$ were defined only at discrete values of time, such as $\{t = kT;\ k \in \mathbb{Z}\}$, then they could be modeled as sample values of sequences of random variables (rvs). These sequences of rvs could then be denoted by $X(t) = \{X(kT);\ k \in \mathbb{Z}\}$ and $Z(t) = \{Z(kT);\ k \in \mathbb{Z}\}$. The case of interest here, however, is where $x(t)$ and $z(t)$ are defined over the continuum of values of t, and thus a continuum of rvs is required. Such a probabilistic model is known as a *random process* or, synonymously, a *stochastic process*. These models behave somewhat similarly to random sequences, but they behave differently in a myriad of small but important ways.

7.2 Random processes

A *random process* $\{Z(t);\ t \in \mathbb{R}\}$ is a collection[1] of rvs, one for each $t \in \mathbb{R}$. The parameter t usually models time, and any given instant in time is often referred to as an *epoch*. Thus there is one rv for each epoch. Sometimes the range of t is restricted to some finite interval, $[a, b]$, and then the process is denoted by $\{Z(t);\ t \in [a, b]\}$.

There must be an underlying sample space Ω over which these rvs are defined. That is, for each epoch $t \in \mathbb{R}$ (or $t \in [a, b]$), the rv $Z(t)$ is a function $\{Z(t, \omega);\ \omega \in \Omega\}$ mapping sample points $\omega \in \Omega$ to real numbers.

A given sample point $\omega \in \Omega$ within the underlying sample space determines the sample values of $Z(t)$ for each epoch t. The collection of all these sample values for a given sample point ω, i.e. $\{Z(t, \omega);\ t \in \mathbb{R}\}$, is called a *sample function* $\{z(t) \colon \mathbb{R} \to \mathbb{R}\}$ of the process.

Thus $Z(t, \omega)$ can be viewed as a function of ω for fixed t, in which case it is the rv $Z(t)$, or it can be viewed as a function of t for fixed ω, in which case it is the sample function $\{z(t) \colon \mathbb{R} \to \mathbb{R}\} = \{Z(t, \omega);\ t \in \mathbb{R}\}$ corresponding to the given ω. Viewed as a function of both t and ω, $\{Z(t, \omega);\ t \in \mathbb{R},\ \omega \in \Omega\}$ is the random process itself; the sample point ω is usually suppressed, leading to the notation $\{Z(t);\ t \in \mathbb{R}\}$

Suppose a random process $\{Z(t);\ t \in \mathbb{R}\}$ models the channel noise and $\{z(t) : \mathbb{R} \to \mathbb{R}\}$ is a sample function of this process. At first this seems inconsistent with the traditional elementary view that a random process or set of random variables models an experimental situation a priori (before performing the experiment) and the sample function models the result a posteriori (after performing the experiment). The trouble here is that the experiment might run from $t = -\infty$ to $t = \infty$, so there can be no "before" for the experiment and "after" for the result.

There are two ways out of this perceived inconsistency. First, the notion of "before and after" in the elementary view is inessential; the only important thing is the view

[1] Since a random variable is a mapping from Ω to $\mathbb{R}$, the sample values of a rv are real and thus the sample functions of a random process are real. It is often important to define objects called complex random variables that map Ω to $\mathbb{C}$. One can then define a complex random process as a process that maps each $t \in \mathbb{R}$ into a complex rv. These complex random processes will be important in studying noise waveforms at baseband.

that a multiplicity of sample functions might occur, but only one actually does. This point of view is appropriate in designing a cellular telephone for manufacture. Each individual phone that is sold experiences its own noise waveform, but the device must be manufactured to work over the multiplicity of such waveforms.

Second, whether we view a function of time as going from $-\infty$ to $+\infty$ or going from some large negative to large positive time is a matter of mathematical convenience. We often model waveforms as persisting from $-\infty$ to $+\infty$, but this simply indicates a situation in which the starting time and ending time are sufficiently distant to be irrelevant.

In order to specify a random process $\{Z(t);\, t \in \mathbb{R}\}$, some kind of rule is required from which joint distribution functions can, at least in principle, be calculated. That is, for all positive integers n, and all choices of n epochs $t_1, t_2, \ldots, t_n$ it must be possible (in principle) to find the joint distribution function,

$$F_{Z(t_1),\ldots,Z(t_n)}(z_1, \ldots, z_n) = \Pr\{Z(t_1) \le z_1, \ldots, Z(t_n) \le z_n\}, \tag{7.1}$$

for all choices of the real numbers $z_1, \ldots, z_n$. Equivalently, if densities exist, it must be possible (in principle) to find the joint density,

$$f_{Z(t_1),\ldots,Z(t_n)}(z_1, \ldots, z_n) = \frac{\partial^n F_{Z(t_1),\ldots,Z(t_n)}(z_1, \ldots, z_n)}{\partial z_1 \cdots \partial z_n}, \tag{7.2}$$

for all real $z_1, \ldots, z_n$. Since n can be arbitrarily large in (7.1) and (7.2), it might seem difficult for a simple rule to specify all these quantities, but a number of simple rules are given in the following examples that specify all these quantities.

7.2.1 Examples of random processes

The following generic example will turn out to be both useful and quite general. We saw earlier that we could specify waveforms by the sequence of coefficients in an orthonormal expansion. In the following example, a random process is similarly specified by a sequence of random variables used as coefficients in an orthonormal expansion.

Example 7.2.1 Let $Z_1, Z_2, \ldots$ be a sequence of random variables (rvs) defined on some sample space Ω and let $\{\phi_1(t)\}, \{\phi_2(t)\}, \ldots$ be a sequence of orthogonal (or orthonormal) real functions. For each $t \in \mathbb{R}$, let the rv $Z(t)$ be defined as $Z(t) = \sum_k Z_k \phi_k(t)$. The corresponding random process is then $\{Z(t);\, t \in \mathbb{R}\}$. For each t, $Z(t)$ is simply a sum of rvs, so we could, in principle, find its distribution function. Similarly, for each n-tuple $t_1, \ldots, t_n$ of epochs, $Z(t_1), \ldots, Z(t_n)$ is an n-tuple of rvs whose joint distribution could be found in principle. Since $Z(t)$ is a countably infinite sum of rvs, $\sum_{k=1}^{\infty} Z_k \phi_k(t)$, there might be some mathematical intricacies in finding, or even defining, its distribution function. Fortunately, as will be seen, such intricacies do not arise in the processes of most interest here.

It is clear that random processes can be defined as in the above example, but it is less clear that this will provide a mechanism for constructing reasonable models of actual physical noise processes. For the case of Gaussian processes, which will be defined shortly, this class of models will be shown to be broad enough to provide a flexible set of noise models.

The following few examples specialize the above example in various ways.

Example 7.2.2 Consider binary PAM, but view the input signals as independent identically distributed (iid) rvs $U_1, U_2, \ldots$ which take on the values ± 1 with probability 1/2 each. Assume that the modulation pulse is $\operatorname{sinc}(t/T)$ so the baseband random process is given by

$$U(t) = \sum_k U_k \operatorname{sinc}\left(\frac{t-kT}{T}\right).$$

At each sampling epoch kT, the rv $U(kT)$ is simply the binary rv U_k. At epochs between the sampling epochs, however, $U(t)$ is a countably infinite sum of binary rvs whose variance will later be shown to be 1, but whose distribution function is quite ugly and not of great interest.

Example 7.2.3 A random variable is said to be zero-mean Gaussian if it has the probability density

$$f_Z(z) = \frac{1}{\sqrt{2\pi\sigma^2}} \exp\left(\frac{-z^2}{2\sigma^2}\right), \tag{7.3}$$

where σ^2 is the variance of Z. A common model for a noise process $\{Z(t);\, t \in \mathbb{R}\}$ arises by letting

$$Z(t) = \sum_k Z_k \operatorname{sinc}\left(\frac{t-kT}{T}\right), \tag{7.4}$$

where $\ldots, Z_{-1}, Z_0, Z_1, \ldots$ is a sequence of iid zero-mean Gaussian rvs of variance σ^2. At each sampling epoch kT, the rv $Z(kT)$ is the zero-mean Gaussian rv Z_k. At epochs between the sampling epochs, $Z(t)$ is a countably infinite sum of independent zero-mean Gaussian rvs, which turns out to be itself zero-mean Gaussian of variance σ^2. Section 7.3 considers sums of Gaussian rvs and their interrelations in detail. The sample functions of this random process are simply sinc expansions and are limited to the baseband $[-1/2T,\, 1/2T]$. This example, as well as Example 7.2.2, brings out the following mathematical issue: the expected energy in $\{Z(t);\, t \in \mathbb{R}\}$ turns out to be infinite. As discussed later, this energy can be made finite either by truncating $Z(t)$ to some finite interval much larger than any time of interest or by similarly truncating the sequence $\{Z_k;\, k \in \mathbb{Z}\}$.

Another slightly disturbing aspect of this example is that this process cannot be "generated" by a sequence of Gaussian rvs entering a generating device that multiplies them by T-spaced sinc functions and adds them. The problem is the same as the problem with sinc functions in Chapter 6: they extend forever, and thus the process cannot be generated with finite delay. This is not of concern here, since we are not trying to generate random processes, only to show that interesting processes can be defined. The approach here will be to define and analyze a wide variety of random processes, and then to see which are useful in modeling physical noise processes.

Example 7.2.4 Let $\{Z(t);\, t \in [-1, 1]\}$ be defined by $Z(t) = tZ$ for all $t \in [-1, 1]$, where Z is a zero-mean Gaussian rv of variance 1. This example shows that random processes can be very degenerate; a sample function of this process is fully specified by the sample value $z(t)$ at $t = 1$. The sample functions are simply straight lines through the origin with random slope. This illustrates that the sample functions of a random process do not necessarily "look" random.

7.2.2 The mean and covariance of a random process

Often the first thing of interest about a random process is the mean at each epoch t and the covariance between any two epochs t and τ. The mean, $\mathsf{E}[Z(t)] = \overline{Z}(t)$, is simply a real-valued function of t, and can be found directly from the distribution function $F_{Z(t)}(z)$ or density $f_{Z(t)}(z)$. It can be verified that $\overline{Z}(t)$ is 0 for all t for Examples 7.2.2, 7.2.3, and 7.2.4. For Example 7.2.1, the mean cannot be specified without specifying more about the random sequence and the orthogonal functions.

The covariance[2] is a real-valued function of the epochs t and τ. It is denoted by $\mathsf{K}_Z(t, \tau)$ and defined by

$$\mathsf{K}_Z(t, \tau) = \mathsf{E}\left[\left(Z(t) - \overline{Z}(t)\right)\left(Z(\tau) - \overline{Z}(\tau)\right)\right]. \tag{7.5}$$

This can be calculated (in principle) from the joint distribution function $F_{Z(t),Z(\tau)}(z_1, z_2)$ or from the density $f_{Z(t),Z(\tau)}(z_1, z_2)$. To make the covariance function look a little simpler, we usually split each random variable $Z(t)$ into its mean, $\overline{Z}(t)$, and its fluctuation, $\widetilde{Z}(t) = Z(t) - \overline{Z}(t)$. The covariance function is then given by

$$\mathsf{K}_Z(t, \tau) = \mathsf{E}\left[\widetilde{Z}(t)\widetilde{Z}(\tau)\right]. \tag{7.6}$$

The random processes of most interest to us are used to model noise waveforms and usually have zero mean, in which case $Z(t) = \widetilde{Z}(t)$. In other cases, it often aids intuition to separate the process into its mean (which is simply an ordinary function) and its fluctuation, which by definition has zero mean.

The covariance function for the generic random process in Example 7.2.1 can be written as follows:

$$\mathsf{K}_Z(t, \tau) = \mathsf{E}\left[\sum_k \widetilde{Z}_k \phi_k(t) \sum_m \widetilde{Z}_m \phi_m(\tau)\right]. \tag{7.7}$$

If we assume that the rvs $Z_1, Z_2, \ldots$ are iid with variance σ^2, then $\mathsf{E}[\widetilde{Z}_k \widetilde{Z}_m] = 0$ for all $k \neq m$ and $\mathsf{E}[\widetilde{Z}_k \widetilde{Z}_m] = \sigma^2$ for $k = m$. Thus, ignoring convergence questions, (7.7) simplifies to

$$\mathsf{K}_Z(t, \tau) = \sigma^2 \sum_k \phi_k(t)\phi_k(\tau). \tag{7.8}$$

[2] This is often called the *autocovariance* to distinguish it from the covariance between two processes; we will not need to refer to this latter type of covariance.

For the sampling expansion, where $\phi_k(t) = \text{sinc}(t/T - k)$, it can be shown (see (7.48)) that the sum in (7.8) is simply $\text{sinc}[(t-\tau)/T]$. Thus for Examples 7.2.2 and 7.2.3, the covariance is given by

$$\mathsf{K}_Z(t, \tau) = \sigma^2 \,\text{sinc}\left(\frac{t-\tau}{T}\right)$$

where $\sigma^2 = 1$ for the binary PAM case of Example 7.2.2. Note that this covariance depends only on $t - \tau$ and not on the relationship between t or τ and the sampling points kT. These sampling processes are considered in more detail later.

7.2.3 Additive noise channels

The communication channels of greatest interest to us are known as *additive noise channels*. Both the channel input and the noise are modeled as random processes, $\{X(t);\ t \in \mathbb{R}\}$ and $\{Z(t);\ t \in \mathbb{R}\}$, both on the same underlying sample space Ω. The channel output is another random process, $\{Y(t);\ t \in \mathbb{R}\}$ and $Y(t) = X(t) + Z(t)$. This means that, for each epoch t, the random variable $Y(t)$ is equal to $X(t) + Z(t)$.

Note that one could always define the noise on a channel as the difference $Y(t) - X(t)$ between output and input. The notion of *additive noise* inherently also includes the assumption that the processes $\{X(t);\ t \in \mathbb{R}\}$ and $\{Z(t);\ t \in \mathbb{R}\}$ are statistically independent.[3]

As discussed earlier, the additive noise model $Y(t) = X(t) + Z(t)$ implicitly assumes that the channel attenuation, propagation delay, and carrier frequency and phase are perfectly known and compensated for. It also assumes that the input waveform is not changed by any disturbances other than the noise $Z(t)$.

Additive noise is most frequently modeled as a Gaussian process, as discussed in Section 7.3. Even when the noise is not modeled as Gaussian, it is often modeled as some modification of a Gaussian process. Many rules of thumb in engineering and statistics about noise are stated without any mention of Gaussian processes, but often are valid only for Gaussian processes.

7.3 Gaussian random variables, vectors, and processes

This section first defines Gaussian random variables (rvs), then jointly Gaussian random vectors (rvs), and finally Gaussian random processes. The covariance function and joint density function for Gaussian rvs are then derived. Finally, several equivalent conditions for rvs to be jointly Gaussian are derived.

A rv W is a *normalized Gaussian* rv, or more briefly a *normal*[4] rv, if it has the probability density

$$f_W(w) = \frac{1}{\sqrt{2\pi}} \exp\left(\frac{-w^2}{2}\right).$$

[3] More specifically, this means that, for all $k > 0$, all epochs $t_1, \ldots, t_k$ and all epochs $\tau_1, \ldots, \tau_k$ the rvs $X(t_1), \ldots, X(t_k)$ are statistically independent of $Z(\tau_1), \ldots, Z(\tau_k)$.

[4] Some people use normal rv as a synonym for Gaussian rv.

This density is symmetric around 0, and thus the mean of W is 0. The variance is 1, which is probably familiar from elementary probability and is demonstrated in Exercise 7.1. A random variable Z is a *Gaussian* rv if it is a scaled and shifted version of a normal rv, i.e. if $Z = \sigma W + \bar{Z}$ for a normal rv W. It can be seen that $\bar{Z}$ is the mean of Z and σ^2 is the variance.[5] The density of Z (for $\sigma^2 > 0$) is given by

$$f_Z(z) = \frac{1}{\sqrt{2\pi\sigma^2}} \exp\left(\frac{-(z-\bar{Z})^2}{2\sigma^2}\right). \tag{7.9}$$

A Gaussian rv Z of mean $\bar{Z}$ and variance σ^2 is denoted by $Z \sim \mathcal{N}(\bar{Z}, \sigma^2)$. The Gaussian rvs used to represent noise almost invariably have zero mean. Such rvs have the density $f_Z(z) = (1/\sqrt{2\pi\sigma^2}) \exp(-z^2/2\sigma^2)$, and are denoted by $Z \sim \mathcal{N}(0, \sigma^2)$.

Zero-mean Gaussian rvs are important in modeling noise and other random phenomena for the following reasons:

- they serve as good approximations to the sum of many independent zero-mean rvs (recall the central limit theorem);
- they have a number of extremal properties – as discussed later, they are, in several senses, the most random rvs for a given variance;
- they are easy to manipulate analytically, given a few simple properties;
- they serve as representative channel noise models, which provide insight about more complex models.

Definition 7.3.1 A set of n random variables $Z_1, \ldots, Z_n$ is *zero-mean jointly Gaussian* if there is a set of iid normal rvs $W_1, \ldots, W_\ell$ such that each Z_k, $1 \le k \le n$, can be expressed as

$$Z_k = \sum_{m=1}^{\ell} a_{km} W_m; \qquad 1 \le k \le n, \tag{7.10}$$

where $\{a_{km};\ 1 \le k \le n, 1 \le m \le \ell\}$ is an array of real numbers. More generally, $Z'_1, \ldots, Z'_n$ are *jointly Gaussian* if $Z'_k = Z_k + \bar{Z}'_k$, where the set $Z_1, \ldots, Z_n$ is zero-mean jointly Gaussian and $\bar{Z}'_1, \ldots, \bar{Z}'_n$ is a set of real numbers.

It is convenient notationally to refer to a set of n random variables $Z_1, \ldots, Z_n$ as a random vector[6] (rv) $\boldsymbol{Z} = (Z_1, \ldots, Z_n)^{\mathsf{T}}$. Letting A be the n by ℓ real matrix with elements $\{a_{km};\ 1 \le k \le n, 1 \le m \le \ell\}$, (7.10) can then be represented more compactly as

$$\boldsymbol{Z} = \mathsf{A}\boldsymbol{W}, \tag{7.11}$$

where $\boldsymbol{W}$ is an ℓ-tuple of iid normal rvs. Similarly, the jointly Gaussian random vector $\boldsymbol{Z}'$ can be represented as $\boldsymbol{Z}' = \mathsf{A}\boldsymbol{Z} + \bar{\boldsymbol{Z}}'$, where $\bar{\boldsymbol{Z}}'$ is an n-vector of real numbers.

[5] It is convenient to define Z to be Gaussian even in the deterministic case where $\sigma = 0$, but then (7.9) is invalid.

[6] The class of random vectors for a given n over a given sample space satisfies the axioms of a vector space, but here the vector notation is used simply as a notational convenience.

In the remainder of this chapter, all random variables, random vectors, and random processes are assumed to be zero-mean unless explicitly designated otherwise. In other words, only the fluctuations are analyzed, with the means added at the end.[7]

It is shown in Exercise 7.2 that any sum $\sum_m a_{km} W_m$ of iid normal rvs $W_1, \ldots, W_n$ is a Gaussian rv, so that each Z_k in (7.10) is Gaussian. Jointly Gaussian means much more than this, however. The random variables $Z_1, \ldots, Z_n$ must also be related as linear combinations of the same set of iid normal variables. Exercises 7.3 and 7.4 illustrate some examples of pairs of random variables which are individually Gaussian but not jointly Gaussian. These examples are slightly artificial, but illustrate clearly that the joint density of jointly Gaussian rvs is much more constrained than the possible joint densities arising from constraining marginal distributions to be Gaussian.

The definition of jointly Gaussian looks a little contrived at first, but is in fact very natural. Gaussian rvs often make excellent models for physical noise processes because noise is often the summation of many small effects. The central limit theorem is a mathematically precise way of saying that the sum of a very large number of independent small zero-mean random variables is approximately zero-mean Gaussian. Even when different sums are statistically dependent on each other, they are different linear combinations of a common set of independent small random variables. Thus, the jointly Gaussian assumption is closely linked to the assumption that the noise is the sum of a large number of small, essentially independent, random disturbances. Assuming that the underlying variables are Gaussian simply makes the model analytically clean and tractable.

An important property of any jointly Gaussian n-dimensional rv $\boldsymbol{Z}$ is the following: for any real m by n real matrix B, the rv $\boldsymbol{Y} = \mathsf{B}\boldsymbol{Z}$ is also jointly Gaussian. To see this, let $\boldsymbol{Z} = \mathsf{A}\boldsymbol{W}$, where $\boldsymbol{W}$ is a normal rv. Then

$$\boldsymbol{Y} = \mathsf{B}\boldsymbol{Z} = \mathsf{B}(\mathsf{A}\boldsymbol{W}) = (\mathsf{BA})\boldsymbol{W}. \tag{7.12}$$

Since BA is a real matrix, $\boldsymbol{Y}$ is jointly Gaussian. A useful application of this property arises when A is diagonal, so $\boldsymbol{Z}$ has arbitrary independent Gaussian components. This implies that $\boldsymbol{Y} = \mathsf{B}\boldsymbol{Z}$ is jointly Gaussian whenever a rv $\boldsymbol{Z}$ has independent Gaussian components.

Another important application is where B is a 1 by n matrix and Y is a random variable. Thus every linear combination $\sum_{k=1}^{n} b_k Z_k$ of a jointly Gaussian rv $\boldsymbol{Z} = (Z_1, \ldots, Z_n)^{\mathsf{T}}$ is Gaussian. It will be shown later in this section that this is an if and only if property; that is, if every linear combination of a rv $\boldsymbol{Z}$ is Gaussian, then $\boldsymbol{Z}$ is jointly Gaussian.

We now have the machinery to define zero-mean Gaussian processes.

Definition 7.3.2 $\{Z(t); t \in \mathbb{R}\}$ is a *zero-mean Gaussian process* if, for all positive integers n and all finite sets of epochs $t_1, \ldots, t_n$, the set of random variables $Z(t_1), \ldots, Z(t_n)$ is a (zero-mean) jointly Gaussian set of random variables.

[7] When studying estimation and conditional probabilities, means become an integral part of many arguments, but these arguments will not be central here.

If the covariance, $\mathsf{K}_Z(t,\tau) = \mathsf{E}[Z(t)Z(\tau)]$, is known for each pair of epochs t, τ, then, for any finite set of epochs $t_1, \ldots, t_n$, $\mathsf{E}[Z(t_k)Z(t_m)]$ is known for each pair (t_k, t_m) in that set. Sections 7.3.1 and 7.3.2 will show that the joint probability density for any such set of (zero-mean) jointly Gaussian rvs depends only on the covariances of those variables. This will show that a zero-mean Gaussian process is specified by its covariance function. A nonzero-mean Gaussian process is similarly specified by its covariance function and its mean.

7.3.1 The covariance matrix of a jointly Gaussian random vector

Let an n-tuple of (zero-mean) rvs $Z_1, \ldots, Z_n$ be represented as a rv $\boldsymbol{Z} = (Z_1, \ldots, Z_n)^{\mathsf{T}}$. As defined earlier, $\boldsymbol{Z}$ is jointly Gaussian if $\boldsymbol{Z} = \mathsf{A}\boldsymbol{W}$, where $\boldsymbol{W} = (W_1, W_2, \ldots, W_\ell)^{\mathsf{T}}$ is a vector of iid normal rvs and A is an n by ℓ real matrix. Each rv Z_k, and all linear combinations of $Z_1, \ldots, Z_n$, are Gaussian.

The covariance of two (zero-mean) rvs Z_1, Z_2 is $\mathsf{E}[Z_1 Z_2]$. For a rv $\boldsymbol{Z} = (Z_1, \ldots Z_n)^{\mathsf{T}}$ the covariance between all pairs of random variables is very conveniently represented by the n by n covariance matrix

$$\mathsf{K}_{\boldsymbol{Z}} = \mathsf{E}[\boldsymbol{Z}\boldsymbol{Z}^{\mathsf{T}}].$$

Appendix 7.11.1 develops a number of properties of covariance matrices (including the fact that they are identical to the class of nonnegative definite matrices). For a vector $\boldsymbol{W} = W_1, \ldots, W_\ell$ of independent normalized Gaussian rvs, $\mathsf{E}[W_j W_m] = 0$ for $j \neq m$ and 1 for $j = m$. Thus

$$\mathsf{K}_{\boldsymbol{W}} = \mathsf{E}[\boldsymbol{W}\boldsymbol{W}^{\mathsf{T}}] = \mathsf{I}_\ell,$$

where I_ℓ is the ℓ by ℓ identity matrix. For a zero-mean jointly Gaussian vector $\boldsymbol{Z} = \mathsf{A}\boldsymbol{W}$, the covariance matrix is thus given by

$$\mathsf{K}_{\boldsymbol{Z}} = \mathsf{E}[\mathsf{A}\boldsymbol{W}\boldsymbol{W}^{\mathsf{T}}\mathsf{A}^{\mathsf{T}}] = \mathsf{A}\mathsf{E}[\boldsymbol{W}\boldsymbol{W}^{\mathsf{T}}]\mathsf{A}^{\mathsf{T}} = \mathsf{A}\mathsf{A}^{\mathsf{T}}. \tag{7.13}$$

7.3.2 The probability density of a jointly Gaussian random vector

The *probability density*, $f_{\boldsymbol{Z}}(\boldsymbol{z})$, of a rv $\boldsymbol{Z} = (Z_1, Z_2, \ldots, Z_n)^{\mathsf{T}}$ is the joint probability density of the components $Z_1, \ldots, Z_n$. An important example is the iid rv $\boldsymbol{W}$, where the components W_k, $1 \le k \le n$, are iid and normal, $W_k \sim \mathcal{N}(0,1)$. By taking the product of the n densities of the individual rvs, the density of $\boldsymbol{W} = (W_1, W_2, \ldots, W_n)^{\mathsf{T}}$ is given by

$$f_{\boldsymbol{W}}(\boldsymbol{w}) = \frac{1}{(2\pi)^{n/2}} \exp\left(\frac{-w_1^2 - w_2^2 - \cdots - w_n^2}{2}\right) = \frac{1}{(2\pi)^{n/2}} \exp\left(\frac{-\|\boldsymbol{w}\|^2}{2}\right). \tag{7.14}$$

This shows that the density of $\boldsymbol{W}$ at a sample value $\boldsymbol{w}$ depends only on the squared distance $\|\boldsymbol{w}\|^2$ of the sample value from the origin. That is, $f_{\boldsymbol{W}}(\boldsymbol{w})$ is spherically symmetric around the origin, and points of equal probability density lie on concentric spheres around the origin.

Consider the transformation $\boldsymbol{Z} = \mathsf{A}\boldsymbol{W}$, where $\boldsymbol{Z}$ and $\boldsymbol{W}$ each have n components and A is n by n. If we let $\boldsymbol{a}_1, \boldsymbol{a}_2, \ldots, \boldsymbol{a}_n$ be the n columns of A, then this means that $\boldsymbol{Z} = \sum_m \boldsymbol{a}_m W_m$. That is, for any sample values $w_1, \ldots, w_n$ for $\boldsymbol{W}$, the corresponding sample value for $\boldsymbol{Z}$ is $\boldsymbol{z} = \sum_m \boldsymbol{a}_m w_m$. Similarly, if we let $\boldsymbol{b}_1, \ldots, \boldsymbol{b}_n$ be the rows of A, then $Z_k = \boldsymbol{b}_k \boldsymbol{W}$.

Let $\mathcal{B}_\delta$ be a cube, δ on a side, of the sample values of $\boldsymbol{W}$ defined by $\mathcal{B}_\delta = \{\boldsymbol{w} : 0 \le w_k \le \delta;\ 1 \le k \le n\}$ (see Figure 7.1). The set $\mathcal{B}'_\delta$ of vectors $\boldsymbol{z} = \mathsf{A}\boldsymbol{w}$ for $\boldsymbol{w} \in \mathcal{B}_\delta$ is a parallelepiped whose sides are the vectors $\delta\boldsymbol{a}_1, \ldots, \delta\boldsymbol{a}_n$. The determinant, $\det(\mathsf{A})$, of A has the remarkable geometric property that its magnitude, $|\det(\mathsf{A})|$, is equal to the volume of the parallelepiped with sides $\boldsymbol{a}_k$; $1 \le k \le n$. Thus the unit cube $\mathcal{B}_\delta$, with volume δ^n, is mapped by A into a parallelepiped of volume $|\det \mathsf{A}|\delta^n$.

Assume that the columns $\boldsymbol{a}_1, \ldots, \boldsymbol{a}_n$ of A are linearly independent. This means that the columns must form a basis for $\mathbb{R}^n$, and thus that every vector $\boldsymbol{z}$ is some linear combination of these columns, i.e. that $\boldsymbol{z} = \mathsf{A}\boldsymbol{w}$ for some vector $\boldsymbol{w}$. The matrix A must then be invertible, i.e. there is a matrix A^{-1} such that $\mathsf{A}\mathsf{A}^{-1} = \mathsf{A}^{-1}\mathsf{A} = \mathsf{I}_n$, where I_n is the n by n identity matrix. The matrix A maps the unit vectors of $\mathbb{R}^n$ into the vectors $\boldsymbol{a}_1, \ldots, \boldsymbol{a}_n$ and the matrix A^{-1} maps $\boldsymbol{a}_1, \ldots, \boldsymbol{a}_n$ back into the unit vectors.

If the columns of A are not linearly independent, i.e. A is not invertible, then A maps the unit cube in $\mathbb{R}^n$ into a subspace of dimension less than n. In terms of Figure 7.1, the unit cube would be mapped into a straight line segment. The area, in 2D space, of a straight line segment is 0, and more generally the volume in n-space of any lower-dimensional set of points is 0. In terms of the determinant, $\det \mathsf{A} = 0$ for any noninvertible matrix A.

Assuming again that A is invertible, let $\boldsymbol{z}$ be a sample value of $\boldsymbol{Z}$ and let $\boldsymbol{w} = \mathsf{A}^{-1}\boldsymbol{z}$ be the corresponding sample value of $\boldsymbol{W}$. Consider the incremental cube $\boldsymbol{w} + \mathcal{B}_\delta$ cornered at $\boldsymbol{w}$. For δ very small, the probability $P_\delta(\boldsymbol{w})$ that $\boldsymbol{W}$ lies in this cube is $f_{\boldsymbol{W}}(\boldsymbol{w})\delta^n$ plus terms that go to zero faster than δ^n as $\delta \to 0$. This cube around $\boldsymbol{w}$ maps into a parallelepiped of volume $\delta^n|\det(\mathsf{A})|$ around $\boldsymbol{z}$, and no other sample value of $\boldsymbol{W}$ maps into this parallelepiped. Thus $P_\delta(\boldsymbol{w})$ is also equal to $f_{\boldsymbol{Z}}(\boldsymbol{z})\delta^n|\det(\mathsf{A})|$

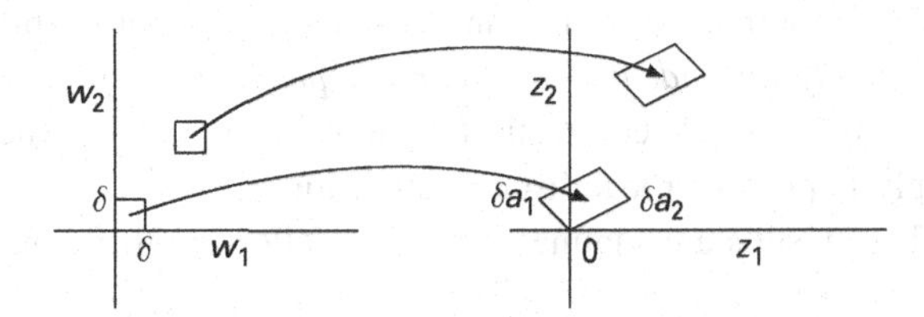

Figure 7.1. Example illustrating how $\boldsymbol{Z} = \mathsf{A}\boldsymbol{W}$ maps cubes into parallelepipeds. Let $Z_1 = -W_1 + 2W_2$ and $Z_2 = W_1 + W_2$. The figure shows the set of sample pairs z_1, z_2 corresponding to $0 \le w_1 \le \delta$ and $0 \le w_2 \le \delta$. It also shows a translation of the same cube mapping into a translation of the same parallelepiped.

plus negligible terms. Going to the limit $\delta \to 0$, we have

$$f_{\mathbf{Z}}(\mathbf{z})|\det(\mathsf{A})| = \lim_{\delta\to 0} \frac{P_\delta(\mathbf{w})}{\delta^n} = f_{\mathbf{W}}(\mathbf{w}). \tag{7.15}$$

Since $\mathbf{w} = \mathsf{A}^{-1}\mathbf{z}$, we obtain the explicit formula

$$f_{\mathbf{Z}}(\mathbf{z}) = \frac{f_{\mathbf{W}}(\mathsf{A}^{-1}\mathbf{z})}{|\det(\mathsf{A})|}. \tag{7.16}$$

This formula is valid for any random vector $\boldsymbol{W}$ with a density, but we are interested in the vector $\boldsymbol{W}$ of iid Gaussian rvs, $\mathcal{N}(0,1)$. Substituting (7.14) into (7.16), we obtain

$$f_{\mathbf{Z}}(\mathbf{z}) = \frac{1}{(2\pi)^{n/2}|\det(\mathsf{A})|}\exp\left(\frac{-\|\mathsf{A}^{-1}\mathbf{z}\|^2}{2}\right), \tag{7.17}$$

$$= \frac{1}{(2\pi)^{n/2}|\det(\mathsf{A})|}\exp\left(-\frac{1}{2}\mathbf{z}^{\mathsf{T}}(\mathsf{A}^{-1})^{\mathsf{T}}\mathsf{A}^{-1}\mathbf{z}\right). \tag{7.18}$$

We can simplify this somewhat by recalling from (7.13) that the covariance matrix of $\boldsymbol{Z}$ is given by $\mathsf{K}_{\mathbf{Z}} = \mathsf{A}\mathsf{A}^{\mathsf{T}}$. Thus, $\mathsf{K}_{\mathbf{Z}}^{-1} = (\mathsf{A}^{-1})^{\mathsf{T}}\mathsf{A}^{-1}$. Substituting this into (7.18) and noting that $\det(\mathsf{K}_{\mathbf{Z}}) = |\det(\mathsf{A})|^2$, we obtain

$$f_{\mathbf{Z}}(\mathbf{z}) = \frac{1}{(2\pi)^{n/2}\sqrt{\det(\mathsf{K}_{\mathbf{Z}})}}\exp\left(-\frac{1}{2}\mathbf{z}^{\mathsf{T}}\mathsf{K}_{\mathbf{Z}}^{-1}\mathbf{z}\right). \tag{7.19}$$

Note that this probability density depends only on the covariance matrix of $\boldsymbol{Z}$ and not directly on the matrix A.

The density in (7.19) does rely, however, on A being nonsingular. If A is singular, then at least one of its rows is a linear combination of the other rows, and thus, for some m, $1 \le m \le n$, Z_m is a linear combination of the other Z_k. The random vector $\boldsymbol{Z}$ is still jointly Gaussian, but the joint probability density does not exist (unless one wishes to view the density of Z_m as a unit impulse at a point specified by the sample values of the other variables). It is possible to write out the distribution function for this case, using step functions for the dependent rvs, but it is not worth the notational mess. It is more straightforward to face the problem and find the density of a maximal set of linearly independent rvs, and specify the others as deterministic linear combinations.

It is important to understand that there is a large difference between rvs being *statistically dependent* and *linearly dependent*. If they are linearly dependent, then one or more are deterministic functions of the others, whereas statistical dependence simply implies a probabilistic relationship.

These results are summarized in the following theorem.

Theorem 7.3.1 (Density for jointly Gaussian rvs) *Let $\boldsymbol{Z}$ be a (zero-mean) jointly Gaussian rv with a nonsingular covariance matrix $\mathsf{K}_{\mathbf{Z}}$. Then the probability density $f_{\mathbf{Z}}(\mathbf{z})$ is given by (7.19). If $\mathsf{K}_{\mathbf{Z}}$ is singular, then $f_{\mathbf{Z}}(\mathbf{z})$ does not exist, but the density in (7.19) can be applied to any set of linearly independent rvs out of $Z_1, \ldots, Z_n$.*

For a zero-mean Gaussian process $Z(t)$, the covariance function $\mathsf{K}_Z(t, \tau)$ specifies $\mathsf{E}[Z(t_k)Z(t_m)]$ for arbitrary epochs t_k and t_m and thus specifies the covariance matrix for any finite set of epochs $t_1, \ldots, t_n$. From Theorem was 7.3.1, this also specifies the joint probability distribution for that set of epochs. Thus the covariance function specifies all joint probability distributions for all finite sets of epochs, and thus specifies the process in the sense[8] of Section 7.2. In summary, we have the following important theorem.

Theorem 7.3.2 (Gaussian process) *A zero-mean Gaussian process is specified by its covariance function* $\mathsf{K}(t, \tau)$.

7.3.3 Special case of a 2D zero-mean Gaussian random vector

The probability density in (7.19) is now written out in detail for the 2D case. Let $\mathsf{E}[Z_1^2] = \sigma_1^2$, $\mathsf{E}[Z_2^2] = \sigma_2^2$, and $\mathsf{E}[Z_1 Z_2] = \kappa_{12}$. Thus

$$\mathsf{K}_Z = \begin{bmatrix} \sigma_1^2 & \kappa_{12} \\ \kappa_{12} & \sigma_2^2 \end{bmatrix}.$$

Let ρ be the *normalized covariance* $\rho = \kappa_{12}/(\sigma_1\sigma_2)$. Then $\det(\mathsf{K}_Z) = \sigma_1^2\sigma_2^2 - \kappa_{12}^2 = \sigma_1^2\sigma_2^2(1-\rho^2)$. Note that ρ must satisfy $|\rho| \le 1$ with strict inequality if K_Z is nonsingular:

$$\mathsf{K}_Z^{-1} = \frac{1}{\sigma_1^2\sigma_2^2 - \kappa_{12}^2}\begin{bmatrix} \sigma_2^2 & -\kappa_{12} \\ -\kappa_{12} & \sigma_1^2 \end{bmatrix} = \frac{1}{1-\rho^2}\begin{bmatrix} 1/\sigma_1^2 & -\rho/(\sigma_1\sigma_2) \\ -\rho/(\sigma_1\sigma_2) & 1/\sigma_2^2 \end{bmatrix};$$

$$\begin{aligned} f_Z(z) &= \frac{1}{2\pi\sqrt{\sigma_1^2\sigma_2^2 - \kappa_{12}^2}} \exp\left(\frac{-z_1^2\sigma_2^2 + 2z_1z_2\kappa_{12} - z_2^2\sigma_1^2}{2(\sigma_1^2\sigma_2^2 - \kappa_{12}^2)}\right) \\ &= \frac{1}{2\pi\sigma_1\sigma_2\sqrt{1-\rho^2}} \exp\left(\frac{-(z_1/\sigma_1)^2 + 2\rho(z_1/\sigma_1)(z_2/\sigma_2) - (z_2/\sigma_2)^2}{2(1-\rho^2)}\right). \end{aligned} \tag{7.20}$$

Curves of equal probability density in the plane correspond to points where the argument of the exponential function in (7.20) is constant. This argument is quadratic, and thus points of equal probability density form an ellipse centered on the origin. The ellipses corresponding to different values of probability density are concentric, with larger ellipses corresponding to smaller densities.

If the normalized covariance ρ is 0, the axes of the ellipse are the horizontal and vertical axes of the plane; if $\sigma_1 = \sigma_2$, the ellipse reduces to a circle; and otherwise the ellipse is elongated in the direction of the larger standard deviation. If $\rho > 0$, the density in the first and third quadrants is increased at the expense of the second

[8] As will be discussed later, focusing on the pointwise behavior of a random process at all finite sets of epochs has some of the same problems as specifying a function pointwise rather than in terms of $\mathcal{L}_2$-equivalence. This can be ignored for the present.

and fourth, and thus the ellipses are elongated in the first and third quadrants. This is reversed, of course, for $\rho < 0$.

The main point to be learned from this example, however, is that the detailed expression for two dimensions in (7.20) is messy. The messiness gets far worse in higher dimensions. Vector notation is almost essential. One should reason directly from the vector equations and use standard computer programs for calculations.

7.3.4 $\mathbf{Z} = \mathsf{A}\mathbf{W}$, where A is orthogonal

An n by n real matrix A for which $\mathsf{A}\mathsf{A}^{\mathsf{T}} = \mathsf{I}_n$ is called an *orthogonal matrix* or *orthonormal matrix* (orthonormal is more appropriate, but orthogonal is more common). For $\mathbf{Z} = \mathsf{A}\mathbf{W}$, where $\mathbf{W}$ is iid normal and A is orthogonal, $\mathsf{K}_{\mathbf{Z}} = \mathsf{A}\mathsf{A}^{\mathsf{T}} = \mathsf{I}_n$. Thus $\mathsf{K}_{\mathbf{Z}}^{-1} = \mathsf{I}_n$ also, and (7.19) becomes

$$f_{\mathbf{Z}}(\mathbf{z}) = \frac{\exp(-(1/2)\mathbf{z}^{\mathsf{T}}\mathbf{z})}{(2\pi)^{n/2}} = \prod_{k=1}^{n} \frac{\exp(-z_k^2/2)}{\sqrt{2\pi}}. \tag{7.21}$$

This means that A transforms $\mathbf{W}$ into a random vector $\mathbf{Z}$ with the same probability density, and thus the components of $\mathbf{Z}$ are still normal and iid. To understand this better, note that $\mathsf{A}\mathsf{A}^{\mathsf{T}} = \mathsf{I}_n$ means that A^{T} is the inverse of A and thus that $\mathsf{A}^{\mathsf{T}}\mathsf{A} = \mathsf{I}_n$. Letting $\boldsymbol{a}_m$ be the mth column of A, the equation $\mathsf{A}^{\mathsf{T}}\mathsf{A} = \mathsf{I}_n$ means that $\boldsymbol{a}_m^{\mathsf{T}}\boldsymbol{a}_j = \delta_{mj}$ for each m, j, $1 \le m, j \le n$, i.e. that the columns of A are orthonormal. Thus, for the 2D example, the unit vectors $\boldsymbol{e}_1, \boldsymbol{e}_2$ are mapped into orthonormal vectors $\boldsymbol{a}_1, \boldsymbol{a}_2$, so that the transformation simply rotates the points in the plane. Although it is difficult to visualize such a transformation in higher-dimensional space, it is still called a rotation, and has the property that $||\mathsf{A}\boldsymbol{w}||^2 = \boldsymbol{w}^{\mathsf{T}}\mathsf{A}^{\mathsf{T}}\mathsf{A}\boldsymbol{w}$, which is just $\boldsymbol{w}^{\mathsf{T}}\boldsymbol{w} = ||\boldsymbol{w}||^2$. Thus, each point $\boldsymbol{w}$ maps into a point $\mathsf{A}\boldsymbol{w}$ at the same distance from the origin as itself.

Not only are the columns of an orthogonal matrix orthonormal, but also the rows, say $\{\boldsymbol{b}_k;\ 1 \le k \le n\}$ are orthonormal (as is seen directly from $\mathsf{A}\mathsf{A}^{\mathsf{T}} = \mathsf{I}_n$). Since $Z_k = \boldsymbol{b}_k\mathbf{W}$, this means that, for any set of orthonormal vectors $\boldsymbol{b}_1, \ldots, \boldsymbol{b}_n$, the random variables $Z_k = \boldsymbol{b}_k\mathbf{W}$ are normal and iid for $1 \le k \le n$.

7.3.5 Probability density for Gaussian vectors in terms of principal axes

This section describes what is often a more convenient representation for the probability density of an n-dimensional (zero-mean) Gaussian rv $\mathbf{Z}$ with a nonsingular covariance matrix $\mathsf{K}_{\mathbf{Z}}$. As shown in Appendix 7.11.1, the matrix $\mathsf{K}_{\mathbf{Z}}$ has n real orthonormal eigenvectors, $\boldsymbol{q}_1, \ldots, \boldsymbol{q}_n$, with corresponding nonnegative (but not necessarily distinct[9]) real eigenvalues, $\lambda_1, \ldots, \lambda_n$. Also, for any vector $\boldsymbol{z}$, it is shown that $\boldsymbol{z}^{\mathsf{T}}\mathsf{K}_{\mathbf{Z}}^{-1}\boldsymbol{z}$ can be

[9] If an eigenvalue λ has multiplicity m, it means that there is an m-dimensional subspace of vectors $\boldsymbol{q}$ satisfying $\mathsf{K}_{\mathbf{Z}}\boldsymbol{q} = \lambda\boldsymbol{q}$; in this case, any orthonormal set of m such vectors can be chosen as the m eigenvectors corresponding to that eigenvalue.

expressed as $\sum_k \lambda_k^{-1}|\langle z, q_k\rangle|^2$. Substituting this in (7.19), we have

$$f_Z(z) = \frac{1}{(2\pi)^{n/2}\sqrt{\det(\mathsf{K}_Z)}} \exp\left(-\sum_k \frac{|\langle z, q_k\rangle|^2}{2\lambda_k}\right). \tag{7.22}$$

Note that $\langle z, q_k\rangle$ is the projection of z in the direction $\mathbf{q}_k$, where $\mathbf{q}_k$ is the kth of n orthonormal directions. The determinant of an n by n real matrix can be expressed in terms of the n eigenvalues (see Appendix 7.11.1) as $\det(\mathsf{K}_Z) = \prod_{k=1}^n \lambda_k$. Thus (7.22) becomes

$$f_Z(z) = \prod_{k=1}^{n} \frac{1}{\sqrt{2\pi\lambda_k}} \exp\left(\frac{-|\langle z, q_k\rangle|^2}{2\lambda_k}\right). \tag{7.23}$$

This is the product of n Gaussian densities. It can be interpreted as saying that the Gaussian rvs $\{\langle \mathbf{Z}, q_k\rangle; 1 \le k \le n\}$ are statistically independent with variances $\{\lambda_k; 1 \le k \le n\}$. In other words, if we represent the rv $\mathbf{Z}$ using $q_1, \ldots, q_n$ as a basis, then the components of $\mathbf{Z}$ in that coordinate system are independent random variables. The orthonormal eigenvectors are called *principal axes* for $\mathbf{Z}$.

This result can be viewed in terms of the contours of equal probability density for $\mathbf{Z}$ (see Figure 7.2). Each such contour satisfies

$$c = \sum_k \frac{|\langle z, q_k\rangle|^2}{2\lambda_k},$$

where c is proportional to the log probability density for that contour. This is the equation of an ellipsoid centered on the origin, where q_k is the kth axis of the ellipsoid and $\sqrt{2c\lambda_k}$ is the length of that axis.

The probability density formulas in (7.19) and (7.23) suggest that, for every covariance matrix K, there is a jointly Gaussian rv that has that covariance, and thus has that probability density. This is in fact true, but to verify it we must demonstrate that for every covariance matrix K there is a matrix A (and thus a rv $\mathbf{Z} = \mathsf{A}\boldsymbol{W}$) such that $\mathsf{K} = \mathsf{A}\mathsf{A}^{\mathsf{T}}$. There are many such matrices for any given K, but a particularly convenient one is given in (7.84). As a function of the eigenvectors and eigenvalues of K, it is $\mathsf{A} = \sum_k \sqrt{\lambda_k} q_k q_k^{\mathsf{T}}$. Thus, for every nonsingular covariance matrix K, there is a jointly Gaussian rv whose density satisfies (7.19) and (7.23).

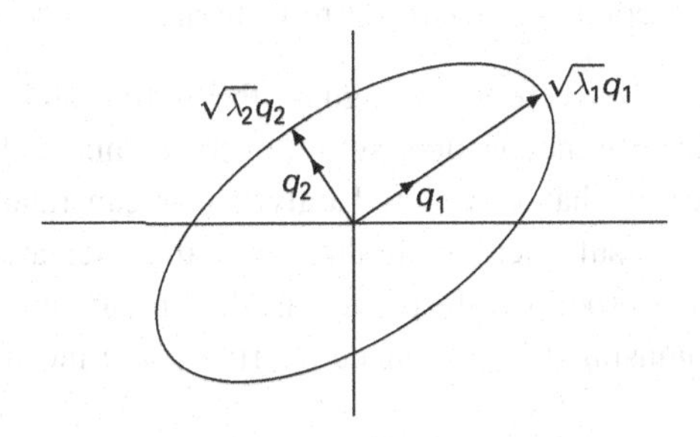

Figure 7.2. Contours of equal probability density. Points z on the q_1 axis are points for which $\langle z, q_2\rangle = 0$ and points on the q_2 axis satisfy $\langle z, q_1\rangle = 0$. Points on the illustrated ellipse satisfy $z^{\mathsf{T}}\mathsf{K}_Z^{-1}z = 1$.

7.3.6 Fourier transforms for joint densities

As suggested in Exercise 7.2, Fourier transforms of probability densities are useful for finding the probability density of sums of independent random variables. More generally, for an n-dimensional rv $\boldsymbol{Z}$, we can define the n-dimensional Fourier transform of $f_{\boldsymbol{Z}}(\boldsymbol{z})$ as follows:

$$\hat{f}_{\boldsymbol{Z}}(\boldsymbol{s}) = \int \cdots \int f_{\boldsymbol{Z}}(\boldsymbol{z}) \exp(-2\pi i \boldsymbol{s}^{\mathsf{T}} \boldsymbol{z})\, dz_1 \cdots dz_n = \mathsf{E}[\exp(-2\pi i \boldsymbol{s}^{\mathsf{T}} \boldsymbol{Z})]. \tag{7.24}$$

If $\boldsymbol{Z}$ is jointly Gaussian, this is easy to calculate. For any given $\boldsymbol{s} \neq \boldsymbol{0}$, let $X = \boldsymbol{s}^{\mathsf{T}}\boldsymbol{Z} = \sum_k s_k Z_k$. Thus X is Gaussian with variance $\mathsf{E}[\boldsymbol{s}^{\mathsf{T}}\boldsymbol{Z}\boldsymbol{Z}^{\mathsf{T}}\boldsymbol{s}] = \boldsymbol{s}^{\mathsf{T}}\mathsf{K}_{\boldsymbol{Z}}\boldsymbol{s}$. From Exercise 7.2,

$$\hat{f}_X(\theta) = \mathsf{E}[\exp(-2\pi i \theta \boldsymbol{s}^{\mathsf{T}}\boldsymbol{Z})] = \exp\left(-\frac{(2\pi\theta)^2 \boldsymbol{s}^{\mathsf{T}}\mathsf{K}_{\boldsymbol{Z}}\boldsymbol{s}}{2} \right). \tag{7.25}$$

Comparing (7.25) for $\theta = 1$ with (7.24), we see that

$$\hat{f}_{\boldsymbol{Z}}(\boldsymbol{s}) = \exp\left(-\frac{(2\pi)^2 \boldsymbol{s}^{\mathsf{T}}\mathsf{K}_{\boldsymbol{Z}}\boldsymbol{s}}{2} \right). \tag{7.26}$$

The above derivation also demonstrates that $\hat{f}_{\boldsymbol{Z}}(\boldsymbol{s})$ is determined by the Fourier transform of each linear combination of the elements of $\boldsymbol{Z}$. In other words, if an arbitrary rv $\boldsymbol{Z}$ has covariance $\mathsf{K}_{\boldsymbol{Z}}$ and has the property that all linear combinations of $\boldsymbol{Z}$ are Gaussian, then the Fourier transform of its density is given by (7.26). Thus, assuming that the Fourier transform of the density uniquely specifies the density, $\boldsymbol{Z}$ must be jointly Gaussian if all linear combinations of $\boldsymbol{Z}$ are Gaussian.

A number of equivalent conditions have now been derived under which a (zero-mean) random vector $\boldsymbol{Z}$ is jointly Gaussian. In summary, each of the following are necessary and sufficient conditions for a rv $\boldsymbol{Z}$ with a nonsingular covariance $\mathsf{K}_{\boldsymbol{Z}}$ to be jointly Gaussian:

- $\boldsymbol{Z} = \mathsf{A}\boldsymbol{W}$, where the components of $\boldsymbol{W}$ are iid normal and $\mathsf{K}_{\boldsymbol{Z}} = \mathsf{A}\mathsf{A}^{\mathsf{T}}$;
- $\boldsymbol{Z}$ has the joint probability density given in (7.19);
- $\boldsymbol{Z}$ has the joint probability density given in (7.23);
- all linear combinations of $\boldsymbol{Z}$ are Gaussian random variables.

For the case where $\mathsf{K}_{\boldsymbol{Z}}$ is singular, the above conditions are necessary and sufficient for any linearly independent subset of the components of $\boldsymbol{Z}$.

This section has considered only zero-mean random variables, vectors, and processes. The results here apply directly to the fluctuation of arbitrary random variables, vectors, and processes. In particular, the probability density for a jointly Gaussian rv $\boldsymbol{Z}$ with a nonsingular covariance matrix $\mathsf{K}_{\boldsymbol{Z}}$ and mean vector $\overline{\boldsymbol{Z}}$ is given by

$$f_{\boldsymbol{Z}}(\boldsymbol{z}) = \frac{1}{(2\pi)^{n/2}\sqrt{\det(\mathsf{K}_{\boldsymbol{Z}})}} \exp\left(-\frac{1}{2}(\boldsymbol{z} - \overline{\boldsymbol{Z}})^{\mathsf{T}}\mathsf{K}_{\boldsymbol{Z}}^{-1}(\boldsymbol{z} - \overline{\boldsymbol{Z}}) \right). \tag{7.27}$$

7.4 Linear functionals and filters for random processes

This section defines the important concept of linear functionals of arbitrary random processes $\{Z(t); t \in \mathbb{R}\}$ and then specializes to Gaussian random processes, where the results of the Section 7.3 can be used. Assume that the sample functions $Z(t, \omega)$ of $Z(t)$ are real $\mathcal{L}_2$ waveforms. These sample functions can be viewed as vectors in the $\mathcal{L}_2$ space of real waveforms. For any given real $\mathcal{L}_2$ waveform $g(t)$, there is an inner product,

$$\langle Z(t, \omega), g(t) \rangle = \int_{-\infty}^{\infty} Z(t, \omega) g(t) \mathrm{d}t.$$

By the Schwarz inequality, the magnitude of this inner product in the space of real $\mathcal{L}_2$ functions is upperbounded by $\|Z(t, \omega)\| \, \|g(t)\|$ and is thus a finite real value for each ω. This then maps sample points ω into real numbers and is thus a random variable,[10] denoted by $V = \int_{-\infty}^{\infty} Z(t) g(t) \mathrm{d}t$. This rv V is called a *linear functional* of the process $\{Z(t); t \in \mathbb{R}\}$.

As an example of the importance of linear functionals, recall that the demodulator for both PAM and QAM contains a filter $q(t)$ followed by a sampler. The output of the filter at a sampling time kT for an input $u(t)$ is $\int u(t) q(kT - t) \mathrm{d}t$. If the filter input also contains additive noise $Z(t)$, then the output at time kT also contains the linear functional $\int Z(t) q(kT - t) \mathrm{d}t$.

Similarly, for any random process $\{Z(t); t \in \mathbb{R}\}$ (again assuming $\mathcal{L}_2$ sample functions) and any real $\mathcal{L}_2$ function $h(t)$, consider the result of passing $Z(t)$ through the filter with impulse response $h(t)$. For any $\mathcal{L}_2$ sample function $Z(t, \omega)$, the filter output at any given time τ is the inner product

$$\langle Z(t, \omega), h(\tau - t) \rangle = \int_{-\infty}^{\infty} Z(t, \omega) h(\tau - t) \mathrm{d}t.$$

For each real τ, this maps sample points ω into real numbers, and thus (aside from measure-theoretic issues)

$$V(\tau) = \int Z(t) h(\tau - t) \mathrm{d}t \tag{7.28}$$

is a rv for each τ. This means that $\{V(\tau); \tau \in \mathbb{R}\}$ is a random process. This is called the *filtered process* resulting from passing $Z(t)$ through the filter $h(t)$. Not much can be said about this general problem without developing a great deal of mathematics, so instead we restrict ourselves to Gaussian processes and other relatively simple examples.

For a Gaussian process, we would hope that a linear functional is a Gaussian random variable. The following examples show that some restrictions are needed even for the class of Gaussian processes.

[10] One should use measure theory over the sample space Ω to interpret these mappings carefully, but this is unnecessary for the simple types of situations here and would take us too far afield.

Example 7.4.1 Let $Z(t) = tX$ for all $t \in \mathbb{R}$, where $X \sim \mathcal{N}(0, 1)$. The sample functions of this Gaussian process have infinite energy with probability 1. The output of the filter also has infinite energy except for very special choices of $h(t)$.

Example 7.4.2 For each $t \in [0, 1]$, let $W(t)$ be a Gaussian rv, $W(t) \sim \mathcal{N}(0, 1)$. Assume also that $\mathsf{E}[W(t)W(\tau)] = 0$ for each $t \neq \tau \in [0, 1]$. The sample functions of this process are discontinuous everywhere.[11] We do not have the machinery to decide whether the sample functions are integrable, let alone whether the linear functionals above exist; we discuss this example further later.

In order to avoid the mathematical issues in Example 7.4.2, along with a host of other mathematical issues, we start with Gaussian processes defined in terms of orthonormal expansions.

7.4.1 Gaussian processes defined over orthonormal expansions

Let $\{\phi_k(t);\ k \geq 1\}$ be a countable set of real orthonormal functions and let $\{Z_k;\ k \geq 1\}$ be a sequence of independent Gaussian random variables, $\{\mathcal{N}(0, \sigma_k^2)\}$. Consider the Gaussian process defined by

$$Z(t) = \sum_{k=1}^{\infty} Z_k \phi_k(t). \tag{7.29}$$

Essentially all zero-mean Gaussian processes of interest can be defined this way, although we will not prove this. Clearly a mean can be added if desired, but zero-mean processes are assumed in what follows. First consider the simple case in which σ_k^2 is nonzero for only finitely many values of k, say $1 \leq k \leq n$. In this case, for each $t \in \mathbb{R}$, $Z(t)$ is a finite sum, given by

$$Z(t) = \sum_{k=1}^{n} Z_k \phi_k(t), \tag{7.30}$$

of independent Gaussian rvs and thus is Gaussian. It is also clear that $Z(t_1), Z(t_2), \ldots, Z(t_\ell)$ are jointly Gaussian for all $\ell, t_1, \ldots, t_\ell$, so $\{Z(t);\ t \in \mathbb{R}\}$ is in fact a Gaussian random process. The energy in any sample function, $z(t) = \sum_k z_k \phi_k(t)$, is $\sum_{k=1}^{n} z_k^2$. This is finite (since the sample values are real and thus finite), so every sample function is $\mathcal{L}_2$. The covariance function is then easily calculated to be

$$\mathsf{K}_Z(t, \tau) = \sum_{k,m} \mathsf{E}[Z_k Z_m] \phi_k(t) \phi_m(\tau) = \sum_{k=1}^{n} \sigma_k^2\, \phi_k(t) \phi_k(\tau). \tag{7.31}$$

Next consider the linear functional $\int Z(t) g(t)\, dt$, where $g(t)$ is a real $\mathcal{L}_2$ function:

$$V = \int_{-\infty}^{\infty} Z(t) g(t) \mathrm{d}t = \sum_{k=1}^{n} Z_k \int_{-\infty}^{\infty} \phi_k(t) g(t) \mathrm{d}t. \tag{7.32}$$

[11] Even worse, the sample functions are not measurable. This process would not even be called a random process in a measure-theoretic formulation, but it provides an interesting example of the occasional need for a measure-theoretic formulation.

Since V is a weighted sum of the zero-mean independent Gaussian rvs $Z_1, \ldots, Z_n$, V is also Gaussian with variance given by

$$\sigma_V^2 = \mathsf{E}[V^2] = \sum_{k=1}^{n} \sigma_k^2 |\langle \boldsymbol{\phi}_k, \boldsymbol{g} \rangle|^2 . \tag{7.33}$$

Next consider the case where n is infinite but $\sum_k \sigma_k^2 < \infty$. The sample functions are still $\mathcal{L}_2$ (at least with probability 1). Equations (7.29) – (7.33) are still valid, and Z is still a Gaussian rv. We do not have the machinery to prove this easily, although Exercise 7.7 provides quite a bit of insight into why these results are true.

Finally, consider a finite set of $\mathcal{L}_2$ waveforms $\{g_m(t);\ 1 \le m \le \ell\}$ and let $V_m = \int_{-\infty}^{\infty} Z(t) g_m(t)\, dt$. By the same argument as above, V_m is a Gaussian rv for each m. Furthermore, since each linear combination of these variables is also a linear functional, it is also Gaussian, so $\{V_1, \ldots, V_\ell\}$ is jointly Gaussian.

7.4.2 Linear filtering of Gaussian processes

We can use the same argument as in Section 7.4.1 to look at the output of a linear filter (see Figure 7.3) for which the input is a Gaussian process $\{Z(t); t \in \mathbb{R}\}$. In particular, assume that $Z(t) = \sum_k Z_k \phi_k(t)$, where $Z_1, Z_2, \ldots$ is an independent sequence $\{Z_k \sim \mathcal{N}(0, \sigma_k^2\}$ satisfying $\sum_k \sigma_k^2 < \infty$ and where $\phi_1(t), \phi_2(t), \ldots$ is a sequence of orthonormal functions.

Assume that the impulse response $h(t)$ of the filter is a real $\mathcal{L}_1$ and $\mathcal{L}_2$ waveform. Then, for any given sample function $Z(t, \omega) = \sum_k Z_k(\omega)\phi_k(t)$ of the input, the filter output at any epoch τ is given by

$$V(\tau, \omega) = \int_{-\infty}^{\infty} Z(t, \omega) h(\tau - t) \mathrm{d}t = \sum_k Z_k(\omega) \int_{-\infty}^{\infty} \phi_k(t) h(\tau - t) \mathrm{d}t. \tag{7.34}$$

Each integral on the right side of (7.34) is an $\mathcal{L}_2$ function of τ (see Exercise 7.5). It follows from this (see Exercise 7.7) that $\int_{-\infty}^{\infty} Z(t, \omega) h(\tau - t) \mathrm{d}t$ is an $\mathcal{L}_2$ waveform with probability 1. For any given epoch τ, (7.34) maps sample points ω to real values, and thus $V(\tau, \omega)$ is a sample value of a random variable $V(\tau)$ defined as

$$V(\tau) = \int_{-\infty}^{\infty} Z(t) h(\tau - t) \mathrm{d}t = \sum_k Z_k \int_{-\infty}^{\infty} \phi_k(t) h(\tau - t) \mathrm{d}t. \tag{7.35}$$

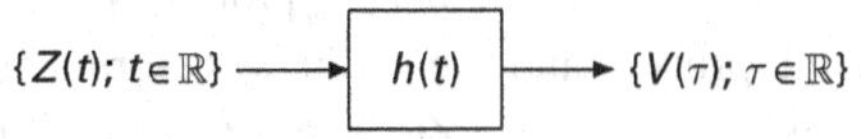

$\{Z(t);\, t \in \mathbb{R}\} \longrightarrow$ $h(t)$ $\longrightarrow \{V(\tau);\, \tau \in \mathbb{R}\}$

Figure 7.3. Filtered random process.

This is a Gaussian rv for each epoch τ. For any set of epochs $\tau_1, \ldots, \tau_\ell$, we see that $V(\tau_1), \ldots, V(\tau_\ell)$ are jointly Gaussian. Thus $\{V(\tau); \tau \in \mathbb{R}\}$ is a Gaussian random process.

We summarize Sections 7.4.1 and 7.4.2 in the following theorem.

Theorem 7.4.1 *Let $\{Z(t); t \in \mathbb{R}\}$ be a Gaussian process $Z(t) = \sum_k Z_k \phi_k(t)$, where $\{Z_k; k \geq 1\}$ is a sequence of independent Gaussian rvs $\mathcal{N}(0, \sigma_k^2)$, where $\sum \sigma_k^2 < \infty$ and $\{\phi_k(t); k \geq 1\}$ is an orthonormal set. Then*

- *for any set of $\mathcal{L}_2$ waveforms $g_1(t), \ldots, g_\ell(t)$, the linear functionals $\{Z_m; 1 \leq m \leq \ell\}$ given by $Z_m = \int_{-\infty}^{\infty} Z(t) g_m(t)\, dt$ are zero-mean jointly Gaussian;*
- *for any filter with real $\mathcal{L}_1$ and $\mathcal{L}_2$ impulse response $h(t)$, the filter output $\{V(\tau); \tau \in \mathbb{R}\}$ given by (7.35) is a zero-mean Gaussian process.*

These are important results. The first, concerning sets of linear functionals, is important when we represent the input to the channel in terms of an orthonormal expansion; the noise can then often be expanded in the same orthonormal expansion. The second, concerning linear filtering, shows that when the received signal and noise are passed through a linear filter, the noise at the filter output is simply another zero-mean Gaussian process. This theorem is often summarized by saying that linear operations preserve Gaussianity.

7.4.3 Covariance for linear functionals and filters

Assume that $\{Z(t); t \in \mathbb{R}\}$ is a random process and that $g_1(t), \ldots, g_\ell(t)$ are real $\mathcal{L}_2$ waveforms. We have seen that if $\{Z(t); t \in \mathbb{R}\}$ is Gaussian, then the linear functionals $V_1, \ldots, V_\ell$ given by $V_m = \int_{-\infty}^{\infty} Z(t) g_m(t) dt$ are jointly Gaussian for $1 \leq m \leq \ell$. We now want to find the covariance for each pair V_j, V_m of these random variables. The result does not depend on the process $Z(t)$ being Gaussian. The computation is quite simple, although we omit questions of limits, interchanges of order of expectation and integration, etc. A more careful derivation could be made by returning to the sampling-theorem arguments before, but this would somewhat obscure the ideas. Assuming that the process $Z(t)$ has zero mean,

$$\mathsf{E}[V_j V_m] = \mathsf{E}\left[\int_{-\infty}^{\infty} Z(t) g_j(t)\, dt \int_{-\infty}^{\infty} Z(\tau) g_m(\tau) d\tau\right] \tag{7.36}$$

$$= \int_{t=-\infty}^{\infty} \int_{\tau=-\infty}^{\infty} g_j(t) \mathsf{E}[Z(t) Z(\tau)] g_m(\tau) dt\, d\tau \tag{7.37}$$

$$= \int_{t=-\infty}^{\infty} \int_{\tau=-\infty}^{\infty} g_j(t) \mathsf{K}_Z(t, \tau) g_m(\tau) dt\, d\tau. \tag{7.38}$$

Each covariance term (including $\mathsf{E}[V_m^2]$ for each m) then depends only on the covariance function of the process and the set of waveforms $\{g_m; 1 \leq m \leq \ell\}$.

The convolution $V(r) = \int Z(t) h(r - t) dt$ is a linear functional at each time r, so the covariance for the filtered output of $\{Z(t); t \in \mathbb{R}\}$ follows in the same way as the results above. The output $\{V(r)\}$ for a filter with a real $\mathcal{L}_2$ impulse response $\boldsymbol{h}$ is given by (7.35), so the covariance of the output can be found as follows:

$$\begin{aligned}\mathsf{K}_V(r,s) &= \mathsf{E}[V(r)V(s)] \\ &= \mathsf{E}\left[\int_{-\infty}^{\infty} Z(t)h(r-t)\mathrm{d}t \int_{-\infty}^{\infty} Z(\tau)h(s-\tau)\mathrm{d}\tau\right] \\ &= \int_{-\infty}^{\infty}\int_{-\infty}^{\infty} h(r-t)\mathsf{K}_Z(t,\tau)h(s-\tau)\mathrm{d}t\,\mathrm{d}\tau. \qquad (7.39)\end{aligned}$$

7.5 Stationarity and related concepts

Many of the most useful random processes have the property that the location of the time origin is irrelevant, i.e. they "behave" the same way at one time as at any other time. This property is called *stationarity*, and such a process is called a *stationary process*.

Since the location of the time origin must be irrelevant for stationarity, random processes that are defined over any interval other than $(-\infty, \infty)$ cannot be stationary. Thus, assume a process that is defined over $(-\infty, \infty)$.

The next requirement for a random process $\{Z(t); t \in \mathbb{R}\}$ to be stationary is that $Z(t)$ must be identically distributed for all epochs $t \in \mathbb{R}$. This means that, for any epochs t and $t+\tau$, and for any real number x, $\Pr\{Z(t) \le x\} = \Pr\{Z(t+\tau) \le x\}$. This does not mean that $Z(t)$ and $Z(t+\tau)$ are the same random variables; for a given sample outcome ω of the experiment, $Z(t,\omega)$ is typically unequal to $Z(t+\tau,\omega)$. It simply means that $Z(t)$ and $Z(t+\tau)$ have the same distribution function, i.e.

$$F_{Z(t)}(x) = F_{Z(t+\tau)}(x) \qquad \text{for all } x. \qquad (7.40)$$

This is still not enough for stationarity, however. The joint distributions over any set of epochs must remain the same if all those epochs are shifted to new epochs by an arbitrary shift τ. This includes the previous requirement as a special case, so we have the following definition.

Definition 7.5.1 A random process $\{Z(t); t \in \mathbb{R}\}$ is *stationary* if, for all positive integers ℓ, for all sets of epochs $t_1, \ldots, t_\ell \in \mathbb{R}$, for all amplitudes $z_1, \ldots, z_\ell$, and for all shifts $\tau \in \mathbb{R}$,

$$F_{Z(t_1),\ldots,Z(t_\ell)}(z_1,\ldots,z_\ell) = F_{Z(t_1+\tau),\ldots,Z(t_\ell+\tau)}(z_1,\ldots,z_\ell). \qquad (7.41)$$

For the typical case where densities exist, this can be rewritten as

$$f_{Z(t_1),\ldots,Z(t_\ell)}(z_1,\ldots,z_\ell) = f_{Z(t_1+\tau),\ldots,Z(t_\ell+\tau)}(z_1,\ldots,z_\ell) \qquad (7.42)$$

for all $z_1, \ldots, z_\ell \in \mathbb{R}$.

For a (zero-mean) Gaussian process, the joint distribution of $Z(t_1), \ldots, Z(t_\ell)$ depends only on the covariance of those variables. Thus, this distribution will be the same as that of $Z(t_1+\tau), \dots, Z(t_\ell+\tau)$ if $\mathsf{K}_Z(t_m, t_j) = \mathsf{K}_Z(t_m+\tau, t_j+\tau)$ for $1 \le m$, $j \le \ell$. This condition will be satisfied for all τ, all ℓ, and all $t_1, \dots, t_\ell$ (verifying that $\{Z(t)\}$ is stationary) if $\mathsf{K}_Z(t_1, t_2) = \mathsf{K}_Z(t_1+\tau, t_2+\tau)$ for all τ and all t_1, t_2. This latter condition will be satisfied if $\mathsf{K}_Z(t_1, t_2) = \mathsf{K}_Z(t_1-t_2, 0)$ for all t_1, t_2. We have thus shown that a zero-mean Gaussian process is stationary if

$$\mathsf{K}_Z(t_1, t_2) = \mathsf{K}_Z(t_1-t_2, 0) \qquad \text{for all } t_1, t_2 \in \mathbb{R}. \tag{7.43}$$

Conversely, if (7.43) is not satisfied for some choice of t_1, t_2, then the joint distribution of $Z(t_1), Z(t_2)$ must be different from that of $Z(t_1-t_2), Z(0)$, and the process is not stationary. The following theorem summarizes this.

Theorem 7.5.1 *A zero-mean Gaussian process $\{Z(t);\ t\in\mathbb{R}\}$ is stationary if and only if (7.43) is satisfied.*

An obvious consequence of this is that a Gaussian process with a nonzero mean is stationary if and only if its mean is constant and its fluctuation satisfies (7.43).

7.5.1 Wide-sense stationary (WSS) random processes

There are many results in probability theory that depend only on the covariances of the random variables of interest (and also the mean if nonzero). For random processes, a number of these classical results are simplified for stationary processes, and these simplifications depend only on the mean and covariance of the process rather than full stationarity. This leads to the following definition.

Definition 7.5.2 A random process $\{Z(t);\ t\in\mathbb{R}\}$ is *wide-sense stationary (WSS)* if $\mathsf{E}[Z(t_1)] = \mathsf{E}[Z(0)]$ and $\mathsf{K}_Z(t_1, t_2) = \mathsf{K}_Z(t_1-t_2, 0)$ for all $t_1, t_2 \in \mathbb{R}$.

Since the covariance function $\mathsf{K}_Z(t+\tau, t)$ of a WSS process is a function of only one variable τ, we will often write the covariance function as a function of one variable, namely $\tilde{\mathsf{K}}_Z(\tau)$ in place of $\mathsf{K}_Z(t+\tau, t)$. In other words, the single variable in the single-argument form represents the difference between the two arguments in two-argument form. Thus, for a WSS process, $\mathsf{K}_Z(t, \tau) = \mathsf{K}_Z(t-\tau, 0) = \tilde{\mathsf{K}}_Z(t-\tau)$.

The random processes defined as expansions of T-spaced sinc functions have been discussed several times. In particular, let

$$V(t) = \sum_k V_k \operatorname{sinc}\left(\frac{t-kT}{T}\right), \tag{7.44}$$

where $\{\dots, V_{-1}, V_0, V_1, \dots\}$ is a sequence of (zero-mean) iid rvs. As shown in (7.8), the covariance function for this random process is given by

$$\mathsf{K}_V(t, \tau) = \sigma_V^2 \sum_k \operatorname{sinc}\left(\frac{t-kT}{T}\right) \operatorname{sinc}\left(\frac{\tau-kT}{T}\right), \tag{7.45}$$

where σ_V^2 is the variance of each V_k. The sum in (7.45), as shown below, is a function only of $t-\tau$, leading to the following theorem.

Theorem 7.5.2 (Sinc expansion) *The random process in (7.44) is WSS. In addition, if the rvs $\{V_k; k \in \mathbb{Z}\}$ are iid Gaussian, the process is stationary. The covariance function is given by*

$$\tilde{\mathsf{K}}_V(t-\tau) = \sigma_V^2 \operatorname{sinc}\left(\frac{t-\tau}{T}\right). \tag{7.46}$$

Proof From the sampling theorem, any $\mathcal{L}_2$ function $u(t)$, baseband-limited to $1/2T$, can be expanded as

$$u(t) = \sum_k u(kT) \operatorname{sinc}\left(\frac{t-kT}{T}\right). \tag{7.47}$$

For any given τ, take $u(t)$ to be $\operatorname{sinc}[(t-\tau)/T]$. Substituting this in (7.47), we obtain

$$\operatorname{sinc}\left(\frac{t-\tau}{T}\right) = \sum_k \operatorname{sinc}\left(\frac{kT-\tau}{T}\right) \operatorname{sinc}\left(\frac{t-kT}{T}\right) = \sum_k \operatorname{sinc}\left(\frac{\tau-kT}{T}\right) \operatorname{sinc}\left(\frac{t-kT}{T}\right). \tag{7.48}$$

Substituting this in (7.45) shows that the process is WSS with the stated covariance. As shown in Section 7.4.1, $\{V(t); t \in \mathbb{R}\}$ is Gaussian if the rvs $\{V_k\}$ are Gaussian. From Theorem 7.5.1, this Gaussian process is stationary since it is WSS. □

Next consider another special case of the sinc expansion in which each V_k is binary, taking values ± 1 with equal probability. This corresponds to a simple form of a PAM transmitted waveform. In this case, $V(kT)$ must be ± 1, whereas, for values of t between the sample points, $V(t)$ can take on a wide range of values. Thus this process is WSS but cannot be stationary. Similarly, any discrete distribution for each V_k creates a process that is WSS but not stationary.

There are not many important models of *noise* processes that are WSS but not stationary,[12] despite the above example and the widespread usage of the term WSS. Rather, the notion of wide-sense stationarity is used to make clear, for some results, that they depend only on the mean and covariance, thus perhaps making it easier to understand them.

The Gaussian sinc expansion brings out an interesting theoretical non sequitur. Assuming that $\sigma_V^2 > 0$, i.e. that the process is not the trivial process for which $V(t) = 0$ with probability 1 for all t, the expected energy in the process (taken over all time) is infinite. It is not difficult to convince oneself that the sample functions of this process have infinite energy with probability 1. Thus, stationary noise models are simple to work with, but the sample functions of these processes do not fit into the $\mathcal{L}_2$ theory of waveforms that has been developed. Even more important than the issue of infinite energy, stationary noise models make unwarranted assumptions about the very distant

[12] An important exception is interference from other users, which, as the above sinc expansion with binary signals shows, can be WSS but not stationary. Even in this case, if the interference is modeled as just part of the noise (rather than specifically as interference), the nonstationarity is usually ignored.

past and future. The extent to which these assumptions affect the results about the present is an important question that must be asked.

The problem here is not with the peculiarities of the Gaussian sinc expansion. Rather it is that stationary processes have constant power over all time, and thus have infinite energy. One practical solution[13] to this is simple and familiar. The random process is simply truncated in any convenient way. Thus, when we say that noise is stationary, we mean that it is stationary within a much longer time interval than the interval of interest for communication. This is not very precise, and the notion of *effective stationarity* is now developed to formalize this notion of a truncated stationary process.

7.5.2 Effectively stationary and effectively WSS random processes

Definition 7.5.3 A (zero-mean) random process is *effectively stationary within* $[-T_0/2, T_0/2]$ if the joint probability assignment for $t_1, \ldots, t_n$ is the same as that for $t_1+\tau, t_2+\tau, \ldots, t_n+\tau$ whenever $t_1, \ldots, t_n$ and $t_1+\tau, t_2+\tau, \ldots, t_n+\tau$ are all contained in the interval $[-T_0/2, T_0/2]$. It is *effectively WSS within* $[-T_0/2, T_0/2]$ if $\mathsf{K}_Z(t, \tau)$ is a function only of $t-\tau$ for $t, \tau \in [-T_0/2, T_0/2]$. A random process with nonzero mean is effectively stationary (effectively WSS) if its mean is constant within $[-T_0/2, T_0/2]$ and its fluctuation is effectively stationary (WSS) within $[-T_0/2, T_0/2]$.

One way to view a stationary (WSS) random process is in the limiting sense of a process that is effectively stationary (WSS) for all intervals $[-T_0/2, T_0/2]$. For operations such as linear functionals and filtering, the nature of this limit as T_0 becomes large is quite simple and natural, whereas, for frequency-domain results, the effect of finite T_0 is quite subtle.

For an effectively WSS process within $[-T_0/2, T_0/2]$, the covariance within $[-T_0/2, T_0/2]$ is a function of a single parameter, $\mathsf{K}_Z(t, \tau) = \tilde{\mathsf{K}}_Z(t-\tau)$ for $t, \tau \in [-T_0/2, T_0/2]$. As illustrated by Figure 7.4, however, that $t-\tau$ can range from $-T_0$ (for $t = -T_0/2, \tau = T_0/2$) to T_0 (for $t = T_0/2, \tau = -T_0/2$).

Since a Gaussian process is determined by its covariance function and mean, it is effectively stationary within $[-T_0/2, T_0/2]$ if it is effectively WSS.

Note that the difference between a stationary and effectively stationary random process for large T_0 is primarily a difference in the model and not in the situation being modeled. If two models have a significantly different behavior over the time intervals of interest, or, more concretely, if noise in the distant past or future has a significant effect, then the entire modeling issue should be rethought.

[13] There is another popular solution to this problem. For any $\mathcal{L}_2$ function $g(t)$, the energy in $g(t)$ outside of $[-T_0/2, T_0/2]$ vanishes as $T_0 \to \infty$, so intuitively the effect of these tails on the linear functional $\int g(t)Z(t)dt$ vanishes as $T_0 \to 0$. This provides a nice intuitive basis for ignoring the problem, but it fails, both intuitively and mathematically, in the frequency domain.

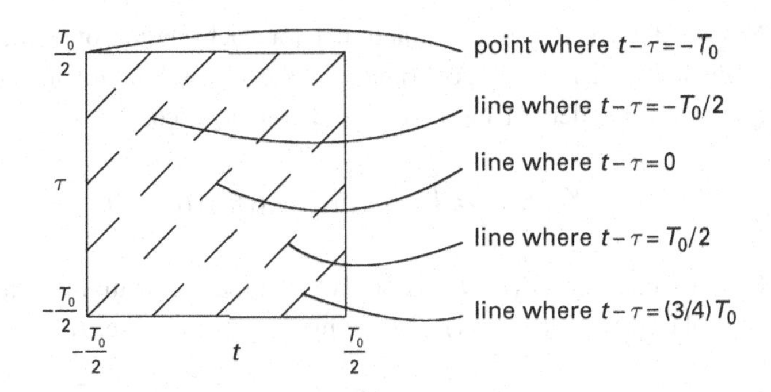

Figure 7.4. Relationship of the two-argument covariance function $\mathsf{K}_Z(t,\,\tau)$ and the one-argument function $\tilde{\mathsf{K}}_Z(t-\tau)$ for an effectively WSS process; $\mathsf{K}_Z(t,\,\tau)$ is constant on each dashed line above. Note that, for example, the line for which $t-\tau=(3/4)T_0$ applies only for pairs $(t,\,\tau)$ where $t \ge T_0/2$ and $\tau \le -T_0/2$. Thus $\tilde{\mathsf{K}}_Z((3/4)T_0)$ is not necessarily equal to $\mathsf{K}_Z((3/4)T_0, 0)$. It can be easily verified, however, that $\tilde{\mathsf{K}}_Z(\alpha T_0) = \mathsf{K}_Z(\alpha T_0, 0)$ for all $\alpha \le 1/2$.

7.5.3 Linear functionals for effectively WSS random processes

The covariance matrix for a set of linear functionals and the covariance function for the output of a linear filter take on simpler forms for WSS or effectively WSS processes than the corresponding forms for general processes derived in Section 7.4.3.

Let $Z(t)$ be a zero-mean WSS random process with covariance function $\tilde{\mathsf{K}}_Z(t-\tau)$ for $t,\tau \in [-T_0/2,\, T_0/2]$, and let $g_1(t), g_2(t), \ldots, g_\ell(t)$ be a set of $\mathcal{L}_2$ functions nonzero only within $[-T_0/2,\, T_0/2]$. For the conventional WSS case, we can take $T_0=\infty$. Let the linear functional V_m be given by $\int_{-T_0/2}^{T_0/2} Z(t) g_m(t)\,\mathrm{d}t$ for $1 \le m \le \ell$. The covariance $\mathsf{E}[V_m V_j]$ is then given by

$$\mathsf{E}[V_m V_j] = \mathsf{E}\left[\int_{-T_0/2}^{T_0/2} Z(t) g_m(t)\mathrm{d}t \int_{-\infty}^{\infty} Z(\tau) g_j(\tau)\mathrm{d}\tau\right]$$

$$= \int_{-T_0/2}^{T_0/2}\int_{-T_0/2}^{T_0/2} g_m(t)\tilde{\mathsf{K}}_Z(t-\tau) g_j(\tau)\mathrm{d}\tau\,\mathrm{d}t. \tag{7.49}$$

Note that this depends only on the covariance where $t,\tau \in [-T_0/2,\, T_0/2]$, i.e. where $\{Z(t)\}$ is effectively WSS. This is not surprising, since we would not expect V_m to depend on the behavior of the process outside of where $g_m(t)$ is nonzero.

7.5.4 Linear filters for effectively WSS random processes

Next consider passing a random process $\{Z(t);\, t \in \mathbb{R}\}$ through a linear time-invariant filter whose impulse response $h(t)$ is $\mathcal{L}_2$. As pointed out in (7.28), the output of the filter is a random process $\{V(\tau);\, \tau \in \mathbb{R}\}$ given by

$$V(\tau) = \int_{-\infty}^{\infty} Z(t_1) h(\tau - t_1)\mathrm{d}t_1.$$

Note that $V(\tau)$ is a linear functional for each choice of τ. The covariance function evaluated at t, τ is the covariance of the linear functionals $V(t)$ and $V(\tau)$. Ignoring questions of orders of integration and convergence,

$$\mathsf{K}_V(t,\tau) = \int_{-\infty}^{\infty}\int_{-\infty}^{\infty} h(t-t_1)\mathsf{K}_Z(t_1,t_2)h(\tau-t_2)\mathrm{d}t_1\,\mathrm{d}t_2. \tag{7.50}$$

First assume that $\{Z(t);\, t\in\mathbb{R}\}$ is WSS in the conventional sense. Then $\mathsf{K}_Z(t_1,t_2)$ can be replaced by $\tilde{\mathsf{K}}_Z(t_1-t_2)$. Replacing t_1-t_2 by s (i.e. t_1 by t_2+s), we have

$$\mathsf{K}_V(t,\tau) = \int_{-\infty}^{\infty}\left[\int_{-\infty}^{\infty} h(t-t_2-s)\tilde{\mathsf{K}}_Z(s)\mathrm{d}s\right] h(\tau-t_2)\mathrm{d}t_2.$$

Replacing t_2 by $\tau+\mu$ yields

$$\mathsf{K}_V(t,\tau) = \int_{-\infty}^{\infty}\left[\int_{-\infty}^{\infty} h(t-\tau-\mu-s)\tilde{\mathsf{K}}_Z(s)\mathrm{d}s\right] h(-\mu)\mathrm{d}\mu. \tag{7.51}$$

Thus $\mathsf{K}_V(t,\tau)$ is a function only of $t-\tau$. This means that $\{V(t);\, t\in\mathbb{R}\}$ is WSS. This is not surprising; passing a WSS random process through a linear time-invariant filter results in another WSS random process.

If $\{Z(t);\, t\in\mathbb{R}\}$ is a Gaussian process, then, from Theorem 7.4.1, $\{V(t);\, t\in\mathbb{R}\}$ is also a Gaussian process. Since a Gaussian process is determined by its covariance function, it follows that if $Z(t)$ is a stationary Gaussian process, then $V(t)$ is also a stationary Gaussian process.

We do not have the mathematical machinery to carry out the above operations carefully over the infinite time interval.[14] Rather, it is now assumed that $\{Z(t);\, t\in\mathbb{R}\}$ is effectively WSS within $[-T_0/2, T_0/2]$. It will also be assumed that the impulse response $h(t)$ above is time-limited in the sense that, for some finite A, $h(t)=0$ for $|t|>A$.

Theorem 7.5.3 *Let $\{Z(t);\, t\in\mathbb{R}\}$ be effectively WSS within $[-T_0/2, T_0/2]$ and have sample functions that are $\mathcal{L}_2$ within $[-T_0/2, T_0/2]$ with probability 1. Let $Z(t)$ be the input to a filter with an $\mathcal{L}_2$ time-limited impulse response $\{h(t)\colon [-A, A]\to\mathbb{R}\}$. Then, for $T_0/2>A$, the output random process $\{V(t);\, t\in\mathbb{R}\}$ is WSS within $[-T_0/2+A, T_0/2-A]$ and its sample functions within $[-T_0/2+A, T_0/2-A]$ are $\mathcal{L}_2$ with probability 1.*

Proof Let $z(t)$ be a sample function of $Z(t)$ and assume $z(t)$ is $\mathcal{L}_2$ within $[-T_0/2, T_0/2]$. Let $v(\tau)=\int z(t)h(\tau-t)\,\mathrm{d}t$ be the corresponding filter output. For each $\tau\in[-T_0/2+A, T_0/2-A]$, $v(\tau)$ is determined by $z(t)$ in the range $t\in[-T_0/2, T_0/2]$.

[14] More important, we have no justification for modeling a process over the infinite time interval. Later, however, after building up some intuition about the relationship of an infinite interval to a very large interval, we can use the simpler equations corresponding to infinite intervals.

Thus, if we replace $z(t)$ by $z_0(t) = z(t)\,\text{rect}[T_0]$, the filter output, say $v_0(\tau)$, will equal $v(\tau)$ for $\tau \in [-T_0/2+A,\ T_0/2-A]$. The time-limited function $z_0(t)$ is $\mathcal{L}_1$ as well as $\mathcal{L}_2$. This implies that the Fourier transform $\hat{z}_0(f)$ is bounded, say by $\hat{z}_0(f) \le B$, for each f. Since $\hat{v}_0(f) = \hat{z}_0(f)\hat{h}(f)$, we see that

$$\int |\hat{v}_0(f)|^2\,\mathrm{d}f = \int |\hat{z}_0(f)|^2|\hat{h}(f)|^2\,\mathrm{d}f \le B^2 \int |\hat{h}(f)|^2\,\mathrm{d}f < \infty.$$

This means that $\hat{v}_0(f)$, and thus also $v_0(t)$, is $\mathcal{L}_2$. Now $v_0(t)$, when truncated to $[-T_0/2+A,\ T_0/2-A]$ is equal to $v(t)$ truncated to $[-T_0/2+A,\ T_0/2-A]$, so the truncated version of $v(t)$ is $\mathcal{L}_2$. Thus the sample functions of $\{V(t)\}$, truncated to $[-T_0/2+A,\ T_0/2-A]$, are $\mathcal{L}_2$ with probability 1.

Finally, since $\{Z(t);\, t \in \mathbb{R}\}$ can be truncated to $[-T_0/2,\ T_0/2]$ with no lack of generality, it follows that $\mathsf{K}_Z(t_1, t_2)$ can be truncated to $t_1, t_2 \in [-T_0/2,\ T_0/2]$. Thus, for $t, \tau \in [-T_0/2+A,\ T_0/2-A]$, (7.50) becomes

$$\mathsf{K}_V(t,\tau) = \int_{-T_0/2}^{T_0/2}\int_{-T_0/2}^{T_0/2} h(t-t_1)\tilde{\mathsf{K}}_Z(t_1-t_2)h(\tau-t_2)\,\mathrm{d}t_1\,\mathrm{d}t_2. \tag{7.52}$$

The argument in (7.50) and (7.51) shows that $V(t)$ is effectively WSS within $[-T_0/2+A,\ T_0/2-A]$. □

Theorem 7.5.3, along with the effective WSS result about linear functionals, shows us that results about WSS processes can be used within finite intervals. The result in the theorem about the interval of effective stationarity being reduced by filtering should not be too surprising. If we truncate a process and then pass it through a filter, the filter spreads out the effect of the truncation. For a finite-duration filter, however, as assumed here, this spreading is limited.

The notion of stationarity (or effective stationarity) makes sense as a modeling tool where T_0 is very much larger than other durations of interest, and in fact where there is no need for explicit concern about how long the process is going to be stationary.

Theorem 7.5.3 essentially tells us that we can have our cake and eat it too. That is, transmitted waveforms and noise processes can be truncated, thus making use of both common sense and $\mathcal{L}_2$ theory, but at the same time insights about stationarity can still be relied upon. More specifically, random processes can be modeled as stationary, without specifying a specific interval $[-T_0/2,\ T_0/2]$ of effective stationarity, because stationary processes can now be viewed as asymptotic versions of finite-duration processes.

Appendices 7.11.2 and 7.11.3 provide a deeper analysis of WSS processes truncated to an interval. The truncated process is represented as a Fourier series with random variables as coefficients. This gives a clean interpretation of what happens as the interval size is increased without bound, and also gives a clean interpretation of the effect of time-truncation in the frequency domain. Another approach to a truncated process is the Karhunen–Loeve expansion, which is discussed in Appendix 7.11.4.

7.6 Stationarity in the frequency domain

Stationary and WSS zero-mean processes, and particularly Gaussian processes, are often viewed more insightfully in the frequency domain than in the time domain. An effectively WSS process over $[-T_0/2, T_0/2]$ has a single-variable covariance function $\tilde{\mathsf{K}}_Z(\tau)$ defined over $[-T_0, T_0]$. A WSS process can be viewed as a process that is effectively WSS for each T_0. The energy in such a process, truncated to $[-T_0/2, T_0/2]$, is linearly increasing in T_0, but the covariance simply becomes defined over a larger and larger interval as $T_0 \to \infty$. We assume in what follows that this limiting covariance is $\mathcal{L}_2$. This does not appear to rule out any but the most pathological processes.

First we look at linear functionals and linear filters, ignoring limiting questions and convergence issues and assuming that T_0 is "large enough." We will refer to the random processes as stationary, while still assuming $\mathcal{L}_2$ sample functions.

For a zero-mean WSS process $\{Z(t); t \in \mathbb{R}\}$ and a real $\mathcal{L}_2$ function $g(t)$, consider the linear functional $V = \int g(t)Z(t)\mathrm{d}t$. From (7.49),

$$\mathsf{E}[V^2] = \int_{-\infty}^{\infty} g(t)\left[\int_{-\infty}^{\infty} \tilde{\mathsf{K}}_Z(t-\tau)g(\tau)\mathrm{d}\tau\right]\mathrm{d}t \tag{7.53}$$

$$= \int_{-\infty}^{\infty} g(t)\left[\tilde{\mathsf{K}}_Z * \boldsymbol{g}\right](t)\mathrm{d}t, \tag{7.54}$$

where $\tilde{\mathsf{K}}_Z * \boldsymbol{g}$ denotes the convolution of the waveforms $\tilde{\mathsf{K}}_Z(t)$ and $g(t)$. Let $S_Z(f)$ be the Fourier transform of $\tilde{\mathsf{K}}_Z(t)$. The function $S_Z(f)$ is called the *spectral density* of the stationary process $\{Z(t); t \in \mathbb{R}\}$. Since $\tilde{\mathsf{K}}_Z(t)$ is $\mathcal{L}_2$, real, and symmetric, its Fourier transform is also $\mathcal{L}_2$, real, and symmetric, and, as shown later, $S_Z(f) \geq 0$. It is also shown later that $S_Z(f)$ at each frequency f can be interpreted as the power per unit frequency at f.

Let $\theta(t) = [\tilde{\mathsf{K}}_Z * \boldsymbol{g}](t)$ be the convolution of $\tilde{\mathsf{K}}_Z$ and $\boldsymbol{g}$. Since $\boldsymbol{g}$ and K_Z are real, $\theta(t)$ is also real, so $\theta(t) = \theta^*(t)$. Using Parseval's theorem for Fourier transforms,

$$\mathsf{E}[V^2] = \int_{-\infty}^{\infty} g(t)\theta^*(t)\mathrm{d}t = \int_{-\infty}^{\infty} \hat{g}(f)\hat{\theta}^*(f)\mathrm{d}f.$$

Since $\theta(t)$ is the convolution of K_Z and $\boldsymbol{g}$, we see that $\hat{\theta}(f) = S_Z(f)\hat{g}(f)$. Thus,

$$\mathsf{E}[V^2] = \int_{-\infty}^{\infty} \hat{g}(f)S_Z(f)\hat{g}^*(f)\mathrm{d}f = \int_{-\infty}^{\infty} |\hat{g}(f)|^2 S_Z(f)\mathrm{d}f. \tag{7.55}$$

Note that $\mathsf{E}[V^2] \geq 0$ and that this holds for all real $\mathcal{L}_2$ functions $g(t)$. The fact that $g(t)$ is real constrains the transform $\hat{g}(f)$ to satisfy $\hat{g}(f) = \hat{g}^*(-f)$, and thus $|\hat{g}(f)| = |\hat{g}(-f)|$ for all f. Subject to this constraint and the constraint that $|\hat{g}(f)|$ be $\mathcal{L}_2$, $|\hat{g}(f)|$ can be chosen as any $\mathcal{L}_2$ function. Stated another way, $\hat{g}(f)$ can be chosen arbitrarily for $f \geq 0$ subject to being $\mathcal{L}_2$.

Since $S_Z(f) = S_Z(-f)$, (7.55) can be rewritten as follows:

$$\mathsf{E}[V^2] = \int_0^\infty 2\,|\hat{g}(f)|^2 S_Z(f)\mathrm{d}f.$$

Since $\mathsf{E}[V^2] \geq 0$ and $|\hat{g}(f)|$ is arbitrary, it follows that $S_Z(f) \geq 0$ for all $f \in \mathbb{R}$.

The conclusion here is that the spectral density of any WSS random process must be nonnegative. Since $S_Z(f)$ is also the Fourier transform of $\tilde{\mathsf{K}}(t)$, this means that a necessary property of any single-variable covariance function is that it have a nonnegative Fourier transform.

Next, let $V_m = \int g_m(t)Z(t)\,\mathrm{d}t$, where the function $g_m(t)$ is real and $\mathcal{L}_2$ for $m = 1, 2$. From (7.49), we have

$$\mathsf{E}[V_1V_2] = \int_{-\infty}^{\infty} g_1(t)\left[\int_{-\infty}^{\infty} \tilde{\mathsf{K}}_Z(t-\tau)g_2(\tau)\mathrm{d}\tau\right]\mathrm{d}t \tag{7.56}$$

$$= \int_{-\infty}^{\infty} g_1(t)\left[\tilde{\mathsf{K}} * \boldsymbol{g}_2\right](t)\mathrm{d}t. \tag{7.57}$$

Let $\hat{g}_m(f)$ be the Fourier transform of $g_m(t)$ for $m = 1, 2$, and let $\theta(t) = [\tilde{\mathsf{K}}_Z(t) * \boldsymbol{g}_2](t)$ be the convolution of $\tilde{\mathsf{K}}_Z$ and $\boldsymbol{g}_2$. Let $\hat{\theta}(f) = S_Z(f)\hat{g}_2(f)$ be its Fourier transform. As before, we have

$$\mathsf{E}[V_1V_2] = \int \hat{g}_1(f)\hat{\theta}^*(f)\mathrm{d}f = \int \hat{g}_1(f)S_Z(f)\hat{g}_2^*(f)\mathrm{d}f. \tag{7.58}$$

There is a remarkable feature in the above expression. If $\hat{g}_1(f)$ and $\hat{g}_2(f)$ have no overlap in frequency, then $\mathsf{E}[V_1V_2] = 0$. In other words, for any stationary process, two linear functionals over different frequency ranges must be uncorrelated. If the process is Gaussian, then the linear functionals are independent. This means in essence that Gaussian noise in different frequency bands must be independent. That this is true simply because of stationarity is surprising. Appendix 7.11.3 helps to explain this puzzling phenomenon, especially with respect to effective stationarity.

Next, let $\{\phi_m(t);\ m \in \mathbb{Z}\}$ be a set of real orthonormal functions and let $\{\hat{\phi}_m(f)\}$ be the corresponding set of Fourier transforms. Letting $V_m = \int Z(t)\phi_m(t)\mathrm{d}t$, (7.58) becomes

$$\mathsf{E}[V_mV_j] = \int \hat{\phi}_m(f)S_Z(f)\hat{\phi}_j^*(f)\mathrm{d}f. \tag{7.59}$$

If the set of orthonormal functions $\{\phi_m(t);\ m \in \mathbb{Z}\}$ is limited to some frequency band, and if $S_Z(f)$ is constant, say with value $N_0/2$ in that band, then

$$\mathsf{E}[V_mV_j] = \frac{N_0}{2}\int \hat{\phi}_m(f)\hat{\phi}_j^*(f)\mathrm{d}f. \tag{7.60}$$

By Parseval's theorem for Fourier transforms, we have $\int \hat{\phi}_m(f)\hat{\phi}_j^*(f)\mathrm{d}f = \delta_{mj}$, and thus

$$\mathsf{E}[V_mV_j] = \frac{N_0}{2}\delta_{mj}. \tag{7.61}$$

The rather peculiar-looking constant $N_0/2$ is explained in Section 7.7. For now, however, it is possible to interpret the meaning of the spectral density of a noise process.

Suppose that $S_Z(f)$ is continuous and approximately constant with value $S_Z(f_c)$ over some narrow band of frequencies around f_c, and suppose that $\phi_1(t)$ is constrained to that narrow band. Then the variance of the linear functional $\int_{-\infty}^{\infty} Z(t)\phi_1(t)\mathrm{d}t$ is approximately $S_Z(f_c)$. In other words, $S_Z(f_c)$ in some fundamental sense describes the energy in the noise per degree of freedom at the frequency f_c. Section 7.7 interprets this further.

7.7 White Gaussian noise

Physical noise processes are very often reasonably modeled as zero-mean, stationary, and Gaussian. There is one further simplification that is often reasonable. This is that the covariance between the noise at two epochs dies out very rapidly as the interval between those epochs increases. The interval over which this covariance is significantly nonzero is often very small relative to the intervals over which the signal varies appreciably. This means that the covariance function $\tilde{\mathsf{K}}_Z(\tau)$ looks like a short-duration pulse around $\tau = 0$.

We know from linear system theory that $\int \tilde{\mathsf{K}}_Z(t-\tau)g(\tau)\mathrm{d}\tau$ is equal to $g(t)$ if $\tilde{\mathsf{K}}_Z(t)$ is a unit impulse. Also, this integral is approximately equal to $g(t)$ if $\tilde{\mathsf{K}}_Z(t)$ has unit area and is a narrow pulse relative to changes in $g(t)$. It follows that, under the same circumstances, (7.56) becomes

$$\mathsf{E}[V_1 V_2^*] = \int_t \int_\tau g_1(t)\tilde{\mathsf{K}}_Z(t-\tau)g_2(\tau)\mathrm{d}\tau\,\mathrm{d}t \approx \int g_1(t)g_2(t)\mathrm{d}t. \tag{7.62}$$

This means that if the covariance function is very narrow relative to the functions of interest, then its behavior relative to those functions is specified by its area. In other words, the covariance function can be viewed as an impulse of a given magnitude. We refer to a zero-mean WSS Gaussian random process with such a narrow covariance function as *white Gaussian noise (WGN)*. The area under the covariance function is called the *intensity* or the *spectral density* of the WGN and is denoted by the symbol $N_0/2$. Thus, for $\mathcal{L}_2$ functions $g_1(t), g_2(t), \ldots$ in the range of interest, and for WGN (denoted by $\{W(t);\, t \in \mathbb{R}\}$) of intensity $N_0/2$, the random variable $V_m = \int W(t)g_m(t)\mathrm{d}t$ has variance given by

$$\mathsf{E}[V_m^2] = (N_0/2)\int g_m^2(t)\mathrm{d}t. \tag{7.63}$$

Similarly, the rvs V_j and V_m have covariance given by

$$\mathsf{E}[V_j V_m] = (N_0/2)\int g_j(t)g_m(t)\mathrm{d}t. \tag{7.64}$$

Also, $V_1, V_2, \ldots$ are jointly Gaussian.

The most important special case of (7.63) and (7.64) is to let $\phi_j(t)$ be a set of orthonormal functions and let $W(t)$ be WGN of intensity $N_0/2$. Let $V_m = \int \phi_m(t)W(t)\mathrm{d}t$. Then, from (7.63) and (7.64),

$$\mathsf{E}[V_j V_m] = (N_0/2)\delta_{jm}. \tag{7.65}$$

This is an important equation. It says that if the noise can be modeled as WGN, then when the noise is represented in terms of *any* orthonormal expansion, the resulting rvs are iid. Thus, we can represent signals in terms of an arbitrary orthonormal expansion, and represent WGN in terms of the same expansion, and the result is iid Gaussian rvs.

Since the coefficients of a WGN process in any orthonormal expansion are iid Gaussian, it is common to also refer to a random vector of iid Gaussian rvs as WGN.

If $\mathsf{K}_W(t)$ is approximated by $(N_0/2)\delta(t)$, then the spectral density is approximated by $S_W(f) = N_0/2$. If we are concerned with a particular band of frequencies, then we are interested in $S_W(f)$ being constant within that band, and in this case $\{W(t);\, t \in \mathbb{R}\}$ can be represented as white noise within that band. If this is the only band of interest, we can model[15] $S_W(f)$ as equal to $N_0/2$ everywhere, in which case the corresponding model for the covariance function is $(N_0/2)\delta(t)$.

The careful reader will observe that WGN has not really been defined. What has been said, in essence, is that if a stationary zero-mean Gaussian process has a covariance function that is very narrow relative to the variation of all functions of interest, or a spectral density that is constant within the frequency band of interest, then we can pretend that the covariance function is an impulse times $N_0/2$, where $N_0/2$ is the value of $S_W(f)$ within the band of interest. Unfortunately, according to the definition of random process, there cannot be any Gaussian random process $W(t)$ whose covariance function is $\tilde{\mathsf{K}}(t) = (N_0/2)\delta(t)$. The reason for this dilemma is that $\mathsf{E}[W^2(t)] = \mathsf{K}_W(0)$. We could interpret $\mathsf{K}_W(0)$ to be either undefined or ∞, but either way $W(t)$ cannot be a random variable (although we could think of it taking on only the values plus or minus ∞).

Mathematicians view WGN as a generalized random process, in the same sense as the unit impulse $\delta(t)$ is viewed as a generalized function. That is, the impulse function $\delta(t)$ is not viewed as an ordinary function taking the value 0 for $t \neq 0$ and the value ∞ at $t = 0$. Rather, it is viewed in terms of its effect on other, better behaved, functions $g(t)$, where $\int_{-\infty}^{\infty} g(t)\delta(t)\,\mathrm{d}t = g(0)$. In the same way, WGN is not viewed in terms of rvs at each epoch of time. Rather, it is viewed as a generalized zero-mean random process for which linear functionals are jointly Gaussian, for which variances and covariances are given by (7.63) and (7.64), and for which the covariance is formally taken to be $(N_0/2)\delta(t)$.

Engineers should view WGN within the context of an overall bandwidth and time interval of interest, where the process is effectively stationary within the time interval and has a constant spectral density over the band of interest. Within that context, the spectral density can be viewed as constant, the covariance can be viewed as an impulse, and (7.63) and (7.64) can be used.

The difference between the engineering view and the mathematical view is that the engineering view is based on a context of given time interval and bandwidth of

[15] This is not as obvious as it sounds, and will be further discussed in terms of the theorem of irrelevance in Chapter 8.

interest, whereas the mathematical view is based on a very careful set of definitions and limiting operations within which theorems can be stated without explicitly defining the context. Although the ability to prove theorems without stating the context is valuable, any application must be based on the context.

When we refer to signal space, what is usually meant is this overall bandwidth and time interval of interest, i.e. the context above. As we have seen, the bandwidth and the time interval cannot both be perfectly truncated, and because of this signal space cannot be regarded as strictly finite-dimensional. However, since the time interval and bandwidth are essentially truncated, visualizing signal space as finite-dimensional with additive WGN is often a reasonable model.

7.7.1 The sinc expansion as an approximation to WGN

Theorem 7.5.2 treated the process $Z(t) = \sum_k Z_k \operatorname{sinc}(t - kT/T)$, where each rv $\{Z_k; k \in \mathbb{Z}\}$ is iid and $\mathcal{N}(0, \sigma^2)$. We found that the process is zero-mean Gaussian and stationary with covariance function $\tilde{\mathsf{K}}_Z(t-\tau) = \sigma^2 \operatorname{sinc}[(t-\tau)/T]$. The spectral density for this process is then given by

$$S_Z(f) = \sigma^2 T \operatorname{rect}(fT). \tag{7.66}$$

This process has a constant spectral density over the baseband bandwidth $\mathsf{W}_b = 1/2T$, so, by making T sufficiently small, the spectral density is constant over a band sufficiently large to include all frequencies of interest. Thus this process can be viewed as WGN of spectral density $N_0/2 = \sigma^2 T$ for any desired range of frequencies $\mathsf{W}_b = 1/2T$ by making T sufficiently small. Note, however, that to approximate WGN of spectral density $N_0/2$, the noise power, i.e. the variance of $Z(t)$ is $\sigma^2 = \mathsf{W}N_0$. In other words, σ^2 must increase with increasing W. This also says that N_0 is the noise power per unit *positive frequency*. The spectral density, $N_0/2$, is defined over both positive and negative frequencies, and so becomes N_0 when positive and negative frequencies are combined, as in the standard definition of bandwidth.[16]

If a sinc process is passed through a linear filter with an arbitrary impulse response $h(t)$, the output is a stationary Gaussian process with spectral density $|\hat{h}(f)|^2 \sigma^2 T \operatorname{rect}(fT)$. Thus, by using a sinc process plus a linear filter, a stationary Gaussian process with any desired nonnegative spectral density within any desired finite bandwith can be generated. In other words, stationary Gaussian processes with arbitrary covariances (subject to $S(f) \geq 0$) can be generated from orthonormal expansions of Gaussian variables.

Since the sinc process is stationary, it has sample waveforms of infinite energy. As explained in Section 7.5.2, this process may be truncated to achieve an effectively stationary process with $\mathcal{L}_2$ sample waveforms. Appendix 7.11.3 provides some insight

[16] One would think that this field would have found a way to be consistent about counting only positive frequencies or positive and negative frequencies. However, the word bandwidth is so widely used among the mathophobic, and Fourier analysis is so necessary for engineers, that one must simply live with such minor confusions.

about how an effectively stationary Gaussian process over an interval T_0 approaches stationarity as $T_0 \to \infty$.

The sinc process can also be used to understand the strange, everywhere uncorrelated, process in Example 7.4.2. Holding $\sigma^2 = 1$ in the sinc expansion as T approaches 0, we get a process whose limiting covariance function is 1 for $t - \tau = 0$ and 0 elsewhere. The corresponding limiting spectral density is 0 everywhere. What is happening is that the power in the process (i.e. $\tilde{K}_Z(0)$) is 1, but that power is being spread over a wider and wider band as $T \to 0$, so the power per unit frequency goes to 0.

To explain this in another way, note that any measurement of this noise process must involve filtering over some very small, but nonzero, interval. The output of this filter will have zero variance. Mathematically, of course, the limiting covariance is $\mathcal{L}_2$-equivalent to 0, so again the mathematics[17] corresponds to engineering reality.

7.7.2 Poisson process noise

The sinc process of Section 7.7.1 is very convenient for generating noise processes that approximate WGN in an easily used formulation. On the other hand, this process is not very believable[18] as a physical process. A model that corresponds better to physical phenomena, particularly for optical channels, is a sequence of very narrow pulses which arrive according to a Poisson distribution in time.

The Poisson distribution, for our purposes, can be simply viewed as a limit of a discrete-time process where the time axis is segmented into intervals of duration Δ and a pulse of width Δ arrives in each interval with probability $\Delta\rho$, independent of every other interval. When such a process is passed through a linear filter, the fluctuation of the output at each instant of time is approximately Gaussian if the filter is of sufficiently small bandwidth to integrate over a very large number of pulses. One can similarly argue that linear combinations of filter outputs tend to be approximately Gaussian, making the process an approximation of a Gaussian process.

We do not analyze this carefully, since our point of view is that WGN, over limited bandwidths, is a reasonable and canonical approximation to a large number of physical noise processes. After understanding how this affects various communication systems, one can go back and see whether the model is appropriate for the given physical noise process. When we study wireless communication, we will find that the major problem is not that the noise is poorly approximated by WGN, but rather that the channel itself is randomly varying.

---

[17] This process also cannot be satisfactorily defined in a measure-theoretic way.

[18] To many people, defining these sinc processes with their easily analyzed properties, but no physical justification, is more troublesome than our earlier use of discrete memoryless sources in studying source coding. In fact, the approach to modeling is the same in each case: first understand a class of easy-to-analyze but perhaps impractical processes, then build on that understanding to understand practical cases. Actually, sinc processes have an advantage here: the bandlimited stationary Gaussian random processes defined this way (although not the method of generation) are widely used as practical noise models, whereas there are virtually no uses of discrete memoryless sources as practical source models.

7.8 Adding noise to modulated communication

Consider the QAM communication problem again. A complex $\mathcal{L}_2$ baseband waveform $u(t)$ is generated and modulated up to passband as a real waveform $x(t) = 2\Re[u(t)e^{2\pi i f_c t}]$. A sample function $w(t)$ of a random noise process $W(t)$ is then added to $x(t)$ to produce the output $y(t) = x(t) + w(t)$, which is then demodulated back to baseband as the received complex baseband waveform $v(t)$.

Generalizing QAM somewhat, assume that $u(t)$ is given by $u(t) = \sum_k u_k \theta_k(t)$, where the functions $\theta_k(t)$ are complex orthonormal functions and the sequence of symbols $\{u_k;\, k \in \mathbb{Z}\}$ are complex numbers drawn from the symbol alphabet and carrying the information to be transmitted. For each symbol u_k, $\Re(u_k)$ and $\Im(u_k)$ should be viewed as sample values of the random variables $\Re(U_k)$ and $\Im(U_k)$. The joint probability distributions of these rvs is determined by the incoming random binary digits and how they are mapped into symbols. The *complex random variable*[19] $\Re(U_k) + i\Im(U_k)$ is then denoted by U_k.

In the same way, $\Re(\sum_k U_k \theta_k(t))$ and $\Im(\sum_k U_k \theta_k(t))$ are random processes denoted, respectively, by $\Re(U(t))$ and $\Im(U(t))$. We then call $U(t) = \Re(U(t)) + i\Im(U(t))$ for $t \in \mathbb{R}$ a *complex random process*. A complex random process $U(t)$ is defined by the joint distribution of $U(t_1), U(t_2), \ldots, U(t_n)$ for all choices of $n, t_1, \ldots, t_n$. This is equivalent to defining both $\Re(U(t))$ and $\Im(U(t))$ as joint processes.

Recall from the discussion of the Nyquist criterion that if the QAM transmit pulse $p(t)$ is chosen to be square root of Nyquist, then $p(t)$ and its T-spaced shifts are orthogonal and can be normalized to be orthonormal. Thus a particularly natural choice here is $\theta_k(t) = p(t - kT)$ for such a $\boldsymbol{p}$. Note that this is a generalization of Chapter 6 in the sense that $\{U_k;\, k \in \mathbb{Z}\}$ is a sequence of complex rvs using random choices from the signal constellation rather than some given sample function of that random sequence. The transmitted passband (random) waveform is then given by

$$X(t) = \sum_k 2\Re\{U_k \theta_k(t) \exp(2\pi i f_c t)\}. \tag{7.67}$$

Recall that the transmitted waveform has twice the power of the baseband waveform. Now define the following:

$$\psi_{k,1}(t) = \Re\{2\theta_k(t) \exp(2\pi i f_c t)\};$$
$$\psi_{k,2}(t) = \Im\{-2\theta_k(t) \exp(2\pi i f_c t)\}.$$

Also, let $U_{k,1} = \Re(U_k)$ and $U_{k,2} = \Im(U_k)$. Then

$$X(t) = \sum_k [U_{k,1}\psi_{k,1}(t) + U_{k,2}\psi_{k,2}(t)].$$

---

[19] Recall that a rv is a mapping from sample points to real numbers, so that a complex rv is a mapping from sample points to complex numbers. Sometimes in discussions involving both rvs and complex rvs, it helps to refer to rvs as real rvs, but the modifier "real" is superfluous.

As shown in Theorem 6.6.1, the set of bandpass functions $\{\psi_{k,\ell};\, k \in \mathbb{Z}, \ell \in \{1,2\}\}$ are orthogonal, and each has energy equal to 2. This again assumes that the carrier frequency f_c is greater than all frequencies in each baseband function $\theta_k(t)$.

In order for $u(t)$ to be $\mathcal{L}_2$, assume that the number of orthogonal waveforms $\theta_k(t)$ is arbitrarily large but finite, say $\theta_1(t), \ldots, \theta_n(t)$. Thus $\{\psi_{k,\ell}\}$ is also limited to $1 \le k \le n$.

Assume that the noise $\{W(t);\, t \in \mathbb{R}\}$ is white over the band of interest and effectively stationary over the time interval of interest, but has $\mathcal{L}_2$ sample functions.[20] Since $\{\psi_{k,l};\, 1 \le k \le n, \ell = 1,2\}$ is a finite real orthogonal set, the projection theorem can be used to express each sample noise waveform $\{w(t);\, t \in \mathbb{R}\}$ as

$$w(t) = \sum_{k=1}^{n} [z_{k,1}\psi_{k,1}(t) + z_{k,2}\psi_{k,2}(t)] + w_{\perp}(t), \tag{7.68}$$

where $w_{\perp}(t)$ is the component of the sample noise waveform perpendicular to the space spanned by $\{\psi_{k,l};\, 1 \le k \le n, \ell = 1,2\}$. Let $Z_{k,\ell}$ be the rv with sample value $z_{k,\ell}$. Then each rv $Z_{k,\ell}$ is a linear functional on $W(t)$. Since $\{\psi_{k,l};\, 1 \le k \le n, \ell = 1,2\}$ is an orthogonal set, the rvs $Z_{k,\ell}$ are iid Gaussian rvs. Let $W_{\perp}(t)$ be the random process corresponding to the sample function $w_{\perp}(t)$ above. Expanding $\{W_{\perp}(t);\, t \in \mathbb{R}\}$ in an orthonormal expansion orthogonal to $\{\psi_{k,l};\, 1 \le k \le n, \ell = 1,2\}$, the coefficients are assumed to be independent of the $Z_{k,\ell}$, at least over the time and frequency band of interest. What happens to these coefficients outside of the region of interest is of no concern, other than assuming that $W_{\perp}(t)$ is independent of $U_{k,\ell}$ and $Z_{k,\ell}$ for $1 \le k \le n$ and $\ell = \{1,2\}$. The received waveform $Y(t) = X(t) + W(t)$ is then given by

$$Y(t) = \sum_{k=1}^{n} \left[(U_{k,1} + Z_{k,1})\psi_{k,1}(t) + (U_{k,2} + Z_{k,2})\psi_{k,2}(t)\right] + W_{\perp}(t).$$

When this is demodulated,[21] the baseband waveform is represented as the complex waveform,

$$V(t) = \sum_{k} (U_k + Z_k)\theta_k(t) + Z_{\perp}(t), \tag{7.69}$$

where each Z_k is a complex rv given by $Z_k = Z_{k,1} + iZ_{k,2}$ and the baseband residual noise $Z_{\perp}(t)$ is independent of $\{U_k, Z_k;\, 1 \le k \le n\}$. The variance of each real rv $Z_{k,1}$ and $Z_{k,2}$ is taken by convention to be $N_0/2$. We follow this convention because we are measuring the input power at baseband; as mentioned many times, the power at passband is scaled to be twice that at baseband. The point here is that N_0 is not a physical constant; rather, it is the noise power per unit positive frequency *in the units used to represent the signal power.*

[20] Since the set of orthogonal waveforms $\theta_k(t)$ is not necessarily time- or frequency-limited, the assumption here is that the noise is white over a much larger time and frequency interval than the nominal bandwidth and time interval used for communication. This assumption is discussed further in Chapter 8.

[21] Some filtering is necessary before demodulation to remove the residual noise that is far out of band, but we do not want to analyze that here.

7.8.1 Complex Gaussian random variables and vectors

Noise waveforms, after demodulation to baseband, are usually complex and are thus represented, as in (7.69), by a sequence of complex random variables which is best regarded as a complex random vector (rv). It is possible to view any such n-dimensional complex rv $\mathbf{Z} = \mathbf{Z}_{\text{re}} + \text{i}\mathbf{Z}_{\text{im}}$ as a $2n$-dimensional real rv

$$\begin{bmatrix} \mathbf{Z}_{\text{re}} \\ \mathbf{Z}_{\text{im}} \end{bmatrix}, \quad \text{where } \mathbf{Z}_{\text{re}} = \Re(\mathbf{Z}) \text{ and } \mathbf{Z}_{\text{im}} = \Im(\mathbf{Z}).$$

For many of the same reasons that it is desirable to work directly with a complex baseband waveform rather than a pair of real passband waveforms, it is often beneficial to work directly with complex rvs.

A complex rv $Z = Z_{\text{re}} + \text{i}Z_{\text{im}}$ is *Gaussian* if Z_{re} and Z_{im} are jointly Gaussian; Z is *circularly symmetric Gaussian*[22] if it is Gaussian and in addition Z_{re} and Z_{im} are iid. In this case (assuming zero mean as usual), the amplitude of Z is a Rayleigh-distributed rv and the phase is uniformly distributed; thus the joint density is circularly symmetric. A circularly symmetric complex Gaussian rv Z is fully described by its mean $\bar{Z}$ (which we continue to assume to be 0 unless stated otherwise) and its variance $\sigma^2 = \mathsf{E}[\tilde{Z}\tilde{Z}^*]$. A circularly symmetric complex Gaussian rv Z of mean $\bar{Z}$ and variance σ^2 is denoted by $Z \sim \mathcal{CN}(\bar{Z}, \sigma^2)$.

A complex random vector $\mathbf{Z}$ is a jointly Gaussian rv if the real and imaginary components of $\mathbf{Z}$ collectively are jointly Gaussian; it is also circularly symmetric if the density of the fluctuation $\tilde{\mathbf{Z}}$ (i.e. the joint density of the real and imaginary parts of the components of $\tilde{\mathbf{Z}}$) is the same[23] as that of $e^{i\theta}\tilde{\mathbf{Z}}$ for all phase angles θ.

An important example of a circularly symmetric Gaussian rv is $\mathbf{Z} = (Z_1, \ldots, Z_n)^{\mathsf{T}}$, where the real and imaginary components collectively are iid and $\mathcal{N}(0, 1)$. Because of the circular symmetry of each Z_k, multiplying $\mathbf{Z}$ by $e^{i\theta}$ simply rotates each Z_k and the probability density does not change. The probability density is just that of $2n$ iid $\mathcal{N}(0, 1)$ rvs, which is

$$f_{\mathbf{Z}}(\mathbf{z}) = \frac{1}{(2\pi)^n} \exp\left(\frac{\sum_{k=1}^{n} -|z_k|^2}{2} \right), \tag{7.70}$$

where we have used the fact that $|z_k|^2 = \Re(z_k)^2 + \Im(z_k)^2$ for each k to replace a sum over $2n$ terms with a sum over n terms.

Another much more general example is to let A be an arbitrary complex n by n matrix and let the complex rv $\boldsymbol{Y}$ be defined by

$$\boldsymbol{Y} = \mathsf{A}\mathbf{Z}, \tag{7.71}$$

---

[22] This is sometimes referred to as complex proper Gaussian.

[23] For a single complex rv Z with Gaussian real and imaginary parts, this phase-invariance property is enough to show that the real and imaginary parts are jointly Gaussian, and thus that Z is circularly symmetric Gaussian. For a random vector with Gaussian real and imaginary parts, phase invariance as defined here is not sufficient to ensure the jointly Gaussian property. See Exercise 7.14 for an example.

where $\boldsymbol{Z}$ has iid real and imaginary normal components as above. The complex **rv** defined in this way has jointly Gaussian real and imaginary parts. To see this, represent (7.71) as the following real linear transformation of $2n$ real space:

$$\begin{bmatrix} \boldsymbol{Y}_{\mathrm{re}} \\ \boldsymbol{Y}_{\mathrm{im}} \end{bmatrix} = \begin{bmatrix} \mathsf{A}_{\mathrm{re}} & -\mathsf{A}_{\mathrm{im}} \\ \mathsf{A}_{\mathrm{im}} & \mathsf{A}_{\mathrm{re}} \end{bmatrix} \begin{bmatrix} \boldsymbol{Z}_{\mathrm{re}} \\ \boldsymbol{Z}_{\mathrm{im}} \end{bmatrix}, \tag{7.72}$$

where $\boldsymbol{Y}_{\mathrm{re}} = \Re(\boldsymbol{Y})$, $\boldsymbol{Y}_{\mathrm{im}} = \Im(\boldsymbol{Y})$, $\mathsf{A}_{\mathrm{re}} = \Re(\mathsf{A})$, and $\mathsf{A}_{\mathrm{im}} = \Im(\mathsf{A})$.

The rv $\boldsymbol{Y}$ is also circularly symmetric.[24] To see this, note that $\mathrm{e}^{\mathrm{i}\theta}\boldsymbol{Y} = \mathrm{e}^{\mathrm{i}\theta}\mathsf{A}\boldsymbol{Z} = \mathsf{A}\mathrm{e}^{\mathrm{i}\theta}\boldsymbol{Z}$. Since $\boldsymbol{Z}$ is circularly symmetric, the density at any given sample value $\boldsymbol{z}$ (i.e. the density for the real and imaginary parts of $\boldsymbol{z}$) is the same as that at $\mathrm{e}^{\mathrm{i}\theta}\boldsymbol{z}$. This in turn implies[25] that the density at $\boldsymbol{y}$ is the same as that at $\mathrm{e}^{\mathrm{i}\theta}\boldsymbol{y}$.

The covariance matrix of a complex **rv** $\boldsymbol{Y}$ is defined as

$$\mathsf{K}_{\boldsymbol{Y}} = \mathsf{E}[\boldsymbol{Y}\boldsymbol{Y}^{\dagger}], \tag{7.73}$$

where $\boldsymbol{Y}^{\dagger}$ is defined as $\boldsymbol{Y}^{\mathsf{T}*}$. For a random vector $\boldsymbol{Y}$ defined by (7.71), $\mathsf{K}_{\boldsymbol{Y}} = \mathsf{A}\mathsf{A}^{\dagger}$.

Finally, for a circularly symmetric complex Gaussian vector as defined in (7.71), the probability density is given by

$$f_{\boldsymbol{Y}}(\boldsymbol{y}) = \frac{1}{(2\pi)^n \det(\mathsf{K}_{\boldsymbol{Y}})} \exp\left(-\boldsymbol{y}^{\dagger}\mathsf{K}_{\boldsymbol{Y}}\boldsymbol{y}\right). \tag{7.74}$$

It can be seen that complex circularly symmetric Gaussian vectors behave quite similarly to (real) jointly Gaussian vectors. Both are defined by their covariance matrices, the properties of the covariance matrices are almost identical (see Appendix 7.11.1), the covariance can be expressed as $\mathsf{A}\mathsf{A}^{\dagger}$, where A describes a linear transformation from iid components, and the transformation A preserves the circularly symmetric Gaussian property in the complex case and the jointly Gaussian property in the real case.

An arbitrary (zero-mean) complex Gaussian rv is not specified by its variance, since $\mathsf{E}[Z_{\mathrm{re}}^2]$ might be different from $\mathsf{E}[Z_{\mathrm{im}}^2]$. Similarly, an arbitrary (zero-mean) complex Gaussian vector is not specified by its covariance matrix. In fact, arbitrary Gaussian complex n-vectors are usually best viewed as $2n$-dimensional real vectors; the simplifications from dealing with complex Gaussian vectors directly are primarily constrained to the circularly symmetric case.

7.9 Signal-to-noise ratio

There are a number of different measures of signal power, noise power, energy per symbol, energy per bit, and so forth, which are defined here. These measures are explained

[24] Conversely, as we will see later, all circularly symmetric jointly Gaussian rvs can be defined this way.

[25] This is not as simple as it appears, and is shown more carefully in the exercises. It is easy to become facile at working in $\mathbb{R}^n$ and $\mathbb{C}^n$, but going back and forth between $\mathbb{R}^{2n}$ and $\mathbb{C}^n$ is tricky and inelegant (witness (7.72) and (7.71)).

in terms of QAM and PAM, but they also apply more generally. In Section 7.8, a fairly general set of orthonormal functions was used, and here a specific set is assumed. Consider the orthonormal functions $p_k(t) = p(t-kT)$ as used in QAM, and use a nominal passband bandwidth $\mathsf{W} = 1/T$. Each QAM symbol U_k can be assumed to be iid with energy $E_s = \mathsf{E}[|U_k|^2]$. This is the signal energy per two real dimensions (i.e. real plus imaginary). The noise energy per two real dimensions is defined to be N_0. Thus the signal-to-noise ratio is defined to be

$$\text{SNR} = \frac{E_s}{N_0} \qquad \text{for QAM.} \tag{7.75}$$

For baseband PAM, using real orthonormal functions satisfying $p_k(t) = p(t-kT)$, the signal energy per signal is $E_s = \mathsf{E}[|U_k|^2]$. Since the signal is 1D, i.e. real, the noise energy per dimension is defined to be $N_0/2$. Thus, the SNR is defined to be

$$\text{SNR} = \frac{2E_s}{N_0} \qquad \text{for PAM.} \tag{7.76}$$

For QAM there are W complex degrees of freedom per second, so the signal power is given by $P = E_s\mathsf{W}$. For PAM at baseband, there are $2\mathsf{W}$ degrees of freedom per second, so the signal power is $P = 2E_s\mathsf{W}$. Thus, in each case, the SNR becomes

$$\text{SNR} = \frac{P}{N_0\mathsf{W}} \qquad \text{for QAM and PAM.} \tag{7.77}$$

We can interpret the denominator here as the overall noise power in the bandwidth W, so SNR is also viewed as the signal power divided by the noise power in the nominal band. For those who like to minimize the number of formulas they remember, all of these equations for SNR follow from a basic definition as the signal energy per degree of freedom divided by the noise energy per degree of freedom.

PAM and QAM each use the same signal energy for each degree of freedom (or at least for each complex pair of degrees of freedom), whereas other systems might use the available degrees of freedom differently. For example, PAM with baseband bandwidth W occupies bandwidth $2\mathsf{W}$ if modulated to passband, and uses only half the available degrees of freedom. For these situations, SNR can be defined in several different ways depending on the context. As another example, frequency-hopping is a technique used both in wireless and in secure communication. It is the same as QAM, except that the carrier frequency f_c changes pseudo-randomly at intervals long relative to the symbol interval. Here the bandwidth W might be taken as the bandwidth of the underlying QAM system, or as the overall bandwidth within which f_c hops. The SNR in (7.77) is quite different in the two cases.

The appearance of W in the denominator of the expression for SNR in (7.77) is rather surprising and disturbing at first. It says that if more bandwidth is allocated to a communication system with the same available power, then SNR *decreases*. This is because the signal energy per degree of freedom decreases when it is spread over more degrees of freedom, but the noise is everywhere. We will see later that the net gain can be made positive.

Another important parameter is the rate R; this is the number of transmitted bits per second, which is the number of bits per symbol, $\log_2|\mathcal{A}|$, times the number of symbols per second. Thus

$$R = \mathsf{W}\log_2|\mathcal{A}| \quad \text{for QAM}; \qquad R = 2\mathsf{W}\log_2|\mathcal{A}| \quad \text{for PAM}. \tag{7.78}$$

An important parameter is the *spectral efficiency* of the system, which is defined as $\rho = R/\mathsf{W}$. This is the transmitted number of bits per second in each unit frequency interval. For QAM and PAM, ρ is given by (7.78) to be

$$\rho = \log_2|\mathcal{A}| \quad \text{for QAM}; \qquad \rho = 2\log_2|\mathcal{A}| \quad \text{for PAM}. \tag{7.79}$$

More generally, the spectral efficiency ρ can be defined as the number of transmitted bits per degree of freedom. From (7.79), achieving a large value of spectral efficiency requires making the symbol alphabet large; note that ρ increases only logarithmically with $|\mathcal{A}|$.

Yet another parameter is the energy per bit E_{b}. Since each symbol contains $\log_2 \mathcal{A}$ bits, E_{b} is given for both QAM and PAM by

$$E_{\mathrm{b}} = \frac{E_{\mathrm{s}}}{\log_2|\mathcal{A}|}. \tag{7.80}$$

One of the most fundamental quantities in communication is the ratio E_{b}/N_0. Since E_{b} is the signal energy per bit and N_0 is the noise energy per two degrees of freedom, this provides an important limit on energy consumption. For QAM, we substitute (7.75) and (7.79) into (7.80), yielding

$$\frac{E_{\mathrm{b}}}{N_0} = \frac{\mathrm{SNR}}{\rho}. \tag{7.81}$$

The same equation is seen to be valid for PAM. This says that achieving a small value for E_{b}/N_0 requires a small ratio of SNR to ρ. We look at this next in terms of channel capacity.

One of Shannon's most famous results was to develop the concept of the capacity C of an additive WGN communication channel. This is defined as the supremum of the number of bits per second that can be transmitted and received with arbitrarily small error probability. For the WGN channel with a constraint W on the bandwidth and a constraint P on the received signal power, he showed that

$$C = \mathsf{W}\log_2\left(1 + \frac{P}{\mathsf{W}N_0}\right). \tag{7.82}$$

Furthermore, arbitrarily small error probability can be achieved at any rate $R < C$ by using channel coding of arbitrarily large constraint length. He also showed, and later results strengthened the fact, that larger rates would lead to large error probabilities. These results will be demonstrated in Chapter 8; they are widely used as a benchmark for comparison with particular systems. Figure 7.5 shows a sketch of C as a function of W. Note that C increases monotonically with W, reaching a limit of $(P/N_0)\log_2 e$ as $\mathsf{W} \to \infty$. This is known as the ultimate Shannon limit on achievable rate. Note also that

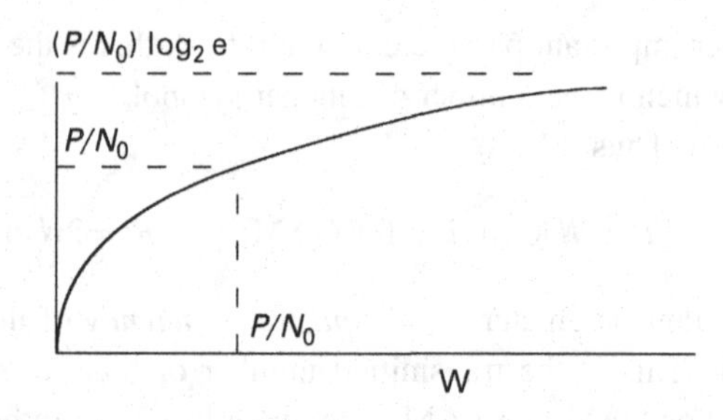

Figure 7.5. Capacity as a function of bandwidth W for fixed P/N_0.

when $\mathsf{W} = P/N_0$, i.e. when the bandwidth is large enough for the SNR to reach 1, then C is within $1/\log_2 \mathrm{e}$, 69% of the ultimate Shannon limit. This is usually expressed as being within 1.6 dB of the ultimate Shannon limit.

Shannon's result showed that the error probability can be made arbitrarily small for any rate $R < C$. Using (7.81) for C, ρ for R/W, and SNR for $P/(\mathsf{W}N_0)$, the inequality $R < C$ becomes

$$\rho < \log_2(1 + \mathrm{SNR}). \tag{7.83}$$

If we substitute this into (7.81), we obtain

$$\frac{E_\mathrm{b}}{N_0} > \frac{\mathrm{SNR}}{\log_2(1 + \mathrm{SNR})}.$$

This is a monotonic increasing function of the single-variable SNR, which in turn is decreasing in W. Thus $(E_\mathrm{b}/N_0)_{\min}$ is monotonically decreasing in W. As $\mathsf{W} \to \infty$ it reaches the limit $\ln 2 = 0.693$, i.e. -1.59 dB. As W decreases, it grows, reaching 0 dB at SNR = 1, and increasing without bound for yet smaller W. The limiting spectral efficiency, however, is C/W. This is also monotonically decreasing in W, going to 0 as $\mathsf{W} \to \infty$. In other words, there is a trade-off between the required E_b/N_0, which is preferably small, and the required spectral efficiency ρ, which is preferably large. This is discussed further in Chapter 8.

7.10 Summary of random processes

The additive noise in physical communication systems is usually best modeled as a random process, i.e. a collection of random variables, one at each real-valued instant of time. A random process can be specified by its joint probability distribution over all finite sets of epochs, but additive noise is most often modeled by the assumption that the rvs are all zero-mean Gaussian and their joint distribution is jointly Gaussian.

These assumptions were motivated partly by the central limit theorem, partly by the simplicity of working with Gaussian processes, partly by custom, and partly by various extremal properties. We found that jointly Gaussian means a great deal more than individually Gaussian, and that the resulting joint densities are determined by the covariance matrix. These densities have ellipsoidal contours of equal probability density whose axes are the eigenfunctions of the covariance matrix.

A sample function $Z(t, \omega)$ of a random process $Z(t)$ can be viewed as a waveform and interpreted as an $\mathcal{L}_2$ vector. For any fixed $\mathcal{L}_2$ function $g(t)$, the inner product $\langle g(t), Z(t, \omega)\rangle$ maps ω into a real number and thus can be viewed over Ω as a random variable. This rv is called a linear function of $Z(t)$ and is denoted by $\int g(t)Z(t)\,dt$.

These linear functionals arise when expanding a random process into an orthonormal expansion and also at each epoch when a random process is passed through a linear filter. For simplicity, these linear functionals and the underlying random processes are not viewed in a measure-theoretic perspective, although the $\mathcal{L}_2$ development in Chapter 4 provides some insight about the mathematical subtleties involved.

Noise processes are usually viewed as being stationary, which effectively means that their statistics do not change in time. This generates two problems: first, the sample functions have infinite energy, and, second, there is no clear way to see whether results are highly sensitive to time regions far outside the region of interest. Both of these problems are treated by defining effective stationarity (or effective wide-sense stationarity) in terms of the behavior of the process over a finite interval. This analysis shows, for example, that Gaussian linear functionals depend only on effective stationarity over the signal space of interest. From a practical standpoint, this means that the simple results arising from the assumption of stationarity can be used without concern for the process statistics outside the time range of interest.

The spectral density of a stationary process can also be used without concern for the process outside the time range of interest. If a process is effectively WSS, it has a single-variable covariance function corresponding to the interval of interest, and this has a Fourier transform which operates as the spectral density over the region of interest. How these results change as the region of interest approaches ∞ is explained in Appendix 7.11.3.

7.11 Appendix: Supplementary topics

7.11.1 Properties of covariance matrices

This appendix summarizes some properties of covariance matrices that are often useful but not absolutely critical to our treatment of random processes. Rather than repeat everything twice, we combine the treatment for real and complex rvs together. On a first reading, however, one might assume everything to be real. Most of the results are the same in each case, although the complex-conjugate signs can be removed in the real case. It is important to realize that the properties developed here apply to non-Gaussian as well as Gaussian rvs. All rvs and rvs here are assumed to be zero-mean.

A square matrix K is a *covariance matrix* if a (real or complex) rv $\mathbf{Z}$ exists such that $\mathsf{K} = \mathsf{E}[\mathbf{Z}\mathbf{Z}^{\mathsf{T}*}]$. The complex conjugate of the transpose, $\mathbf{Z}^{\mathsf{T}*}$, is called the *Hermitian transpose* and is denoted by $\mathbf{Z}^{\dagger}$. If $\mathbf{Z}$ is real, of course, $\mathbf{Z}^{\dagger} = \mathbf{Z}^{\mathsf{T}}$. Similarly, for a matrix K, the Hermitian conjugate, denoted by $\mathsf{K}^{\dagger}$, is $\mathsf{K}^{\mathsf{T}*}$. A matrix is *Hermitian* if $\mathsf{K} = \mathsf{K}^{\dagger}$. Thus a real Hermitian matrix (a Hermitian matrix containing all real terms) is a symmetric matrix.

An n by n square matrix K with real or complex terms is *nonnegative definite* if it is Hermitian and if, for all $\boldsymbol{b} \in \mathbb{C}^n$, $\boldsymbol{b}^\dagger \mathsf{K} \boldsymbol{b}$ is real and nonnegative. It is *positive definite* if, in addition, $\boldsymbol{b}^\dagger \mathsf{K} \boldsymbol{b} > 0$ for $\boldsymbol{b} \neq 0$. We now list some of the important relationships between nonnegative definite, positive definite, and covariance matrices and state some other useful properties of covariance matrices.

(1) Every covariance matrix K is nonnegative definite. To see this, let $\boldsymbol{Z}$ be a **rv** such that $\mathsf{K} = \mathsf{E}[\boldsymbol{Z}\boldsymbol{Z}^\dagger]$; K is Hermitian since $\mathsf{E}[Z_k Z_m^*] = \mathsf{E}[Z_m^* Z_k]$ for all k, m. For any $\boldsymbol{b} \in \mathbb{C}^n$, let $X = \boldsymbol{b}^\dagger \boldsymbol{Z}$. Then $0 \le \mathsf{E}[|X|^2] = \mathsf{E}\left[(\boldsymbol{b}^\dagger \boldsymbol{Z})(\boldsymbol{b}^\dagger \boldsymbol{Z})^*\right] = \mathsf{E}\left[\boldsymbol{b}^\dagger \boldsymbol{Z}\boldsymbol{Z}^\dagger \boldsymbol{b}\right] = \boldsymbol{b}^\dagger \mathsf{K} \boldsymbol{b}$.

(2) For any complex n by n matrix A, the matrix $\mathsf{K} = \mathsf{A}\mathsf{A}^\dagger$ is a covariance matrix. In fact, let $\boldsymbol{Z}$ have n independent unit-variance elements so that $\mathsf{K}_{\boldsymbol{Z}}$ is the identity matrix I_n. Then $\boldsymbol{Y} = \mathsf{A}\boldsymbol{Z}$ has the covariance matrix $\mathsf{K}_{\boldsymbol{Y}} = \mathsf{E}[(\mathsf{A}\boldsymbol{Z})(\mathsf{A}\boldsymbol{Z})^\dagger] = \mathsf{E}[\mathsf{A}\boldsymbol{Z}\boldsymbol{Z}^\dagger \mathsf{A}^\dagger] = \mathsf{A}\mathsf{A}^\dagger$. Note that if A is real and $\boldsymbol{Z}$ is real, then $\boldsymbol{Y}$ is real and, of course, $\mathsf{K}_{\boldsymbol{Y}}$ is real. It is also possible for A to be real and $\boldsymbol{Z}$ complex, and in this case $\mathsf{K}_{\boldsymbol{Y}}$ is still real but $\boldsymbol{Y}$ is complex.

(3) A covariance matrix K is positive definite if and only if K is nonsingular. To see this, let $\mathsf{K} = \mathsf{E}[\boldsymbol{Z}\boldsymbol{Z}^\dagger]$ and note that, if $\boldsymbol{b}^\dagger \mathsf{K} \boldsymbol{b} = 0$ for some $\boldsymbol{b} \neq 0$, then $X = \boldsymbol{b}^\dagger \boldsymbol{Z}$ has zero variance, and therefore is 0 with probability 1. Thus $\mathsf{E}[X\boldsymbol{Z}^\dagger] = 0$, so $\boldsymbol{b}^\dagger \mathsf{E}[\boldsymbol{Z}\boldsymbol{Z}^\dagger] = 0$. Since $\boldsymbol{b} \neq 0$ and $\boldsymbol{b}^\dagger \mathsf{K} = 0$, K must be singular. Conversely, if K is singular, there is some $\boldsymbol{b}$ such that $\mathsf{K}\boldsymbol{b} = 0$, so $\boldsymbol{b}^\dagger \mathsf{K} \boldsymbol{b}$ is also 0.

(4) A complex number λ is an *eigenvalue* of a square matrix K if $\mathsf{K}\boldsymbol{q} = \lambda \boldsymbol{q}$ for some nonzero vector $\boldsymbol{q}$; the corresponding $\boldsymbol{q}$ is an *eigenvector* of K. The following results about the eigenvalues and eigenvectors of positive (nonnegative) definite matrices K are standard linear algebra results (see, for example, Strang (1976), sect 5.5).

All eigenvalues of K are positive (nonnegative). If K is real, the eigenvectors can be taken to be real. Eigenvectors of different eigenvalues are orthogonal, and the eigenvectors of any one eigenvalue form a subspace whose dimension is called the *multiplicity* of that eigenvalue. If K is n by n, then n orthonormal eigenvectors $\boldsymbol{q}_1, \dots, \boldsymbol{q}_n$ can be chosen. The corresponding list of eigenvalues $\lambda_1, \dots, \lambda_n$ need not be distinct; specifically, the number of repetitions of each eigenvalue equals the multiplicity of that eigenvalue. Finally, $\det(\mathsf{K}) = \prod_{k=1}^n \lambda_k$.

(5) If K is nonnegative definite, let Q be the matrix with the orthonormal columns $\boldsymbol{q}_1, \dots, \boldsymbol{q}_n$ defined in item (4) above. Then Q satisfies $\mathsf{K}\mathsf{Q} = \mathsf{Q}\Lambda$, where $\Lambda = \operatorname{diag}(\lambda_1, \dots, \lambda_n)$. This is simply the vector version of the eigenvector/eigenvalue relationship above. Since $\boldsymbol{q}_k^\dagger \boldsymbol{q}_m = \delta_{nm}$, Q also satisfies $\mathsf{Q}^\dagger \mathsf{Q} = \mathsf{I}_n$. We then also have $\mathsf{Q}^{-1} = \mathsf{Q}^\dagger$ and thus $\mathsf{Q}\mathsf{Q}^\dagger = \mathsf{I}_n$; this says that the rows of Q are also orthonormal. Finally, by post-multiplying $\mathsf{K}\mathsf{Q} = \mathsf{Q}\Lambda$ by $\mathsf{Q}^\dagger$, we see that $\mathsf{K} = \mathsf{Q}\Lambda\mathsf{Q}^\dagger$. The matrix Q is called *unitary* if complex and *orthogonal* if real.

(6) If K is positive definite, then $\mathsf{K}\boldsymbol{b} \neq 0$ for $\boldsymbol{b} \neq 0$. Thus K can have no zero eigenvalues and Λ is nonsingular. It follows that K can be inverted as $\mathsf{K}^{-1} = \mathsf{Q}\Lambda^{-1}\mathsf{Q}^\dagger$. For any n-vector $\boldsymbol{b}$,

$$\boldsymbol{b}^\dagger \mathsf{K}^{-1} \boldsymbol{b} = \sum_k \lambda_k^{-1} |\langle \boldsymbol{b}, \boldsymbol{q}_k \rangle|^2.$$

To see this, note that $\boldsymbol{b}^{\dagger}\mathsf{K}^{-1}\boldsymbol{b} = \boldsymbol{b}^{\dagger}Q\Lambda^{-1}\mathsf{Q}^{\dagger}\boldsymbol{b}$. Letting $\boldsymbol{v} = \mathsf{Q}^{\dagger}\boldsymbol{b}$ and using the fact that the rows of Q^{T} are the orthonormal vectors $\boldsymbol{q}_k$, we see that $\langle \boldsymbol{b}, \boldsymbol{q}_k \rangle$ is the kth component of $\boldsymbol{v}$. We then have $\boldsymbol{v}^{\dagger}\Lambda^{-1}\boldsymbol{v} = \sum_k \lambda_k^{-1}|v_k|^2$, which is equivalent to the desired result. Note that $\langle \boldsymbol{b}, \boldsymbol{q}_k \rangle$ is the projection of $\boldsymbol{b}$ in the direction of $\boldsymbol{q}_k$.

(7) We have $\det \mathsf{K} = \prod_{k=1}^{n} \lambda_k$, where $\lambda_1, \ldots, \lambda_n$ are the eigenvalues of K repeated according to their multiplicity. Thus, if K is positive definite, $\det \mathsf{K} > 0$, and, if K is nonnegative definite, $\det \mathsf{K} \geq 0$.

(8) If K is a positive definite (semi-definite) matrix, then there is a unique positive definite (semi-definite) square root matrix R satisfying $\mathsf{R}^2 = \mathsf{K}$. In particular, R is given by

$$\mathsf{R} = \mathsf{Q}\Lambda^{1/2}\mathsf{Q}^{\dagger}, \quad \text{where } \Lambda^{1/2} = \text{diag}\left(\sqrt{\lambda_1}, \sqrt{\lambda_2}, \ldots, \sqrt{\lambda_n}\right). \tag{7.84}$$

(9) If K is nonnegative definite, then K is a covariance matrix. In particular, K is the covariance matrix of $\boldsymbol{Y} = \mathsf{R}\mathbf{Z}$, where R is the square root matrix in (7.84) and $\mathsf{K}_{\mathsf{Z}} = \mathsf{I}_m$. This shows that zero-mean jointly Gaussian rvs exist with any desired covariance matrix; the definition of jointly Gaussian here as a linear combination of normal rvs does not limit the possible set of covariance matrices.

For any given covariance matrix K, there are usually many choices for A satisfying $\mathsf{K} = \mathsf{A}\mathsf{A}^{\dagger}$. The square root matrix R is simply a convenient choice. Some of the results in this section are summarized in the following theorem.

Theorem 7.11.1 An n by n matrix K is a covariance matrix if and only if it is nonnegative definite. Also K is a covariance matrix if and only if $\mathsf{K} = \mathsf{A}\mathsf{A}^{\dagger}$ for an n by n matrix A. One choice for A is the square root matrix R in (7.84).

7.11.2 The Fourier series expansion of a truncated random process

Consider a (real zero-mean) random process that is effectively WSS over some interval $[-T_0/2, T_0/2]$ where T_0 is viewed intuitively as being very large. Let $\{Z(t); |t| \leq T_0/2\}$ be this process truncated to the interval $[-T_0/2, T_0/2]$. The objective of this and Section 7.11.3 is to view this truncated process in the frequency domain and discover its relation to the spectral density of an untruncated WSS process. A second objective is to interpret the statistical independence between different frequencies for stationary Gaussian processes in terms of a truncated process.

Initially assume that $\{Z(t); |t| \leq T_0/2\}$ is arbitrary; the effective WSS assumption will be added later. Assume the sample functions of the truncated process are $\mathcal{L}_2$ real functions with probability 1. Each $\mathcal{L}_2$ sample function, say $\{Z(t, \omega); |t| \leq T_0/2\}$ can then be expanded in a Fourier series, as follows:

$$Z(t, \omega) = \sum_{k=-\infty}^{\infty} \hat{Z}_k(\omega) e^{2\pi i k t/T_0}, \qquad |t| \leq \frac{T_0}{2}. \tag{7.85}$$

The orthogonal functions here are complex and the coefficients $\hat{Z}_k(\omega)$ can be similarly complex. Since the sample functions $\{Z(t,\omega);\ |t| \le T_0/2\}$ are real, $\hat{Z}_k(\omega) = \hat{Z}^*_{-k}(\omega)$ for each k. This also implies that $\hat{Z}_0(\omega)$ is real. The inverse Fourier series is given by

$$\hat{Z}_k(\omega) = \frac{1}{T_0}\int_{-T_0/2}^{T_0/2} Z(t,\omega)\mathrm{e}^{-2\pi i k t/T_0}\,\mathrm{d}t. \tag{7.86}$$

For each sample point ω, $\hat{Z}_k(\omega)$ is a complex number, so $\hat{Z}_k$ is a complex rv, i.e. $\Re(\hat{Z}_k)$ and $\Im(\hat{Z}_k)$ are both rvs. Also, $\Re(\hat{Z}_k) = \Re(\hat{Z}_{-k})$ and $\Im(\hat{Z}_k) = -\Im(\hat{Z}_{-k})$ for each k. It follows that the truncated process $\{Z(t);\ |t| \le T_0/2\}$ defined by

$$Z(t) = \sum_{k=-\infty}^{\infty} \hat{Z}_k \mathrm{e}^{2\pi i k t/T_0}, \qquad -\frac{T_0}{2} \le t \le \frac{T_0}{2}, \tag{7.87}$$

is a (real) random process and the complex rvs $\hat{Z}_k$ are complex linear functionals of $Z(t)$ given by

$$\hat{Z}_k = \frac{1}{T_0}\int_{-T_0/2}^{T_0/2} Z(t)\mathrm{e}^{-2\pi i k t/T_0}\,\mathrm{d}t. \tag{7.88}$$

Thus (7.87) and (7.88) are a Fourier series pair between a random process and a sequence of complex rvs. The sample functions satisfy

$$\frac{1}{T_0}\int_{-T_0/2}^{T_0/2} Z^2(t,\omega)\,\mathrm{d}t = \sum_{k\in\mathbb{Z}} |\hat{Z}_k(\omega)|^2,$$

so that

$$\frac{1}{T_0}\mathsf{E}\left[\int_{t=-T_0/2}^{T_0/2} Z^2(t)\,\mathrm{d}t\right] = \sum_{k\in\mathbb{Z}} \mathsf{E}\left[|\hat{Z}_k|^2\right]. \tag{7.89}$$

The assumption that the sample functions are $\mathcal{L}_2$ with probability 1 can be seen to be equivalent to the assumption that

$$\sum_{k\in\mathbb{Z}} S_k < \infty, \qquad \text{where } S_k = \mathsf{E}[|\hat{Z}_k|^2]. \tag{7.90}$$

This is summarized in the following theorem.

Theorem 7.11.2 *If a zero-mean (real) random process is truncated to* $[-T_0/2, T_0/2]$, *and the truncated sample functions are* $\mathcal{L}_2$ *with probability 1, then the truncated process is specified by the joint distribution of the complex Fourier-coefficient random variables* $\{\hat{Z}_k\}$. *Furthermore, any joint distribution of* $\{\hat{Z}_k;\ k \in \mathbb{Z}\}$ *that satisfies (7.90) specifies such a truncated process.*

The covariance function of a truncated process can be calculated from (7.87) as follows:

$$\begin{aligned}\mathsf{K}_Z(t,\tau) = \mathsf{E}[Z(t)Z^*(\tau)] &= \mathsf{E}\left[\sum_k \hat{Z}_k \mathrm{e}^{2\pi i k t/T_0} \sum_m \hat{Z}^*_m \mathrm{e}^{-2\pi i m \tau/T_0}\right] \\ &= \sum_{k,m} \mathsf{E}[\hat{Z}_k \hat{Z}^*_m]\mathrm{e}^{2\pi i k t/T_0}\mathrm{e}^{-2\pi i m\tau/T_0}, \qquad \text{for } -\frac{T_0}{2} \le t,\ \tau \le \frac{T_0}{2}.\end{aligned} \tag{7.91}$$

Note that if the function on the right of (7.91) is extended over all $t, \tau \in \mathbb{R}$, it becomes periodic in t with period T_0 for each τ, and periodic in τ with period T_0 for each t.

Theorem 7.11.2 suggests that virtually any truncated process can be represented as a Fourier series. Such a representation becomes far more insightful and useful, however, if the Fourier coefficients are uncorrelated. Sections 7.11.3 and 7.11.4 look at this case and then specialize to Gaussian processes, where uncorrelated implies independent.

7.11.3 Uncorrelated coefficients in a Fourier series

Consider the covariance function in (7.91) under the additional assumption that the Fourier coefficients $\{\tilde{Z}_k; k \in \mathbb{Z}\}$ are uncorrelated, i.e. that $\mathsf{E}[\hat{Z}_k \hat{Z}_m^*] = 0$ for all k, m such that $k \neq m$. This assumption also holds for $m = -k \neq 0$, and, since $Z_k = Z_{-k}^*$ for all k, implies both that $\mathsf{E}[(\Re(Z_k))^2] = \mathsf{E}[(\Im(Z_k))^2]$ and $\mathsf{E}[\Re(Z_k)\Im(Z_k)] = 0$ (see Exercise 7.10). Since $\mathsf{E}[\hat{Z}_k \hat{Z}_m^*] = 0$ for $k \neq m$, (7.91) simplifies to

$$\mathsf{K}_Z(t, \tau) = \sum_{k \in \mathbb{Z}} S_k e^{2\pi i k (t-\tau)/T_0}, \qquad \text{for } -\frac{T_0}{2} \leq t, \; \tau \leq \frac{T_0}{2}. \tag{7.92}$$

This says that $\mathsf{K}_Z(t, \tau)$ is a function only of $t - \tau$ over $-T_0/2 \leq t, \tau \leq T_0/2$, i.e. that $\mathsf{K}_Z(t, \tau)$ is effectively WSS over $[-T_0/2, T_0/2]\}$. Thus $\mathsf{K}_Z(t, \tau)$ can be denoted by $\tilde{\mathsf{K}}_Z(t - \tau)$ in this region, and

$$\tilde{\mathsf{K}}_Z(\tau) = \sum_k S_k e^{2\pi i k \tau / T_0}. \tag{7.93}$$

This means that the variances S_k of the sinusoids making up this process are the Fourier series coefficients of the covariance function $\tilde{\mathsf{K}}_Z(r)$.

In summary, the assumption that a truncated (real) random process has uncorrelated Fourier series coefficients over $[-T_0/2, T_0/2]$ implies that the process is WSS over $[-T_0/2, T_0/2]$ and that the variances of those coefficients are the Fourier coefficients of the single-variable covariance. This is intuitively plausible since the sine and cosine components of each of the corresponding sinusoids are uncorrelated and have equal variance.

Note that $\mathsf{K}_Z(t, \tau)$ in the above example is defined for all $t, \tau \in [-T_0/2, T_0/2]$ and thus $t - \tau$ ranges from $-T_0$ to T_0 and $\tilde{\mathsf{K}}_Z(r)$ must satisfy (7.93) for $-T_0 \leq r \leq T_0$. From (7.93), $\tilde{\mathsf{K}}_Z(r)$ is also periodic with period T_0, so the interval $[-T_0, T_0]$ constitutes two periods of $\tilde{\mathsf{K}}_Z(r)$. This means, for example, that $\mathsf{E}[Z(-\varepsilon)Z^*(\varepsilon)] = \mathsf{E}[Z(T_0/2 - \varepsilon)Z^*(-T_0/2 + \varepsilon)]$. More generally, the periodicity of $\tilde{\mathsf{K}}_Z(r)$ is reflected in $\mathsf{K}_Z(t, \tau)$, as illustrated in Figure 7.6.

We have seen that essentially any random process, when truncated to $[-T_0/2, T_0/2]$, has a Fourier series representation, and that, if the Fourier series coefficients are uncorrelated, then the truncated process is WSS over $[-T_0/2, T_0/2]$ and has a covariance function which is periodic with period T_0. This proves the first half of the following theorem.

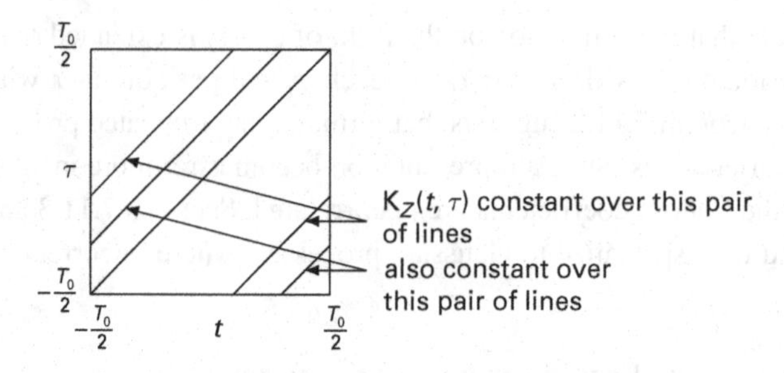

Figure 7.6. Constraint on $\mathsf{K}_Z(t, \tau)$ imposed by periodicity of $\tilde{\mathsf{K}}_Z(t-\tau)$.

Theorem 7.11.3 *Let $\{Z(t);\ t \in [-T_0/2, T_0/2]\}$ be a finite-energy zero-mean (real) random process over $[-T_0/2, T_0/2]$ and let $\{\hat{Z}_k;\ k \in \mathbb{Z}\}$ be the Fourier series rvs of (7.87) and (7.88).*

- *If $\mathsf{E}[Z_k Z_m^*] = S_k \delta_{k,m}$ for all $k, m \in \mathbb{Z}$, then $\{Z(t); t \in [-T_0/2, T_0/2]\}$ is effectively WSS within $[-T_0/2, T_0/2]$ and satisfies (7.93).*
- *If $\{Z(t);\ t \in [-T_0/2, T_0/2]\}$ is effectively WSS within $[-T_0/2, T_0/2]$ and if $\tilde{\mathsf{K}}_Z(t-\tau)$ is periodic with period T_0 over $[-T_0, T_0]$, then $\mathsf{E}[Z_k Z_m^*] = S_k \delta_{k,m}$ for some choice of $S_k \geq 0$ and for all $k, m \in \mathbb{Z}$.*

Proof To prove the second part of the theorem, note from (7.88) that

$$\mathsf{E}[\hat{Z}_k \hat{Z}_m^*] = \frac{1}{T_0^2} \int_{-T_0/2}^{T_0/2} \int_{-T_0/2}^{T_0/2} \mathsf{K}_Z(t, \tau) \mathrm{e}^{-2\pi i k t/T_0} \mathrm{e}^{2\pi i m \tau/T_0} \, \mathrm{d}t \, \mathrm{d}\tau. \tag{7.94}$$

By assumption, $\mathsf{K}_Z(t, \tau) = \tilde{\mathsf{K}}_Z(t-\tau)$ for $t, \tau \in [-T_0/2, T_0/2]$ and $\tilde{\mathsf{K}}_Z(t-\tau)$ is periodic with period T_0. Substituting $s = t - \tau$ for t as a variable of integration, (7.94) becomes

$$\mathsf{E}[Z_k Z_m^*] = \frac{1}{T_0^2} \int_{-T_0/2}^{T_0/2} \left(\int_{-T_0/2-\tau}^{T_0/2-\tau} \tilde{\mathsf{K}}_Z(s) \mathrm{e}^{-2\pi i k s/T_0} \, \mathrm{d}s \right) \mathrm{e}^{-2\pi i k \tau/T_0} \mathrm{e}^{2\pi i m \tau/T_0} \, \mathrm{d}\tau. \tag{7.95}$$

The integration over s does not depend on τ because the interval of integration is one period and $\tilde{\mathsf{K}}_Z$ is periodic. Thus, this integral is only a function of k, which we denote by $T_0 S_k$. Thus

$$\mathsf{E}[Z_k Z_m^*] = \frac{1}{T_0} \int_{-T_0/2}^{T_0/2} S_k \mathrm{e}^{-2\pi i (k-m)\tau/T_0} \, \mathrm{d}\tau = \begin{cases} S_k & \text{for } m = k; \\ 0 & \text{otherwise.} \end{cases} \tag{7.96}$$

This shows that the Z_k are uncorrelated, completing the proof. □

The next issue is to find the relationship between these processes and processes that are WSS over all time. This can be done most cleanly for the case of Gaussian processes. Consider a WSS (and therefore stationary) zero-mean Gaussian random

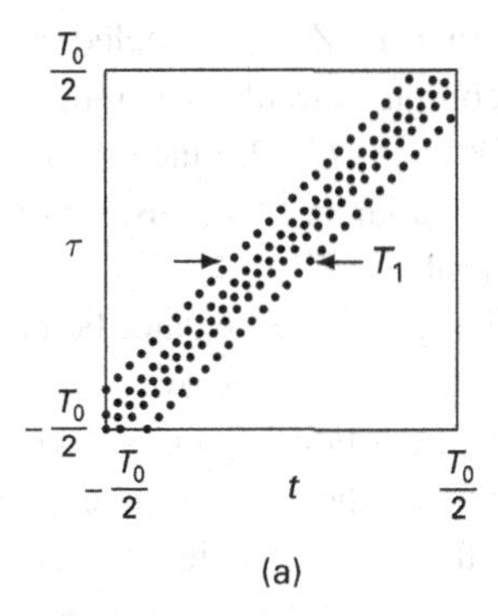

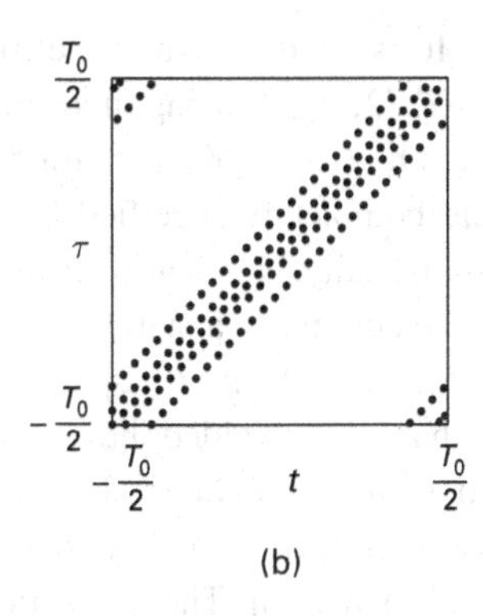

Figure 7.7. (a) $\mathsf{K}_{Z'}(t, \tau)$ over the region $-T_0/2 \leq t, \tau \leq T_0/2$ for a stationary process Z' satisfying $\tilde{\mathsf{K}}_{Z'}(\tau) = 0$ for $|\tau| > T_1/2$. (b) $\mathsf{K}_Z(t, \tau)$ for an approximating process Z comprising independent sinusoids, spaced by $1/T_0$ and with uniformly distributed phase. Note that the covariance functions are identical except for the anomalous behavior at the corners where t is close to $T_0/2$ and τ is close to $-T_0/2$ or vice versa.

process[26] $\{Z'(t); t \in \mathbb{R}\}$ with covariance function $\tilde{\mathsf{K}}_{Z'}(\tau)$ and assume a limited region of nonzero covariance; i.e.

$$\tilde{\mathsf{K}}_{Z'}(\tau) = 0 \qquad \text{for} \quad |\tau| > \frac{T_1}{2}.$$

Let $S_{Z'}(f) \geq 0$ be the spectral density of Z' and let T_0 satisfy $T_0 > T_1$. The Fourier series coefficients of $\tilde{\mathsf{K}}_{Z'}(\tau)$ over the interval $[-T_0/2, T_0/2]$ are then given by $S_k = S_{Z'}(k/T_0)/T_0$. Suppose this process is approximated over the interval $[-T_0/2, T_0/2]$ by a truncated Gaussian process $\{Z(t); t \in [-T_0/2, T_0/2]\}$ composed of independent Fourier coefficients $\hat{Z}_k$, i.e.

$$Z(t) = \sum_k \hat{Z}_k e^{2\pi i k t/T_0}, \qquad -\frac{T_0}{2} \leq t \leq \frac{T_0}{2},$$

where

$$\mathsf{E}[\hat{Z}_k \hat{Z}_m^*] = S_k \delta_{k,m}, \qquad \text{for all } k, m \in \mathbb{Z}.$$

By Theorem 7.11.3, the covariance function of $Z(t)$ is $\tilde{\mathsf{K}}_Z(\tau) = \sum_k S_k e^{2\pi i k t/T_0}$. This is periodic with period T_0 and for $|\tau| \leq T_0/2$, $\tilde{\mathsf{K}}_Z(\tau) = \tilde{\mathsf{K}}_{Z'}(\tau)$. The original process $Z'(t)$ and the approximation $Z(t)$ thus have the same covariance for $|\tau| \leq T_0/2$. For $|\tau| > T_0/2$, $\tilde{\mathsf{K}}_{Z'}(\tau) = 0$, whereas $\tilde{\mathsf{K}}_Z(\tau)$ is periodic over all τ. Also, of course, Z' is stationary, whereas Z is effectively stationary within its domain $[-T_0/2, T_0/2]$. The difference between Z and Z' becomes more clear in terms of the two-variable covariance function, illustrated in Figure 7.7.

[26] Equivalently, one can assume that Z' is effectively WSS over some interval much larger than the intervals of interest here.

It is evident from the figure that if Z' is modeled as a Fourier series over $[-T_0/2, T_0/2]$ using independent complex circularly symmetric Gaussian coefficients, then $\mathsf{K}_{Z'}(t, \tau) = \mathsf{K}_Z(t, \tau)$ for $|t|, |\tau| \le (T_0 - T_1)/2$. Since zero-mean Gaussian processes are completely specified by their covariance functions, this means that Z' and Z are statistically identical over this interval.

In summary, a stationary Gaussian process Z' cannot be perfectly modeled over an interval $[-T_0/2, T_0/2]$ by using a Fourier series over that interval. The anomalous behavior is avoided, however, by using a Fourier series over an interval large enough to include the interval of interest plus the interval over which $\tilde{\mathsf{K}}_{Z'}(\tau) \ne 0$. If this latter interval is unbounded, then the Fourier series model can only be used as an approximation. The following theorem has been established.

Theorem 7.11.4 *Let $Z'(t)$ be a zero-mean stationary Gaussian random process with spectral density $S(f)$ and covariance $\tilde{\mathsf{K}}_{Z'}(\tau) = 0$ for $|\tau| \ge T_1/2$. Then for $T_0 > T_1$, the truncated process $Z(t) = \sum_k Z_k e^{2\pi ikt/T_0}$ for $|t| \le T_0/2$, where the Z_k are independent and $Z_k \sim \mathcal{CN}(S(k/T_0)/T_0)$ for all $k \in \mathbb{Z}$ is statistically identical to $Z'(t)$ over $[-(T_0 - T_1)/2, (T_0 - T_1)/2]$.*

The above theorem is primarily of conceptual use, rather than as a problem-solving tool. It shows that, aside from the anomalous behavior discussed above, stationarity can be used over the region of interest without concern for how the process behaves far outside the interval of interest. Also, since T_0 can be arbitrarily large, and thus the sinusoids arbitrarily closely spaced, we see that the relationship between stationarity of a Gaussian process and independence of frequency bands is quite robust and more than something valid only in a limiting sense.

7.11.4 The Karhunen–Loeve expansion

There is another approach, called the Karhunen–Loeve expansion, for representing a random process that is truncated to some interval $[-T_0/2, T_0/2]$ by an orthonormal expansion. The objective is to choose a set of orthonormal functions such that the coefficients in the expansion are uncorrelated.

We start with the covariance function $\mathsf{K}(t, \tau)$ defined for $t, \tau \in [-T_0/2, T_0/2]$. The basic facts about these time-limited covariance functions are virtually the same as the facts about covariance matrices in Appendix 7.11.1. That is, $\mathsf{K}(t, \tau)$ is nonnegative definite in the sense that for all $\mathcal{L}_2$ functions $g(t)$,

$$\int_{-T_0/2}^{T_0/2} \int_{-T_0/2}^{T_0/2} g(t) \mathsf{K}_Z(t, \tau) g(\tau) \, dt \, d\tau \ge 0$$

Note that K_Z also has real-valued orthonormal eigenvectors defined over $[-T_0/2, T_0/2]$ and nonnegative eigenvalues. That is,

$$\int_{-T_0/2}^{T_0/2} \mathsf{K}_Z(t, \tau) \phi_m(\tau) d\tau = \lambda_m \phi_m(t); \quad t \in \left[-\frac{T_0}{2}, \frac{T_0}{2} \right],$$

where $\langle \boldsymbol{\phi}_m, \boldsymbol{\phi}_k \rangle = \delta_{m,k}$. These eigenvectors span the $\mathcal{L}_2$ space of real functions over $[-T_0/2, T_0/2]$. By using these eigenvectors as the orthonormal functions of $Z(t) = \sum_m Z_m \phi_m(t)$, it is easy to show that $\mathsf{E}[Z_m Z_k] = \lambda_m \delta_{m,k}$. In other words, given an arbitrary covariance function over the truncated interval $[-T_0/2, T_0/2]$, we can find a particular set of orthonormal functions so that $Z(t) = \sum_m Z_m \phi_m(t)$ and $\mathsf{E}[Z_m Z_k] = \lambda_m \delta_{m,k}$. This is called the Karhunen–Loeve expansion.

These equations for the eigenvectors and eigenvalues are well known integral equations and can be calculated by computer. Unfortunately, they do not provide a great deal of insight into the frequency domain.

7.12 Exercises

7.1 (a) Let X, Y be iid rvs, each with density $f_X(x) = \alpha \exp(-x^2/2)$. In part (b), we show that α must be $1/\sqrt{2\pi}$ in order for $f_X(x)$ to integrate to 1, but in this part we leave α undetermined. Let $S = X^2 + Y^2$. Find the probability density of S in terms of α. [Hint. Sketch the contours of equal probability density in the X, Y plane.]

(b) Prove from part (a) that α must be $1/\sqrt{2\pi}$ in order for S, and thus X and Y, to be rvs. Show that $\mathsf{E}[X] = 0$ and that $\mathsf{E}[X^2] = 1$.

(c) Find the probability density of $R = \sqrt{S}$ (R is called a *Rayleigh* rv).

7.2 (a) Let $X \sim \mathcal{N}(0, \sigma_X^2)$ and $Y \sim \mathcal{N}(0, \sigma_Y^2)$ be independent zero-mean Gaussian rvs. By convolving their densities, find the density of $X + Y$. [Hint. In performing the integration for the convolution, you should do something called "completing the square" in the exponent. This involves multiplying and dividing by $\mathrm{e}^{\alpha y^2/2}$ for some α, and you can be guided in this by knowing what the answer is. This technique is invaluable in working with Gaussian rvs.]

(b) The Fourier transform of a probability density $f_X(x)$ is $\hat{f}_X(\theta) = \int f_X(x)\mathrm{e}^{-2\pi \mathrm{i} x\theta}\,\mathrm{d}x = \mathsf{E}[\mathrm{e}^{-2\pi \mathrm{i} X\theta}]$. By scaling the basic Gaussian transform in (4.48), show that, for $X \sim \mathcal{N}(0, \sigma_X^2)$,

$$\hat{f}_X(\theta) = \exp\left(-\frac{(2\pi\theta)^2 \sigma_X^2}{2}\right).$$

(c) Now find the density of $X + Y$ by using Fourier transforms of the densities.

(d) Using the same Fourier transform technique, find the density of $V = \sum_{k=1}^{n} \alpha_k W_k$, where $W_1, \ldots, W_k$ are independent normal rvs.

7.3 In this exercise you will construct two rvs that are individually Gaussian but not jointly Gaussian. Consider the nonnegative random variable X with density given by

$$f_X(x) = \sqrt{\frac{2}{\pi}} \exp\left(\frac{-x^2}{2}\right), \qquad \text{for } x \geq 0.$$

Let U be binary, ± 1, with $p_U(1) = p_U(-1) = 1/2$.

(a) Find the probability density of $Y_1 = UX$. Sketch the density of Y_1 and find its mean and variance.

(b) Describe two normalized Gaussian rvs, say Y_1 and Y_2, such that the joint density of Y_1, Y_2 is *zero* in the second and fourth quadrants of the plane. It is nonzero in the first and third quadrants where it has the density $(1/\pi)\exp(-y_1^2/2 - y_2^2/2)$. Are Y_1, Y_2 jointly Gaussian? [Hint. Use part (a) for Y_1 and think about how to construct Y_2.]

(c) Find the covariance $\mathsf{E}[Y_1 Y_2]$. [Hint. First find the mean of the rv X above.]

(d) Use a variation of the same idea to construct two normalized Gaussian rvs V_1, V_2 whose probability is concentrated on the diagonal axes $v_1 = v_2$ and $v_1 = -v_2$, i.e. for which $\Pr(V_1 \neq V_2 \text{ and } V_1 \neq -V_2) = 0$. Are V_1, V_2 jointly Gaussian?

7.4 Let $W_1 \sim \mathcal{N}(0,1)$ and $W_2 \sim \mathcal{N}(0,1)$ be independent normal rvs. Let $X = \max(W_1, W_2)$ and $Y = \min(W_1, W_2)$.

(a) Sketch the transformation from sample values of W_1, W_2 to sample values of X, Y. Which sample pairs w_1, w_2 of W_1, W_2 map into a given sample pair x, y of X, Y?

(b) Find the probability density $f_{XY}(x,y)$ of X, Y. Explain your argument briefly but work from your sketch rather than equations.

(c) Find $f_S(s)$, where $S = X + Y$.

(d) Find $f_D(d)$, where $D = X - Y$.

(e) Let U be a random variable taking the values ± 1 with probability 1/2 each and let U be statistically independent of W_1, W_2. Are S and UD jointly Gaussian?

7.5 Let $\phi(t)$ be an $\mathcal{L}_1$ and $\mathcal{L}_2$ function of energy 1 and let $h(t)$ be $\mathcal{L}_1$ and $\mathcal{L}_2$. Show that $\int_{-\infty}^{\infty} \phi(t) h(\tau - t)\,\mathrm{d}t$ is an $\mathcal{L}_2$ function of τ. [Hint. Consider the Fourier transform of $\phi(t)$ and $h(t)$.]

7.6 (a) Generalize the random process of (7.30) by assuming that the Z_k are arbitrarily correlated. Show that every sample function is still $\mathcal{L}_2$.

(b) For this same case, show that $\iint |\mathsf{K}_Z(t,\tau)|^2\,\mathrm{d}t\,\mathrm{d}\tau < \infty$.

7.7 (a) Let $Z_1, Z_2, \ldots$ be a sequence of independent Gaussian rvs, $Z_k \sim \mathcal{N}(0, \sigma_k^2)$, and let $\{\phi_k(t) : \mathbb{R} \to \mathbb{R}\}$ be a sequence of orthonormal functions. Argue from fundamental definitions that, for each t, $Z(t) = \sum_{k=1}^{n} Z_k \phi_k(t)$ is a Gaussian rv. Find the variance of $Z(t)$ as a function of t.

(b) For any set of epochs $t_1, \ldots, t_\ell$, let $Z(t_m) = \sum_{k=1}^{n} Z_k \phi_k(t_m)$ for $1 \le m \le \ell$. Explain carefully from the basic definitions why $\{Z(t_1), \ldots, Z(t_\ell)\}$ are jointly Gaussian and specify their covariance matrix. Explain why $\{Z(t);\ t \in \mathbb{R}\}$ is a Gaussian random process.

(c) Now let $n = \infty$ in the definition of $Z(t)$ in part (a) and assume that $\sum_k \sigma_k^2 < \infty$. Also assume that the orthonormal functions are bounded for all k and t in the sense that, for some constant A, $|\phi_k(t)| \le A$ for all k and t. Consider the linear combination of rvs:

$$Z(t) = \sum_k Z_k \phi_k(t) = \lim_{n\to\infty} \sum_{k=1}^{n} Z_k \phi_k(t).$$

Let $Z^{(n)}(t) = \sum_{k=1}^{n} Z_k \phi_k(t)$. For any given t, find the variance of $Z^{(j)}(t) - Z^{(n)}(t)$ for $j > n$. Show that, for all $j > n$, this variance approaches 0 as $n \to \infty$. Explain intuitively why this indicates that $Z(t)$ is a Gaussian rv. Note: $Z(t)$ is, in fact, a Gaussian rv, but proving this rigorously requires considerable background; $Z(t)$ is a limit of a sequence of rvs, and each rv is a function of a sample space – the issue here is the same as that of a sequence of functions going to a limit function, where we had to invoke the Riesz–Fischer theorem.

(d) For the above Gaussian random process $\{Z(t); t \in \mathbb{R}\}$, let $z(t)$ be a sample function of $Z(t)$ and find its energy, i.e. $\|z\|^2$, in terms of the sample values $z_1, z_2, \ldots$ of $Z_1, Z_2, \ldots$ Find the expected energy in the process, $\mathsf{E}[\|\{Z(t); t \in \mathbb{R}\}\|^2]$.

(e) Find an upperbound on $\Pr\{\|\{Z(t); t \in \mathbb{R}\}\|^2 > \alpha\}$ that goes to zero as $\alpha \to \infty$. [Hint. You might find the Markov inequality useful. This says that for a nonnegative rv Y, $\Pr\{Y \geq \alpha\} \leq \mathsf{E}[Y]/\alpha$.] Explain why this shows that the sample functions of $\{Z(t)\}$ are $\mathcal{L}_2$ with probability 1.

7.8 Consider a stochastic process $\{Z(t); t \in \mathbb{R}\}$ for which each sample function is a sequence of rectangular pulses as in Figure 7.8. Analytically, $Z(t) = \sum_{k=-\infty}^{\infty} Z_k \operatorname{rect}(t-k)$, where $\ldots Z_{-1}, Z_0, Z_1, \ldots$ is a sequence of iid normal variables, $Z_k \sim \mathcal{N}(0, 1)$.

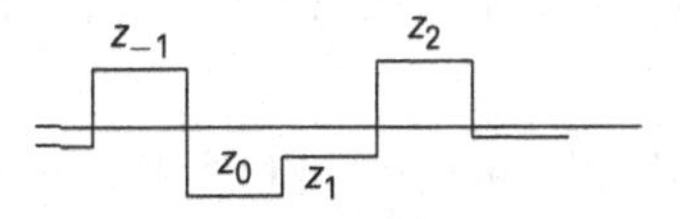

Figure 7.8.

(a) Is $\{Z(t); t \in \mathbb{R}\}$ a Gaussian random process? Explain why or why not carefully.

(b) Find the covariance function of $\{Z(t); t \in \mathbb{R}\}$.

(c) Is $\{Z(t); t \in \mathbb{R}\}$ a stationary random process? Explain carefully.

(d) Now suppose the stochastic process is modified by introducing a random time shift Φ which is uniformly distributed between 0 and 1. Thus, the new process $\{V(t); t \in \mathbb{R}\}$ is defined by $V(t) = \sum_{k=-\infty}^{\infty} Z_k \operatorname{rect}(t - k - \Phi)$. Find the conditional distribution of $V(0.5)$ conditional on $V(0) = v$.

(e) Is $\{V(t); t \in \mathbb{R}\}$ a Gaussian random process? Explain why or why not carefully.

(f) Find the covariance function of $\{V(t); t \in \mathbb{R}\}$.

(g) Is $\{V(t); t \in \mathbb{R}\}$ a stationary random process? It is easier to explain this than to write a lot of equations.

7.9 Consider the Gaussian sinc process, $V(t) = \sum_k V_k \operatorname{sinc}[(t-kT)/T]$, where $\{\ldots, V_{-1}, V_0, V_1, \ldots,\}$ is a sequence of iid rvs, $V_k \sim \mathcal{N}(0, \sigma^2)$.

(a) Find the probability density for the linear functional $\int V(t)\,\text{sinc}(t/T)\text{d}t$.
(b) Find the probability density for the linear functional $\int V(t)\,\text{sinc}(\alpha t/T)\text{d}t$ for $\alpha > 1$.
(c) Consider a linear filter with impulse response $h(t) = \text{sinc}(\alpha t/T)$, where $\alpha > 1$. Let $\{Y(t)\}$ be the output of this filter when $V(t)$ is the input. Find the covariance function of the process $\{Y(t)\}$. Explain why the process is Gaussian and why it is stationary.
(d) Find the probability density for the linear functional $Y(\tau) = \int V(t)\,\text{sinc}(\alpha(t-\tau)/T)\text{d}t$ for $\alpha \geq 1$ and arbitrary τ.
(e) Find the spectral density of $\{Y(t);\, t \in \mathbb{R}\}$.
(f) Show that $\{Y(t);\, t \in \mathbb{R}\}$ can be represented as $Y(t) = \sum_k Y_k\,\text{sinc}[(t-kT)/T]$ and characterize the rvs $\{Y_k;\, k \in Z\}$.
(g) Repeat parts (c), (d), and (e) for $\alpha < 1$.
(h) Show that $\{Y(t)\}$ in the $\alpha < 1$ case can be represented as a Gaussian sinc process (like $\{V(t)\}$ but with an appropriately modified value of T).
(i) Show that if any given process $\{Z(t);\, t \in \mathbb{R}\}$ is stationary, then so is the process $\{Y(t);\, t \in \mathbb{R}\}$, where $Y(t) = Z^2(t)$ for all $t \in \mathbb{R}$.

7.10 (Complex random variables)

(a) Suppose the zero-mean complex random variables X_k and X_{-k} satisfy $X^*_{-k} = X_k$ for all k. Show that if $\mathsf{E}[X_k X^*_{-k}] = 0$ then $\mathsf{E}[(\Re(X_k))^2] = \mathsf{E}[(\Im(X_k))^2]$ and $\mathsf{E}[\Re(X_k)\Im(X_{-k})] = 0$.
(b) Use this to show that if $\mathsf{E}[X_k X^*_m] = 0$ then $\mathsf{E}[\Re(X_k)\Re(X_m)] = 0$, $\mathsf{E}[\Re(X_k)\Im(X_m)] = 0$, and $\mathsf{E}[\Im(X_k)\Im(X_m)] = 0$ for all m not equal to either k or $-k$.

7.11 Explain why the integral in (7.58) must be real for $g_1(t)$ and $g_2(t)$ real, but the integrand $\hat{g}_1(f)S_Z(f)\hat{g}^*_2(f)$ need not be real.

7.12 (Filtered white noise) Let $\{Z(t)\}$ be a WGN process of spectral density $N_0/2$.

(a) Let $Y = \int_0^T Z(t)\,\text{d}t$. Find the probability density of Y.
(b) Let $Y(t)$ be the result of passing $Z(t)$ through an ideal baseband filter of bandwidth W whose gain is adjusted so that its impulse response has unit energy. Find the joint distribution of $Y(0)$ and $Y(1/4\mathsf{W})$.
(c) Find the probability density of

$$V = \int_0^\infty \text{e}^{-t} Z(t)\text{d}t.$$

7.13 (Power spectral density)

(a) Let $\{\phi_k(t)\}$ be any set of real orthonormal $\mathcal{L}_2$ waveforms whose transforms are limited to a band B, and let $\{W(t)\}$ be WGN with respect to B with power spectral density $S_W(f) = N_0/2$ for $f \in B$. Let the orthonormal expansion of $W(t)$ with respect to the set $\{\phi_k(t)\}$ be defined by

$$\tilde{W}(t) = \sum_k W_k \phi_k(t),$$

where $W_k = \langle W(t), \phi_k(t) \rangle$. Show that $\{W_k\}$ is an iid Gaussian sequence, and give the probability distribution of each W_k.

(b) Let the band $B = [-1/2T, 1/2T]$, and let $\phi_k(t) = (1/\sqrt{T})$ $\text{sinc}[(t - kT)/T]$, $k \in \mathbb{Z}$. Interpret the result of part (a) in this case.

7.14 (Complex Gaussian vectors)

(a) Give an example of a 2D complex rv $\mathbf{Z} = (Z_1, Z_2)$, where $Z_k \sim \mathcal{CN}(0, 1)$ for $k = 1, 2$ and where $\mathbf{Z}$ has the same joint probability distribution as $e^{i\phi}\mathbf{Z}$ for all $\phi \in [0, 2\pi]$, but where $\mathbf{Z}$ is not jointly Gaussian and thus not circularly symmetric Gaussian. [Hint. Extend the idea in part (d) of Exercise 7.3.]

(b) Suppose a complex rv $Z = Z_{\text{re}} + iZ_{\text{im}}$ has the properties that Z_{re} and Z_{im} are individually Gaussian and that Z has the same probability density as $e^{i\phi}Z$ for all $\phi \in [0, 2\pi]$. Show that Z is complex circularly symmetric Gaussian.

8 Detection, coding, and decoding

8.1 Introduction

Chapter 7 showed how to characterize noise as a random process. This chapter uses that characterization to retrieve the signal from the noise-corrupted received waveform. As one might guess, this is not possible without occasional errors when the noise is unusually large. The objective is to retrieve the data while minimizing the effect of these errors. This process of retrieving data from a noise-corrupted version is known as *detection*.

Detection, decision making, hypothesis testing, and decoding are synonyms. The word *detection* refers to the effort to detect whether some phenomenon is present or not on the basis of observations. For example, a radar system uses observations to detect whether or not a target is present; a quality control system attempts to *detect* whether a unit is defective; a medical test *detects* whether a given disease is present. The meaning of detection has been extended in the digital communication field from a yes/no decision to a decision at the receiver between a finite set of possible transmitted signals. Such a decision between a set of possible transmitted signals is also called *decoding*, but here the possible set is usually regarded as the set of codewords in a code rather than the set of signals in a signal set.[1] *Decision making* is, again, the process of deciding between a number of mutually exclusive alternatives. *Hypothesis testing* is the same, but here the mutually exclusive alternatives are called hypotheses. We use the word hypotheses for the possible choices in what follows, since the word conjures up the appropriate intuitive image of making a choice between a set of alternatives, where only one alternative is correct and there is a possibility of erroneous choice.

These problems will be studied initially in a purely probabilistic setting. That is, there is a probability model within which each hypothesis is an event. These events are mutually exclusive and collectively exhaustive; i.e., the sample outcome of the experiment lies in one and only one of these events, which means that in each performance of the experiment, one and only one hypothesis is correct. Assume there are M hypotheses,[2] labeled $a_0, \ldots, a_{M-1}$. The sample outcome of the experiment will

[1] As explained more fully later, there is no fundamental difference between a code and a signal set.

[2] The principles here apply essentially without change for a countably infinite set of hypotheses; for an uncountably infinite set of hypotheses, the process of choosing a hypothesis from an observation is called *estimation*. Typically, the probability of choosing correctly in this case is 0, and the emphasis is on making an estimate that is close in some sense to the correct hypothesis.

be one of these M events, and this defines a random symbol U which, for each m, takes the value a_m when event a_m occurs. The marginal probability $p_U(a_m)$ of hypothesis a_m is denoted by p_m and is called the *a-priori probability* of a_m. There is also a random variable (rv) V, called the observation. A sample value v of V is observed, and on the basis of that observation the detector selects one of the possible M hypotheses. The observation could equally well be a complex random variable, a random vector, a random process, or a random symbol; these generalizations are discussed in what follows.

Before discussing how to make decisions, it is important to understand when and why decisions must be made. For a binary example, assume that the conditional probability of hypothesis a_0 given the observation is 2/3 and that of hypothesis a_1 is 1/3. Simply deciding on hypothesis a_0 and forgetting about the probabilities throws away the information about the probability that the decision is correct. However, actual decisions sometimes must be made. In a communication system, the user usually wants to receive the message (even partly garbled) rather than a set of probabilities. In a control system, the controls must occasionally take action. Similarly, managers must occasionally choose between courses of action, between products, and between people to hire. In a sense, it is by making decisions that we return from the world of mathematical probability models to the world being modeled.

There are a number of possible criteria to use in making decisions. Initially assume that the criterion is to maximize the probability of correct choice. That is, when the experiment is performed, the resulting experimental outcome maps into both a sample value a_m for U and a sample value v for V. The decision maker observes v (but not a_m) and maps v into a decision $\tilde{u}(v)$. The decision is correct if $\tilde{u}(v) = a_m$. In principle, maximizing the probability of correct choice is almost trivially simple. Given v, calculate $p_{U|V}(a_m|v)$ for each possible hypothesis a_m. This *is* the probability that a_m is the correct hypothesis conditional on v. Thus the rule for maximizing the probability of being correct is to choose $\tilde{u}(v)$ to be that a_m for which $p_{U|V}(a_m|v)$ is maximized. For each possible observation v, this is denoted by

$$\tilde{u}(v) = \arg\max_m [p_{U|V}(a_m|v)] \qquad \text{(MAP rule)}, \tag{8.1}$$

where $\arg\max_m$ means the argument m that maximizes the function. If the maximum is not unique, the probability of being correct is the same no matter which maximizing m is chosen, so, to be explicit, the smallest such m will be chosen.[3] Since the rule (8.1) applies to each possible sample output v of the random variable V, (8.1) also defines the selected hypothesis as a random symbol $\tilde{U}(V)$. The conditional probability $p_{U|V}$ is called an *a-posteriori probability*. This is in contrast to the *a-priori probability* p_U of the hypothesis before the observation of V. The decision rule in (8.1) is thus called the maximum a-posteriori probability (MAP) rule.

[3] As discussed in Appendix 8.10, it is sometimes desirable to choose randomly among the maximum a-posteriori choices when the maximum in (8.1) is not unique. There are often situations (such as with discrete coding and decoding) where nonuniqueness occurs with positive probability.

An important consequence of (8.1) is that the MAP rule depends only on the conditional probability $p_{U|V}$ and thus is completely determined by the joint distribution of U and V. Everything else in the probability space is irrelevant to making a MAP decision.

When distinguishing between different decision rules, the MAP decision rule in (8.1) will be denoted by $\tilde{u}_{\text{MAP}}(v)$. Since the MAP rule maximizes the probability of correct decision for each sample value v, it also maximizes the probability of correct decision averaged over all v. To see this analytically, let $\tilde{u}_D(v)$ be an arbitrary decision rule. Since $\tilde{u}_{\text{MAP}}$ maximizes $p_{U|V}(m \mid v)$ over m,

$$p_{U|V}(\tilde{u}_{\text{MAP}}(v)|v) - p_{U|V}(\tilde{u}_D(v)|v) \geq 0; \qquad \text{for each rule } D \text{ and observation } v. \quad (8.2)$$

Taking the expected value of the first term on the left over the observation V, we get the probability of correct decision using the MAP decision rule. The expected value of the second term on the left for any given D is the probability of correct decision using that rule. Thus, taking the expected value of (8.2) over V shows that the MAP rule maximizes the probability of correct decision over the observation space. The above results are very simple, but also important and fundamental. They are summarized in the following theorem.

Theorem 8.1.1 *The MAP rule, given in (8.1), maximizes the probability of a correct decision, both for each observed sample value v and as an average over V. The MAP rule is determined solely by the joint distribution of U and V.*

Before discussing the implications and use of the MAP rule, the above assumptions are reviewed. First, a probability model was assumed in which all probabilities are known, and in which, for each performance of the experiment, one and only one hypothesis is correct. This conforms very well to the communication model in which a transmitter sends one of a set of possible signals and the receiver, given signal plus noise, makes a decision on the signal actually sent. It does not always conform well to a scientific experiment attempting to verify the existence of some new phenomenon; in such situations, there is often no sensible way to model a-priori probabilities. Detection in the absence of known a-priori probabilities is discussed in Appendix 8.10.

The next assumption was that maximizing the probability of correct decision is an appropriate decision criterion. In many situations, the cost of a wrong decision is highly asymmetric. For example, when testing for a treatable but deadly disease, making an error when the disease is present is far more costly than making an error when the disease is not present. As shown in Exercise 8.1, it is easy to extend the theory to account for relative costs of errors.

With the present assumptions, the detection problem can be stated concisely in the following probabilistic terms. There is an underlying sample space Ω, a probability measure, and two rvs U and V of interest. The corresponding experiment is performed, an observer sees the sample value v of rv V, but does not observe anything else, particularly not the sample value of U, say a_m. The observer uses a detection rule, $\tilde{u}(v)$, which is a function mapping each possible value of v to a possible value of U.

If $\tilde{v}(v) = a_m$, the detection is correct; otherwise an error has been made. The above MAP rule maximizes the probability of correct detection conditional on each v and also maximizes the unconditional probability of correct detection. Obviously, the observer must know the conditional probability assignment $p_{U|V}$ in order to use the MAP rule.

Sections 8.2 and 8.3 are restricted to the case of binary hypotheses where $M = 2$. This allows us to understand most of the important ideas, but simplifies the notation considerably. This is then generalized to an arbitrary number of hypotheses; fortunately, this extension is almost trivial.

8.2 Binary detection

Assume a probability model in which the correct hypothesis U is a binary random variable with possible values $\{a_0, a_1\}$ and a-priori probabilities p_0 and p_1. In the communication context, the a-priori probabilities are usually modeled as equiprobable, but occasionally there are multistage detection processes in which the result of the first stage can be summarized by a new set of a-priori probabilities. Thus let p_0 and $p_1 = 1 - p_0$ be arbitrary. Let V be a rv with a conditional probability density $f_{V|U}(v \mid a_m)$ that is finite and nonzero for all $v \in \mathbb{R}$ and $m \in \{0, 1\}$. The modifications for zero densities, discrete V, complex V, or vector V are relatively straightforward and are discussed later.

The conditional densities $f_{V|U}(v \mid a_m)$, $m \in \{0, 1\}$, are called *likelihoods* in the jargon of hypothesis testing. The marginal density of V is given by $f_V(v) = p_0 f_{V|U}(v \mid a_0) + p_1 f_{V|U}(v \mid a_1)$. The a-posteriori probability of U, for $m = 0$ or 1, is given by

$$p_{U|V}(a_m|v) = \frac{p_m f_{V|U}(v \mid a_m)}{f_V(v)}. \tag{8.3}$$

Writing out (8.1) explicitly for this case, we obtain

$$\frac{p_0 f_{V|U}(v \mid a_0)}{f_V(v)} \begin{array}{c} \geq^{\tilde{U}=a_0} \\ <_{\tilde{U}=a_1} \end{array} \frac{p_1 f_{V|U}(v \mid a_1)}{f_V(v)}. \tag{8.4}$$

This "equation" indicates that the MAP decision is a_0 if the left side is greater than or equal to the right, and is a_1 if the left side is less than the right. Choosing the decision $\tilde{U} = a_0$ when equality holds in (8.4) is an arbitrary choice and does not affect the probability of being correct. Canceling $f_V(v)$ and rearranging, we obtain

$$\Lambda(v) = \frac{f_{V|U}(v \mid a_0)}{f_{V|U}(v \mid a_1)} \begin{array}{c} \geq^{\tilde{U}=a_0} \\ <_{\tilde{U}=a_1} \end{array} \frac{p_1}{p_0} = \eta. \tag{8.5}$$

The ratio $\Lambda(v) = f_{V|U}(v \mid a_0)/f_{V|U}(v \mid a_1)$ is called the *likelihood ratio*, and is a function only of v. The ratio $\eta = p_1/p_0$ is called the *threshold* and depends only on the a-priori probabilities. The binary MAP rule (or MAP test, as it is usually called) then compares the likelihood ratio to the threshold, and decides on hypothesis a_0 if the threshold is reached, and on hypothesis a_1 otherwise. Note that if the a-priori probability p_0 is increased, the threshold decreases, and the set of v for which hypothesis a_0 is chosen increases; this corresponds to our intuition – the more certain we are initially that U is 0, the stronger the evidence required to make us change our minds. As shown in Exercise 8.1, the only effect of minimizing over costs rather than error probability is to change the threshold η in (8.5).

An important special case of (8.5) is that in which $p_0 = p_1$. In this case $\eta = 1$, and the rule chooses $\tilde{U}(v) = a_0$ for $f_{V|U}(v|a_0) \geq f_{V|U}(v|a_1)$ and chooses $\tilde{U}(v) = 1$ otherwise. This is called a *maximum likelihood (ML) rule* or *test.* In the communication case, as mentioned above, the a-priori probabilities are usually equal, so MAP then reduces to ML. The maximum likelihood test is also often used when p_0 and p_1 are unknown.

The *probability of error*, i.e. one minus the probability of choosing correctly, is now derived for MAP detection. First we find the probability of error conditional on each hypothesis, $\Pr\{e|U = a_1\}$ and $\Pr\{e|U = a_0\}$. The overall probability of error is then given by

$$\Pr\{e\} = p_0 \Pr\{e|U = a_0\} + p_1 \Pr\{e|U = a_1\}.$$

In the radar field, $\Pr\{e|U = a_0\}$ is called the probability of false alarm, and $\Pr\{e|U = a_1\}$ is called the probability of a miss. Also $1 - \Pr\{e|U = a_1\}$ is called the probability of detection. In statistics, $\Pr\{e|U = a_1\}$ is called the probability of error of the second kind, and $\Pr\{e|U = a_0\}$ is the probability of error of the first kind. These terms are not used here.

Note that (8.5) partitions the space of observed sample values into two regions: $R_0 = \{v : \Lambda(v) \geq \eta\}$ is the region for which $\tilde{U} = a_0$ and $R_1 = \{v : \Lambda(v) < \eta\}$ is the region for which $\tilde{U} = a_1$. For $U = a_1$, an error occurs if and only if v is in R_0, and for $U = a_0$ an error occurs if and only if v is in R_1. Thus,

$$\Pr\{e|U = a_0\} = \int_{R_1} f_{V|U}(v|a_0)\mathrm{d}v; \tag{8.6}$$

$$\Pr\{e|U = a_1\} = \int_{R_0} f_{V|U}(v|a_1)\mathrm{d}v. \tag{8.7}$$

Another, often simpler, approach is to work directly with the likelihood ratio. Since $\Lambda(v)$ is a function of the observed sample value v, the random variable, $\Lambda(V)$, also called a likelihood ratio, is defined as follows: for every sample point ω, $V(\omega)$ is the corresponding sample value v, and $\Lambda(V)$ is then shorthand for $\Lambda(V(\omega))$. In the same way, $\tilde{U}(V)$ (or more briefly $\tilde{U}$) is the decision random variable. In these terms, (8.5) states that

$$\tilde{U} = a_0 \quad \text{if and only if} \quad \Lambda(V) \geq \eta. \tag{8.8}$$

Thus, for MAP detection with a threshold η,

$$\Pr\{e|U = a_0\} = \Pr\{\tilde{U} = a_1|U = a_0\} = \Pr\{\Lambda(V) < \eta|U = a_0\}; \tag{8.9}$$

$$\Pr\{e|U = a_1\} = \Pr\{\tilde{U} = a_0|U = a_1\} = \Pr\{\Lambda(V) \geq \eta|U = a_1\}. \tag{8.10}$$

A *sufficient statistic* is defined as any function of the observation v from which the likelihood ratio can be calculated. As examples, v itself, $\Lambda(v)$, and any one-to-one function of $\Lambda(v)$ are sufficient statistics. Note that $\Lambda(v)$, and functions of $\Lambda(v)$, are often simpler to work with than v, since $\Lambda(v)$ is simply a real number, whereas v could be a vector or a waveform.

We have seen that the MAP rule (and thus also the ML rule) is a threshold test on the likelihood ratio. Similarly the min-cost rule (see Exercise 8.1) and the Neyman–Pearson test (which, as shown in Appendix 8.10, makes no assumptions about a-priori probabilities) are threshold tests on the likelihood ratio. Not only are all these binary decision rules based only on threshold tests on the likelihood ratio, but the properties of these rules, such as the conditional error probabilities in (8.9) and (8.10), are based only on $\Lambda(V)$ and η. In fact, it is difficult to imagine any sensible binary decision procedure, especially in the digital communication context, that is not a threshold test on the likelihood ratio. Thus, once a sufficient statistic has been calculated from the observed vector, that observed vector has no further value in any decision rule of interest here.

The log likelihood ratio, $\mathrm{LLR}(V) = \ln[\Lambda(V)]$, is an important sufficient statistic which is often easier to work with than the likelihood ratio itself. As seen in Section 8.3, the LLR is particularly convenient for use with Gaussian noise statistics.

8.3 Binary signals in white Gaussian noise

This section first treats standard 2-PAM, then 2-PAM with an offset, then binary signals with vector observations, and finally binary signals with waveform observations.

8.3.1 Detection for PAM antipodal signals

Consider PAM antipodal modulation (i.e. 2-PAM), as illustrated in Figure 8.1. The correct hypothesis U is either $a_0 = a$ or $a_1 = -a$. Let $Z \sim \mathcal{N}(0, N_0/2)$ be a Gaussian noise rv of mean 0 and variance $N_0/2$, independent of U. That is,

$$f_Z(z) = \frac{1}{\sqrt{2\pi N_0/2}} \exp\left(\frac{-z^2}{N_0}\right).$$

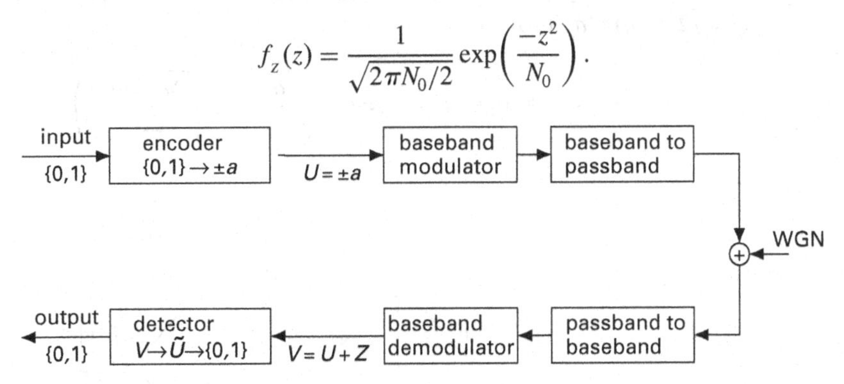

Figure 8.1. The source produces a binary digit, which is mapped into $U = \pm a$. This is modulated into a waveform, WGN is added, and the resultant waveform is demodulated and sampled, resulting in a noisy received value $V = U + Z$. From Section 7.8, $Z \sim \mathcal{N}(0, N_0/2)$. This is explained more fully later. Based on this observation, the receiver makes a decision $\tilde{U}$ and maps this back to the binary output, which is the hypothesized version of the binary input.

Assume that 2-PAM is simplified by sending only a single binary symbol (rather than a sequence over time) and by observing only the single sample value v corresponding to that input. As seen later, these simplifications are unnecessary, but they permit the problem to be viewed in the simplest possible context. The observation V (i.e. the channel output prior to detection) is $a+Z$ or $-a+Z$, depending on whether $U=a$ or $-a$. Thus, conditional on $U=a$, $V\sim\mathcal{N}(a, N_0/2)$ and, conditional on $U=-a$, $V\sim\mathcal{N}(-a, N_0/2)$:

$$f_{V|U}(v|a)=\frac{1}{\sqrt{\pi N_0}}\exp\left(\frac{-(v-a)^2}{N_0}\right); \qquad f_{V|U}(v|-a)=\frac{1}{\sqrt{\pi N_0}}\exp\left(\frac{-(v+a)^2}{N_0}\right).$$

The likelihood ratio is the ratio of these likelihoods, and is given by

$$\Lambda(v)=\exp\left(\frac{-(v-a)^2+(v+a)^2}{N_0}\right)=\exp\left(\frac{4av}{N_0}\right). \tag{8.11}$$

Substituting this into (8.5), we obtain

$$\exp\left(\frac{4av}{N_0}\right) \begin{matrix} \geq^{\tilde{U}=a} \\ <_{\tilde{U}=-a} \end{matrix} \frac{p_1}{p_0}=\eta. \tag{8.12}$$

This is further simplified by taking the logarithm, yielding

$$\mathrm{LLR}(v)=\frac{4av}{N_0} \begin{matrix} \geq^{\tilde{U}=a} \\ <_{\tilde{U}=-a} \end{matrix} \ln\eta; \tag{8.13}$$

$$v \begin{matrix} \geq^{\tilde{U}=a} \\ <_{\tilde{U}=-a} \end{matrix} \frac{N_0\ln\eta}{4a}. \tag{8.14}$$

Figure 8.2 interprets this decision rule.

The probability of error, given $U=-a$, is seen to be the probability that the noise value is greater than $a+N_0\ln\eta/4a$. Since the noise has variance $N_0/2$, this is the probability that the normalized Gaussian rv $Z/\sqrt{N_0/2}$ exceeds $a/\sqrt{N_0/2}+\sqrt{N_0/2}\,\ln(\eta)/2a$. Thus,

$$\Pr\{e\,|\,U=-a\}=Q\left(\frac{a}{\sqrt{N_0/2}}+\frac{\sqrt{N_0/2}\ln\eta}{2a}\right), \tag{8.15}$$

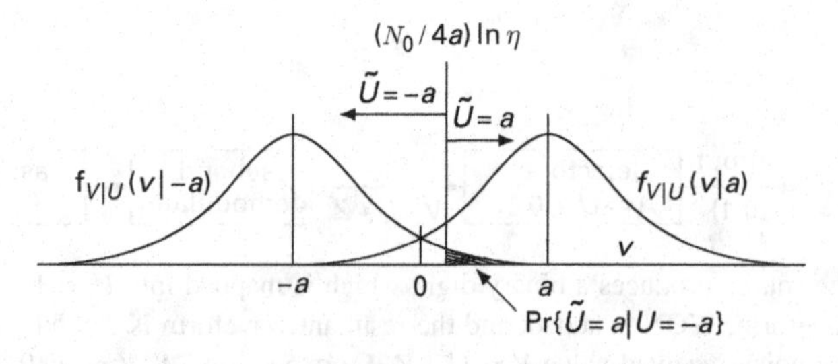

Figure 8.2. Binary hypothesis testing for antipodal signals, $0\to a$, $1\to -a$. The a-priori probabilities are p_0 and p_1, the threshold is $\eta=p_0/p_1$, and the noise is $\mathcal{N}(0, N_0/2)$.

where $Q(x)$, the complementary distribution function of $\mathcal{N}(0,1)$, is given by

$$Q(x) = \int_x^\infty \frac{1}{\sqrt{2\pi}} \exp\left(\frac{-z^2}{2}\right) \mathrm{d}z.$$

The probability of error given $U = a$ is calculated the same way, but is the probability that $-Z$ is greater than or equal to $a - N_0 \ln \eta / 4a$. Since $-Z$ has the same distribution as Z,

$$\Pr\{e|U = a\} = Q\left(\frac{a}{\sqrt{N_0/2}} - \frac{\sqrt{N_0/2} \ln \eta}{2a}\right). \tag{8.16}$$

It is more insightful to express $a/\sqrt{N_0/2}$ as $\sqrt{2a^2/N_0}$. As seen before, a^2 can be viewed as the energy per bit, E_b, so that (8.15) and (8.16) become

$$\Pr\{e|U = -a\} = Q\left(\sqrt{\frac{2E_b}{N_0}} + \frac{\ln \eta}{2\sqrt{2E_b/N_0}}\right), \tag{8.17}$$

$$\Pr\{e|U = a\} = Q\left(\sqrt{\frac{2E_b}{N_0}} - \frac{\ln \eta}{2\sqrt{2E_b/N_0}}\right). \tag{8.18}$$

Note that these formulas involve only the ratio E_b/N_0 rather than E_b or N_0 separately. If the signal, observation, and noise had been measured on a different scale, then both E_b and N_0 would change by the same factor, helping to explain why only the ratio is relevant. In fact, the scale could be normalized so that either the noise has variance 1 or the signal has variance 1.

The hypotheses in these communication problems are usually modeled as equiprobable, $p_0 = p_1 = 1/2$. In this case, $\ln \eta = 0$ and the MAP rule is equivalent to the ML rule. Equations (8.17) and (8.18) then simplify to the following:

$$\Pr\{e\} = \Pr\{e|U = -a\} = \Pr\{e|U = a\} = Q\left(\sqrt{\frac{2E_b}{N_0}}\right). \tag{8.19}$$

In terms of Figure 8.2, this is the tail of either Gaussian distribution from the point 0 where they cross. This equation keeps reappearing in different guises, and it will soon seem like a completely obvious result for a variety of Gaussian detection problems.

8.3.2 Detection for binary nonantipodal signals

Next consider the slightly more complex case illustrated in Figure 8.3. Instead of mapping 0 to $+a$ and 1 to $-a$, 0 is mapped to an arbitrary number b_0 and 1 to an arbitrary number b_1. To analyze this, let c be the midpoint between b_0 and b_1, $c = (b_0 + b_1)/2$. Assuming $b_1 < b_0$, let $a = b_0 - c = c - b_1$. Conditional on $U = b_0$, the observation is $V = c + a + Z$; conditional on $U = b_1$, it is $V = c - a + Z$. In other words, this more general case is simply the result of shifting the previous signals by the constant c.

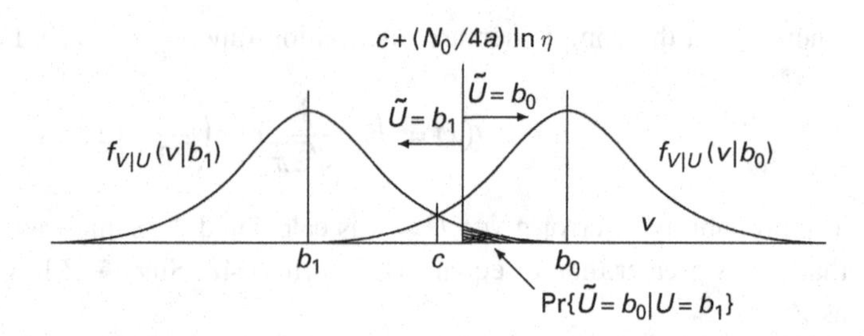

Figure 8.3. Binary hypothesis testing for arbitrary signals, $0 \to b_0, 1 \to b_1$, for $b_0 > b_1$. With $c = (b_0+b_1)/2$ and $a = |b_0 - b_1|/2$, this is the same as Figure 8.2 shifted by c. For $b_0 < b_1$, the picture must be reversed, but the answer is the same.

Define $\tilde{V} = V - c$ as the result of shifting the observation by $-c$; $\tilde{V}$ is a sufficient statistic and $\tilde{V} = \pm a + Z$. This is the same as the antipodal signal case in Section 8.3.1, so the error probability is again given by (8.15) and (8.16).

The energy used in achieving this error probability has changed from the antipodal case. The energy per bit (assuming equal a-priori probabilities) is now $(b_0^2 + b_1{}^2)/2 = a^2 + c^2$. A center value c is frequently used as a "pilot tone" in communication for tracking the channel. We see that E_b is then the sum of the energy used for the actual binary transmission (a^2) plus the energy used for the pilot tone (c^2). The fraction of energy E_b used for the signal is $\gamma = a^2/(a^2+c^2)$, and (8.17) and (8.18) are changed as follows:

$$\Pr\{e|U=b_1\} = Q\left(\sqrt{\frac{2\gamma E_b}{N_0}} + \frac{\ln\eta}{2\sqrt{2\gamma E_b/N_0}}\right), \tag{8.20}$$

$$\Pr\{e|U=b_0\} = Q\left(\sqrt{\frac{2\gamma E_b}{N_0}} - \frac{\ln\eta}{2\sqrt{2\gamma E_b/N_0}}\right). \tag{8.21}$$

For example, a common binary communication technique called *on-off keying* uses the binary signals 0 and $2a$. In this case, $\gamma = 1/2$ and there is an energy loss of 3 dB from the antipodal case. For the ML rule, the probability of error then becomes $Q(\sqrt{E_b/N_0})$.

8.3.3 Detection for binary real vectors in WGN

Next consider the vector version of the Gaussian detection problem. Suppose the observation is a random n-vector $\boldsymbol{V} = \boldsymbol{U} + \boldsymbol{Z}$. The noise $\boldsymbol{Z}$ is a random n-vector $(Z_1, Z_2, \ldots, Z_n)^{\mathsf{T}}$, independent of $\boldsymbol{U}$, with iid components given by $Z_k \sim \mathcal{N}(0, N_0/2)$. The input $\boldsymbol{U}$ is a random n-vector with M possible values (hypotheses). The mth hypothesis, $0 \le m \le M-1$, is denoted by $\boldsymbol{a}_m = (a_{m1}, a_{m2}, \ldots, a_{mn})^{\mathsf{T}}$. A sample value $\boldsymbol{v}$ of $\boldsymbol{V}$ is observed, and the problem is to make a MAP decision, denoted by $\tilde{\boldsymbol{U}}$, about $\boldsymbol{U}$.

Initially assume the binary antipodal case, where $\boldsymbol{a}_1 = -\boldsymbol{a}_0$. For notational simplicity, let $\boldsymbol{a}_0$ be denoted by $\boldsymbol{a} = (a_1, a_2, \ldots, a_n)^{\mathsf{T}}$. Thus the two hypotheses are $\boldsymbol{U} = \boldsymbol{a}$

and $U = -a$, and the observation is either $a + Z$ or $-a + Z$. The likelihoods are then given by

$$f_{V|U}(\boldsymbol{v}\,|\,\boldsymbol{a}) = \frac{1}{(\pi N_0)^{n/2}} \exp \sum_{k=1}^{n} \frac{-(v_k - a_k)^2}{N_0} = \frac{1}{(\pi N_0)^{n/2}} \exp\left(\frac{-\|\boldsymbol{v}-\boldsymbol{a}\|^2}{N_0}\right),$$

$$f_{V|U}(\boldsymbol{v}\,|\,-\boldsymbol{a}) = \frac{1}{(\pi N_0)^{n/2}} \exp \sum_{k=1}^{n} \frac{-(v_k + a_k)^2}{N_0} = \frac{1}{(\pi N_0)^{n/2}} \exp\left(\frac{-\|\boldsymbol{v}+\boldsymbol{a}\|^2}{N_0}\right).$$

The log likelihood ratio is thus given by

$$\mathrm{LLR}(\boldsymbol{v}) = \frac{-\|\boldsymbol{v}-\boldsymbol{a}\|^2 + \|\boldsymbol{v}+\boldsymbol{a}\|^2}{N_0} = \frac{4\langle \boldsymbol{v}, \boldsymbol{a}\rangle}{N_0}, \tag{8.22}$$

and the MAP test is

$$\mathrm{LLR}(\boldsymbol{v}) = \frac{4\langle \boldsymbol{v}, \boldsymbol{a}\rangle}{N_0} \begin{array}{c} \geq^{\tilde{U}=\boldsymbol{a}} \\ <_{\tilde{U}=-\boldsymbol{a}} \end{array} \ln \frac{p_1}{p_0} = \ln \eta.$$

This can be restated as follows:

$$\frac{\langle \boldsymbol{v}, \boldsymbol{a}\rangle}{\|\boldsymbol{a}\|} \begin{array}{c} \geq^{\tilde{U}=\boldsymbol{a}} \\ <_{\tilde{U}=-\boldsymbol{a}} \end{array} \frac{N_0 \ln \eta}{4\|\boldsymbol{a}\|}. \tag{8.23}$$

The projection of the observation $\boldsymbol{v}$ onto the signal $\boldsymbol{a}$ is $(\langle \boldsymbol{v}, \boldsymbol{a}\rangle / \|\boldsymbol{a}\|)(\boldsymbol{a}/\|\boldsymbol{a}\|)$. Thus the left side of (8.23) is the component of $\boldsymbol{v}$ in the direction of $\boldsymbol{a}$, showing that the decision is based solely on that component of $\boldsymbol{v}$. This result is rather natural; the noise is independent in different orthogonal directions, and only the noise in the direction of the signal should be relevant in detecting the signal.

The geometry of the situation is particularly clear in the ML case (see Figure 8.4). The noise is spherically symmetric around the origin, and the likelihoods depend only

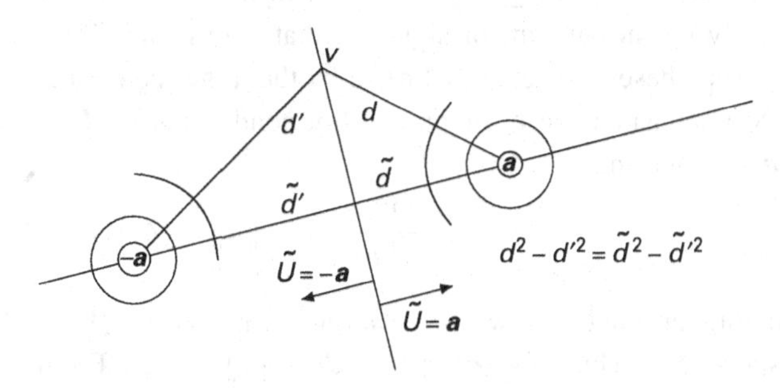

Figure 8.4. ML decision regions for binary signals in WGN. A vector $\boldsymbol{v}$ on the threshold boundary is shown. The distance from $\boldsymbol{v}$ to $\boldsymbol{a}$ is $d = \|\boldsymbol{v}-\boldsymbol{a}\|$. Similarly, the distance to $-\boldsymbol{a}$ is $d' = \|\boldsymbol{v}+\boldsymbol{a}\|$. As shown algebraically in (8.22), any point at which $d^2 - d'^2 = 0$ is a point at which $\langle \boldsymbol{v}, \boldsymbol{a}\rangle = 0$, and thus at which the LLR is 0. Geometrically, from the Pythagorean theorem, however, $d^2 - d'^2 = \tilde{d}^2 - \tilde{d}'^2$, where $\tilde{d}$ and $\tilde{d}'$ are the distances from $\boldsymbol{a}$ and $-\boldsymbol{a}$ to the projection of $\boldsymbol{v}$ on the straight line generated by $\boldsymbol{a}$. This demonstrates geometrically why it is only the projection of $\boldsymbol{v}$ onto $\boldsymbol{a}$ that is relevant.

on the distance from the origin. The ML detection rule is then equivalent to choosing the hypothesis closest to the received point. The set of points equidistant from the two hypotheses, as illustrated in Figure 8.4, is the perpendicular bisector between them; this bisector is the set of $\boldsymbol{v}$ satisfying $\langle \boldsymbol{v}, \boldsymbol{a}\rangle = 0$. The set of points closer to $\boldsymbol{a}$ is on the $\boldsymbol{a}$ side of this perpendicular bisector; it is determined by $\langle \boldsymbol{v}, \boldsymbol{a}\rangle > 0$ and is mapped into $\boldsymbol{a}$ by the ML rule. Similarly, the set of points closer to $-\boldsymbol{a}$ is determined by $\langle \boldsymbol{v}, \boldsymbol{a}\rangle < 0$, and is mapped into $-\boldsymbol{a}$. In the general MAP case, the region mapped into $\boldsymbol{a}$ is again separated from the region mapped into $-\boldsymbol{a}$ by a perpendicular to $\boldsymbol{a}$, but in this case it is the perpendicular defined by $\langle \boldsymbol{v}, \boldsymbol{a}\rangle = N_0 \ln(\eta)/4$.

Another way of interpreting (8.23) is to view it in a different coordinate system. That is, choose $\boldsymbol{\phi}_1 = \boldsymbol{a}/\|\boldsymbol{a}\|$ as one element of an orthonormal basis for the n-vectors, and choose another $n-1$ orthonormal vectors by the Gram–Schmidt procedure. In this new coordinate system, $\boldsymbol{v}$ can be expressed as $(v_1', v_2', \ldots, v_n')^{\mathsf{T}}$, where, for $1 \le k \le n$, $v_k' = \langle \boldsymbol{v}, \boldsymbol{\phi}_k\rangle$. Since $\langle \boldsymbol{v}, \boldsymbol{a}\rangle = \|\boldsymbol{a}\|\langle \boldsymbol{v}, \boldsymbol{\phi}_1\rangle = \|\boldsymbol{a}\| v_1'$, the left side of (8.23) is simply v_1', i.e. the size of the projection of $\boldsymbol{v}$ onto $\boldsymbol{a}$. Thus (8.23) becomes

$$v_1' \;\mathop{\gtrless}_{\tilde{U}=1}^{\tilde{U}=0}\; \frac{N_0 \ln \eta}{4\|\boldsymbol{a}\|}.$$

This is the same as the scalar MAP test in (8.14). In other words, the vector problem is the same as the scalar problem when the appropriate coordinate system is chosen. Actually, the derivation of (8.23) has shown something more, namely that v_1' is a sufficient statistic. The components $v_2', \ldots, v_n'$, which contain only noise, cancel out in (8.22) if (8.22) is expressed in the new coordinate system. The fact that the coordinates of $\boldsymbol{v}$ in directions orthogonal to the signal do not affect the LLR is sometimes called the *theorem of irrelevance*. A generalized form of this theorem is stated later as Theorem 8.4.2.

Some additional insight into (8.23) (in the original coordinate system) can be gained by writing $\langle \boldsymbol{v}, \boldsymbol{a}\rangle$ as $\sum_k v_k a_k$. This says that the MAP test weights each coordinate linearly by the amount of signal in that coordinate. This is not surprising, since the two hypotheses are separated more by the larger components of $\boldsymbol{a}$ than by the smaller.

Next consider the error probability conditional on $\boldsymbol{U} = -\boldsymbol{a}$. Given $\boldsymbol{U} = -\boldsymbol{a}$, $\boldsymbol{V} = -\boldsymbol{a} + \boldsymbol{Z}$, and thus

$$\frac{\langle \boldsymbol{V}, \boldsymbol{a}\rangle}{\|\boldsymbol{a}\|} = -\|\boldsymbol{a}\| + \langle \boldsymbol{Z}, \boldsymbol{\phi}_1\rangle.$$

Conditional on $\boldsymbol{U} = -\boldsymbol{a}$, the mean and variance of $\langle \boldsymbol{V}, \boldsymbol{a}\rangle/\|\boldsymbol{a}\|$ are $-\|\boldsymbol{a}\|$ and $N_0/2$, respectively. Thus, $\langle \boldsymbol{V}, \boldsymbol{a}\rangle/\|\boldsymbol{a}\| \sim \mathcal{N}(-\|\boldsymbol{a}\|, N_0/2)$. From (8.23), the probability of error, given $\boldsymbol{U} = -\boldsymbol{a}$, is the probability that $\mathcal{N}(-\|\boldsymbol{a}\|, N_0/2)$ exceeds $N_0 \ln(\eta)/4\,\|\boldsymbol{a}\|$. This is the probability that Z is greater than $\|\boldsymbol{a}\| + N_0 \ln(\eta)/(4\,\|\boldsymbol{a}\|)$. Normalizing as in Section 8.3.1, we obtain

$$\Pr\{e | \boldsymbol{U} = -\boldsymbol{a}\} = Q\left(\sqrt{\frac{2\|\boldsymbol{a}\|^2}{N_0}} + \frac{\ln \eta}{2\sqrt{2\|\boldsymbol{a}\|^2/N_0}}\right). \tag{8.24}$$

By the same argument,

$$\Pr\{e|U=\boldsymbol{a}\} = Q\left(\sqrt{\frac{2\|\boldsymbol{a}\|^2}{N_0}} - \frac{\ln\eta}{2\sqrt{2\|\boldsymbol{a}\|^2/N_0}}\right). \tag{8.25}$$

It can be seen that this is the same answer as given by (8.15) and (8.16) when the problem is converted to a coordinate system where a is collinear with a coordinate vector. The energy per bit is $E_b = \|\boldsymbol{a}\|^2$, so that (8.17) and (8.18) follow as before. This is not surprising, of course, since this vector decision problem is identical to the scalar problem when the appropriate basis is used.

For most communication problems, the a-priori probabilities are assumed to be equal, so that $\eta = 1$. Thus, as in (8.19),

$$\Pr\{e\} = Q\left(\sqrt{\frac{2E_b}{N_0}}\right). \tag{8.26}$$

This gives us a useful sanity check – the probability of error does not depend on the orthonormal coordinate basis.

Now suppose that the binary hypotheses correspond to nonantipodal vector signals, say $\boldsymbol{b}_0$ and $\boldsymbol{b}_1$. We analyze this in the same way as the scalar case. Namely, let $\boldsymbol{c} = (\boldsymbol{b}_0 + \boldsymbol{b}_1)/2$ and $\boldsymbol{a} = \boldsymbol{b}_0 - \boldsymbol{c}$. Then the two signals are $\boldsymbol{b}_0 = \boldsymbol{a} + \boldsymbol{c}$ and $\boldsymbol{b}_1 = -\boldsymbol{a} + \boldsymbol{c}$. As before, converting the observation $\boldsymbol{V}$ to $\tilde{\boldsymbol{V}} = \boldsymbol{V} - \boldsymbol{c}$ shifts the midpoint and converts the problem back to the antipodal case. The error probability depends only on the distance $2\|\boldsymbol{a}\|$ between the signals, but the energy per bit, assuming equiprobable inputs, is $E_b = \|\boldsymbol{a}\|^2 + \|\boldsymbol{c}\|^2$. Thus the center point $\boldsymbol{c}$ contributes to the energy, but not to the error probability. In the important special case where $\boldsymbol{b}_1$ and $\boldsymbol{b}_0$ are orthogonal and have equal energy, $\|\boldsymbol{a}\| = \|\boldsymbol{c}\|$ and

$$\Pr(e) = Q(\sqrt{E_b/N_0}). \tag{8.27}$$

It is often more convenient, especially when dealing with $M > 2$ hypotheses, to express the LLR for the nonantipodal case directly in terms of $\boldsymbol{b}_0$ and $\boldsymbol{v}_1$. Using (8.22) for the shifted vector $\tilde{\boldsymbol{V}}$, the LLR can be expressed as follows:

$$\mathrm{LLR}(\boldsymbol{v}) = \frac{-\|\boldsymbol{v} - \boldsymbol{b}_0\|^2 + \|\boldsymbol{v} - \boldsymbol{b}_1^2\|^2}{N_0}. \tag{8.28}$$

For ML detection, this is simply the minimum-distance rule, and for MAP the interpretation is the same as for the antipodal case.

8.3.4 Detection for binary complex vectors in WGN

Next consider the complex vector version of the same problem. Assume the observation is a complex random n-vector $\boldsymbol{V} = \boldsymbol{U} + \boldsymbol{Z}$. The noise, $\boldsymbol{Z} = (Z_1, \ldots, Z_n)^{\mathsf{T}}$, is a complex random vector of n zero-mean complex iid Gaussian rvs with iid real and imaginary parts, each $\mathcal{N}(0, N_0/2)$. Thus each Z_k is circularly symmetric and denoted by $\mathcal{CN}(0, N_0)$. The input $\boldsymbol{U}$ is independent of $\boldsymbol{Z}$ and binary, taking on value $\boldsymbol{a}$ with probability p_0 and $-a$ with probability p_1, where $\boldsymbol{a} = (a_1, \ldots, a_n)^{\mathsf{T}}$ is an arbitrary complex n-vector.

This problem can be reduced to that of Section 8.3.3 by letting $\boldsymbol{Z}'$ be the $2n$-dimensional real random vector with components $\Re(Z_k)$ and $\Im(Z_k)$ for $1 \le k \le n$. Similarly, let $\boldsymbol{a}'$ be the $2n$-dimensional real vector with components $\Re(a_k)$ and $\Im(a_k)$ for $1 \le k \le n$ and let $\boldsymbol{U}'$ be the real random vector that takes on values $\boldsymbol{a}'$ or $-\boldsymbol{a}'$. Finally, let $\boldsymbol{V}' = \boldsymbol{U}' + \boldsymbol{Z}'$.

Recalling that probability densities for complex random variables or vectors are equal to the joint probability densities for the real and imaginary parts, we have

$$f_{V|U}(\boldsymbol{v}|\boldsymbol{a}) = f_{V'|U'}(\boldsymbol{v}'|\boldsymbol{a}') = \frac{1}{(\pi N_0)^n} \exp \sum_{k=1}^{n} \frac{-\Re(v_k - a_k)^2 - \Im(v_k - a_k)^2}{N_0};$$

$$f_{V|U}(\boldsymbol{v}|-\boldsymbol{a}) = f_{V'|U'}(\boldsymbol{v}'|-\boldsymbol{a}') = \frac{1}{(\pi N_0)^n} \exp \sum_{k=1}^{n} \frac{-\Re(v_k + a_k)^2 - \Im(v_k + a_k)^2}{N_0}.$$

The LLR is then given by

$$\mathrm{LLR}(\boldsymbol{v}) = \frac{-\|\boldsymbol{v} - \boldsymbol{a}\|^2 + \|\boldsymbol{v} + \boldsymbol{a}\|^2}{N_0}. \tag{8.29}$$

Note that

$$\|\boldsymbol{v} - \boldsymbol{a}\|^2 = \|\boldsymbol{v}\|^2 - \langle \boldsymbol{v}, \boldsymbol{a}\rangle - \langle \boldsymbol{a}, \boldsymbol{v}\rangle + \|\boldsymbol{a}\|^2 = \|\boldsymbol{v}\|^2 - 2\Re(\langle \boldsymbol{v}, \boldsymbol{a}\rangle) + \|\boldsymbol{a}\|^2.$$

Using this and the analagous expression for $\|\boldsymbol{v} + \boldsymbol{a}\|^2$, (8.29) becomes

$$\mathrm{LLR}(\boldsymbol{v}) = \frac{4\Re(\langle \boldsymbol{v}, \boldsymbol{a}\rangle)}{N_0}. \tag{8.30}$$

The MAP test can now be stated as follows:

$$\frac{\Re(\langle \boldsymbol{v}, \boldsymbol{a}\rangle)}{\|\boldsymbol{a}\|} \mathop{\gtrless}_{\tilde{U}=-\boldsymbol{a}}^{\tilde{U}=\boldsymbol{a}} \frac{N_0 \ln \eta}{4\|\boldsymbol{a}\|}. \tag{8.31}$$

Note that the value of the LLR and the form of the MAP test are the same as the real vector case except for the real part of $\langle \boldsymbol{v}, \boldsymbol{a}\rangle$. The significance of this real-part operation is now discussed.

In the n-dimensional complex vector space, $\langle \boldsymbol{v}, \boldsymbol{a}\rangle / \|\boldsymbol{a}\|$ is the complex value of the projection of $\boldsymbol{v}$ in the direction of $\boldsymbol{a}$. In order to understand this projection better, consider an orthonormal basis in which $\boldsymbol{a} = (1, 0, 0, \ldots, 0)^{\mathsf{T}}$. Then $\langle \boldsymbol{v}, \boldsymbol{a}\rangle / \|\boldsymbol{a}\| = v_1$. Thus $\Re(v_1) = \pm 1 + \Re(z_1)$ and $\Im(v_1) = \Im(z_1)$. Clearly, only $\Re(v_1)$ is relevant to the binary decision. Using $\Re(\langle \boldsymbol{v}, \boldsymbol{a}\rangle / \|\boldsymbol{a}\|)$ in (8.31) is simply the general way of stating this elementary idea. If the complex plane is viewed as a 2D real space, then taking the real part of $\langle \boldsymbol{v}, \boldsymbol{a}\rangle$ is equivalent to taking the further projection of this 2D real vector in the direction of $\boldsymbol{a}$ (see Exercise 8.12).

The other results and interpretations of Section 8.3.3 remain unchanged. In particular, since $\|a'\| = \|a\|$, the error probability results are given by

$$\Pr\{e|U=-a\} = Q\left(\sqrt{\frac{2\|a\|^2}{N_0}} + \frac{\ln\eta}{2\sqrt{2\|a\|^2/N_0}}\right); \tag{8.32}$$

$$\Pr\{e|U=a\} = Q\left(\sqrt{\frac{2\|a\|^2}{N_0}} - \frac{\ln\eta}{2\sqrt{2\|a\|^2/N_0}}\right). \tag{8.33}$$

For the ML case, recognizing that $\|a\|^2 = E_b$, we have the following familiar result:

$$\Pr\{e\} = Q\left(\sqrt{\frac{2E_b}{N_0}}\right). \tag{8.34}$$

Finally, for the nonantipodal case with hypotheses $\boldsymbol{b}_0$ and $\boldsymbol{b}_1$, the LLR is again given by (8.28).

8.3.5 Detection of binary antipodal waveforms in WGN

This section extends the vector case of Sections 8.3.3 and 8.3.4 to the waveform case. It will be instructive to do this simultaneously for both passband real random processes and baseband complex random processes. Let $U(t)$ be the baseband modulated waveform. As before, the situation is simplified by transmitting a single bit rather than a sequence of bits, so, for some arbitrary, perhaps complex, baseband waveform $a(t)$, the binary input 0 is mapped into $U(t) = a(t)$ and 1 is mapped into $U(t) = -a(t)$; the a-priori probabilities are denoted by p_0 and p_1. Let $\{\theta_k(t); k \in \mathbb{Z}\}$ be a complex orthonormal expansion covering the baseband region of interest, and let $a(t) = \sum_k a_k \theta_k(t)$.

Assume $U(t) = \pm a(t)$ is modulated onto a carrier f_c larger than the baseband bandwidth. The resulting bandpass waveform is denoted by $X(t) = \pm b(t)$, where, from Section 7.8, the modulated form of $a(t)$, denoted by $b(t)$, can be represented as

$$b(t) = \sum_k b_{k,1}\psi_{k,1}(t) + b_{k,2}\psi_{k,2}(t),$$

where

$$b_{k,1} = \Re(a_k); \qquad \psi_{k,1}(t) = \Re\{2\theta_k(t)\exp[2\pi i f_c t]\};$$
$$b_{k,2} = \Im(a_k); \qquad \psi_{k,2}(t) = -\Im\{2\theta_k(t)\exp[2\pi i f_c t]\}.$$

From Theorem 6.6.1, the set of waveforms $\{\psi_{k,j}(t); k \in \mathbb{Z}, j \in \{1,2\}\}$ are orthogonal, each with energy 2. Let $\{\phi_m(t); m \in \mathbb{Z}\}$ be a set of orthogonal functions, each of energy 2 and each orthogonal to each of the $\psi_{k,j}(t)$. Assume that $\{\phi_m(t); m \in \mathbb{Z}\}$, together with the $\psi_{k,j}(t)$, span $\mathcal{L}_2$.

The noise $W(t)$, by assumption, is WGN. It can be represented as

$$W(t) = \sum_k \left(Z_{k,1}\psi_{k,1}(t) + Z_{k,2}\psi_{k,2}(t)\right) + \sum_m W_m \phi_m(t),$$

where $\{Z_{k,m};\, k \in \mathbb{Z}, m \in \{1,2\}\}$ is the set of scaled linear functionals of the noise in the $\mathcal{L}_2$ vector space spanned by the $\psi_{k,m}(t)$, and $\{W_m;\, m \in \mathbb{Z}\}$ is the set of linear functionals of the noise in the orthogonal complement of the space. As will be seen shortly, the joint distribution of the W_m makes no difference in choosing between $a(t)$ and $-a(t)$, so long as the W_m are independent of the $Z_{k,j}$ and the transmitted binary digit. The observed random process at passband is then $Y(t) = X(t) + W(t)$:

$$Y(t) = \sum_k \left[Y_{k,1}\psi_{k,1}(t) + Y_{k,2}\psi_{k,2}(t)\right] + \sum_m W_m \phi_m(t),$$

where

$$Y_{k,1} = \left(\pm b_{k,1} + Z_{k,1}\right); \qquad Y_{k,2} = \left(\pm b_{k,2} + Z_{k,2}\right).$$

First assume that a finite number n of orthonormal functions are used to represent $a(t)$. This is no loss of generality, since the single function $a(t)/\|a(t)\|$ would be sufficient. Suppose also, initially, that only a finite number, say $W_1, \ldots, W_\ell$, of the orthogonal noise functionals are observed. Assume also that the noise variables, $Z_{k,j}$ and W_m, are independent and are each[4] $\mathcal{N}(0, N_0/2)$. Then the likelihoods are given by

$$f_{Y|X}(y\,|\,b) = \frac{1}{(\pi N_0)^n} \exp\left(\sum_{k=1}^{n}\sum_{j=1}^{2} \frac{-(y_{k,j} - b_{k,j})^2}{N_0} + \sum_{m=1}^{\ell} \frac{-w_m^2}{N_0} \right),$$

$$f_{Y|X}(y\,|\,-b) = \frac{1}{(\pi N_0)^n} \exp\left(\sum_{k=1}^{n}\sum_{j=1}^{2} \frac{-(y_{k,j} + b_{k,j})^2}{N_0} + \sum_{m=1}^{\ell} \frac{-w_m^2}{N_0} \right).$$

The log likelihood ratio is thus given by

$$\mathrm{LLR}(y) = \sum_{k=1}^{n}\sum_{j=1}^{2} \frac{-(y_{k,j} - b_{k,j})^2 + (y_{k,j} + b_{k,j})^2}{N_0}$$

$$= \frac{-\|\boldsymbol{y} - \boldsymbol{b}\|^2 + \|\boldsymbol{y} + \boldsymbol{b}\|^2}{N_0} \tag{8.35}$$

$$= \sum_{k=1}^{n}\sum_{j=1}^{2} \frac{4 y_{k,j} b_{k,j}}{N_0} = \frac{4\langle \boldsymbol{y}, \boldsymbol{b}\rangle}{N_0}, \tag{8.36}$$

and the MAP test is

$$\frac{\langle \boldsymbol{y}, \boldsymbol{b}\rangle}{\|\boldsymbol{b}\|} \;\begin{matrix} \geq^{\tilde{X}=b} \\ <_{\tilde{X}=-b} \end{matrix}\; \frac{N_0 \ln \eta}{4\|\boldsymbol{b}\|}.$$

[4] Recall that $N_0/2$ is the noise variance using the same scale as used for the input, and the use of orthogonal functions of energy 2 at passband corresponds to orthonormal functions at baseband. Thus, since the input energy is measured at baseband, the noise is also; at passband, then, both the signal energy and the spectral density of the noise are multiplied by 2.

This is the same as the real vector case analyzed in Section 8.3.3. In fact, the only difference is that the observation here includes noise in the degrees of freedom orthogonal to the range of interest, and the derivation of the LLR shows clearly why these noise variables do not appear in the LLR. In fact, the number ℓ of rvs W_m can be taken to be arbitrarily large, and they can have any joint density. So long as they are independent of the $Z_{k,j}$ (and of $X(t)$), they cancel out in the LLR. In other words, WGN should be viewed as noise that is iid Gaussian over a large enough space to represent the signal, and is independent of the signal and noise elsewhere.

The preceding argument leading to (8.35) and (8.36) is not entirely satisfying mathematically, since it is based on the slightly vague notion of the signal space of interest, but in fact it is just this feature that makes it useful in practice, since physical noise characteristics do change over large changes in time and frequency.

The inner product in (8.36) is the inner product over the $\mathcal{L}_2$ space of real sequences. Since these sequences are coefficients in an orthogonal (rather than orthonormal) expansion, the conversion to an inner product over the corresponding functions (see Exercise 8.5) is given by

$$\sum_{k,j} y_{k,j} b_{k,j} = \frac{1}{2}\int y(t)b(t)\mathrm{d}t. \tag{8.37}$$

This shows that the LLR is independent of the basis, and that this waveform problem reduces to the single-dimensional problem if $b(t)$ is a multiple of one of the basis functions. Also, if a countably infinite basis for the signal space of interest is used, (8.37) is still valid.

Next consider what happens when $Y(t) = \pm b(t) + W(t)$ is demodulated to the baseband waveform $V(t)$. The component $\sum_m W_m(t)$ of $Y(t)$ extends to frequencies outside the passband, and thus $Y(t)$ is filtered before demodulation, preventing an aliasing-like effect between $\sum_m W_m(t)$ and the signal part of $Y(t)$ (see Exercise 6.11). Assuming that this filtering does not affect $b(t)$, $b(t)$ maps back into $a(t) = \sum_k a_k\theta_k(t)$, where $a_k = b_{k,1} + \mathrm{i}b_{k,2}$. Similarly, $W(t)$ maps into

$$Z(t) = \sum_k Z_k\theta_k(t) + Z_\perp(t),$$

where $Z_k = Z_{k,1} + \mathrm{i}Z_{k,2}$ and $Z_\perp(t)$ is the result of filtering and frequency demodulation on $\sum_m W_m\phi_m(t)$. The received baseband complex process is then given by

$$V(t) = \sum_k V_k\theta_k(t) + Z_\perp(t), \qquad \text{where} \quad V_k = \pm a_k + Z_k. \tag{8.38}$$

By the filtering assumption above, the sample functions of $Z_\perp(t)$ are orthogonal to the space spanned by the $\theta_k(t)$, and thus the sequence $\{V_k;\, k \in \mathbb{Z}\}$ is determined from $V(t)$. Since $V_k = Y_{k,1} + \mathrm{i}Y_{k,2}$, the sample value LLR$(\boldsymbol{y})$ in (8.36) is determined as follows by the sample values of $\{v_k;\, k \in \mathbb{Z}\}$:

$$\mathrm{LLR}(\boldsymbol{y}) = \frac{4\langle \boldsymbol{y}, \boldsymbol{b}\rangle}{N_0} = \frac{4\Re(\langle \boldsymbol{v}, \boldsymbol{a}\rangle)}{N_0}. \tag{8.39}$$

Thus $\{v_k; k \in \mathbb{Z}\}$ is a sufficient statistic for $y(t)$, and thus the MAP test based on $y(t)$ can be performed using $v(t)$. Now an implementation that first finds the sample function $v(t)$ from $y(t)$ and then does a MAP test on $v(t)$ is simply a particular kind of test on $y(t)$, and thus cannot achieve a smaller error probability than the MAP test on $\boldsymbol{y}$. Finally, since $\{v_k; k \in \mathbb{Z}\}$ is a sufficient statistic for $y(t)$, it is also a sufficient statistic for $v(t)$ and thus the orthogonal noise $Z_\perp(t)$ is irrelevant.

Note that the LLR in (8.39) is the same as the complex vector result in (8.30). One could repeat the argument there, adding in an orthogonal expansion for $Z_\perp(t)$ to verify the argument that $Z_\perp(t)$ is irrelevant. Since $Z_\perp(t)$ could take on virtually any form, the argument above, based on the fact that $Z_\perp(t)$ is a function of $\sum_m W_m \phi_m(t)$, which is independent of the signal and noise in the signal space, is more insightful.

To summarize this subsection, the detection of a single bit, sent by generating antipodal signals at baseband and modulating to passband, has been analyzed. After adding WGN, the received waveform is demodulated to baseband and then the single bit is detected. The MAP detector at passband is a threshold test on $\int y(t)b(t)\mathrm{d}t$. This is equivalent to a threshold test at baseband on $\Re(\int v(t)a^*(t)\mathrm{d}t)$. This shows that no loss of optimality occurs by demodulating to baseband and also shows that detection can be done either at passband or at baseband. In the passband case, the result is an immediate extension of binary detection for real vectors, and at baseband it is an immediate extension of binary detection of complex vectors.

The results of this section can now be interpreted in terms of PAM and QAM, while still assuming a "one-shot" system in which only one binary digit is actually sent. Recall that for both PAM and QAM modulation, the modulation pulse $p(t)$ is orthogonal to its T-spaced time shifts if $|\hat{p}(f)|^2$ satisfies the Nyquist criterion. Thus, if the corresponding received baseband waveform is passed through a matched filter (a filter with impulse response $p^*(t)$) and sampled at times kT, the received samples will have no intersymbol interference. For a single bit transmitted at discrete time 0, $u(t) = \pm a(t) = \pm a p(t)$. The output of the matched filter at receiver time 0 is then given by

$$\int v(t)p^*(t)\mathrm{d}t = \frac{\Re[\langle \boldsymbol{v}, a\rangle]}{a},$$

which is a scaled version of the LLR. Thus the receiver from Chapter 6 that avoids intersymbol interference also calculates the LLR, from which a threshold test yields the MAP detection.

Section 8.4 shows that this continues to provide MAP tests on successive signals. It should be noted also that sampling the output of the matched filter at time 0 yields the MAP test whether or not $p(t)$ has been chosen to avoid intersymbol interference.

It is important to note that the performance of binary antipodal communication in WGN depends only on the energy of the transmitted waveform. With ML detection, the error probability is the familiar expression $Q(\sqrt{2E_\mathrm{b}/N_0})$, where $E_\mathrm{b} = \int |a(t)|^2\, \mathrm{d}t$ and the variance of the noise in each real degree of freedom in the region of interest is $N_0/2$.

This completes the analysis of binary detection in WGN, including the relationship between the vector case and waveform case and that between complex waveforms or vectors at baseband and real waveforms or vectors at passband.

The following sections analyze M-ary detection. The relationships between vector and waveform and between real and complex is the same as above, so the following sections each assume whichever of these cases is most instructive without further discussion of these relationships.

8.4 M-ary detection and sequence detection

The analysis in Section 8.3 was limited in several ways. First, only binary signal sets were considered, and second only the "one-shot" problem where a single bit rather than a sequence of bits was considered. In this section, M-ary signal sets for arbitrary M will be considered, and this will then be used to study the transmission of a sequence of signals and to study arbitrary modulation schemes.

8.4.1 M-ary detection

Going from binary to M-ary hypothesis testing is a simple extension. To be specific, this will be analyzed for the complex random vector case. Let the observation be a complex random n-vector $\boldsymbol{V}$ and let the complex random n-vector $\boldsymbol{U}$ to be detected take on a value from the set $\{\boldsymbol{a}_0, \dots, \boldsymbol{a}_{M-1}\}$ with a-priori probabilities $p_0, \dots, p_{M-1}$. Denote the a-posteriori probabilities by $p_{\boldsymbol{U}|\boldsymbol{V}}(\boldsymbol{a}_m|\boldsymbol{v})$. The MAP rule (see Section 8.1) then chooses $\tilde{\boldsymbol{U}}(\boldsymbol{v}) = \arg\max_m p_{\boldsymbol{U}|\boldsymbol{V}}(\boldsymbol{a}_m|\boldsymbol{v})$. Assuming that the likelihoods can be represented as probability densities $f_{\boldsymbol{V}|\boldsymbol{U}}$, the MAP rule can be expressed as

$$\tilde{\boldsymbol{U}}(\boldsymbol{v}) = \arg\max_m p_m f_{\boldsymbol{V}|\boldsymbol{U}}(\boldsymbol{v}|\boldsymbol{a}_m).$$

Usually, the simplest approach to this M-ary rule is to consider multiple binary hypothesis testing problems. That is, $\tilde{\boldsymbol{U}}(v)$ is that $\boldsymbol{a}_m$ for which

$$\Lambda_{m,m'}(\boldsymbol{v}) = \frac{f_{\boldsymbol{V}|\boldsymbol{U}}(\boldsymbol{v}|\boldsymbol{a}_m)}{f_{\boldsymbol{V}|\boldsymbol{U}}(\boldsymbol{v}|\boldsymbol{a}_{m'})} \geq \frac{p_{m'}}{p_m}$$

for all m'. In the case of ties, it makes no difference which of the maximizing hypotheses are chosen.

For the complex vector additive WGN case, the observation is $\boldsymbol{V} = \boldsymbol{U} + \boldsymbol{Z}$, where $\boldsymbol{Z}$ is complex Gaussian noise with iid real and imaginary components. As derived in (8.29), the log likelihood ratio (LLR) between each pair of hypotheses $\boldsymbol{a}_m$ and $\boldsymbol{a}_{m'}$ is given by

$$\mathrm{LLR}_{m,m'}(\boldsymbol{v}) = \frac{-\|\boldsymbol{v} - \boldsymbol{a}_m\|^2 + \|\boldsymbol{v} - \boldsymbol{a}_{m'}\|^2}{N_0}. \tag{8.40}$$

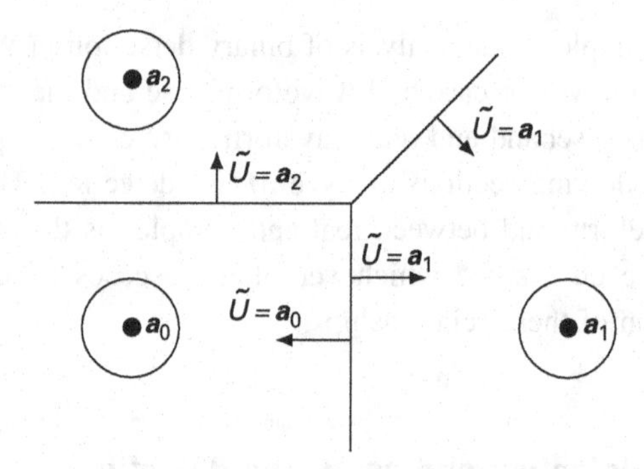

Figure 8.5. Decision regions for a 3-ary alphabet of vector signals in iid Gaussian noise. For ML detection, the decision regions are Voronoi regions, i.e. regions separated by perpendicular bisectors between the signal points.

Thus, each binary test separates the observation space[5] into two regions separated by the perpendicular bisector between the two points. With M hypotheses, the space is separated into the Voronoi regions of points closest to each of the signals (hypotheses) (see Figure 8.5). If the a-priori probabilities are unequal, then these perpendicular bisectors are shifted, remaining perpendicular to the axis joining the two signals, but no longer being bisectors.

The probability that noise carries the observation across one of these perpendicular bisectors is given in (8.31). The only new problem that arises with M-ary hypothesis testing is that the error probability, given $\boldsymbol{U} = m$, is the union of $M-1$ events, namely crossing the corresponding perpendicular to each other point. This can be found exactly by integrating over the n-dimensional vector space, but is usually upperbounded and approximated by the union bound, where the probability of crossing each perpendicular is summed over the $M-1$ incorrect hypotheses. This is usually a good approximation (if M is not too large), because the Gaussian density decreases so rapidly with distance; thus, in the ML case, most errors are made when observations occur roughly halfway between the transmitted and the detected signal points.

8.4.2 Successive transmissions of QAM signals in WGN

This subsection extends the "one-shot" analysis of detection for QAM and PAM in the presence of WGN to the case in which an n-tuple of successive independent symbols are transmitted. We shall find that, under many conditions, both the detection rule and the corresponding probability of symbol error can be analyzed by looking at one symbol at a time.

First consider a QAM modulation system using a modulation pulse $p(t)$. Assume that $p(t)$ has unit energy and is orthonormal to its T-spaced shifts $\{p(t-kT);\ k \in \mathbb{Z}\}$,

[5] For an n-dimensional complex vector space, it is simplest to view the observation space as the corresponding $2n$-dimensional real vector space.

i.e. that $\{p(t-kT);\ k \in \mathbb{Z}\}$ is a set of orthonormal functions. Let $\mathcal{A} = \{a_0, \ldots, a_{M-1}\}$ be the alphabet of complex input signals and denote the input waveform over an arbitrary n-tuple of successive input signals by

$$u(t) = \sum_{k=1}^{n} u_k p(t-kT),$$

where each u_k is a selection from the input alphabet $\mathcal{A}$.

Let $\{\phi_k(t);\ k \geq 1\}$ be an orthonormal basis of complex $\mathcal{L}_2$ waveforms such that the first n waveforms in that basis are given by $\phi_k(t) = p(t-kT)$, $1 \leq k \leq n$. The received baseband waveform is then given by

$$V(t) = \sum_{k=1}^{\infty} V_k \phi_k(t) = \sum_{k=1}^{n} (u_k + Z_k) p(t-kT) + \sum_{k>n} Z_k \phi_k(t). \tag{8.41}$$

We now compare two different detection schemes. In the first, a single ML decision between the M^n hypotheses for all possible joint values of $U_1, \ldots, U_n$ is made based on $V(t)$. In the second scheme, for each k, $1 \leq k \leq n$, an ML decision between the M possible hypotheses $a_0, \ldots, a_{M-1}$ is made for input U_k based on the observation $V(t)$. Thus in this scheme, n separate M-ary decisions are made, one for each of the n successive inputs.

For the first alternative, each hypothesis corresponds to an n-dimensional vector of inputs, $\boldsymbol{u} = (u_1, \ldots, u_n)^{\mathsf{T}}$. As in Section 8.3.5, the sample value $v(t) = \sum_k v_k \phi_k(t)$ of the received waveform can be taken as an ℓ-tuple $\boldsymbol{v} = (v_1, v_2, \ldots, v_\ell)^{\mathsf{T}}$ with $\ell \geq n$. The likelihood of $\boldsymbol{v}$ conditional on $\boldsymbol{u}$ is then given by

$$f_{V|U}(\boldsymbol{v}|\boldsymbol{u}) = \prod_{k=1}^{n} f_Z(v_k - u_k) \prod_{k=n+1}^{\ell} f_Z(v_k).$$

For any two hypotheses, say $\boldsymbol{u} = (u_1, \ldots, u_n)^{\mathsf{T}}$ and $\boldsymbol{u}' = (u'_1, \ldots, u'_n)^{\mathsf{T}}$, the likelihood ratio and LLR are given by

$$\Lambda_{\boldsymbol{u},\boldsymbol{u}'}(\boldsymbol{v}) = \prod_{k=1}^{n} \frac{f_Z(v_k - u_k)}{f_Z(v_k - u'_k)}, \tag{8.42}$$

$$\mathrm{LLR}_{\boldsymbol{u},\boldsymbol{u}'}(\boldsymbol{v}) = \frac{-\|\boldsymbol{v}-\boldsymbol{u}\|^2 + \|\boldsymbol{v}-\boldsymbol{u}'\|^2}{N_0}. \tag{8.43}$$

Note that for each $k > n$, v_k does not appear in this likelihood ratio. Thus this likelihood ratio is still valid[6] in the limit $\ell \to \infty$, but the only relevant terms in the decision are $v_1, \ldots, v_n$. Therefore, let $\boldsymbol{v} = (v_1, \ldots, v_n)^{\mathsf{T}}$ in what follows. From (8.43), this likelihood ratio is positive if and only if $\|\boldsymbol{v}-\boldsymbol{u}\| < \|\boldsymbol{v}-\boldsymbol{u}'\|$. The conclusion is that, for M^n-ary detection, done jointly on $u_1, \ldots, u_n$, the ML decision is the vector $\boldsymbol{u}$ that minimizes the distance $\|\boldsymbol{v}-\boldsymbol{u}\|$.

[6] In fact, these final $\ell - n$ components do not have to be independent or equally distributed, they must simply be independent of the signals and noise for $1 \leq k \leq n$.

Consider how to minimize $\|\boldsymbol{v}-\boldsymbol{u}\|$. Note that

$$\|\boldsymbol{v}-\boldsymbol{u}\|^2 = \sum_{k=1}^{n}(v_k - u_k)^2. \tag{8.44}$$

Suppose that $\tilde{\boldsymbol{u}} = (\tilde{u}_1, \ldots, \tilde{u}_n)^{\mathsf{T}}$ minimizes this sum. Then for each k, $\tilde{u}_k$ minimizes $(v_k - u_k)^2$ over the M choices for u_k (otherwise some $a_m \neq \tilde{u}_k$ could be substituted for $\tilde{u}_k$ to reduce $(v_k - u_k)^2$ and therefore reduce the sum in (8.44)). Thus the ML sequence detector with M^n hypotheses detects each U_k by minimizing $(v_k - u_k)^2$ over the M hypotheses for that U_k.

Next consider the second alternative above. For a given sample observation $\boldsymbol{v} = (v_1, \ldots, v_\ell)^{\mathsf{T}}$ and a given k, $1 \le k \le n$, the likelihood of $\boldsymbol{v}$ conditional on $U_k = u_k$ is given by

$$f_{\boldsymbol{V}|U_k}(\boldsymbol{v}|u_k) = f_Z(v_k - u_k) \prod_{j\neq k, 1\le j\le n} f_{V_j}(v_j) \prod_{j=n+1}^{\ell} f_Z(v_j),$$

where $f_{V_j}(v_j) = \sum_m p_m f_{V_j|U_j}(v_j|a_m)$ is the marginal probability of V_j. The likelihood ratio of $\boldsymbol{v}$ between the hypotheses $U_k = a_m$ and $U_k = a_{m'}$ is then

$$\Lambda^{(k)}_{m,m'}(\boldsymbol{v}) = \frac{f_Z(v_k - a_m)}{f_Z(v_k - a_{m'})}.$$

This is the familiar 1D nonantipodal Gaussian detection problem, and the ML decision is to choose $\tilde{u}_k$ as the a_m closest to u_k. Thus, given the sample observation $v(t)$, the vector $(\tilde{u}_1, \ldots, \tilde{u}_n)^{\mathsf{T}}$ of individual M-ary ML detectors for each U_k is the same as the M^n-ary ML sequence detector for the sequence $\boldsymbol{U} = (U_1, \ldots, U_n)^{\mathsf{T}}$. Moreover, each such detector is equivalent to a vector of ML decisions on each U_k based solely on the observation V_k.

Summarizing, we have proved the following theorem.

Theorem 8.4.1 *Let $U(t) = \sum_{k=1}^{n} U_k p(t-kT)$ be a QAM (or PAM) baseband input to a WGN channel and assume that $\{p(t-kT); 1 \le k \le n\}$ is an orthonormal sequence. Then the M^n-ary ML decision on $\boldsymbol{U} = (U_1, \ldots, U_n)^{\mathsf{T}}$ is equivalent to making separate M-ary ML decisions on each U_k, $1 \le k \le n$, where the decision on each U_k can be based either on the observation $v(t)$ or the observation of v_k.*

Note that the theorem states that the same decision is made for both sequence detection and separate detection for each signal. It *does not* say that the probability of an error within the sequence is the same as the error for a single signal. Letting P be the probability of error for a single signal, the probability of error for the sequence is $1-(1-P)^n$.

Theorem 8.4.1 makes no assumptions about the probabilities of the successive inputs, although the use of ML detection would not minimize the probability of error if the inputs were not independent and equally likely. If coding is used between the n input signals, then not all of these M^n n-tuples are possible. In this case, ML detection on the *possible encoded* sequences (as opposed to all M^n sequences) is different from separate detection on each signal. As an example, if the transmitter always repeats each signal, with $u_1 = u_2$, $u_3 = u_4$, etc., then the detection of u_1 should be based on both v_1 and v_2. Similarly, the detection of u_3 should be based on v_3 and v_4, etc.

When coding is used, it is possible to make ML decisions on each signal separately, and then to use the coding constraints to correct errors in the detected sequence. These individual signal decisions are then called *hard decisions*. It is also possible, for each k, to save a sufficient statistic (such as v_k) for the decision on U_k. This is called a *soft decision* since it saves all the relevant information needed for an ML decision between the set of possible codewords. Since the ML decision between possible encoded sequences minimizes the error probability (assuming equiprobable codewords), soft decisions yield smaller error probabilities than hard decisions.

Theorem 8.4.1 can be extended to MAP detection if the input signals are statistically independent of each other (see Exercise 8.15). One can see this intuitively by drawing the decision boundaries for the 2D real case; these decision boundaries are then horizontal and vertical lines.

A nice way to interpret Theorem 8.4.1 is to observe that the detection of each signal U_k depends only on the corresponding received signal V_k; all other components of the received vector are irrelevant to the decision on U_k. Section 8.4.3 generalizes from QAM to arbitrary modulation schemes and also generalizes this notion of irrelevance.

8.4.3 Detection with arbitrary modulation schemes

The previous sections have concentrated on detection of PAM and QAM systems, using real hypotheses $\mathcal{A} = \{a_0, \dots, a_{M-1}\}$ for PAM and complex hypotheses $\mathcal{A} = \{a_0, \dots, a_{M-1}\}$ for QAM. In each case, a sequence $\{u_k; k \in \mathbb{Z}\}$ of signals from $\mathcal{A}$ is modulated into a baseband waveform $u(t) = \sum_k u_k p(t-kT)$. The PAM waveform is then either transmitted or first modulated to passband. The complex QAM waveform is necessarily modulated to a real passband waveform.

This is now generalized by considering a signal set $\mathcal{A}$ to be an M-ary alphabet, $\{\boldsymbol{a}_0, \dots, \boldsymbol{a}_{M-1}\}$, of real n-tuples. Thus each $\boldsymbol{a}_m$ is an element of $\mathbb{R}^n$. The n components of the mth signal vector are denoted by $\boldsymbol{a}_m = (a_{m,1}, \dots, a_{m,n})^{\mathsf{T}}$. The selected signal vector $\boldsymbol{a}_m$ is then modulated into a signal waveform $\boldsymbol{b}_m(t) = \sum_{k=1}^{n} a_{m,k}\phi_k(t)$, where $\{\phi_1(t), \dots, \phi_n(t)\}$ is a set of n orthonormal waveforms.

The above provides a general scenario for mapping the symbols 0 to $M-1$ into a set of signal waveforms $\boldsymbol{b}_0(t)$ to $\boldsymbol{b}_{M-1}(t)$. A provision must also be made for transmitting a sequence of such M-ary symbols. If these symbols are to be transmitted at T-spaced intervals, the most straightforward way of accomplishing this is to choose the orthonormal waveforms $\phi_1(t), \dots, \phi_n(t)$ in such a way that $\phi_k(t-\ell T)$ and $\phi_j(t-\ell' T)$ are orthonormal for all j, k, $1 \le j, k \le n$, and all integers ℓ, ℓ'. In this case, a sequence of symbols, say $s_1, s_2, \dots$, each drawn from the alphabet $\{0, \dots, M-1\}$, could be mapped into a sequence of waveforms $\boldsymbol{b}_{s_1}(t), \boldsymbol{b}_{s_2}(t-T), \dots$ The transmitted waveform would then be $\sum_\ell \boldsymbol{b}_{s_\ell}(t-\ell T)$.

Note that PAM is a special case of this scenario where the dimension n is 1. The function $\phi_1(t)$ in this case is the real modulation pulse $p(t)$ for baseband transmission and $\sqrt{2}\,p(t)\cos(2\pi f_c t)$ for passband transmission; QAM is another special case where n is 2 at passband. In this case, the complex signals a_m are viewed as 2D real signals. The orthonormal waveforms (assuming real $p(t)$) are $\phi_1(t) = \sqrt{2}\,p(t)\cos(2\pi f_c t)$ and $\phi_2(t) = \sqrt{2}\,p(t)\sin(2\pi f_c t)$.

More generally, it is not necessary to start at baseband and shift to passband,[7] and it is not necessary for successive signals to be transmitted as time shifts of a basic waveform set. For example, in frequency-hopping systems, successive n-dimensional signals can be modulated to different carrier frequencies. What is important is that the successive transmitted signal waveforms are all orthogonal to each other.

Let $X(t)$ be the first signal waveform in such a sequence of successive waveforms. Then $X(t)$ is a choice from the set of M waveforms, $\boldsymbol{b}_0(t), \dots, \boldsymbol{b}_{M-1}(t)$. We can represent $X(t)$ as $\sum_{k=1}^{n} X_k \phi_k(t)$, where, under hypothesis m, $X_k = a_{m,k}$ for $1 \le k \le n$. Let $\phi_{n+1}(t), \phi_{n+2}(t), \dots$ be an additional set of orthonormal functions such that the entire set $\{\phi_k(t); k \ge 1\}$ spans the space of real $\mathcal{L}_2$ waveforms. The subsequence $\phi_{n+1}(t), \phi_{n+2}(t), \dots$ might include the successive time shifts of $\phi_1(t), \dots, \phi_n(t)$ for the example above, but in general can be arbitrary. We do assume, however, that successive signal waveforms are orthogonal to $\phi_1(t), \dots, \phi_n(t)$, and thus that they can be expanded in terms of $\phi_{n+1}(t), \phi_{n+2}(t), \dots$ The received random waveform $Y(t)$ is assumed to be the sum of $X(t)$, the WGN $Z(t)$, and contributions of signal waveforms other than X. These other waveforms could include successive signals from the given channel input and also signals from other users. This sum can be expanded over an arbitrarily large number, say ℓ, of these orthonormal functions:

$$Y(t) = \sum_{k=1}^{\ell} Y_k \phi_k(t) = \sum_{k=1}^{n} (X_k + Z_k)\phi_k(t) + \sum_{k=n+1}^{\ell} Y_k \phi_k(t). \tag{8.45}$$

Note that in (8.45) the random process $\{Y(t);\ t \in \mathbb{R}\}$ specifies the random variables $Y_1, \dots, Y_\ell$. Assuming that the sample waveforms of $Y(t)$ are $\mathcal{L}_2$, it also follows that the limit as $\ell \to \infty$ of $Y_1, \dots, Y_\ell$ specifies $Y(t)$ in the $\mathcal{L}_2$ sense. Thus we consider $Y_1, \dots, Y_\ell$ to be the observation at the channel output. It is convenient to separate these terms into two vectors, $\boldsymbol{Y} = (Y_1, \dots, Y_n)^{\mathsf{T}}$ and $\boldsymbol{Y}' = (Y_{n+1}, \dots, Y_\ell)^{\mathsf{T}}$.

Similarly, the WGN $Z(t) = \sum_k Z_k \phi_k(t)$ can be represented by $\boldsymbol{Z} = (Z_1, \dots, Z_n)^{\mathsf{T}}$ and $\boldsymbol{Z}' = (Z_{n+1}, \dots, Z_\ell)^{\mathsf{T}}$, and $X(t)$ can be represented as $\boldsymbol{X} = (X_1, \dots, X_n)^{\mathsf{T}}$. Finally, let $V(t) = \sum_{k>n} V_k \phi_k(t)$ be the contributions from other users and successive signals from the given user. Since these terms are orthogonal to $\phi_1(t), \dots, \phi_n(t)$, $V(t)$ can be represented by $\boldsymbol{V}' = (V_{n+1}, \dots, V_\ell)^{\mathsf{T}}$. With these changes, (8.45) becomes

$$\boldsymbol{Y} = \boldsymbol{X} + \boldsymbol{Z}; \qquad \boldsymbol{Y}' = \boldsymbol{Z}' + \boldsymbol{V}'. \tag{8.46}$$

[7] It seems strange at first that the real vector and real waveform case here is more general than the complex case, but the complex case is used for notational and conceptual simplifications at baseband, where the baseband waveform will be modulated to passsband and converted to a real waveform.

The observation is a sample value of $(\boldsymbol{Y}, \boldsymbol{Y}')$, and the detector must choose the MAP value of $\boldsymbol{X}$. Assuming that $\boldsymbol{X}, \boldsymbol{Z}, \boldsymbol{Z}'$, and $\boldsymbol{V}'$ are statistically independent, the likelihoods can be expressed as follows:

$$f_{\boldsymbol{YY}'|\boldsymbol{X}}(\boldsymbol{yy}'|\boldsymbol{a}_m) = f_{\boldsymbol{Z}}(\boldsymbol{y}-\boldsymbol{a}_m)f_{\boldsymbol{Y}'}(\boldsymbol{y}').$$

The likelihood ratio between hypotheses $\boldsymbol{a}_m$ and $\boldsymbol{a}_{m'}$ is then given by

$$\Lambda_{m,m'}(\boldsymbol{y}) = \frac{f_{\boldsymbol{Z}}(\boldsymbol{y}-\boldsymbol{a}_m)}{f_{\boldsymbol{Z}}(\boldsymbol{y}-\boldsymbol{a}_{m'})}. \tag{8.47}$$

The important thing here is that all the likelihood ratios (for $0 \le m, m' \le M-1$) depend only on $\boldsymbol{Y}$, and thus $\boldsymbol{Y}$ is a sufficient statistic for a MAP decision on $\boldsymbol{X}$. Note that $\boldsymbol{Y}'$ is irrelevant to the decision, and thus its probability density is irrelevant (other than the need to assume that $\boldsymbol{Y}'$ is statistically independent of $(\boldsymbol{Z}, \boldsymbol{X})$). This also shows that the size of ℓ is irrelevant. This is summarized (and slightly generalized by dropping the Gaussian noise assumption) in the following theorem.

Theorem 8.4.2 (Theorem of irrelevance) *Let $\{\phi_k(t); k \ge 1\}$ be a set of real orthonormal functions. Let $X(t) = \sum_{k=1}^{n} X_k\phi_k(t)$ and $Z(t) = \sum_{k=1}^{n} Z_k\phi_k(t)$ be the input to a channel and the corresponding noise, respectively, where $\boldsymbol{X} = (X_1, \dots, X_n)^{\mathsf{T}}$ and $\boldsymbol{Z} = (Z_1, \dots, Z_n)^{\mathsf{T}}$ are random vectors. Let $Y'(t) = \sum_{k>n} Y_k\phi_k(t)$, where, for each $\ell > n$, $\boldsymbol{Y}' = (Y_{n+1}, \dots, Y_\ell)^{\mathsf{T}}$ is a random vector that is statistically independent of the pair $\boldsymbol{X}$, $\boldsymbol{Z}$. Let $\boldsymbol{Y} = \boldsymbol{X}+\boldsymbol{Z}$. Then the LLR and the MAP detection of $\boldsymbol{X}$ from the observation of $\boldsymbol{Y}, \boldsymbol{Y}'$ depends only on $\boldsymbol{Y}$. That is, the observed sample value of $\boldsymbol{Y}'$ is irrelevant.*

The orthonormal set $\{\phi_1(t), \dots, \phi_n(t)\}$ chosen above appears to have a more central importance than it really has. What is important is the existence of an n-dimensional subspace of real $\mathcal{L}_2$ that includes the signal set and has the property that the noise and signals orthogonal to this subspace are independent of the noise and signal within the subspace. In the usual case, we choose this subspace to be the space spanned by the signal set, but there are also cases where the subspace must be somewhat larger to provide for the independence between the subspace and its complement.

The irrelevance theorem does not specify how to do MAP detection based on the observed waveform, but rather shows how to reduce the problem to a finite-dimensional problem. Since the likelihood ratios specify both the decision regions and the error probability for MAP detection, it is clear that the choice of orthonormal set cannot influence either the error probability or the mapping of received waveforms to hypotheses.

One important constraint in the above analysis is that both the noise and the interference (from successive transmissions and from other users) are additive. The other important constraint is that the interference is both orthogonal to the signal $X(t)$ and also statistically independent of $X(t)$. The orthogonality is why $\boldsymbol{Y} = \boldsymbol{X}+\boldsymbol{Z}$, with no contribution from the interference. The statistical independence is what makes $\boldsymbol{Y}'$ irrelevant.

If the interference is orthogonal but not independent, then a MAP decision could still be made based on $\boldsymbol{Y}$ alone; the error probability would be the same as if $\boldsymbol{Y}, \boldsymbol{Y}'$ were independent, but using the dependence at the decoder could lead to a smaller error probability.

On the other hand, if the interference is nonorthogonal but independent, then $\boldsymbol{Y}$ would include both noise and a contribution from the interference, and the error probability would typically be larger, but never smaller, than in the orthogonal case. As a rule of thumb, then, nonorthogonal interference tends to increase error probability, whereas dependence (if the receiver makes use of it) tends to reduce error probability.

If successive statistically independent signals, $X_1, X_2, \ldots,$ are modulated onto distinct sets of orthonormal waveforms (i.e. if X_1 is modulated onto the orthonormal waveforms $\phi_1(t)$ to $\phi_n(t)$, X_2 is modulated onto $\phi_{n+1}(t)$ to $\phi_{2n}(t)$, etc.), then it also follows, as in Section 8.4.2, that ML detection on a sequence $X_1, \ldots, X_\ell$ is equivalent to separate ML decisions on each input signal X_j, $1 \le j \le \ell$. The details are omitted since the only new feature in this extension is the more complicated notation.

The higher-dimensional mappings allowed in this subsection are sometimes called *channel codes*, and are sometimes simply viewed as more complex forms of modulation. The coding field is very large, but the following sections provide an introduction.

8.5 Orthogonal signal sets and simple channel coding

An orthogonal signal set is a set $\boldsymbol{a}_0, \ldots, \boldsymbol{a}_{M-1}$ of M real orthogonal M-vectors, each with the same energy E. Without loss of generality we choose a basis for $\mathbb{R}^M$ in which the mth basis vector is $\boldsymbol{a}_m/\sqrt{E}$. In this basis, $\boldsymbol{a}_0 = (\sqrt{E}, 0, 0, \ldots, 0)^\mathsf{T}$, $\boldsymbol{a}_1 = (0, \sqrt{E}, 0, \ldots, 0)^\mathsf{T}$, etc. Modulation onto an orthonormal set $\{\phi_m(t)\}$ of waveforms then maps hypothesis $\boldsymbol{a}_m$ $(0 \le m \le M-1)$ into the waveform $\sqrt{E}\phi_m(t)$. After addition of WGN, the sufficient statistic for detection is a sample value $\boldsymbol{y}$ of $\boldsymbol{Y} = \boldsymbol{A} + \boldsymbol{Z}$, where $\boldsymbol{A}$ takes on the values $\boldsymbol{a}_0, \ldots, \boldsymbol{a}_{M-1}$ with equal probability and $\boldsymbol{Z} = (Z_0, \ldots, Z_{M-1})^\mathsf{T}$ has iid components $\mathcal{N}(0, N_0/2)$. It can be seen that the ML decision is to decide on that m for which y_m is largest.

The major case of interest for orthogonal signals is where M is a power of 2, say $M = 2^b$. Thus the signal set can be used to transmit b binary digits, so the energy per bit is $E_\mathrm{b} = E/b$. The number of required degrees of freedom for the signal set, however, is $M = 2^b$, so the spectral efficiency ρ (the number of bits per pair of degrees of freedom) is then $\rho = b/2^{b-1}$. As b gets large, ρ gets small at almost an exponential rate. It will be shown, however, that for large enough E_b, as b gets large holding E_b constant, the ML error probabiliity goes to 0. In particular, for any $E_\mathrm{b}/N_0 < \ln 2 = 0.693$, the error probability goes to 0 exponentially as $b \to \infty$. Recall that $\ln 2 = 0.693$, i.e. -1.59 dB, is the Shannon limit for reliable communication on a WGN channel with unlimited bandwidth. Thus the derivation to follow will establish the Shannon theorem for WGN and unlimited bandwidth. Before doing that, however, two closely related types of signal sets are discussed.

8.5.1 Simplex signal sets

Consider the random vector $\boldsymbol{A}$ with orthogonal equiprobable sample values $\boldsymbol{a}_0, \ldots, \boldsymbol{a}_{M-1}$ as described above. The mean value of $\boldsymbol{A}$ is then given by

$$\overline{\boldsymbol{A}} = \left(\frac{\sqrt{E}}{M}, \frac{\sqrt{E}}{M}, \ldots, \frac{\sqrt{E}}{M} \right)^{\mathsf{T}}.$$

We have seen that if a signal set is shifted by a constant vector, the Voronoi detection regions are also shifted and the error probability remains the same. However, such a shift can change the expected energy of the random signal vector. In particular, if the signals are shifted to remove the mean, then the signal energy is reduced by the energy (squared norm) of the mean. In this case, the energy of the mean is E/M. A *simplex signal set* is an orthogonal signal set with the mean removed. That is,

$$\boldsymbol{S} = \boldsymbol{A} - \overline{\boldsymbol{A}}; \quad \boldsymbol{s}_m = \boldsymbol{a}_m - \overline{\boldsymbol{A}}; \quad 0 \le m \le M-1.$$

In other words, the mth component of $\boldsymbol{s}_m$ is $\sqrt{E}\,(M-1)/M$ and each other component is $-\sqrt{E}/M$. Each simplex signal has energy $E(M-1)/M$, so the simplex set has the same error probability as the related orthogonal set, but requires less energy by a factor of $(M-1)/M$. The simplex set of size M has dimensionality $M-1$, as can be seen from the fact that the sum of all the signals is 0, so the signals are linearly dependent. Figure 8.6 illustrates the orthogonal and simplex sets for $M = 2$ and 3.

For small M, the simplex set is a substantial improvement over the orthogonal set. For example, for $M = 2$, it has a 3 dB energy advantage (it is simply the antipodal 1D set). Also, it uses one fewer dimension than the orthogonal set. For large M, however, the improvement becomes almost negligible.

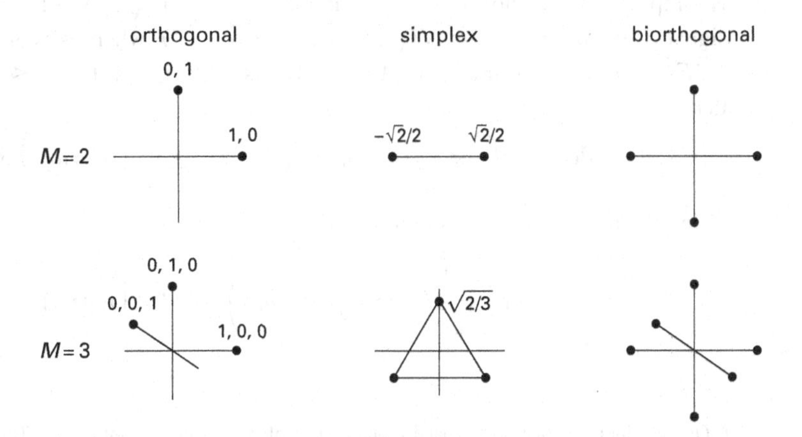

Figure 8.6. Orthogonal, simplex, and biorthogonal signal constellations.

8.5.2 Biorthogonal signal sets

If $\boldsymbol{a}_0, \ldots, \boldsymbol{a}_{M-1}$ is a set of orthogonal signals, we call the set of $2M$ signals consisting of $\pm\boldsymbol{a}_0, \ldots, \pm\boldsymbol{a}_{M-1}$ a *biorthogonal signal set*; 2D and 3D examples of biorthogonal signals sets are given in Figure 8.6.

It can be seen by the same argument used for orthogonal signal sets that the ML detection rule for such a set is to first choose the dimension m for which $|y_m|$ is largest, and then choose $\boldsymbol{a}_m$ or $-\boldsymbol{a}_m$ depending on whether y_m is positive or negative. Orthogonal signal sets and simplex signal sets each have the property that each signal is equidistant from every other signal. For biorthogonal sets, each signal is equidistant from all but one of the other signals. The exception, for the signal $\boldsymbol{a}_m$, is the signal $-\boldsymbol{a}_m$.

The biorthogonal signal set of M dimensions contains twice as many signals as the orthogonal set (thus sending one extra bit per signal), but has the same minimum distance between signals. It is hard to imagine[8] a situation where we would prefer an orthogonal signal set to a biorthogonal set, since one extra bit per signal is achieved at essentially no cost. However, for the limiting argument to follow, an orthogonal set is used in order to simplify the notation. As M gets very large, the advantage of biorthogonal signals becomes smaller, which is why, asymptotically, the two are equivalent.

8.5.3 Error probability for orthogonal signal sets

Since the signals differ only by the ordering of the coordinates, the probability of error does not depend on which signal is sent; thus $\Pr(e) = \Pr(e \,|\, A = \boldsymbol{a}_0)$. Conditional on $A = \boldsymbol{a}_0$, Y_0 is $\mathcal{N}(\sqrt{E}, N_0/2)$ and Y_m is $\mathcal{N}(0, N_0/2)$ for $1 \le m \le M-1$. Note that if $A = \boldsymbol{a}_0$ and $Y_0 = y_0$, then an error is made if $Y_m \ge y_0$ for any m, $1 \le m \le M-1$. Thus

$$\Pr(e) = \int_{-\infty}^{\infty} f_{Y_0|\boldsymbol{A}}(y_0 \,|\, \boldsymbol{a}_0) \Pr\left(\bigcup_{m=1}^{M-1} (Y_m \ge y_0 \,|\, \boldsymbol{A} = \boldsymbol{a}_0) \right) \mathrm{d}y_0. \tag{8.48}$$

The rest of the derivation of $\Pr(e)$, and its asymptotic behavior as M gets large, is simplified if we normalize the outputs to $W_m = Y_m\sqrt{2/N_0}$. Then, conditional on signal $\boldsymbol{a}_0$ being sent, W_0 is $\mathcal{N}(\sqrt{2E/N_0}, 1) = \mathcal{N}(\alpha, 1)$, where α is an abbreviation for $\sqrt{2E/N_0}$. Also, conditional on $A = \boldsymbol{a}_0$, W_m is $\mathcal{N}(0, 1)$ for $1 \le m \le M-1$. It follows that

$$\Pr(e) = \int_{-\infty}^{\infty} f_{W_0|\boldsymbol{A}}(w_0 \,|\, \boldsymbol{a}_0) \Pr\left(\bigcup_{m=1}^{M-1} (W_m \ge w_0 \,|\, \boldsymbol{A} = \boldsymbol{a}_0) \right) \mathrm{d}w_0. \tag{8.49}$$

Using the union bound on the union above,

$$\Pr\left(\bigcup_{m=1}^{M-1} (W_m \ge w_0 \,|\, A = \boldsymbol{a}_0) \right) \le (M-1)Q(w_0). \tag{8.50}$$

[8] One possibility is that at passband a phase error of π can turn $\boldsymbol{a}_m$ into $-\boldsymbol{a}_m$. Thus with biorthogonal signals it is necessary to track phase or use differential phase.

The union bound is quite tight when applied to independent quantitities that have small aggregate probability. Thus, this bound will be quite tight when w_0 is large and M is not too large. When w_0 is small, however, the bound becomes loose. For example, for $w_0 = 0$, $Q(w_0) = 1/2$ and the bound in (8.50) is $(M-1)/2$, much larger than the obvious bound of 1 for any probability. Thus, in the analysis to follow, the left side of (8.50) will be upperbounded by 1 for small w_0 and by $(M-1)Q(w_0)$ for large w_0. Since both 1 and $(M-1)Q(w_0)$ are valid upperbounds for all w_0, the dividing point γ between small and large can be chosen arbitrarily. It is chosen in what follows to satisfy

$$\exp(-\gamma^2/2) = 1/M; \qquad \gamma = \sqrt{2\ln M}. \tag{8.51}$$

It might seem more natural in light of (8.50) to replace γ above by the γ_1 that satisfies $(M-1)Q(\gamma_1) = 1$, and that turns out to be the natural choice in the lowerbound to $\Pr(e)$ developed in Exercise 8.10. It is not hard to see, however, that γ/γ_1 goes to 1 as $M \to \infty$, so the difference is not of major importance. Splitting the integral in (8.49) into $w_0 \le \gamma$ and $w_0 > \gamma$, we obtain

$$\Pr(e) \le \int_{-\infty}^{\gamma} f_{W_0|A}(w_0 \mid \boldsymbol{a}_0)\mathrm{d}w_0 + \int_{\gamma}^{\infty} f_{W_0|A}(w_0 \mid \boldsymbol{a}_0)(M-1)Q(w_0)\mathrm{d}w_0 \tag{8.52}$$

$$\le Q(\alpha-\gamma) + \int_{\gamma}^{\infty} f_{W_0|A}(w_0 \mid a_0)(M-1)Q(\gamma)\exp\left(\frac{\gamma^2}{2} - \frac{w_0^2}{2}\right)\mathrm{d}w_0 \tag{8.53}$$

$$\le Q(\alpha-\gamma) + \int_{\gamma}^{\infty} \frac{1}{\sqrt{2\pi}}\exp\left(\frac{-(w_0-\alpha)^2 + \gamma^2 - w_0^2}{2}\right)\mathrm{d}w_0 \tag{8.54}$$

$$= Q(\alpha-\gamma) + \int_{\gamma}^{\infty} \frac{1}{\sqrt{2\pi}}\exp\left(\frac{-2(w_0-\alpha/2)^2 + \gamma^2 - \alpha^2/2}{2}\right)\mathrm{d}w_0 \tag{8.55}$$

$$= Q(\alpha-\gamma) + \frac{1}{\sqrt{2}}Q\left(\sqrt{2}\left(\gamma - \frac{\alpha}{2}\right)\right)\exp\left(\frac{\gamma^2}{2} - \frac{\alpha^2}{4}\right). \tag{8.56}$$

The first term on the right side of (8.52) is the lower tail of the distribution of W_0, and is the probability that the *negative* of the fluctuation of W_0 exceeds $\alpha - \gamma$, i.e. $Q(\alpha-\gamma)$. In the second term, $Q(w_0)$ is upperbounded using Exercise 8.7(c), thus resulting in (8.53). This is simplified by $(M-1)Q(\gamma) \le M\exp(-\gamma^2/2) = 1$, resulting in (8.54). The exponent is then manipulated to "complete the square" in (8.55), leading to an integral of a Gaussian density, as given in (8.56).

The analysis now breaks into three special cases: the first where $\alpha \le \gamma$; the second where $\alpha/2 \le \gamma < \alpha$; and the third where $\gamma \le \alpha/2$. We explain the significance of these cases after completing the bounds.

Case (1) ($\alpha \le \gamma$) The argument of the first Q function in (8.55) is less than or equal to 0, so its value lies between 1/2 and 1. This means that $\Pr(e) \le 1$, which is a useless result. As seen later, this is the case where the rate is greater than or equal to capacity. It is also shown in Exercise 8.10 that the error probability must be large in this case.

Case (2) ($\alpha/2 \le \gamma < \alpha$) Each Q function in (8.56) has a non-negative argument, so the bound $Q(x) \le (1/2)\exp(-x^2/2)$ applies (see Exercise 8.7(b)):

$$\Pr(e) \le \frac{1}{2}\exp\left(\frac{-(\alpha-\gamma)^2}{2}\right) + \frac{1}{2\sqrt{2}}\exp\left(\frac{-\alpha^2}{4} + \frac{\gamma^2}{2} - (\gamma - \alpha/2)^2\right) \quad (8.57)$$

$$\le \left(\frac{1}{2} + \frac{1}{2\sqrt{2}}\right)\exp\left(\frac{-(\alpha-\gamma)^2}{2}\right) \le \exp\left(\frac{-(\alpha-\gamma)^2}{2}\right). \quad (8.58)$$

Note that (8.58) follows (8.57) from combining the terms in the exponent of the second term. The fact that the exponents are equal is not too surprising, since γ was chosen to equalize approximately the integrands in (8.52) at $w_0 = \gamma$.

Case (3) ($\gamma \le \alpha/2$) The argument of the second Q function in (8.56) is less than or equal to 0, so its value lies between 1/2 and 1 and is upperbounded by 1, yielding

$$\Pr(e) \le \frac{1}{2}\exp\left(\frac{-(\alpha-\gamma)^2}{2}\right) + \frac{1}{2\sqrt{2}}\exp\left(\frac{-\alpha^2}{4} + \frac{\gamma^2}{2}\right) \quad (8.59)$$

$$\le \exp\left(\frac{-\alpha^2}{4} + \frac{\gamma^2}{2}\right). \quad (8.60)$$

Since the two exponents in (8.57) are equal, the first exponent in (8.59) must be smaller than the second, leading to (8.60). This is essentially the union bound derived in Exercise 8.8.

The lowerbound in Exercise 8.10 shows that these bounds are quite tight, but the sense in which they are tight will be explained later.

We now explore what α and γ are in terms of the number of codewords M and the energy per bit E_b. Recall that $\alpha = \sqrt{2E/N_0}$. Also $\log_2 M = b$, where b is the number of bits per signal. Thus $\alpha = \sqrt{2bE_b/N_0}$. From (8.51), $\gamma^2 = 2\ln M = 2b\ln 2$. Thus,

$$\alpha - \gamma = \sqrt{2b}\left[\sqrt{E_b/N_0} - \sqrt{\ln 2}\right].$$

Substituting these values into (8.58) and (8.60), we obtain

$$\Pr(e) \le \exp\left[-b\left(\sqrt{E_b/N_0} - \sqrt{\ln 2}\right)^2\right] \qquad \text{for } \frac{E_b}{4N_0} \le \ln 2 < \frac{E_b}{N_0}; \quad (8.61)$$

$$\Pr(e) \le \exp\left[-b\left(\frac{E_b}{2N_0} - \ln 2\right)\right] \qquad \text{for } \ln 2 < \frac{E_b}{4N_0}. \quad (8.62)$$

We see from this that for fixed $E_b/N_0 > \ln 2$, $\Pr(e) \to 0$ as $b \to \infty$.

Recall that in (7.82) we stated that the capacity (in bits per second) of a WGN channel of bandwidth W, noise spectral density $N_0/2$, and power P is given by

$$C = W\log\left(1 + \frac{P}{WN_0}\right). \quad (8.63)$$

With no bandwidth constraint, i.e. in the limit $\mathsf{W} \to \infty$, the ultimate capacity is $C = P/(N_0 \ln 2)$. This means that, according to Shannon's theorem, for any rate $R < C = P/(N_0 \ln 2)$, there are codes of rate R bits per second for which the error probability is arbitrarily close to 0. Now $P/R = E_\mathrm{b}$, so Shannon says that if $E_b/(N_0 \ln 2) > 1$, then codes exist with arbitrarily small probability of error.

The orthogonal codes provide a concrete proof of this ultimate capacity result, since (8.61) shows that $\Pr(e)$ can be made arbitrarily small (by increasing b) so long as $E_b/(N_0 \ln 2) > 1$. Shannon's theorem also says that the error probability cannot be made small if $E_b/(N_0 \ln 2) < 1$. We have not quite proven that here, although Exercise 8.10 shows that the error probability cannot be made arbitrarily small for an orthogonal code[9] if $E_b/(N_0 \ln 2) < 1$.

The limiting operation here is slightly unconventional. As b increases, E_b is held constant. This means that the energy E in the signal increases linearly with b, but the size of the constellation increases exponentially with b. Thus the bandwidth required for this scheme is infinite in the limit, and going to infinity very rapidly. This means that this is not a practical scheme for approaching capacity, although sets of 64 or even 256 biorthogonal waveforms are used in practice.

The point of the analysis, then, is first to show that this infinite bandwidth capacity can be approached, but second to show also that using large but finite sets of orthogonal (or biorthogonal or simplex) waveforms does decrease error probability for fixed signal-to-noise ratio, and decreases it as much as desired (for rates below capacity) if enough bandwidth is used.

The different forms of solution in (8.61) and (8.62) are interesting, and not simply consequences of the upperbounds used. For case (2), which leads to (8.61), the typical way that errors occur is when $w_0 \approx \gamma$. In this situation, the union bound is on the order of 1, which indicates that, conditional on $y_0 \approx \gamma$, it is quite likely that an error will occur. In other words, the typical error event involves an unusually large negative value for w_0 rather than any unusual values for the other noise terms. In case (3), which leads to (8.62), the typical way for errors to occur is when $w_0 \approx \alpha/2$ and when some other noise term is also at about $\alpha/2$. In this case, an unusual event is needed both in the signal direction and in some other direction.

A more intuitive way to look at this distinction is to visualize what happens when E/N_0 is held fixed and M is varied. Case 3 corresponds to small M, case 2 to larger M, and case 1 to very large M. For small M, one can visualize the Voronoi region around the transmitted signal point. Errors occur when the noise carries the signal point outside the Voronoi region, and that is most likely to occur at the points in the Voronoi surface closest to the transmitted signal, i.e. at points halfway between the transmitted point and some other signal point. As M increases, the number of these

[9] Since a simplex code has the same error probability as the corresponding orthogonal code, but differs in energy from the orthogonal code by a vanishingly small amount as $M \to \infty$, the error probability for simplex codes also cannot be made arbitrarily small for any given $E_b/(N_0 \ln 2) < 1$. It is widely believed, but never proven, that simplex codes are optimal in terms of ML error probability whenever the error probability is small. There is a known example (Steiner, 1994), for all $M \geq 7$, where the simplex is nonoptimal, but in this example the signal-to-noise ratio is very small and the error probability is very large.

midway points increases until one of them is almost certain to cause an error when the noise in the signal direction becomes too large.

8.6 Block coding

This section provides a brief introduction to the subject of coding for error correction on noisy channels. Coding is a major topic in modern digital communication, certainly far more important than suggested by the length of this introduction. In fact, coding is a topic that deserves its own text and its own academic subject in any serious communication curriculum. Suggested texts are Forney (2005) and Lin and Costello (2004). Our purpose here is to provide enough background and examples to understand the role of coding in digital communication, rather than to prepare the student for coding research. We start by viewing orthogonal codes as block codes using a binary alphabet. This is followed by the Reed–Muller codes, which provide considerable insight into coding for the WGN channel. This then leads into Shannon's celebrated noisy-channel coding theorem.

A *block code* is a code for which the incoming sequence of binary digits is segmented into blocks of some given length m and then these binary m-tuples are mapped into *codewords*. There are thus 2^m codewords in the code; these codewords might be binary n-tuples of some *block length* $n > m$, or they might be vectors of signals, or waveforms. Successive codewords then pass through the remaining stages of modulation before transmission. There is no fundamental difference between coding and modulation; for example, the orthogonal code above with $M = 2^m$ codewords can be viewed either as modulation with a large signal set or coding using binary m-tuples as input.

8.6.1 Binary orthogonal codes and Hadamard matrices

When orthogonal codewords are used on a WGN channel, any orthogonal set is equally good from the standpoint of error probability. One possibility, for example, is the use of orthogonal sine waves. From an implementation standpoint, however, there are simpler choices than orthogonal sine waves. Conceptually, also, it is helpful to see that orthogonal codewords can be constructed from binary codewords. This digital approach will turn out to be conceptually important in the study of fading channels and diversity in Chapter 9. It also helps in implementation, since it postpones the point at which digital hardware gives way to analog waveforms.

One digital approach to generating a large set of orthogonal waveforms comes from first generating a set of M binary codewords, each of length M and each distinct pair differing in exactly $M/2$ places. Each binary digit can then be mapped into an antipodal signal, $0 \to +a$ and $1 \to -a$. This yields an M-tuple of real-valued antipodal signals, $s_1, \dots, s_M$, which is then mapped into the waveform $\sum_j s_j \phi_j(t)$, where $\{\phi_j(t); 1 \le j \le M\}$ is an orthonormal set (such as sinc functions or Nyquist pulses). Since each pair of binary codewords differs in $M/2$ places, the corresponding pair of waveforms are orthogonal and each waveform has equal energy. A binary code with the above properties is called a *binary orthogonal code*.

There are many ways to generate binary orthogonal codes. Probably the simplest is from a *Hadamard matrix*. For each integer $m \geq 1$, there is a 2^m by 2^m Hadamard matrix H_m. Each distinct pair of rows in the Hadamard matrix H_m differs in exactly 2^{m-1} places, so the 2^m rows of H_m constitute a binary orthogonal code with 2^m codewords.

It turns out that there is a simple algorithm for generating the Hadamard matrices. The Hadamard matrix H_1 is defined to have the rows 00 and 01, which trivially satisfy the condition that each pair of distinct rows differ in half the positions. For any integer $m > 1$, the Hadamard matrix H_{m+1} of order 2^{m+1} can be expressed as four 2^m by 2^m submatrices. Each of the upper two submatrices is H_m, and the lower two submatrices are H_m and $\overline{H}_m$, where $\overline{H}_m$ is the complement of H_m. This is illustrated in Figure 8.7.

Note that each row of each matrix in Figure 8.7, other than the all-zero row, contains half 0s and half 1s. To see that this remains true for all larger values of m, we can use induction. Thus assume, for given m, that H_m contains a single row of all 0s and $2^m - 1$ rows, each with exactly half 1s. To prove the same for H_{m+1}, first consider the first 2^m rows of H_{m+1}. Each row has twice the length and twice the number of 1s as the corresponding row in H_m. Next consider the final 2^m rows. Note that $\overline{H}_m$ has all 1s in the first row and 2^{m-1} 1s in each other row. Thus the first row in the second set of 2^m rows of H_{m+1} has no 1s in the first 2^m positions and 2^m 1s in the final 2^m positions, yielding 2^m 1s in 2^{m+1} positions. Each remaining row has 2^{m-1} 1s in the first 2^m positions and 2^{m-1} 1s in the final 2^m positions, totaling 2^m 1s as required.

By a similar inductive argument (see Exercise 8.18), the mod-2 sum[10] of any two rows of H_m is another row of H_m. Since the mod-2 sum of two rows gives the positions in which the rows differ, and only the mod-2 sum of a codeword with itself gives the all-zero codeword, this means that the set of rows is a binary orthogonal set.

The fact that the mod-2 sum of any two rows is another row makes the corresponding code a special kind of binary code called a *linear code*, *parity-check code*, or *group code* (these are all synonyms). Binary M-tuples can be regarded as vectors in a vector space over the binary scalar field. It is not necessary here to be precise about what a field is; so far it has been sufficient to consider vector spaces defined over the real or complex fields. However, the binary numbers, using mod-2 addition and ordinary

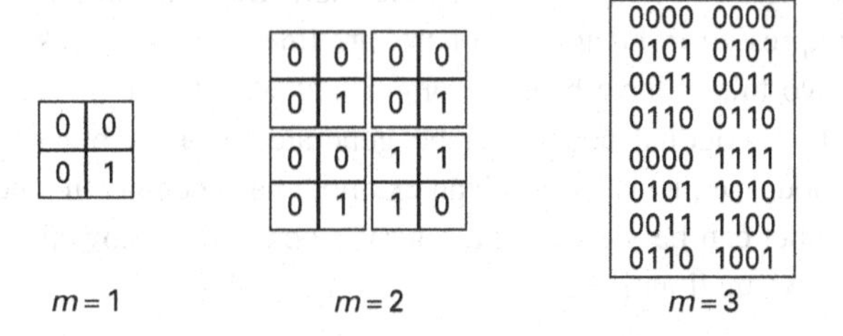

Figure 8.7. Hadamard matrices.

[10] The mod-2 sum of two binary numbers is defined by $0 \oplus 0 = 0$, $0 \oplus 1 = 1$, $1 \oplus 0 = 1$, and $1 \oplus 1 = 0$. The mod-2 sum of two rows (or vectors) or binary numbers is the component-wise row (or vector) of mod-2 sums.

multiplication, also form the field called $\mathbb{F}_2$, and the familiar properties of vector spaces, using $\{0, 1\}$ as scalars, apply here also.

Since the set of codewords in a linear code is closed under mod-2 sums (and also closed under scalar multiplication by 1 or 0), a linear code is a binary vector subspace of the binary vector space of binary M-tuples. An important property of such a subspace, and thus of a linear code, is that the set of positions in which two codewords differ is the set of positions in which the mod-2 sum of those codewords contains the binary digit 1. This means that the distance between two codewords (i.e. the number of positions in which they differ) is equal to the weight (the number of positions containing the binary digit 1) of their mod-2 sum. This means, in turn, that, for a linear code, the minimum distance $d_{\min}$ taken between all distinct pairs of codewords is equal to the minimum weight (minimum number of 1s) of any nonzero codeword.

Another important property of a linear code (other than the trivial code consisting of all binary M-tuples) is that some components x_k of each codeword $\boldsymbol{x} = (x_1, \dots, x_M)^{\mathsf{T}}$ can be represented as mod-2 sums of other components. For example, in the $m = 3$ case of Figure 8.7, $x_4 = x_2 \oplus x_3$, $x_6 = x_2 \oplus x_5$, $x_7 = x_3 \oplus x_5$, $x_8 = x_4 \oplus x_5$, and $x_1 = 0$. Thus only three of the components can be independently chosen, leading to a 3D binary subspace. Since each component is binary, such a 3D subspace contains $2^3 = 8$ vectors. The components that are mod-2 combinations of previous components are called "parity checks" and often play an important role in decoding. The first component, x_1, can be viewed as a parity check since it cannot be chosen independently, but its only role in the code is to help achieve the orthogonality property. It is irrelevant in decoding.

It is easy to modify a binary orthogonal code generated by a Hadamard matrix to generate a binary simplex code, i.e. a binary code which, after the mapping $0 \to a, 1 \to -a$, forms a simplex in Euclidean space. The first component of each binary codeword is dropped, turning the code into M codewords over $M-1$ dimensions. Note that in terms of the antipodal signals generated by the binary digits, dropping the first component converts the signal $+a$ (corresponding to the first binary component 0) into the signal 0 (which corresponds neither to the binary 0 or 1). The generation of the binary biorthogonal code is equally simple; the rows of H_m yield half of the codewords and the rows of $\overline{H}_m$ yield the other half. Both the simplex and the biorthogonal code, as expressed in binary form here, are linear binary block codes.

Two things have been accomplished with this representation of orthogonal codes. First, orthogonal codes can be generated from a binary sequence mapped into an antipodal sequence; second, an example has been given where modulation over a large alphabet can be viewed as a binary block code followed by modulation over a binary or very small alphabet.

8.6.2 Reed–Muller codes

Orthogonal codes (and the corresponding simplex and biorthgonal codes) use enormous bandwidth for large M. The Reed–Muller codes constitute a class of binary linear block codes that include large bandwidth codes (in fact, they include the binary biorthogonal codes), but also allow for much smaller bandwidth expansion, i.e. they allow for binary codes with M codewords, where $\log M$ is much closer to the number of dimensions used by the code.

The Reed–Muller codes are specified by two integer parameters, $m \geq 1$ and $0 \leq r \leq m$; a binary linear block code, denoted by $\mathrm{RM}(r, m)$, exists for each such choice. The parameter m specifies the block length to be $n = 2^m$. The minimum distance $d_{\min}(r, m)$ of the code and the number of binary information digits $k(r, m)$ required to specify a codeword are given by

$$d_{\min}(r, m) = 2^{m-r}; \qquad k(r, m) = \sum_{j=0}^{r} \binom{m}{j}, \tag{8.64}$$

where $\binom{m}{j} = \frac{m!}{j!(m-j)!}$. Thus these codes, like the binary orthogonal codes, exist only for block lengths equal to a power of 2. While there is only one binary orthogonal code (as defined through H_m) for each m, there is a range of RM codes for each m, ranging from large $d_{\min}$ and small k to small $d_{\min}$ and large k as r increases.

For each m, these codes are trivial for $r = 0$ and $r = m$. For $r = 0$ the code consists of two codewords selected by a single bit, so $k(0, m) = 1$; one codeword is all 0s and the other is all 1s, leading to $d_{\min}(0, m) = 2^m$. For $r = m$, the code is the set of all binary 2^m-tuples, leading to $d_{\min}(m, m) = 1$ and $k(m, m) = 2^m$. For $m = 1$, then, there are two RM codes: $\mathrm{RM}(0, 1)$ consists of the two codewords (0,0) and (1,1), and $\mathrm{RM}(1, 1)$ consists of the four codewords (0,0), (0,1), (1,0), and (1,1).

For $m > 1$ and intermediate values of r, there is a simple algorithm, much like that for Hadamard matrices, that specifies the set of codewords. The algorithm is recursive, and, for each $m > 1$ and $0 < r < m$, specifies $\mathrm{RM}(r, m)$ in terms of $\mathrm{RM}(r, m-1)$ and $\mathrm{RM}(r-1, m-1)$. Specifically, $\boldsymbol{x} \in \mathrm{RM}(r, m)$ if $\boldsymbol{x}$ is the concatenation of $\boldsymbol{u}$ and $\boldsymbol{u} \oplus \boldsymbol{v}$, denoted by $\boldsymbol{x} = (\boldsymbol{u}, \boldsymbol{u} \oplus \boldsymbol{v})$, for some $\boldsymbol{u} \in RM(r, m-1,)$ and $\boldsymbol{v} \in RM(r-1, m-1)$. More formally, for $0 < r < m$,

$$\mathrm{RM}(r, m) = \{(\boldsymbol{u}, \boldsymbol{u} \oplus \boldsymbol{v}) \mid \boldsymbol{u} \in \mathrm{RM}(r, m-1), \boldsymbol{v} \in \mathrm{RM}(r-1, m-1)\}. \tag{8.65}$$

The analogy with Hadamard matrices is that $\boldsymbol{x}$ is a row of H_m if $\boldsymbol{u}$ is a row of H_{m-1} and $\boldsymbol{v}$ is either all 1s or all 0s.

The first thing to observe about this definition is that if $\mathrm{RM}(r, m-1)$ and $\mathrm{RM}(r-1, m-1)$ are linear codes, then $\mathrm{RM}(r, m)$ is also. To see this, let $\boldsymbol{x} = (\boldsymbol{u}, \boldsymbol{u} \oplus \boldsymbol{v})$ and $\boldsymbol{x}' = (\boldsymbol{u}', \boldsymbol{u}' \oplus \boldsymbol{v}')$. Then

$$\boldsymbol{x} \oplus \boldsymbol{x}' = (\boldsymbol{u} \oplus \boldsymbol{u}', \boldsymbol{u} \oplus \boldsymbol{u}' \oplus \boldsymbol{v} \oplus \boldsymbol{v}') = (\boldsymbol{u}'', \boldsymbol{u}'' \oplus \boldsymbol{v}''),$$

where $\boldsymbol{u}'' = \boldsymbol{u} \oplus \boldsymbol{u}' \in \mathrm{RM}(r, m-1)$ and $\boldsymbol{v}'' = \boldsymbol{v} \oplus \boldsymbol{v}' \in \mathrm{RM}(r-1, m-1)$. This shows that $\boldsymbol{x} \oplus \boldsymbol{x}' \in \mathrm{RM}(r, m)$, and it follows that $\mathrm{RM}(r, m)$ is a linear code if $\mathrm{RM}(r, m-1)$ and $\mathrm{RM}(r-1, m-1)$ are. Since both $\mathrm{RM}(0, m)$ and $\mathrm{RM}(m, m)$ are linear for all $m \geq 1$, it follows by induction on m that all the Reed–Muller codes are linear.

Another observation is that different choices of the pair $\boldsymbol{u}$ and $\boldsymbol{v}$ cannot lead to the same value of $\boldsymbol{x} = (\boldsymbol{u}, \boldsymbol{u} \oplus \boldsymbol{v})$. To see this, let $\boldsymbol{x}' = (\boldsymbol{u}', \boldsymbol{v}')$. Then, if $\boldsymbol{u} \neq \boldsymbol{u}'$, it follows that the first half of $\boldsymbol{x}$ differs from that of $\boldsymbol{x}'$. Similarly, if $\boldsymbol{u} = \boldsymbol{u}'$ and $\boldsymbol{v} \neq \boldsymbol{v}'$, then the second half of $\boldsymbol{x}$ differs from that of $\boldsymbol{x}'$. Thus, $\boldsymbol{x} = \boldsymbol{x}'$ only if both $\boldsymbol{u} = \boldsymbol{u}'$ and $\boldsymbol{v} = \boldsymbol{v}'$. As a consequence of this, the number of information bits required to specify a codeword in RM(r, m), denoted by $k(r, m)$, is equal to the number required to specify a codeword in RM$(r, m-1)$ plus that to specify a codeword in RM$(r-1, m-1)$, i.e., for $0 < r < m$,

$$k(r, m) = k(r, m-1) + k(r-1, m-1).$$

Exercise 8.19 shows that this relationship implies the explicit form for $k(r, m)$ given in (8.64). Finally, Exercise 8.20 verifies the explicit form for $d_{\min}(r, m)$ in (8.64).

The RM$(1, m)$ codes are the binary biorthogonal codes, and one can view the construction in (8.65) as being equivalent to the Hadamard matrix algorithm by replacing the M by M matrix H_m in the Hadamard algorithm by the $2M$ by M matrix $\begin{bmatrix} H_m \\ G_m \end{bmatrix}$, where $G_m = \overline{H}_m$.

Another interesting case is the RM$(m-2, m)$ codes. These have $d_{\min}(m-2, m) = 4$ and $k(m-2, m) = 2^m - m - 1$ information bits. In other words, they have $m+1$ parity checks. As explained below, these codes are called *extended Hamming codes*.

A property of all RM codes is that all codewords have an even number[11] of 1s and thus the last component in each codeword can be viewed as an overall parity check which is chosen to ensure that the codeword contains an even number of 1s. If this final parity check is omitted from RM$(m-2, m)$ for any given m, the resulting code is still linear and must have a minimum distance of at least 3, since only one component has been omitted. This code is called the Hamming code of block length $2^m - 1$ with m parity checks. It has the remarkable property that every binary $(2^m - 1)$-tuple is either a codeword in this code or distance 1 from a codeword.[12]

The Hamming codes are not particularly useful in practice for the following reasons. If one uses a Hamming code at the input to a modulator and then makes hard decisions on the individual bits before decoding, then a block decoding error is made whenever two or more bit errors occur. This is a small improvement in reliability at a very substantial cost in transmission rate. On the other hand, if soft decisions are made, the use of the extended Hamming code (i.e. RM$(m-2, m)$) extends $d_{\min}$ from 3 to 4, significantly decreasing the error probability with a marginal cost in added redundant bits.

8.7 Noisy-channel coding theorem

Sections 8.5 and 8.6 provided a brief introduction to coding. Several examples were discussed showing that the use of binary codes could accomplish the same thing, for

[11] This property can be easily verified by induction.

[12] To see this, note that there are $2^{2^m - 1 - m}$ codewords, and each codeword has $2^m - 1$ neighbors; these are distinct from the neighbors of other codewords since $d_{\min}$ is at least 3. Adding the codewords and the neighbors, we get the entire set of $2^{2^m - 1}$ vectors. This also shows that the minimum distance is exactly 3.

example as the use of large sets of orthogonal, simplex, or biorthogonal waveforms. There was an ad hoc nature to the development, however, illustrating a number of schemes with various interesting properties, but little in the way of general results.

The earlier results on $\Pr(e)$ for orthogonal codes were more fundamental, showing that $\Pr(e)$ could be made arbitrarily small for a WGN channel with no bandwidth constraint if E_b/N_0 is greater than ln 2. This constituted a special case of the noisy-channel coding theorem, saying that arbitrarily small $\Pr(e)$ can be achieved for that very special channel and set of constraints.

8.7.1 Discrete memoryless channels

This section states and proves the noisy-channel coding theorem for another special case, that of discrete memoryless channels (DMCs). This may seem a little peculiar after all the emphasis in this and the preceding chapter on WGN. There are two major reasons for this choice. The first is that the argument is particularly clear in the DMC case, particularly after studying the AEP for discrete memoryless sources. The second is that the argument can be generalized easily, as will be discussed later. A DMC has a discrete input sequence $\boldsymbol{X} = X_1, \dots, X_k, \dots$ At each discrete time k, the input to the channel belongs to a finite alphabet $\mathcal{X}$ of symbols. For example, in Section 8.6, the input alphabet could be viewed as the signals $\pm a$. The question of interest would then be whether it is possible to communicate reliably over a channel when the decision to use the alphabet $\mathcal{X} = \{a, -a\}$ has already been made. The channel would then be regarded as the part of the channel from signal selection to an output sequence from which detection would be done. In a more general case, the signal set could be an arbitrary QAM set.

A DMC is also defined to have a discrete output sequence $\boldsymbol{Y} = Y_1, \dots, Y_k, \dots,$ where each output Y_k in the output sequence is a selection from a finite alphabet $\mathcal{Y}$ and is a probabilistic function of the input and noise in a way to be described shortly. In the example above, the output alphabet could be chosen as $\mathcal{Y} = \{a, -a\}$, corresponding to the case in which hard decisions are made on each signal at the receiver. The channel would then include the modulation and detection as an internal part, and the question of interest would be whether coding at the input and decoding from the single-letter hard decisions at the output could yield reliable communication.

Another choice would be to use the pre-decision outputs, first quantized to satisfy the finite alphabet constraint. Another almost identical choice would be a detector that produced a quantized LLR as opposed to a decision.

In summary, the choice of discrete memoryless channel alphabets depends on what part of the overall communication problem is being addressed.

In general, a channel is described not only by the input and output alphabets, but also by the probabilistic description of the outputs conditional on the inputs (the probabilistic description of the inputs is selected by the channel user). Let $\boldsymbol{X}^n = (X_1, X_2, \dots X_n)^{\mathsf{T}}$ be the channel input, here viewed either over the lifetime of the channel or any time greater than or equal to the duration of interest. Similarly, the output is denoted by

$Y^n = (Y_1, \dots, Y_n)^{\mathsf{T}}$. For a DMC, the probability of the output n-tuple, conditional on the input n-tuple, is defined to satisfy

$$p_{Y^n|X^n}(y_1, \dots, y_n \mid x_1, \dots, x_n) = \prod_{k=1}^{n} p_{Y_k|X_k}(y_k|x_k), \tag{8.66}$$

where $p_{Y_k|X_k}(y_k = j|x_k = i)$, for each $j \in \mathcal{Y}$ and $i \in \mathcal{X}$, is a function only of i and j and not of the time k. Thus, conditional on the inputs, the outputs are independent and have the same conditional distribution at all times. This conditional distribution is denoted by $P_{i,j}$ for all $i \in \mathcal{X}$ and $j \in \mathcal{Y}$, i.e. $p_{Y_k|X_k}(y_k = j|x_k = i) = P_{i,j}$. Thus the channel is completely described by the input alphabet, the output alphabet, and the conditional distribution function $P_{i,j}$. The conditional distribution function is usually called the *transition function* or *transition matrix*.

The most intensely studied DMC over the past 60 years has been the binary symmetric channel (BSC), which has $\mathcal{X} = \{0, 1\}, \mathcal{Y} = \{0, 1\}$, and satisfies $P_{0,1} = P_{1,0}$. The single number $P_{0,1}$ thus specifies the BSC. The WGN channel with antipodal inputs and ML hard decisions at the output is an example of the BSC. Despite the intense study of the BSC and its inherent simplicity, the question of optimal codes of long block length (optimal in the sense of minimum error probability) is largely unanswered. Thus, the noisy-channel coding theorem, which describes various properties of the achievable error probability through coding plays a particularly important role in coding.

8.7.2 Capacity

This section defines the capacity C of a DMC. Section 8.7.3, after defining the rate R at which information enters the modulator, shows that reliable communication is impossible on a channel if $R > C$. This is known as the converse to the noisy-channel coding theorem, and is in contrast to Section 8.7.4, which shows that arbitrarily reliable communication is possible for any $R < C$. As in the analysis of orthogonal codes, communication at rates below capacity can be made increasingly reliable with increasing block length, while this is not possible for $R > C$.

The capacity is defined in terms of various entropies. For a given DMC and given sequence length n, let $p_{Y^n|X^n}(y^n|x^n)$ be given by (8.66) and let $p_{X^n}(x^n)$ denote an arbitrary probability mass function chosen by the user on the input $X_1, \dots, X_n$. This leads to a joint entropy $\mathsf{H}[X^nY^n]$. From (2.37), this can be broken up as follows:

$$\mathsf{H}[X^nY^n] = \mathsf{H}[X^n] + \mathsf{H}[Y^n|X^n], \tag{8.67}$$

where $\mathsf{H}[Y^n|X^n] = \mathsf{E}[-\log p_{Y^n|X^n}(Y^n|X^n)]$. Note that because $\mathsf{H}[Y^n|X^n]$ is defined as an expectation over both X^n and Y^n, $\mathsf{H}[Y^n|X^n]$ depends on the distribution of X^n as well as the conditional distribution of Y^n given X^n. The joint entropy $\mathsf{H}[X^nY^n]$ can also be broken up the opposite way as follows:

$$\mathsf{H}[X^nY^n] = \mathsf{H}[Y^n] + \mathsf{H}[X^n|Y^n]. \tag{8.68}$$

Combining (8.67) and (8.68), it is seen that $\mathsf{H}[\boldsymbol{X}^n]-\mathsf{H}[\boldsymbol{X}^n|\boldsymbol{Y}^n]=\mathsf{H}[\boldsymbol{Y}^n]-\mathsf{H}[\boldsymbol{Y}^n|\boldsymbol{X}^n]$. This difference of entropies is called the *mutual information* between $\boldsymbol{X}^n$ and $\boldsymbol{Y}^n$ and is denoted by $\mathsf{I}[\boldsymbol{X}^n;\boldsymbol{Y}^n]$, so

$$\mathsf{I}[\boldsymbol{X}^n;\boldsymbol{Y}^n]=\mathsf{H}[\boldsymbol{X}^n]-\mathsf{H}[\boldsymbol{X}^n|\boldsymbol{Y}^n]=\mathsf{H}[\boldsymbol{Y}^n]-\mathsf{H}[\boldsymbol{Y}^n|\boldsymbol{X}^n]. \tag{8.69}$$

The first expression for $\mathsf{I}[\boldsymbol{X}^n;\boldsymbol{Y}^n]$ has a nice intuitive interpretation. From source coding, $\mathsf{H}[\boldsymbol{X}^n]$ represents the number of bits required to represent the channel input. If we look at a particular sample value $\boldsymbol{y}^n$ of the output, $\mathsf{H}[\boldsymbol{X}^n|\boldsymbol{Y}^n=\boldsymbol{y}^n]$ can be interpreted as the number of bits required to represent $\boldsymbol{X}^n$ after observing the output sample value $\boldsymbol{y}^n$. Note that $\mathsf{H}[\boldsymbol{X}^n|\boldsymbol{Y}^n]$ is the expected value of this over $\boldsymbol{Y}^n$. Thus $\mathsf{I}[\boldsymbol{X}^n;\boldsymbol{Y}^n]$ can be interpreted as the reduction in uncertainty, or number of required bits for specification, after passing through the channel. This intuition will lead to the converse to the noisy-channel coding theorem in Section 8.7.3.

The second expression for $\mathsf{I}[\boldsymbol{X}^n;\boldsymbol{Y}^n]$ is the one most easily manipulated. Taking the log of the expression in (8.66), we obtain

$$\mathsf{H}[\boldsymbol{Y}^n|\boldsymbol{X}^n]=\sum_{k=1}^{n}\mathsf{H}[Y_k|X_k]. \tag{8.70}$$

Since the entropy of a sequence of random symbols is upperbounded by the sum of the corresponding terms (see Exercise 2.19),

$$\mathsf{H}[\boldsymbol{Y}^n]\le\sum_{k=1}^{n}\mathsf{H}[Y_k]. \tag{8.71}$$

Substituting this and (8.70) in (8.69) yields

$$\mathsf{I}[\boldsymbol{X}^n;\boldsymbol{Y}^n]\le\sum_{k=1}^{n}\mathsf{I}[X_k;Y_k]. \tag{8.72}$$

If the inputs are independent, then the outputs are also, and (8.71) and (8.72) are satisfied with equality. The mutual information $\mathsf{I}[X_k;Y_k]$ at each time k is a function only of the pmf for X_k, since the output probabilities conditional on the input are determined by the channel. Thus, each mutual information term in (8.72) is upperbounded by the maximum of the mutual information over the input distribution. This maximum is defined as the *capacity* of the DMC, given by

$$C=\max_{\boldsymbol{p}}\sum_{i\in\mathcal{X}}\sum_{j\in\mathcal{Y}}p_iP_{i,j}\log\frac{P_{i,j}}{\sum_{\ell\in\mathcal{X}}p_\ell P_{\ell,j}}, \tag{8.73}$$

where $\boldsymbol{p}=\{p_0,p_1,\ldots,p_{|\mathcal{X}|-1}\}$ is the set (over the alphabet $\mathcal{X}$) of input probabilities. The maximum is over this set of input probabilities, subject to $p_i\ge 0$ for each $i\in\mathcal{X}$ and $\sum_{i\in\mathcal{X}}p_i=1$. The above function is concave in $\boldsymbol{p}$, and thus the maximization is straightforward. For the BSC, for example, the maximum is at $p_0=p_1=1/2$ and

$C = 1 + P_{0,1} \log P_{0,1} + P_{0,0} \log P_{0,0}$. Since C upperbounds $\mathsf{I}[X_k; Y_k]$ for each k, with equality if the distribution for X_k is the maximizing distribution,

$$\mathsf{I}[\boldsymbol{X}^n; \boldsymbol{Y}^n] \leq nC \tag{8.74}$$

with equality if all inputs are independent and chosen with the maximizing probabilities in (8.73).

8.7.3 Converse to the noisy-channel coding theorem

Define the rate R for the DMC above as the number of iid equiprobable binary source digits that enter the channel per channel use. More specifically, assume that nR bits enter the source and are transmitted over the n channel uses under discussion. Assume also that these bits are mapped into the channel input $\boldsymbol{X}^n$ in a one-to-one way. Thus $\mathsf{H}[\boldsymbol{X}^n] = nR$ and $\boldsymbol{X}^n$ can take on $M = 2^{nR}$ equiprobable values. The following theorem now bounds $\Pr(e)$ away from 0 if $R > C$.

Theorem 8.7.1 *Consider a DMC with capacity C. Assume that the rate R satisfies $R > C$. Then, for any block length n, the ML probability of error, i.e. the probability that the decoded n-tuple $\tilde{\boldsymbol{X}}^n$ is unequal to the transmitted n-tuple $\boldsymbol{X}^n$, is lowerbounded by*

$$R - C \leq H_{\mathrm{b}}(\Pr(e)) + R \Pr(e), \tag{8.75}$$

where $H_{\mathrm{b}}(\alpha)$ is the binary entropy, $-\alpha \log \alpha - (1-\alpha)\log(1-\alpha)$.

Remark Note that the right side of (8.75) is 0 at $\Pr(e) = 0$ and is increasing for $\Pr(e) \leq 1/2$, so (8.75) provides a lowerbound to $\Pr(e)$ that depends only on C and R.

Proof Note that $\mathsf{H}[\boldsymbol{X}^n] = nR$ and, from (8.72) and (8.69), $\mathsf{H}(\boldsymbol{X}^n) - \mathsf{H}(\boldsymbol{X}^n|\boldsymbol{Y}^n) \leq nC$. Thus

$$\mathsf{H}(\boldsymbol{X}^n|\boldsymbol{Y}^n) \geq nR - nC. \tag{8.76}$$

For each sample value $\boldsymbol{y}^n$ of $\boldsymbol{Y}^n$, $\mathsf{H}(\boldsymbol{X}^n \,|\, \boldsymbol{Y}^n = \boldsymbol{y}^n)$ is an ordinary entropy. The received $\boldsymbol{y}^n$ is decoded into some $\tilde{\boldsymbol{x}}^n$, and the corresonding probability of error is $\Pr(\boldsymbol{X}^n \neq \tilde{\boldsymbol{x}}^n \,|\, \boldsymbol{Y}^n = \boldsymbol{y}^n)$. The Fano inequality (see Exercise 2.20) states that the entropy $\mathsf{H}(\boldsymbol{X}^n \,|\, \boldsymbol{Y}^n = \boldsymbol{y}^n)$ can be upperbounded as the sum of two terms: first the binary entropy of whether or not $\boldsymbol{X}^n = \tilde{\boldsymbol{x}}^n$, and second the entropy of all $M-1$ possible errors in the case $\boldsymbol{X}^n \neq \tilde{\boldsymbol{x}}^n$, i.e.

$$\mathsf{H}(\boldsymbol{X}^n \,|\, \boldsymbol{Y}^n = \boldsymbol{y}^n) \leq H_{\mathrm{b}}(\Pr(e|\boldsymbol{y}^n)) + \Pr(e|\boldsymbol{y}^n)\log(M-1).$$

Upperbounding $\log(M-1)$ by $\log M = nR$ and averaging over $\boldsymbol{Y}^n$ yields

$$\mathsf{H}(\boldsymbol{X}^n|\boldsymbol{Y}^n) \leq H_{\mathrm{b}}(\Pr(e)) + nR\Pr(e). \tag{8.77}$$

Combining (8.76) and (8.77), we obtain

$$R - C \leq \frac{H_b(\Pr(e))}{n} + R\Pr(e),$$

and upperbounding $1/n$ by 1 yields (8.75). □

Theorem 8.7.1 is not entirely satisfactory, since it shows that the probability of block error cannot be made negligible at rates above capacity, but it does not rule out the possibility that each block error causes only one bit error, say, and thus the probability of bit error might go to 0 as $n \to \infty$. As shown in Gallager (1968, theorem 4.3.4), this cannot happen, but the proof does not add much insight and will be omitted here.

8.7.4 Noisy-channel coding theorem, forward part

There are two critical ideas in the forward part of the coding theorem. The first is to use the AEP on the joint ensemble $\boldsymbol{X}^n\boldsymbol{Y}^n$. The second, however, is what shows the true genius of Shannon. His approach, rather than an effort to find and analyze good codes, was simply to choose each codeword of a code randomly, choosing each letter in each codeword to be iid with the capacity yielding input distribution.

One would think initially that the codewords should be chosen to be maximally different in some sense, but Shannon's intuition said that independence would be enough. Some initial sense of why this might be true comes from looking at the binary orthogonal codes. Here each codeword of length n differs from each other codeword in $n/2$ positions, which is equal to the average number of differences with random choice. Another initial intuition comes from the fact that mutual information between input and output n-tuples is maximized by iid inputs. Truly independent inputs do not allow for coding constraints, but choosing a limited number of codewords using an iid distribution is at least a plausible approach. In any case, the following theorem proves that this approach works.

It clearly makes no sense for the encoder to choose codewords randomly if the decoder does not know what those codewords are, so we visualize the designer of the modem as choosing these codewords and building them into both transmitter and receiver. Presumably the designer is smart enough to test a code before shipping a million copies around the world, but we won't worry about that. We simply average the performance over all random choices. Thus the probability space consists of M independent iid codewords of block length n, followed by a randomly chosen message m, $0 \leq m \leq M-1$, that enters the encoder. The corresponding sample value $\boldsymbol{x}_m^n$ of the mth randomly chosen codeword is transmitted and combined with noise to yield a received sample sequence $\boldsymbol{y}^n$. The decoder then compares $\boldsymbol{y}^n$ with the M possible randomly chosen messages (the decoder knows $\boldsymbol{x}_0^n, \ldots, \boldsymbol{x}_{M-1}^n$, but not m) and chooses the most likely of them. It appears that a simple problem has been replaced by a complex problem, but since there is so much independence between all the random symbols, the new problem is surprisingly simple.

These randomly chosen codewords and channel outputs are now analyzed with the help of the AEP. For this particular problem, however, it is simpler to use a slightly different form of AEP, called the *strong AEP*, than that of Chapter 2. The strong AEP was analyzed in Exercise 2.28 and is reviewed here. Let $\boldsymbol{U}^n = U_1, \ldots, U_n$ be an n-tuple of iid discrete random symbols with alphabet $\mathcal{U}$ and letter probabilities p_j for each $j \in \mathcal{U}$. Then, for any $\varepsilon > 0$, the strongly typical set $S_\varepsilon(\boldsymbol{U}^n)$ of sample n-tuples is defined as follows:

$$S_\varepsilon(\boldsymbol{U}^n) = \left\{ \boldsymbol{u}^n : p_j(1-\varepsilon) < \frac{N_j(\boldsymbol{u}^n)}{n} < p_j(1+\varepsilon); \quad \text{for all } j \in \mathcal{U} \right\}, \tag{8.78}$$

where $N_j(\boldsymbol{u}^n)$ is the number of appearances of letter j in the n-tuple $\boldsymbol{u}^n$. The double inequality in (8.78) will be abbreviated as $N_j(\boldsymbol{u}^n) \in np_j(1 \pm \varepsilon)$, so (8.78) becomes

$$S_\varepsilon(\boldsymbol{U}^n) = \{\boldsymbol{u}^n : N_j(\boldsymbol{u}^n) \in np_j(1 \pm \varepsilon); \quad \text{for all } j \in \mathcal{U}\}. \tag{8.79}$$

Thus, the strongly typical set is the set of n-tuples for which each letter appears with approximately the right relative frequency. For any given ε, the law of large numbers says that $\lim_{n\to\infty} \Pr(N_j(\boldsymbol{U}^n) \in p_j(1 \pm \varepsilon)) = 1$ for each j. Thus (see Exercise 2.28),

$$\lim_{n\to\infty} \Pr(\boldsymbol{U}^n \in S_\varepsilon(\boldsymbol{U}^n)) = 1. \tag{8.80}$$

Next consider the probability of n-tuples in $S_\varepsilon(\boldsymbol{U}^n)$. Note that $p_{\boldsymbol{U}^n}(\boldsymbol{u}^n) = \prod_j p_j^{N_j(\boldsymbol{u}^n)}$. Taking the log of this, we see that

$$\begin{aligned} \log p_{\boldsymbol{U}^n}(\boldsymbol{u}^n) &= \sum_j N_j(\boldsymbol{u}^n) \log p_j \\ &\in \sum_j p_j(1 \pm \varepsilon) \log p_j; \\ \log p_{\boldsymbol{U}^n}(\boldsymbol{u}^n) &\in -n\mathsf{H}(U)(1 \pm \varepsilon), \qquad \text{for } \boldsymbol{u}^n \in S_\varepsilon(\boldsymbol{U}^n). \end{aligned} \tag{8.81}$$

Thus the strongly typical set has the same basic properties as the typical set defined in Chapter 2. Because of the requirement that each letter has a typical number of appearances, however, it has additional properties that are useful in the coding theorem that follows.

Consider an n-tuple of channel input/output pairs, $\boldsymbol{X}^n\boldsymbol{Y}^n = (X_1Y_1), (X_2Y_2), \ldots, (X_nY_n)$, where successive pairs are iid. For each pair XY, let X have the pmf $\{p_i; i \in \mathcal{X}\}$, which achieves capacity in (8.73). Let the pair XY have the pmf $\{p_iP_{i,j}; i \in \mathcal{X}, j \in \mathcal{Y}\}$, where $P_{i,j}$ is the channel transition probability from input i to output j. This is the joint pmf for the randomly chosen codeword that is transmitted and the corresponding received sequence.

The strongly typical set $S_\varepsilon(\boldsymbol{X}^n\boldsymbol{Y}^n)$ is then given by (8.79) as follows:

$$S_\varepsilon(\boldsymbol{X}^n\boldsymbol{Y}^n) = \{\boldsymbol{x}^n\boldsymbol{y}^n : N_{ij}(\boldsymbol{x}^n\boldsymbol{y}^n) \in n\, p_iP_{i,j}(1 \pm \varepsilon); \quad \text{for all } i \in \mathcal{X}, j \in \mathcal{Y}\}, \tag{8.82}$$

where $N_{ij}(\boldsymbol{x}^n\boldsymbol{y}^n)$ is the number of xy pairs in $((x_1y_1), (x_2y_2), \ldots, (x_ny_n))$ for which $x = i$ and $y = j$. Using the same argument as in (8.80), the transmitted codeword $\boldsymbol{X}^n$ and the received n-tuple $\boldsymbol{Y}^n$ jointly satisfy

$$\lim_{n\to\infty} \Pr[(\boldsymbol{X}^n\boldsymbol{Y}^n) \in S_\varepsilon(\boldsymbol{X}^n\boldsymbol{Y}^n)] = 1. \tag{8.83}$$

Applying the same argument as in (8.81) to the pair $x^n y^n$, we obtain

$$\log p_{X^nY^n}(x^n y^n) \in -n\mathsf{H}(XY)(1 \pm \varepsilon), \qquad \text{for } (x^n y^n) \in S_\varepsilon(X^n Y^n). \tag{8.84}$$

The nice feature about strong typicality is that if $x^n y^n$ is in the set $S_\varepsilon(X^n Y^n)$ for a given pair $x^n y^n$, then the given x^n must be in $S_\varepsilon(X^n)$ and the given Y must be in $S_\varepsilon(Y^n)$. To see this, assume that $(x^n y^n) \in S_\varepsilon(X^n Y^n)$. Then, by definition, $N_{ij}(x^n y^n) \in np_i P_{ij}(1 \pm \varepsilon)$ for all i, j. Thus,

$$\begin{aligned} N_i(x^n) &= \sum_j N_{ij}(x^n y^n) \\ &\in \sum_j np_i P_{ij}(1 \pm \varepsilon) = np_i(1 \pm \varepsilon), \qquad \text{for all } i. \end{aligned}$$

Thus $x^n \in S_\varepsilon(X^n)$. By the same argument, $y^n \in S_\varepsilon(Y^n)$.

Theorem 8.7.2 *Consider a DMC with capacity C and let R be any fixed rate $R < C$. Then, for any $\delta > 0$ and all sufficiently large block lengths n, there exist block codes with $M \geq 2^{nR}$ equiprobable codewords such that the ML error probability satisfies* $\Pr(e) \leq \delta$.

Proof As suggested in the preceding discussion, we consider the error probability averaged over the random selection of codes defined above, where, for given block length n and rate R, the number of codewords will be $M = \lceil 2^{nR} \rceil$. Since at least one code must be as good as the average, the theorem can be proved by showing that $\Pr(e) \leq \delta$.

The decoding rule to be used will be different from maximum likelihood, but since the ML rule is optimum for equiprobable messages, proving that $\Pr(e) \leq \delta$ for any decoding rule will prove the theorem. The rule to be used is strong typicality. That is, the decoder, after observing the received sequence y^n, chooses a codeword x_m^n for which $x_m^n y^n$ is jointly typical, i.e. for which $x_m^n y^n \in S_\varepsilon(X^n Y^n)$ for some ε to be determined later. An error is said to be made if either $x_m^n \notin S_\varepsilon(X^n Y^n)$ for the message m actually transmitted or if $x_{m'}^n y^n \in S_\varepsilon(X^n Y^n)$ for some $m' \neq m$. The probability of error given message m is then upperbounded by two terms: $\Pr[X^n Y^n \notin S_\varepsilon(X^n Y^n)]$, where $X^n Y^n$ is the transmitted/received pair, and the probability that some other codeword is jointly typical with Y^n. The other codewords are independent of Y and each are chosen with iid symbols using the same pmf as the transmitted codeword. Let $\overline{X}^n$ be any one of these codewords. Using the union bound,

$$\Pr(e) \leq \Pr[(X^n Y^n) \notin S_\varepsilon(X^n Y^n)] + (M-1)\Pr\left[(\overline{X}^n Y^n) \in S_\varepsilon(X^n Y^n)\right]. \tag{8.85}$$

For any large enough n, (8.83) shows that the first term is at most $\delta/2$. Also $M - 1 \leq 2^{nR}$. Thus

$$\Pr(e) \leq \frac{\delta}{2} + 2^{nR}\Pr\left[(\overline{X}^n Y^n) \in S_\varepsilon(X^n Y^n)\right]. \tag{8.86}$$

To analyze (8.86), define $F(\boldsymbol{y}^n)$ as the set of input sequences $\boldsymbol{x}^n$ that are jointly typical with any given $\boldsymbol{y}^n$. This set is empty if $\boldsymbol{y}^n \notin S_\varepsilon(\boldsymbol{Y}^n)$. Note that, for $\boldsymbol{y}^n \in S_\varepsilon(\boldsymbol{Y}^n)$,

$$p_{\boldsymbol{Y}^n}(\boldsymbol{y}^n) \geq \sum_{\boldsymbol{x}^n \in F(\boldsymbol{y}^n)} p_{\boldsymbol{X}^n\boldsymbol{Y}^n}(\boldsymbol{x}^n\boldsymbol{y}^n) \geq \sum_{\boldsymbol{x}^n \in F(\boldsymbol{y}^n)} 2^{-n\mathsf{H}(XY)(1+\varepsilon)},$$

where the final inequality comes from (8.84). Since $p_{\boldsymbol{Y}^n}(\boldsymbol{y}^n) \leq 2^{-n\mathsf{H}(Y)(1-\varepsilon)}$ for $\boldsymbol{y}^n \in S_\varepsilon(\boldsymbol{Y}^n)$, the conclusion is that the number of n-tuples in $F(\boldsymbol{y}^n)$ for any typical $\boldsymbol{y}^n$ satisfies

$$|F(\boldsymbol{y}^n)| \leq 2^{n[\mathsf{H}(XY)(1+\varepsilon)-\mathsf{H}(Y)(1-\varepsilon)]}. \tag{8.87}$$

This means that the probability that $\overline{\boldsymbol{X}}^n$ lies in $F(\boldsymbol{y}^n)$ is at most the size $|F(\boldsymbol{y}^n)|$ times the maximum probability of a typical $\overline{\boldsymbol{X}}^n$ (recall that $\overline{\boldsymbol{X}}^n$ is independent of $\boldsymbol{Y}^n$ but has the same marginal distribution as $\boldsymbol{X}^n$). Thus,

$$\Pr\left[(\overline{\boldsymbol{X}}^n\boldsymbol{Y}^n) \in S_\varepsilon(\boldsymbol{X}^n\boldsymbol{Y}^n)\right] \leq 2^{-n[\mathsf{H}(X)(1-\varepsilon)+\mathsf{H}(Y)(1-\varepsilon)-\mathsf{H}(XY)(1+\varepsilon)]}$$
$$= 2^{-n\{C-\varepsilon[\mathsf{H}(X)+\mathsf{H}(Y)+\mathsf{H}(XY)]\}},$$

where we have used the fact that $C = \mathsf{H}(X) - \mathsf{H}(X|Y) = \mathsf{H}(X) + \mathsf{H}(Y) - \mathsf{H}(XY)$. Substituting this into (8.86) yields

$$\Pr(e) \leq \frac{\delta}{2} + 2^{n(R-C+\varepsilon\alpha)},$$

where $\alpha = \mathsf{H}(X) + \mathsf{H}(Y) + \mathsf{H}(XY)$. Finally, choosing $\varepsilon = (C-R)/(2\alpha)$,

$$\Pr(e) \leq \frac{\delta}{2} + 2^{-n(C-R)/2} \leq \delta$$

for sufficiently large n. □

The above proof is essentially the original proof given by Shannon, with a little added explanation of details. It will be instructive to explain the essence of the proof without any of the epsilons or deltas. The transmitted and received n-tuple pair $(\boldsymbol{X}^n\boldsymbol{Y}^n)$ is typical with high probability and the typical pairs essentially have probability $2^{-n\mathsf{H}(XY)}$ (including both the random choice of $\boldsymbol{X}^n$ and the random noise). Each typical output $\boldsymbol{y}^n$ essentially has a marginal probability $2^{-n\mathsf{H}(Y)}$. For each typical $\boldsymbol{y}^n$, there are essentially $2^{n\mathsf{H}(X|Y)}$ input n-tuples that are jointly typical with $\boldsymbol{y}^n$ (this is the nub of the proof). An error occurs if any of these are selected to be codewords (other than the actual transmitted codeword). Since there are about $2^{n\mathsf{H}(X)}$ typical input n-tuples altogether, a fraction $2^{-n\mathsf{I}[X;Y]} = 2^{-nC}$ of them are jointly typical with the given received $\boldsymbol{y}^n$.

More recent proofs of the noisy-channel coding theorem also provide much better upperbounds on error probability. These bounds are exponentially decreasing with n, with a rate of decrease that typically becomes vanishingly small as $R \to C$.

The error probability in the theorem is the block error probability averaged over the codewords. This clearly upperbounds the error probability per transmitted binary digit. The theorem can also be easily modified to apply uniformly to each codeword in the code. One simply starts with twice as many codewords as required and eliminates the ones with greatest error probability. The ε and δ in the theorem can still be made arbitrarily small. Usually encoders contain a scrambling operation between input and output to provide privacy for the user, so a uniform bound on error probability is usually unimportant.

8.7.5 The noisy-channel coding theorem for WGN

The coding theorem for DMCs can be easily extended to discrete-time channels with arbitrary real or complex input and output alphabets, but doing this with mathematical generality and precision is difficult with our present tools.

This extension is carried out for the discrete-time Gaussian channel, which will make clear the conditions under which this generalization is easy. Let X_k and Y_k be the input and output to the channel at time k, and assume that $Y_k = X_k + Z_k$, where $Z_k \sim \mathcal{N}(0, N_0/2)$ is independent of X_k and independent of the signal and noise at all other times. Assume the input is constrained in second moment to $\mathsf{E}[X_k^2] \leq E$, so $\mathsf{E}[Y^2] \leq E + N_0/2$.

From Exercise 3.8, the differential entropy of Y is then upperbounded by

$$\mathsf{h}(Y) \leq \frac{1}{2} \log[2\pi \mathrm{e}(E + N_0/2)]. \tag{8.88}$$

This is satisfied with equality if Y is $\mathcal{N}(0, E + N_0/2)$, and thus if X is $\mathcal{N}(0, E)$. For any given input x, $\mathsf{h}(Y|X = x) = (1/2)\log(2\pi \mathrm{e} N_0/2)$, so averaging over the input space yields

$$\mathsf{h}(Y|X) = \frac{1}{2} \log(2\pi \mathrm{e} N_0/2). \tag{8.89}$$

By analogy with the DMC case, let the capacity C (in bits per channel use) be defined as the maximum of $\mathsf{h}(Y) - \mathsf{h}(Y|X)$ subject to the second moment constraint E. Thus, combining (8.88) and (8.89), we have

$$C = \frac{1}{2} \log\left(1 + \frac{2E}{N_0}\right). \tag{8.90}$$

Theorem 8.7.2 applies quite simply to this case. For any given rate R in bits per channel use such that $R < C$, one can quantize the channel input and output space finely enough such that the corresponding discrete capacity is arbitrarily close to C and in particular larger than R. Then Theorem 8.7.2 applies, so rates arbitrarily close to C can be transmitted with arbitrarily high reliability. The converse to the coding theorem can be extended in a similar way.

For a discrete-time WGN channel using $2W$ degrees of freedom per second and a power constraint P, the second moment constraint on each degree of freedom[13]

[13] We were careless in not specifying whether the constraint must be satisfied for each degree of freedom or overall as a time average. It is not hard to show, however, that the mutual information is maximized when the same energy is used in each degree of freedom.

becomes $E = P/2W$ and the capacity C_t in bits per second becomes Shannon's famous formula:

$$C_t = W \log\left(1 + \frac{P}{WN_0}\right). \tag{8.91}$$

This is then the capacity of a WGN channel with input power constrained to P and degrees of freedom per second constrained to $2W$.

With some careful interpretation, this is also the capacity of a continuous-time channel constrained in bandwidth to W and in power to P. The problem here is that if the input is strictly constrained in bandwidth, no information at all can be transmitted. That is, if a single bit is introduced into the channel at time 0, the difference in the waveform generated by symbol 1 and that generated by symbol 0 must be zero before time 0, and thus, by the Paley–Wiener theorem, cannot be nonzero and strictly bandlimited. From an engineering perspective, this does not seem to make sense, but the waveforms used in all engineering systems have negligible but nonzero energy outside the nominal bandwidth.

Thus, to use (8.91) for a bandlimited input, it is necessary to start with the constraint that, for any given $\eta > 0$, at least a fraction $(1-\eta)$ of the energy must lie within a bandwidth W. Then reliable communication is possible at all rates R_t in bits per second less than C_t as given in (8.91). Since this is true for all $\eta > 0$, no matter how small, it makes sense to call this the capacity of the bandlimited WGN channel. This is not an issue in the design of a communication system, since filters must be used, and it is widely recognized that they cannot be entirely bandlimited.

8.8 Convolutional codes

The theory of coding, and particularly of coding theorems, concentrates on block codes, but convolutional codes are also widely used and have essentially no block structure. These codes can be used whether bandwidth is highly constrained or not. We give an example below where there are two output bits for each input bit. Such a code is said to have rate $1/2$ (in input bits per channel bit). More generally, such codes produce an m-tuple of output bits for each b-tuple of input bits for arbitrary integers $0 < b < m$. These codes are said to have rate b/m.

A convolutional code looks very much like a discrete filter. Instead of having a single input and output stream, however, we have b input streams and m output streams. For the example of a convolutional code in Figure 8.8, the number of input streams is $b = 1$ and the number of output streams is $m = 2$, thus producing two output bits per input bit. There is another difference between a convolutional code and a discrete filter; the inputs and outputs for a convolutional code are binary and the addition is modulo 2.

As indicated in Figure 8.8, the outputs for this convolutional code are given by

$$\begin{aligned} U_{k,1} &= D_k \oplus D_{k-1} \oplus D_{k-2}, \\ U_{k,2} &= D_k \qquad\quad\;\; \oplus D_{k-2}. \end{aligned}$$

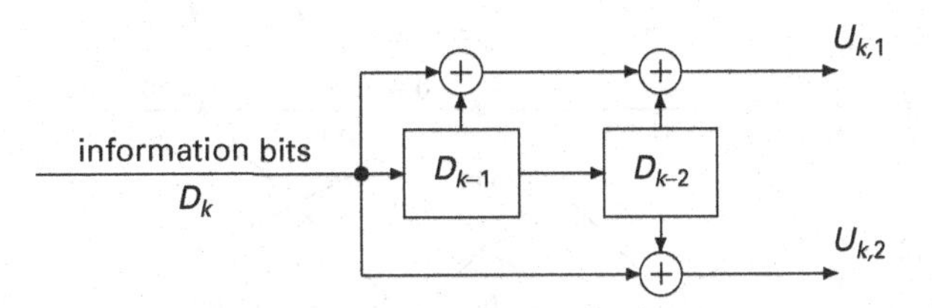

Figure 8.8. Example of a convolutional code.

Thus, each of the two output streams are linear modulo 2 convolutions of the input stream. This encoded pair of binary streams can now be mapped into a pair of signal streams such as antipodal signals $\pm a$. This pair of signal streams can then be interleaved and modulated by a single stream of Nyquist pulses at twice the rate. This baseband waveform can then be modulated to passband and transmitted.

The structure of this code can be most easily visualized by a "trellis" diagram as illustrated in Figure 8.9.

To understand this trellis diagram, note from Figure 8.8 that the encoder is characterized at any epoch k by the previous binary digits, D_{k-1} and D_{k-2}. Thus the encoder has four possible states, corresponding to the four possible values of the pair D_{k-1}, D_{k-2}. Given any of these four states, the encoder output and the next state depend only on the current binary input. Figure 8.9 shows these four states arranged vertically and shows time horizontally. We assume the encoder starts at epoch 0 with $D_{-1} = D_{-2} = 0$.

In the convolutional code of Figure 8.8 and 8.9, the output at epoch k depends on the current input and the previous two inputs. In this case, the *constraint length* of the code is 2. In general, the output could depend on the input and the previous n inputs, and the constraint length is then defined to be n. If the constraint length is n (and a single binary digit enters the encoder at each epoch k), then there are 2^n possible states, and the trellis diagram contains 2^n rather than 4 nodes at each time instant.

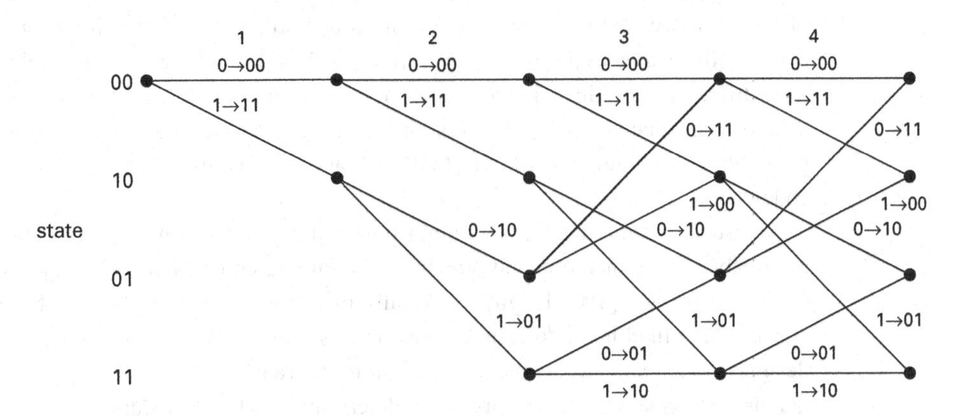

Figure 8.9. Trellis diagram; each transition is labeled with the input and corresponding output.

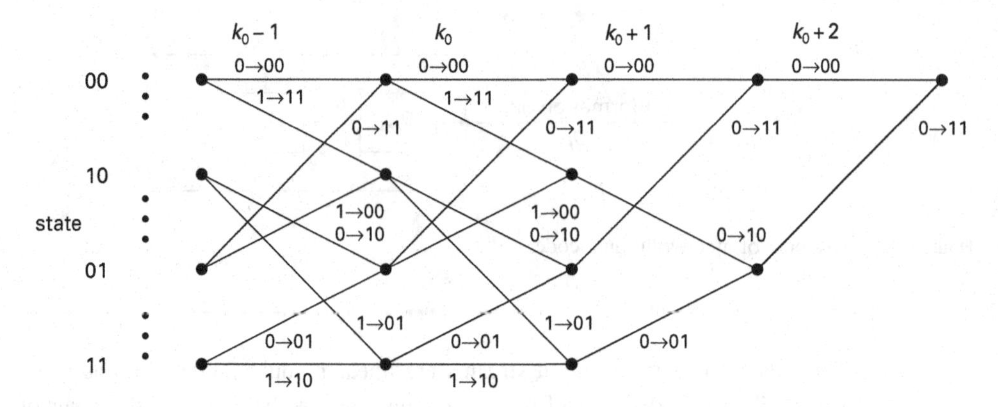

Figure 8.10. Trellis termination.

As we have described convolutional codes above, the encoding starts at time 1 and then continues forever. In practice, because of packetization of data and various other reasons, the encoding usually comes to an end after some large number, say k_0, of binary digits have been encoded. After D_{k_0} enters the encoder, two final 0s enter the encoder, at epochs k_0+1 and k_0+2, and four final encoded digits come out of the encoder. This restores the state of the encoder to state 0, which, as we see later, is very useful for decoding. For the more general case with a constraint length of n, we need n final 0s to restore the encoder to state 0. Altogether, k_0 inputs lead to $2(k_0+n)$ outputs, for a code rate of $k_0/[2(k_0+n)]$. This is referred to as a terminated rate 1/2 code. Figure 8.10 shows the part of the trellis diagram corresponding to this termination.

8.8.1 Decoding of convolutional codes

Decoding a convolutional code is essentially the same as using detection theory to choose between each pair of codewords, and then choosing the best overall (the same as done for the orthogonal code). There is one slight conceptual difference in that, in principle, the encoding continues forever. When the code is terminated, however, this problem does not exist, and in principle one takes the maximum likelihood (ML) choice of all the (finite length) possible codewords.

As usual, assume that the incoming binary digits are iid and equiprobable. This is reasonable if the incoming bit stream has been source encoded. This means that the codewords of any given length are equally likely, which then justifies ML decoding.

Maximum likelihood detection is also used so that codes for error correction can be designed independently of the source data to be transmitted.

Another issue, given iid inputs, is determining what is meant by probability of error. In all of the examples discussed so far, given a received sequence of symbols, we have attempted to choose the codeword that minimizes the probability of error for the entire codeword. An alternative would have been to minimize the probability of error individually for each binary information digit. It turns out to be easier to minimize the sequence error probability than the bit error probability. This, in fact, is what happens when we use ML detection between codewords, as suggested above.

In decoding for error correction, the objective is almost invariably to minimize the sequence probability of error. Along with the convenience suggested here, a major reason is that a binary input is usually a source-coded version of some other source sequence or waveform, and thus a single output error is often as serious as multiple errors within a codeword. Note that ML detection on sequences is assumed in what follows.

8.8.2 The Viterbi algorithm

The Viterbi algorithm is an algorithm for performing ML detection for convolutional codes. Assume for the time being that the code is terminated as in Figure 8.10. It will soon be seen that whether or not the code is terminated is irrelevant. The algorithm will now be explained for the convolutional code in Figure 8.8 and for the assumption of WGN; the extension to arbitrary convolutional codes will be obvious except for the notational complexity of the general case. For any given input $d_1, \dots, d_{k_0}$, let the encoded sequence be $u_{1,1}, u_{1,2}, u_{2,1}, u_{2,2}, \dots, u_{k_0+2,2}$, and let the channel output, after modulation, addition of WGN, and demodulation, be $v_{1,1}, v_{1,2}, v_{2,1}, v_{2,2}, \dots, v_{k_0+2,2}$.

There are 2^{k_0} possible codewords, corresponding to the 2^{k_0} possible binary k_0-tuples $d_1, \dots, d_{k_0}$, so a naive approach to decoding would be to compare the likelihood of each of these codewords. For large k_0, even with today's technology, such an approach would be prohibitive. It turns out, however, that by using the trellis structure of Figure 8.9, this decoding effort can be greatly simplified.

Each input $d_1, \dots, d_{k_0}$ (i.e. each codeword) corresponds to a particular path through the trellis from epoch 1 to k_0+2, and each path, at each epoch k, corresponds to a particular trellis state.

Consider two paths $d_1, \dots, d_{k_0}$ and $d'_1, \dots, d'_{k_0}$ through the trellis that pass through the same state at time k^+ (i.e. at the time immediately after the input and state change at epoch k) and remain together thereafter. Thus, $d_{k+1}, \dots, d_{k_0} = d'_{k+1}, \dots, d'_{k_0}$. For example, from Figure 8.8, we see that both $(0, \dots, 0)$ and $(1, 0, \dots, 0)$ are in state 00 at 3^+ and both remain in the same state thereafter. Since the two paths are in the same state at k^+ and have the same inputs after this time, they both have the same encoder outputs after this time. Thus $u_{k+1,i}, \dots, u_{k_0+2,i} = u'_{k+1,i}, \dots, u'_{k_0+2,i}$ for $i = 1, 2$.

Since each channel output rv $V_{k,i}$ is given by $V_{k,i} = U_{k,i} + Z_{k,i}$ and the Gaussian noise variables $Z_{k,i}$ are independent, this means that, for any channel output $v_{1,1}, \dots, v_{k_0+2,2}$,

$$\frac{f(v_{1,1}, \dots, v_{k_0+2,2} | d_1, \dots, d_{k_0})}{f(v_{1,1}, \dots, v_{k_0+2,2} | d'_1, \dots, d'_{k_0})} = \frac{f(v_{1,1}, \dots, v_{k,2} | d_1, \dots, d_{k_0})}{f(v_{1,1}, \dots, v_{k,2} | d'_1, \dots, d'_{k_0})}.$$

In plain English, this says that if two paths merge at time k^+ and then stay together, the likelihood ratio depends on only the first k output pairs. Thus if the right side exceeds 1, then the path $d_1, \dots, d_{k_0}$ is more likely than the path $d'_1, \dots, d'_{k_0}$. This conclusion holds no matter how the final inputs $d_{k+1}, \dots, d_{k_0}$ are chosen.

We then see that when two paths merge at a node, no matter what the remainder of the path is, the most likely of the paths is the one that is most likely at the point of the merger. Thus, whenever two paths merge, the least likely of the paths can be eliminated at that point. Doing this elimination successively from the smallest k for which paths merge (3 in our example), there is only one survivor for each state at each epoch.

To be specific, let $h(d_1, \dots, d_k)$ be the state at time k^+ with input $d_1, \dots, d_k$. For our example, $h(d_1, \dots, d_k) = (d_{k-1}, d_k)$. Let

$$f_{\max}(k, s) = \max_{h(d_1,\dots,d_k)=s} f(v_{1,1}, \dots, v_{k,2} | d_1, \dots, d_k).$$

These quantities can then be calculated iteratively for each state and each time k by the following iteration:

$$f_{\max}(k+1, s) = \max_{r:r\to s} f_{\max}(k, r) \cdot f(v_{k,1}|u_1(r\to s)) f(v_{k,2}|u_2(r\to s)), \tag{8.92}$$

where the maximization is over the set of states r that have a transition to state s in the trellis and $u_1(r\to s)$ and $u_2(r\to s)$ are the two outputs from the encoder corresponding to a transition from r to s.

This expression is simplified (for WGN) by taking the log, which is proportional to the negative squared distance between $\boldsymbol{v}$ and $\boldsymbol{u}$. For the antipodal signal case in the example, this may be further simplified by simply taking the dot product between $\boldsymbol{v}$ and $\boldsymbol{u}$. Letting $L(k, s)$ be this dot product,

$$L(k+1, s) = \max_{r:r\to s} L(k, r) + v_{k,1}u_1(r\to s) + v_{k,2}u_2(r\to s). \tag{8.93}$$

What this means is that at each epoch $(k+1)$, it is necessary to calculate the inner product in (8.93) for each link in the trellis going from k to $k+1$. These must be maximized over r for each state s at epoch $(k+1)$. The maximum must then be saved as $L(k+1, s)$ for each s. One must, of course, also save the paths taken in arriving at each merging point.

Those familiar with dynamic programming will recognize this recursive algoriothm as an example of the dynamic programming principle.

The complexity of the entire computation for decoding a block of k_0 information bits is proportional to $4(k_0+2)$. In the more general case, where the constraint length of the convolutional coder is n rather than 2, there are 2^n states and the computational complexity is proportional to $2^n(k_0+n)$. The Viterbi algorithm is usually used in cases where the constraint length is moderate, say 6–12, and in these situations the computation is quite moderate, especially compared with 2^{k_0}.

Usually one does not wait until the end of the block to start decoding. When the above computation is performed at epoch k, all the paths up to k' have merged for k' a few constraint lengths less than k. In this case, one can decode without any bound on k_0, and the error probability is viewed in terms of "error events" rather than block error.

8.9 Summary of detection, coding, and decoding

This chapter analyzed the last major segment of a general point-to-point communication system in the presence of noise, namely how to detect the input signals from the noisy version presented at the output. Initially the emphasis was on detection alone; i.e., the assumption was that the rest of the system had been designed and the only question remaining was how to extract the signals.

At a very general level, the problem of detection in this context is trivial. That is, under the assumption that the statistics of the input and the noise are known, the sensible rule is maximum a-posteriori probability decoding: find the a-posteriori probability of all the hypotheses and choose the largest. This is somewhat complicated by questions of whether to carry out sequence detection or bit detection, but these questions are details in a sense.

At a more specific level, however, the detection problem led to many interesting insights and simplifications, particularly for WGN channels. A particularly important simplification is the principle of irrelevance, which says that components of the received waveform in degrees of freedom not occupied by the signal of interest (or statistically related signals) can be ignored in detection of those signals. Looked at in another way, this says that matched filters could be used to extract the degrees of freedom of interest.

The last part of the chapter discussed coding and decoding. The focus changed here to the question of how coding can change the input waveforms so as to make the decoding more effective. In other words, a MAP detector can be designed for any signal structure, but the real problem is to design both the signal structure and detection for effective performance.

At this point, the noisy-channel coding theorem comes into the picture. If $R < C$, then the probability of error can be reduced arbitrarily, meaning that finding the optimal code at a given constraint length is slightly artificial. What is needed is a good trade-off between error probability and the delay and complexity caused by longer constraint lengths.

Thus the problem is not only to overcome the noise, but also to do this with reasonable delay and complexity. Chapter 9 considers some of these problems in the context of wireless communication.

8.10 Appendix: Neyman–Pearson threshold tests

We have seen in the preceding sections that any binary MAP test can be formulated as a comparison of a likelihood ratio with a threshold. It turns out that many other detection rules can also be viewed as threshold tests on likelihood ratios. One of the most important binary detection problems for which a threshold test turns out to be essentially optimum is the *Neyman–Pearson test*. This is often used in those situations in which there is no sensible way to choose a-priori probabilities. In the Neyman–Pearson test, an acceptable value α is established for $\Pr\{e|U=1\}$, and, subject to

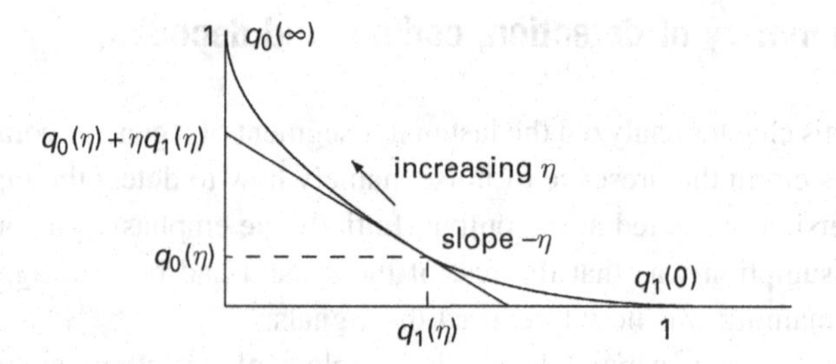

Figure 8.11. The error curve; $q_1(\eta)$ and $q_0(\eta)$ are plotted as parametric functions of η.

the constraint $\Pr\{e|U=1\} \le \alpha$, the Neyman–Pearson test minimizes $\Pr\{e|U=0\}$. We shall show in what follows that such a test is essentially a threshold test. Before demonstrating this, we need some terminology and definitions.

Define $q_0(\eta)$ to be $\Pr\{e|U=0\}$ for a threshold test with threshold η, $0<\eta<\infty$, and similarly define $q_1(\eta)$ as $\Pr\{e|U=1\}$. Thus, for $0<\eta<\infty$,

$$q_0(\eta)=\Pr\{\Lambda(V){<}\eta|U=0\}; \qquad q_1(\eta)=\Pr\{\Lambda(V){\ge}\eta|U=1\}. \tag{8.94}$$

Define $q_0(0)$ as $\lim_{\eta\to 0} q_0(\eta)$ and $q_1(0)$ as $\lim_{\eta\to 0} q_1(\eta)$. Clearly, $q_0(0)=0$, and in typical situations $q_1(0)=1$. More generally, $q_1(0)=\Pr\{\Lambda(V){>}0|U=1\}$. In other words, $q_1(0)<1$ if there is some set of observations that are impossible under $U=0$ but have positive probability under $U=1$. Similarly, define $q_0(\infty)$ as $\lim_{\eta\to\infty} q_0(\eta)$ and $q_1(\infty)$ as $\lim_{\eta\to\infty} q_1(\eta)$. We have $q_0(\infty)=\Pr\{\Lambda(V)<\infty\}$ and $q_1(\infty)=0$.

Finally, for an arbitrary test A, threshold or not, denote $\Pr\{e\,|\,U=0\}$ as $q_0(A)$ and $\Pr\{e|U=1\}$ as $q_1(A)$.

Using (8.94), we can plot $q_0(\eta)$ and $q_1(\eta)$ as parametric functions of η; we call this the *error curve*.[14] Figure 8.11 illustrates this error curve for a typical detection problem such as (8.17) and (8.18) for antipodal binary signalling. We have already observed that, as the threshold η is increased, the set of v mapped into $\tilde{U}=0$ decreases. Thus $q_0(\eta)$ is an increasing function of η and $q_1(\eta)$ is decreasing. Thus, as η increases from 0 to ∞, the curve in Figure 8.11 moves from the lower right to the upper left.

Figure 8.11 also shows a straight line of slope $-\eta$ through the point $(q_1(\eta), q_0(\eta))$ on the error curve. The following lemma shows why this line is important.

Lemma 8.10.1 *For each η, $0<\eta<\infty$, the line of slope $-\eta$ through the point $(q_1(\eta), q_0(\eta))$ lies on or beneath all other points $(q_1(\eta'), q_0(\eta'))$ on the error curve, and also lies beneath $(q_1(A), q_0(A))$ for all tests A.*

Before proving this lemma, we give an example of the error curve for a discrete observation space.

Example 8.10.1 (Discrete observations) Figure 8.12 shows the error curve for an example in which the hypotheses 0 and 1 are again mapped $0\to +a$ and $1\to -a$.

[14] In the radar field, one often plots $1-q_0(\eta)$ as a function of $q_1(\eta)$. This is called the receiver operating characteristic (ROC). If one flips the error curve vertically around the point $1/2$, the ROC results.

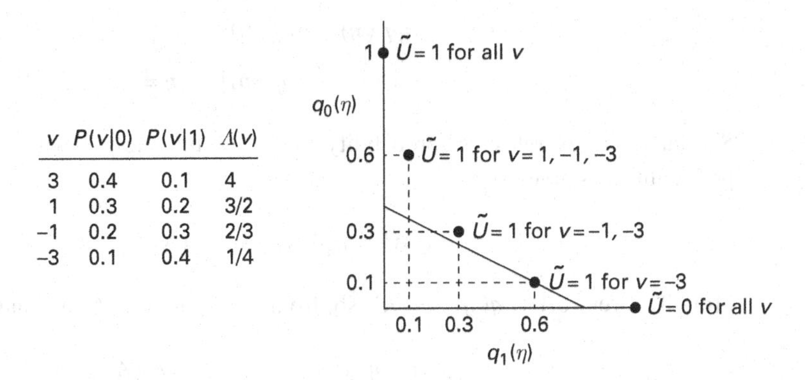

v	$P(v\|0)$	$P(v\|1)$	$\Lambda(v)$
3	0.4	0.1	4
1	0.3	0.2	3/2
−1	0.2	0.3	2/3
−3	0.1	0.4	1/4

Figure 8.12. Error curve for a discrete observation space. There are only five points making up the "curve," one corresponding to each of the five distinct threshold rules. For example, the threshold rule $\tilde{U} = 1$ only for $v = -3$ yields $(q_1(\eta), q_0(\eta)) = (0.6, 0.1)$ for all η in the range 1/4 to 2/3. A straight line of slope $-\eta$ through that point is also shown for $\eta = 1/2$. Lemma 8.10.1 asserts that this line lies on or beneath each point of the error curve and each point $(q_1(A), q_0(A)$ for any other test. Note that as η increases or decreases, this line will rotate around the point $(0.6, 0.1)$ until η becomes larger than 2/3 or smaller than 1/4; it then starts to rotate around the next point in the error curve.

Assume that the observation V can take on only four discrete values $+3, +1, -1, -3$. The probabilities of each of these values, conditional on $U = 0$ and $U = 1$, are given in Figure 8.12. As indicated, the likelihood ratio $\Lambda(v)$ then takes the values 4, 3/2, 2/3, and 1/4, corresponding, respectively, to $v = 3, 1, -1$, and -3.

A threshold test at η decides $\tilde{U} = 0$ if and only if $\Lambda(V) \geq \eta$. Thus, for example, for any $\eta \leq 1/4$, all possible values of v are mapped into $\tilde{U} = 0$. In this range, $q_1(\eta) = 1$ since $U = 1$ always causes an error. Also $q_0(\eta) = 0$ since $U = 0$ never causes an error. In the range $1/4 < \eta \leq 2/3$, since $\Lambda(-3) = 1/4$, the value -3 is mapped into $\tilde{U} = 1$ and all other values into $\tilde{U} = 0$. In this range, $q_1(\eta) = 0.6$, since, when $U = 1$, an error occurs unless $V = -3$.

In the same way, all threshold tests with $2/3 < \eta \leq 3/2$ give rise to the decision rule that maps -1 and -3 into $\tilde{U} = 1$ and 1 and 3 into $\tilde{U} = 0$. In this range, $q_1(\eta) = q_0(\eta) = 0.3$. As shown, there is another decision rule for $3/2 < \eta \leq 4$ and a final decision rule for $\eta > 4$.

The point of this example is that a finite observation space leads to an error curve that is simply a finite set of points. It is also possible for a continuously varying set of outputs to give rise to such an error curve when there are only finitely many possible likelihood ratios. Figure 8.12 illustrates what Lemma 8.10.1 means for error curves consisting only of a finite set of points.

Proof of Lemma 8.10.1 Consider the line of slope $-\eta$ through the point $(q_1(\eta), q_0(\eta))$. From plane geometry, as illustrated in Figure 8.11, we see that the vertical axis intercept of this line is $q_0(\eta) + \eta q_1(\eta)$. To interpret this line, define p_0 and p_1 as a-priori probabilities such that $\eta = p_1/p_0$. The overall error probability for the corresponding MAP test is then given by

$$q(\eta) = p_0 q_0(\eta) + p_1 q_1(\eta)$$
$$= p_0[q_0(\eta) + \eta q_1(\eta)]; \quad \eta = p_1/p_0. \tag{8.95}$$

Similarly, the overall error probability for an arbitrary test A with the same a-priori probabilities is given by

$$q(A) = p_0[q_0(A) + \eta q_1(A)]. \tag{8.96}$$

From Theorem 8.1.1, $q(\eta) \leq q(A)$, so, from (8.95) and (8.96), we have

$$q_0(\eta) + \eta\, q_1(\eta) \leq q_0(A) + \eta\, q_1(A). \tag{8.97}$$

We have seen that the left side of (8.97) is the vertical axis intercept of the line of slope $-\eta$ through $(q_1(\eta), q_0(\eta))$. Similarly, the right side is the vertical axis intercept of the line of slope $-\eta$ through $(q_1(A), q_0(A))$. This says that the point $(q_1(A), q_0(A))$ lies on or above the line of slope $-\eta$ through $(q_1(\eta), q_0(\eta))$. This applies to every test A, which includes every threshold test. □

Lemma 8.10.1 shows that if the error curve gives $q_0(\eta)$ as a differentiable function of $q_1(\eta)$ (as in the case of Figure 8.11), then the line of slope $-\eta$ through $(q_1(\eta), q_0(\eta))$ is a tangent, at point $(q_1(\eta), q_0(\eta))$, to the error curve. Thus in what follows we call this line the *η-tangent* to the error curve. Note that the error curve of Figure 8.12 is not really a curve, but rather a discrete set of points. Each η-tangent, as defined above and illustrated in the figure for $\eta = 2/3$, still lies on or beneath all of these discrete points. Each η-tangent has also been shown to lie below all achievable points $(q_1(A), q_0(A))$, for each arbitrary test A.

Since for each test A the point $(q_1(A), q_0(A))$ lies on or above each η-tangent, it also lies on or above the supremum of these η-tangents over $0 < \eta < \infty$. It also follows, then, that, for each η', $0 < \eta' < \infty$, $(q_1(\eta'), q_0(\eta'))$ lies on or above this supremum. Since $(q_1(\eta'), q_0(\eta'))$ also lies on the η'-tangent, it lies on or beneath the supremum, and thus must lie on the supremum. We conclude that each point of the error curve lies on the supremum of the η-tangents.

Although all points of the error curve lie on the supremum of the η-tangents, all points of the supremum are not necessarily points of the error curve, as seen from Figure 8.12. We shall see shortly, however, that all points on the supremum are achievable by a simple extension of threshold tests. Thus we call this supremum the *extended error curve*.

For the example in Figure 8.11, the extended error curve is the same as the error curve itself. For the discrete example in Figure 8.12, the extended error curve is shown in Figure 8.13.

To understand the discrete case better, assume that the extended error function has a straight line portion of slope $-\eta^*$ and horizontal extent γ. This implies that the distribution function of $\Lambda(V)$ given $U = 1$ has a discontinuity of magnitude γ at η^*. Thus there is a set $\mathcal{V}^*$ of one or more v with $\Lambda(v) = \eta^*$, $\Pr\{\mathcal{V}^* | U = 1\} = \gamma$, and $\Pr\{\mathcal{V}^* | U = 0\} = \eta^*\gamma$. For a MAP test with threshold η^*, the overall error probability

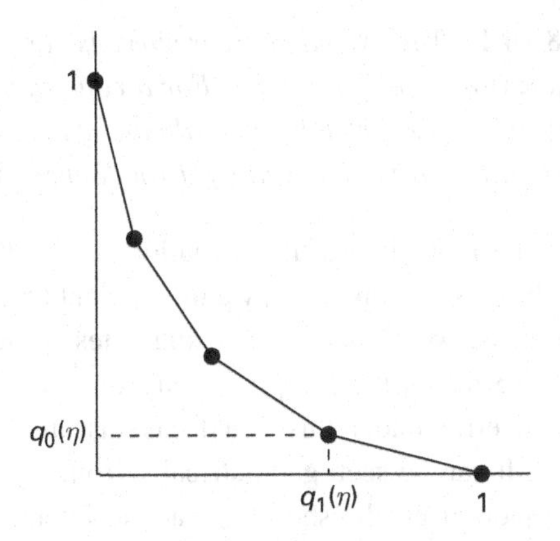

Figure 8.13. Extended error curve for the discrete observation example of Figure 8.12. From Lemma 8.10.1, for each slope $-\eta$, the η-tangent touches the error curve. Thus, the line joining two adjacent points on the error curve must be an η-tangent for its particular slope, and therefore must lie on the extended error curve.

is not affected by whether $v \in \mathcal{V}^*$ is detected as $\tilde{U}=0$ or $\tilde{U}=1$. Our convention is to detect $v \in \mathcal{V}^*$ as $\tilde{U}=0$, which corresponds to the lower right point on the straight line portion of the extended error curve. The opposite convention, detecting $v \in \mathcal{V}^*$ as $\tilde{U}=1$ reduces the error probability given $U=1$ by γ and increases the error probability given $U=0$ by $\eta^*\gamma$, i.e. it corresponds to the upper left point on the straight line portion of the extended error curve.

Note that when we were interested in MAP detection, it made no difference how $v \in \mathcal{V}^*$ was detected for the threshold η^*. For the Neyman–Pearson test, however, it makes a great deal of difference since $q_0(\eta^*)$ and $q_1(\eta^*)$ are changed. In fact, we can achieve any point on the straight line in question by detecting $v \in \mathcal{V}^*$ randomly, increasing the probability of choosing $\tilde{U}=0$ to approach the lower right endpoint. In other words, the extended error curve is the curve relating q_1 to q_0 using a randomized threshold test. For a given η^*, of course, only those $v \in \mathcal{V}^*$ are detected randomly.

To summarize, the Neyman–Pearson test is a randomized threshold test. For a constraint α on $\Pr\{e|U=1\}$, we choose the point α on the abscissa of the extended error curve and achieve the corresponding ordinate as the minimum $\Pr\{e|U=1\}$. If that point on the extended error curve lies within a straight line segment of slope η^*, a randomized test is used for those observations with likelihood ratio η^*.

Since the extended error curve is a supremum of straight lines, it is a convex function. Since these straight lines all have negative slope, it is a monotonic decreasing[15] function. Thus, Figures 8.11 and 8.13 represent the general behavior of extended error curves, with the slight possible exception mentioned above that the endpoints need not have one of the error probabilities equal to 1.

The following theorem summarizes the results obtained for Neyman–Pearson tests.

15 To be more precise, it is strictly decreasing between the endpoints $(q_1(\infty), q_0(\infty))$ and $(q_1(0), q_0(0))$.

Theorem 8.10.1 *The extended error curve is convex and strictly decreasing between* $(q_1(\infty), q_0(\infty))$ *and* $(q_1(0), q_0(0))$. *For a constraint* α *on* $\Pr\{e|U=1\}$, *the minimum value of* $\Pr\{e|U=0\}$ *is given by the ordinate of the extended error curve corresponding to the abscissa* α *and is achieved by a randomized threshold test.*

There is one more interesting variation on the theme of threshold tests. If the a-priori probabilities are unknown, we might want to minimize the maximum probability of error. That is, we visualize choosing a test followed by nature choosing a-priori probabilities to maximize the probability of error. Our objective is to minimize the probability of error under this worst case assumption. The resulting test is called a minmax test. It can be seen geometrically from Figures 8.11 or 8.13 that the minmax test is the randomized threshold test at the intersection of the extended error curve with a 45° line from the origin.

If there is symmetry between $U=0$ and $U=1$ (as in the Gaussian case), then the extended error curve will be symmetric around the 45° degree line, and the threshold will be at $\eta=1$ (i.e. the ML test is also the minmax test). This is an important result for Gaussian communication problems, since it says that ML detection, i.e. minimum distance detection, is robust in the sense of not depending on the input probabilities. If we know the a-priori probabilities, we can do better than the ML test, but we can do no worse.

8.11 Exercises

8.1 (Binary minimum cost detection)

(a) Consider a binary hypothesis testing problem with a-priori probabilities p_0, p_1 and likelihoods $f_{V|U}(v|i)$, $i=0,1$. Let C_{ij} be the cost of deciding on hypothesis j when i is correct. Conditional on an observation $V=v$, find the expected cost (over $U=0,1$) of making the decision $\tilde{U}=j$ for $j=0,1$. Show that the decision of minimum expected cost is given by

$$\tilde{U}_{\text{mincost}} = \arg\min_j \left[C_{0j} p_{U|V}(0|v) + C_{1j} p_{U|V}(1|v) \right].$$

(b) Show that the min cost decision above can be expressed as the following threshold test:

$$\Lambda(v) = \frac{f_{V|U}(v|0)}{f_{V|U}(v|1)} \begin{array}{c} \geq^{\tilde{U}=0} \\ <_{\tilde{U}=1} \end{array} \frac{p_1(C_{10}-C_{11})}{p_0(C_{01}-C_{00})} = \eta.$$

(c) Interpret the result above as saying that the only difference between a MAP test and a minimum cost test is an adjustment of the threshold to take account of the costs; i.e., a large cost of an error of one type is equivalent to having a large a-priori probability for that hypothesis.

8.2 Consider the following two equiprobable hypotheses:

$$U = 0 : V_1 = a\cos\Theta + Z_1, \quad V_2 = a\sin\Theta + Z_2;$$
$$U = 1 : V_1 = -a\cos\Theta + Z_1, \quad V_2 = -a\sin\Theta + Z_2.$$

Assume that Z_1 and Z_2 are iid $\mathcal{N}(0, \sigma^2)$ and that Θ takes on the values $\{-\pi/4, 0, \pi/4\}$, each with probability 1/3. Find the ML decision rule when V_1, V_2 are observed. [Hint. Sketch the possible values of V_1, V_2 for $\mathbf{Z} = 0$ given each hypothesis. Then, without doing any calculations, try to come up with a good intuitive decision rule. Then try to verify that it is optimal.]

8.3 Let

$$V_j = S_j X_j + Z_j, \qquad \text{for } 1 \le j \le 4,$$

where $\{X_j; 1 \le j \le 4\}$ are iid $\mathcal{N}(0, 1)$ and $\{Z_j; 1 \le j \le 4\}$ are iid $\mathcal{N}(0, \sigma^2)$ and independent of $\{X_j; 1 \le j \le 4\}$. Assume that $\{V_j; 1 \le j \le 4\}$ are observed at the output of a communication system and the input is a single binary random variable U which is independent of $\{Z_j; 1 \le j \le 4\}$ and $\{X_j; 1 \le j \le 4\}$. Assume that $S_1, \dots, S_4$ are functions of U, with $S_1 = S_2 = U \oplus 1$ and $S_3 = S_4 = U$.

(a) Find the log likelihood ratio

$$\mathrm{LLR}(\boldsymbol{v}) = \ln\left(\frac{f_{V|U}(\boldsymbol{v}|0)}{f_{V|U}(\boldsymbol{x}^n|1)}\right).$$

(b) Let $\mathcal{E}_a = |V_1|^2 + |V_2|^2$ and $\mathcal{E}_b = |V_3|^2 + |V_4|^2$. Explain why $\{\mathcal{E}_a, \mathcal{E}_b\}$ form a sufficient statistic for this problem and express the log likelihood ratio in terms of the sample values of $\{\mathcal{E}_a, \mathcal{E}_b\}$.

(c) Find the threshold for ML detection.

(d) Find the probability of error. [Hint. Review Exercise 6.1.] Note: we will later see that this corresponds to binary detection in Rayleigh fading.

8.4 Consider binary antipodal MAP detection for the real vector case. Modify the picture and argument in Figure 8.4 to verify the algebraic relation between the squared energy difference and the inner product in (8.22).

8.5 Derive (8.37), i.e. that $\sum_{k,j} y_{k,j} b_{k,j} = (1/2) \int y(t)b(t)\mathrm{d}t$. Explain the factor of 1/2.

8.6 In this problem, you will derive the inequalities

$$\left(1 - \frac{1}{x^2}\right)\frac{1}{x\sqrt{2\pi}}\mathrm{e}^{-x^2/2} \le Q(x) \le \frac{1}{x\sqrt{2\pi}}\mathrm{e}^{-x^2/2}, \qquad \text{for } x > 0, \tag{8.98}$$

where $Q(x) = (2\pi)^{-1/2}\int_x^\infty \exp(-z^2/2)\mathrm{d}z$ is the "tail" of the Normal distribution. The purpose of this is to show that, when x is large, the right side of this inequality is a very tight upperbound on $Q(x)$.

(a) By using a simple change of variable, show that

$$Q(x) = \frac{1}{\sqrt{2\pi}}\mathrm{e}^{-x^2/2}\int_0^\infty \exp\left(-y^2/2 - xy\right)\mathrm{d}y.$$

(b) Show that

$$1 - y^2/2 \le \exp(-y^2/2) \le 1.$$

(c) Use parts (a) and (b) to establish (8.98).

8.7 (Other bounds on $Q(x)$)

(a) Show that the following bound holds for any γ and η such that $0 \le \gamma$ and $0 \le \eta$:

$$Q(\gamma+\eta) \le Q(\gamma)\exp[-\eta\gamma - \eta^2/2].$$

[Hint. Start with $Q(\gamma+\eta) = (1/\sqrt{2\pi})\int_{\gamma+\eta}\exp[-x^2/2]\mathrm{d}x$ and use the change of variable $y = x - \eta$.]

(b) Use part (a) to show that, for all $\eta \ge 0$,

$$Q(\eta) \le \frac{1}{2}\exp[-\eta^2/2].$$

(c) Use part (a) to show that, for all $0 \le \gamma \le w$,

$$\frac{Q(w)}{\exp[-w^2/2]} \le \frac{Q(\gamma)}{\exp[-\gamma^2/2]}.$$

Note: equation (8.98) shows that $Q(w)$ goes to 0 with increasing w as a slowly varying coefficient times $\exp[-w^2/2]$. This demonstrates that the coefficient is decreasing for $w \ge 0$.

8.8 (Orthogonal signal sets) An *orthogonal signal set* is a set $\mathcal{A} = \{\boldsymbol{a}_m,\ 0 \le m \le M-1\}$ of M orthogonal vectors in $\mathbb{R}^M$ with equal energy E; i.e., $\langle \boldsymbol{a}_m, \boldsymbol{a}_j \rangle = E\delta_{mj}$.

(a) Compute the spectral efficiency ρ of $\mathcal{A}$ in bits per two dimensions. Compute the average energy E_b per information bit.

(b) Compute the minimum squared distance $d^2_{\min}(\mathcal{A})$ between these signal points. Show that every signal has $M-1$ nearest neighbors.

(c) Let the noise variance be $N_0/2$ per dimension. Describe a ML detector on this set of M signals. [Hint. Represent the signal set in an orthonormal expansion where each vector is collinear with one coordinate. Then visualize making binary decisions between each pair of possible signals.]

8.9 (Orthogonal signal sets; continuation of Exercise 8.8) Consider a set $\mathcal{A} = \{\boldsymbol{a}_m, 0 \le m \le M-1\}$ of M orthogonal vectors in $\mathbb{R}^M$ with equal energy E.

(a) Use the union bound to show that $\Pr(e)$, using ML detection, is bounded by

$$\Pr(e) \le (M-1)Q(\sqrt{E/N_0}).$$

(b) Let $M \to \infty$ with $E_b = E/\log M$ held constant. Using the upperbound for $Q(x)$ in Exercise 8.7(b), show that if $E_b/N_0 > 2\ln 2$, then $\lim_{M\to\infty}\Pr(e) = 0$. How close is this to the ultimate Shannon limit on E_b/N_0? What is the limit of the spectral efficiency ρ?

8.10 (Lowerbound to $\Pr(e)$ for orthogonal signals)

(a) Recall the exact expression for error probability for orthogonal signals in WGN from (8.49):

$$\Pr(e) = \int_{-\infty}^{\infty} f_{W_0|A}(w_0|\boldsymbol{a}_0) \Pr\left(\bigcup_{m=1}^{M-1} (W_m \geq w_0 | A = \boldsymbol{a}_0)\right) dw_0.$$

Explain why the events $W_m \geq w_0$ for $1 \leq m \leq M-1$ are iid conditional on $A = \boldsymbol{a}_0$ and $W_0 = w_0$.

(b) Demonstrate the following two relations for any w_0:

$$\Pr\left(\bigcup_{m=1}^{M-1} (W_m \geq w_0 | A = \boldsymbol{a}_0)\right) = 1 - [1 - Q(w_0)]^{M-1}$$
$$\geq (M-1)Q(w_0) - \frac{[(M-1)Q(w_0)]^2}{2}.$$

(c) Define γ_1 by $(M-1)Q(\gamma_1) = 1$. Demonstrate the following:

$$\Pr\left(\bigcup_{m=1}^{M-1} (W_m \geq w_0 | A = \boldsymbol{a}_0)\right) \geq \begin{cases} \dfrac{(M-1)Q(w_0)}{2} & \text{for} \quad w_0 > \gamma_1; \\ \dfrac{1}{2} & \text{for} \quad w_0 \leq \gamma_1. \end{cases}$$

(d) Show that

$$\Pr(e) \geq \frac{1}{2} Q(\alpha - \gamma_1).$$

(e) Show that $\lim_{M\to\infty} \gamma_1/\gamma = 1$, where $\gamma = \sqrt{2 \ln M}$. Use this to compare the lowerbound in part (d) to the upperbounds for cases (1) and (2) in Section 8.5.3. In particular, show that $\Pr(e) \geq 1/4$ for $\gamma_1 > \alpha$ (the case where capacity is exceeded).

(f) Derive a tighter lowerbound on $\Pr(e)$ than part (d) for the case where $\gamma_1 \leq \alpha$. Show that the ratio of the log of your lowerbound and the log of the upperbound in Section 8.5.3 approaches 1 as $M \to \infty$. Note: this is much messier than the bounds above.

8.11 Section 8.3.4 discusses detection for binary complex vectors in WGN by viewing complex n-dimensional vectors as $2n$-dimensional real vectors. Here you will treat the vectors directly as n-dimensional complex vectors. Let $\boldsymbol{Z} = (Z_1, \ldots, Z_n)^{\mathsf{T}}$ be a vector of complex iid Gaussian rvs with iid real and imaginary parts, each $\mathcal{N}(0, N_0/2)$. The input $\boldsymbol{U}$ is binary antipodal, taking on values $\boldsymbol{a}$ or $-\boldsymbol{a}$. The observation $\boldsymbol{V}$ is $\boldsymbol{U}+\boldsymbol{Z}$.

(a) The probability density of $\boldsymbol{Z}$ is given by

$$f_{\boldsymbol{Z}}(\boldsymbol{z}) = \frac{1}{(\pi N_0)^n} \exp \sum_{j=1}^{n} \frac{-|z_j|^2}{N_0} = \frac{1}{(\pi N_0)^n} \exp \frac{-\|\boldsymbol{z}\|^2}{N_0}.$$

Explain what this probability density represents (i.e. probability per unit what?)

(b) Give expressions for $f_{V|U}(v|a)$ and $f_{V|U}(v|-a)$.

(c) Show that the log likelihood ratio for the observation v is given by

$$\mathrm{LLR}(v) = \frac{-\|v-a\|^2 + \|v+a\|^2}{N_0}.$$

(d) Explain why this implies that ML detection is minimum distance detection (defining the distance between two complex vectors as the norm of their difference).

(e) Show that $\mathrm{LLR}(v)$ can also be written as $4\Re(\langle v, a\rangle)/N_0$.

(f) The appearance of the real part, $\Re(\langle v, a\rangle)$, in part (e) is surprising. Point out why log likelihood ratios must be real. Also explain why replacing $\Re(\langle v, a\rangle)$ by $|\langle v, a\rangle|$ in the above expression would give a nonsensical result in the ML test.

(g) Does the set of points $\{v : \mathrm{LLR}(v) = 0\}$ form a complex vector space?

8.12 Let D be the function that maps vectors in $\mathcal{C}^n$ into vectors in $\mathcal{R}^{2n}$ by the mapping

$$a = (a_1, a_2, \ldots, a_n) \to (\Re a_1, \Re a_2, \ldots, \Re a_n, \Im a_1, \Im a_2, \ldots, \Im a_n) = D(a).$$

(a) Explain why $a \in \mathcal{C}^n$ and $\mathrm{i}a$ ($\mathrm{i} = \sqrt{-1}$) are contained in the 1D complex subspace of $\mathcal{C}^n$ spanned by a.

(b) Show that $D(a)$ and $D(\mathrm{i}a)$ are orthogonal vectors in $\mathcal{R}^{2n}$.

(c) For $v, a \in \mathcal{C}^n$, the projection of v on a is given by $v_{|a} = (\langle v, a\rangle/\|a\|)(a/\|a\|)$. Show that $D(v_{|a})$ is the projection of $D(v)$ onto the subspace of $\mathcal{R}^{2n}$ spanned by $D(a)$ and $D(\mathrm{i}a)$.

(d) Show that $D((\Re(\langle v, a\rangle)/\|a\|)(a/\|a\|))$ is the further projection of $D(v)$ onto $D(a)$.

8.13 Consider 4-QAM with the four signal points $u = \pm a \pm \mathrm{i}a$. Assume Gaussian noise with spectral density $N_0/2$ per dimension.

(a) Sketch the signal set and the ML decision regions for the received complex sample value y. Find the exact probability of error (in terms of the Q function) for this signal set using ML detection.

(b) Consider 4-QAM as two 2-PAM systems in parallel. That is, a ML decision is made on $\Re(u)$ from $\Re(v)$ and a decision is made on $\Im(u)$ from $\Im(v)$. Find the error probability (in terms of the Q function) for the ML decision on $\Re(u)$ and similarly for the decision on $\Im(u)$.

(c) Explain the difference between what has been called an error in part (a) and what has been called an error in part (b).

(d) Derive the QAM error probability directly from the PAM error probability.

8.14 Consider two 4-QAM systems with the same 4-QAM constellation:

$$s_0 = 1+\mathrm{i}, \quad s_1 = -1+\mathrm{i}, \quad s_2 = -1-\mathrm{i}, \quad s_3 = 1-\mathrm{i}.$$

For each system, a pair of bits is mapped into a signal, but the two mappings are different:

$$\text{mapping 1:} \quad 00 \to s_0, \quad 01 \to s_1, \quad 10 \to s_2, \quad 11 \to s_3;$$

$$\text{mapping 2:} \quad 00 \to s_0, \quad 01 \to s_1, \quad 11 \to s_2, \quad 10 \to s_3.$$

The bits are independent, and 0s and 1s are equiprobable, so the constellation points are equally likely in both systems. Suppose the signals are decoded by the minimum distance decoding rule and the signal is then mapped back into the two binary digits. Find the error probability (in terms of the Q function) for each bit in each of the two systems.

8.15 Re-state Theorem 8.4.1 for the case of MAP detection. Assume that the inputs $U_1, \ldots, U_n$ are independent and each have the a-priori distribution $p_0, \ldots, p_{M-1}$. [Hint. Start with (8.43) and (8.44), which are still valid here.]

8.16 The following problem relates to a digital modulation scheme called minimum shift keying (MSK). Let

$$s_0(t) = \begin{cases} \sqrt{\frac{2E}{T}} \cos(2\pi f_0 t) & \text{if } 0 \le t \le T; \\ 0 & \text{otherwise,} \end{cases}$$

and

$$s_1(t) = \begin{cases} \sqrt{\frac{2E}{T}} \cos(2\pi f_1 t) & \text{if } 0 \le t \le T; \\ 0 & \text{otherwise.} \end{cases}$$

(a) Compute the energy of the signals $s_0(t)$, $s_1(t)$. You may assume that $f_0 T \gg 1$ and $f_1 T \gg 1$.

(b) Find conditions on the frequencies f_0, f_1 and the duration T to ensure both that the signals $s_0(t)$ and $s_1(t)$ are orthogonal and that $s_0(0) = s_0(T) = s_1(0) = s_1(T)$. Why do you think a system with these parameters is called minimum shift keying?

(c) Assume that the parameters are chosen as in part (b). Suppose that, under $U = 0$, the signal $s_0(t)$ is transmitted and, under $U = 1$, the signal $s_1(t)$ is transmitted. Assume that the hypotheses are equally likely. Let the observed signal be equal to the sum of the transmitted signal and a white Gaussian process with spectral density $N_0/2$. Find the optimal detector to minimize the probability of error. Draw a block diagram of a possible implementation.

(d) Compute the probability of error of the detector you have found in part (c).

8.17 Consider binary communication to a receiver containing k_0 antennas. The transmitted signal is $\pm a$. Each antenna has its own demodulator, and the received signal after demodulation at antenna k, $1 \le k \le k_0$, is given by

$$V_k = U g_k + Z_k,$$

where U is $+a$ for $U=0$ and $-a$ for $U=1$. Also, g_k is the gain of antenna k and $Z_k \sim \mathcal{N}(0, \sigma^2)$ is the noise at antenna k; everything is real and $U, Z_1, Z_2, \ldots, Z_{k_0}$ are independent. In vector notation, $\boldsymbol{V} = U\boldsymbol{g}+\boldsymbol{Z}$, where $\boldsymbol{V} = (v_1, \ldots, v_{k_0})^{\mathsf{T}}$, etc.

(a) Suppose that the signal at each receiving antenna k is weighted by an arbitrary real number q_k and the signals are combined as $Y = \sum_k V_k q_k = \langle \boldsymbol{V}, \boldsymbol{q} \rangle$. What is the ML detector for U given the observation Y?
(b) What is the probability of error, $\Pr(e)$, for this detector?
(c) Let $\beta = \langle \boldsymbol{g}, \boldsymbol{q} \rangle / \|\boldsymbol{g}\| \|\boldsymbol{q}\|$. Express $\Pr(e)$ in a form where $\boldsymbol{q}$ does not appear except for its effect on β.
(d) Give an intuitive explanation why changing $\boldsymbol{q}$ to $c\boldsymbol{q}$ for some nonzero scalar c does not change $\Pr(e)$.
(e) Minimize $\Pr(e)$ over all choices of $\boldsymbol{q}$. [Hint. Use part (c).]
(f) Is it possible to reduce $\Pr(e)$ further by doing ML detection on $V_1, \ldots, V_{k_0}$ rather than restricting ourselves to a linear combination of those variables?
(g) Redo part (b) under the assumption that the noise variables have different variances, i.e. $Z_k \sim \mathcal{N}(0, \sigma_k^2)$. As before, $U, Z_1, \ldots, Z_{k_0}$ are independent.
(h) Minimize $\Pr(e)$ in part (g) over all choices of $\boldsymbol{q}$.

8.18 (a) The Hadamard matrix H_1 has the rows 00 and 01. Viewed as binary codewords, this is rather foolish since the first binary digit is always 0 and thus carries no information at all. Map the symbols 0 and 1 into the signals a and $-a$, respectively, $a > 0$, and plot these two signals on a 2D plane. Explain the purpose of the first bit in terms of generating orthogonal signals.
(b) Assume that the mod-2 sum of each pair of rows of H_b is another row of H_b for any given integer $b \geq 1$. Use this to prove the same result for H_{b+1}. [Hint. Look separately at the mod-2 sum of two rows in the first half of the rows, two rows in the second half, and two rows in different halves.]

8.19 (RM codes)

(a) Verify the following combinatorial identity for $0 < r < m$:

$$\sum_{j=0}^{r} \binom{m}{j} = \sum_{j=0}^{r-1} \binom{m-1}{j} + \sum_{j=0}^{r} \binom{m-1}{j}.$$

[Hint. Note that the first term above is the number of binary m-tuples with r or fewer 1s. Consider separately the number of these that end in 1 and end in 0.]
(b) Use induction on m to show that $k(r, m) = \sum_{j=0}^{r} \binom{m}{j}$. Be careful how you handle $r=0$ and $r=m$.

8.20 (RM codes) This exercise first shows that $\mathrm{RM}(r, m) \subset \mathrm{RM}(r+1, m)$ for $0 \leq r < m$. It then shows that $d_{\min}(r, m) = 2^{m-r}$.

(a) Show that if $\mathrm{RM}(r-1, m-1) \subset \mathrm{RM}(r, m-1)$ for all r, $0 < r < m$, then

$$\mathrm{RM}(r-1, m) \subset \mathrm{RM}(r, m) \qquad \text{for all } r,\ 0 < r \leq m.$$

Note: be careful about $r=1$ and $r=m$.

(b) Let $\boldsymbol{x} = (\boldsymbol{u}, \boldsymbol{u} \oplus \boldsymbol{v})$, where $\boldsymbol{u} \in \mathrm{RM}(r, m-1)$ and $\boldsymbol{v} \in \mathrm{RM}(r-1, m-1)$. Assume that $d_{\min}(r, m-1) \leq 2^{m-1-r}$ and $d_{\min}(r-1, m-1) \leq 2^{m-r}$. Show that if $\boldsymbol{x}$ is nonzero, it has at least 2^{m-r} 1s. [Hint (1). For a linear code, $d_{\min}$ is equal to the weight (number of 1s) in the minimum-weight nonzero codeword.] [Hint (2). First consider the case $\boldsymbol{v} = 0$, then the case $\boldsymbol{u} = 0$. Finally use part (a) in considering the case $\boldsymbol{u} \neq 0$, $\boldsymbol{v} \neq 0$, under the subcases $\boldsymbol{u} = \boldsymbol{v}$ and $\boldsymbol{u} \neq \boldsymbol{v}$.]

(c) Use induction on m to show that $d_{\min} = 2^{m-r}$ for $0 \leq r \leq m$.

9 Wireless digital communication

9.1 Introduction

This chapter provides a brief treatment of wireless digital communication systems. More extensive treatments are found in many texts, particularly Tse and Viswanath (2005) and Goldsmith (2005). As the name suggests, wireless systems operate via transmission through space rather than through a wired connection. This has the advantage of allowing users to make and receive calls almost anywhere, including while in motion. Wireless communication is sometimes called mobile communication, since many of the new technical issues arise from motion of the transmitter or receiver.

There are two major new problems to be addressed in wireless that do not arise with wires. The first is that the communication channel often varies with time. The second is that there is often interference between multiple users. In previous chapters, modulation and coding techniques have been viewed as ways to combat the noise on communication channels. In wireless systems, these techniques must also combat time-variation and interference. This will cause major changes both in the modeling of the channel and the type of modulation and coding.

Wireless communication, despite the hype of the popular press, is a field that has been around for over 100 years, starting around 1897 with Marconi's successful demonstrations of wireless telegraphy. By 1901, radio reception across the Atlantic Ocean had been established, illustrating that rapid progress in technology has also been around for quite a while. In the intervening decades, many types of wireless systems have flourished, and often later disappeared. For example, television transmission, in its early days, was broadcast by wireless radio transmitters, which is increasingly being replaced by cable or satellite transmission. Similarly, the point-to-point microwave circuits that formerly constituted the backbone of the telephone network are being replaced by optical fiber. In the first example, wireless technology became outdated when a wired distribution network was installed; in the second, a new wired technology (optical fiber) replaced the older wireless technology. The opposite type of example is occurring today in telephony, where cellular telephony is partially replacing wireline telephony, particularly in parts of the world where the wired network is not well developed. The point of these examples is that there are many situations in which there is a choice between wireless and wire technologies, and the choice often changes when new technologies become available.

Cellular networks will be emphasized in this chapter, both because they are of great current interest and also because they involve a relatively simple architecture within which most of the physical layer communication aspects of wireless systems can be studied. A cellular network consists of a large number of wireless subscribers with cellular telephones (cell phones) that can be used in cars, buildings, streets, etc. There are also a number of fixed base stations arranged to provide wireless electromagnetic communication with arbitrarily located cell phones.

The area covered by a base station, i.e. the area from which incoming calls can reach that base station, is called a cell. One often pictures a cell as a hexagonal region with the base station in the middle. One then pictures a city or region as being broken up into a hexagonal lattice of cells (see Figure 9.1(a)). In reality, the base stations are placed somewhat irregularly, depending on the location of places such as building tops or hill tops that have good communication coverage and that can be leased or bought (see Figure 9.1(b)). Similarly, the base station used by a particular cell phone is selected more on the basis of communication quality than of geographic distance.

Each cell phone, when it makes a call, is connected (via its antenna and electromagnetic radiation) to the base station with the best apparent communication path. The base stations in a given area are connected to a *mobile telephone switching office* (MTSO) by high-speed wire, fiber, or microwave connections. The MTSO is connected to the public wired telephone network. Thus an incoming call from a cell phone is first connected to a base station and from there to the MTSO and then to the wired network. From there the call goes to its destination, which might be another cell phone, or an ordinary wire line telephone, or a computer connection. Thus, we see that a cellular network is not an independent network, but rather an appendage to the wired network. The MTSO also plays a major role in coordinating which base station will handle a call to or from a cell phone and when to hand-off a cell phone conversation from one base station to another.

When another telephone (either wired or wireless) places a call to a given cell phone, the reverse process takes place. First the cell phone is located and an MTSO and nearby base station are selected. Then the call is set up through the MTSO and base station. The wireless link from a base station to a cell phone is called the

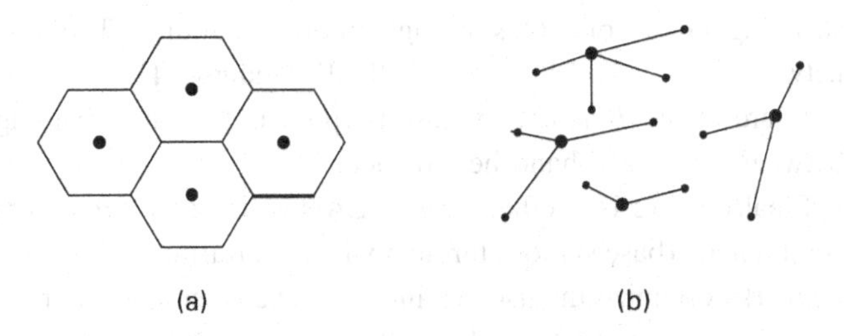

Figure 9.1. Cells and base stations for a cellular network. (a) Oversimplified view in which each cell is hexagonal. (b) More realistic case in which base stations are irregularly placed and cell phones choose the best base station.

downlink (or forward) channel, and the link from a cell phone to a base station is called the uplink (or reverse) channel. There are usually many cell phones connected to a single base station. Thus, for downlink communication, the base station multiplexes the signals intended for the various connected cell phones and broadcasts the resulting single waveform, from which each cell phone can extract its own signal. This set of downlink channels from a base station to multiple cell phones is called a *broadcast channel*. For the uplink channels, each cell phone connected to a given base station transmits its own waveform, and the base station receives the sum of the waveforms from the various cell phones plus noise. The base station must then separate and detect the signals from each cell phone and pass the resulting binary streams to the MTSO. This set of uplink channels to a given base station is called a *multiaccess channel*.

Early cellular systems were analog. They operated by directly modulating a voice waveform on a carrier and transmitting it. Different cell phones in the same cell were assigned different carrier frequencies, and adjacent cells used different sets of frequencies. Cells sufficiently far away from each other could reuse the same set of frequencies with little danger of interference.

All of the newer cellular systems are digital (i.e. use a binary interface), and thus, in principle, can be used for voice or data. Since these cellular systems, and their standards, originally focused on telephony, the current data rates and delays in cellular systems are essentially determined by voice requirements. At present, these systems are still mostly used for telephony, but both the capability to send data and the applications for data are rapidly increasing. Also the capabilities to transmit data at higher rates than telephony rates are rapidly being added to cellular systems.

As mentioned above, there are many kinds of wireless systems other than cellular. First there are the broadcast systems, such as AM radio, FM radio, TV, and paging systems. All of these are similar to the broadcast part of cellular networks, although the data rates, the size of the areas covered by each broadcasting node, and the frequency ranges are very different.

In addition, there are wireless LANs (local area networks). These are designed for much higher data rates than cellular systems, but otherwise are somewhat similar to a single cell of a cellular system. These are designed to connect PCs, shared peripheral devices, large computers, etc. within an office building or similar local environment. There is little mobility expected in such systems, and their major function is to avoid stringing a maze of cables through an office building. The principal standards for such networks are the 802.11 family of IEEE standards. There is a similar even smaller-scale standard called *Bluetooth* whose purpose is to reduce cabling and simplify transfers between office and hand-held devices.

Finally, there is another type of LAN called an *ad hoc network*. Here, instead of a central node (base station) through which all traffic flows, the nodes are all alike. These networks organize themselves into links between various pairs of nodes and develop routing tables using these links. The network layer issues of routing, protocols, and shared control are of primary concern for ad hoc networks; this is somewhat disjoint from our focus here on physical-layer communication issues.

One of the most important questions for all of these wireless systems is that of standardization. Some types of standardization are mandated by the Federal Communication Commission (FCC) in the USA and corresponding agencies in other countries. This has limited the available bandwidth for conventional cellular communication to three frequency bands, one around 0.9 GHz, another around 1.9 GHz, and the other around 5.8 GHz. Other kinds of standardization are important since users want to use their cell phones over national and international areas. There are three well established, mutually incompatible, major types of digital cellular systems. One is the GSM system,[1] which was standardized in Europe and is now used worldwide; another is a TDM (time division modulation) standard developed in the USA; and a third is CDMA (code division multiple access). All of these are evolving and many newer systems with a dizzying array of new features are constantly being introduced. Many cell phones can switch between multiple modes as a partial solution to these incompatibility issues.

This chapter will focus primarily on CDMA, partly because so many newer systems are using this approach, and partly because it provides an excellent medium for discussing communication principles. GSM and TDM will be discussed briefly, but the issues of standardization are so centered on nontechnological issues and so rapidly changing that they will not be discussed further.

In thinking about wireless LANs and cellular telephony, an obvious question is whether they will some day be combined into one network. The use of data rates compatible with voice rates already exists in the cellular network, and the possibility of much higher data rates already exists in wireless LANs, so the question is whether very high data rates are commercially desirable for standardized cellular networks. The wireless medium is a much more difficult medium for communication than the wired network. The spectrum available for cellular systems is quite limited, the interference level is quite high, and rapid growth is increasing the level of interference. Adding higher data rates will exacerbate this interference problem. In addition, the display on hand-held devices is small, limiting the amount of data that can be presented, which suggests that many applications of such devices do not need very high data rates. Thus it is questionable whether very high-speed data for cellular networks is necessary or desirable in the near future. On the other hand, there is intense competition between cellular providers, and each strives to distinguish their service by new features requiring increased data rates.

Subsequent sections in this chapter introduce the study of the technological aspects of wireless channels, focusing primarily on cellular systems. Section 9.2 looks briefly at the electromagnetic properties that propagate signals from transmitter to receiver. Section 9.3 then converts these detailed electromagnetic models into simpler input/output descriptions of the channel. These input/output models can be characterized most simply as linear time-varying filter models.

[1] GSM stands for groupe speciale mobile or global systems for mobile communication, but the acronym is far better known and just as meaningful as the words.

The input/output model views the input, the channel properties, and the output at passband. Section 9.4 then finds the baseband equivalent for this passband view of the channel. It turns out that the channel can then be modeled as a complex baseband linear time-varying filter. Finally, in Section 9.5, this deterministic baseband model is replaced by a stochastic model.

The remainder of the chapter then introduces various issues of communication over such a stochastic baseband channel. Along with modulation and detection in the presence of noise, we also discuss channel measurement, coding, and diversity. The chapter ends with a brief case study of the CDMA cellular standard, IS95.

9.2 Physical modeling for wireless channels

Wireless channels operate via electromagnetic radiation from transmitter to receiver. In principle, one could solve Maxwell's equations for the given transmitted signal to find the electromagnetic field at the receiving antenna. This would have to account for the reflections from nearby buildings, vehicles, and bodies of land and water. Objects in the line of sight between transmitter and receiver would also have to be accounted for.

The wavelength $\Lambda(f)$ of electromagnetic radiation at any given frequency f is given by $\Lambda = c/f$, where $c = 3 \times 10^8$ m/s is the velocity of light. The wavelength in the bands allocated for cellular communication thus lies between 0.05 and 0.3 m. To calculate the electromagnetic field at a receiver, the locations of the receiver and the obstructions would have to be known within submeter accuracies. The electromagnetic field equations therefore appear to be unreasonable to solve, especially on the fly for moving users. Thus, electromagnetism cannot be used to characterize wireless channels in detail, but it will provide understanding about the underlying nature of these channels.

One important question is where to place base stations, and what range of power levels are then necessary on the downlinks and uplinks. To a great extent, this question must be answered experimentally, but it certainly helps to have a sense of what types of phenomena to expect. Another major question is what types of modulation techniques and detection techniques look promising. Here again, a sense of what types of phenomena to expect is important, but the information will be used in a different way. Since cell phones must operate under a wide variety of different conditions, it will make sense to view these conditions probabilistically. Before developing such a stochastic model for channel behavior, however, we first explore the gross characteristics of wireless channels by looking at several highly idealized models.

9.2.1 Free space, fixed transmitting and receiving antennas

First, consider a fixed antenna radiating into free space. In the far field,[2] the electric field and magnetic field at any given location $\boldsymbol{d}$ are perpendicular both to each other

[2] The far field is the field many wavelengths away from the antenna, and (9.1) is the limiting form as this number of wavelengths increases. It is a safe assumption that cellular receivers are in the far field.

and to the direction of propagation from the antenna. They are also proportional to each other, so we focus on only the electric field (just as we normally consider only the voltage or only the current for electronic signals). The electric field at $\boldsymbol{d}$ is, in general, a vector with components in the two co-ordinate directions perpendicular to the line of propagation. If one of these two components is zero, then the electric field at $\boldsymbol{d}$ can be viewed as a real-valued function of time. For simplicity, we look only at this case. The electric waveform is usually a passband waveform modulated around a carrier, and we focus on the complex positive frequency part of the waveform. The electric far-field response at point $\boldsymbol{d}$ to a transmitted complex sinusoid, $\exp(2\pi \mathrm{i} ft)$, can then be expressed as follows:

$$E(f, t, \boldsymbol{d}) = \frac{\alpha_{\mathrm{s}}(\theta, \psi, f) \exp\{2\pi \mathrm{i} f(t - r/c)\}}{r}. \tag{9.1}$$

Here (r, θ, ψ) represents the point $\boldsymbol{d}$ in space at which the electric field is being measured; r is the distance from the transmitting antenna to $\boldsymbol{d}$; and (θ, ψ) represents the vertical and horizontal angles from the antenna to $\boldsymbol{d}$. The radiation pattern of the transmitting antenna at frequency f in the direction (θ, ψ) is denoted by the complex function $\alpha_{\mathrm{s}}(\theta, \psi, f)$. The magnitude of α_{s} includes antenna losses; the phase of α_{s} represents the phase change due to the antenna. The phase of the field also varies with fr/c, corresponding to the delay r/c caused by the radiation traveling at the speed of light c.

We are not concerned here with actually finding the radiation pattern for any given antenna, but only with recognizing that antennas have radiation patterns, and that the free-space far field depends on that pattern as well as on the $1/r$ attenuation and r/c delay.

The reason why the electric field goes down with $1/r$ in free space can be seen by looking at concentric spheres of increasing radius r around the antenna. Since free space is lossless, the total power radiated through the surface of each sphere remains constant. Since the surface area is increasing with r^2, the power radiated per unit area must go down as $1/r^2$, and thus E must go down as $1/r$. This does not imply that power is radiated uniformly in all directions – the radiation pattern is determined by the transmitting antenna. As will be seen later, this r^{-2} reduction of power with distance is sometimes invalid when there are obstructions to free space propagation.

Next, suppose there is a fixed receiving antenna at location $\boldsymbol{d} = (r, \theta, \psi)$. The received waveform at the antenna terminals (in the absence of noise) in response to $\exp(2\pi \mathrm{i} ft)$ is then given by

$$\frac{\alpha(\theta, \psi, f) \exp\{2\pi \mathrm{i} f(t - r/c)\}}{r}, \tag{9.2}$$

where $\alpha(\theta, \psi, f)$ is the product of α_{s} (the antenna pattern of the transmitting antenna) and the antenna pattern of the receiving antenna; thus the losses and phase changes of both antennas are accounted for in $\alpha(\theta, \psi, f)$. The explanation for this response is that the receiving antenna causes only local changes in the electric field, and thus alters neither the r/c delay nor the $1/r$ attenuation.

For the given input and output, a system function $\hat{h}(f)$ can be defined as

$$\hat{h}(f) = \frac{\alpha(\theta, \psi, f)\exp\{-2\pi i f r/c\}}{r}. \tag{9.3}$$

Substituting this in (9.2), the response to $\exp(2\pi i f t)$ is $\hat{h}(f)\exp\{2\pi i f t\}$.

Electromagnetic radiation has the property that the response is linear in the input. Thus the response at the receiver to a superposition of transmitted sinusoids is simply the superposition of responses to the individual sinusoids. The response to an arbitrary input $x(t) = \int \hat{x}(f)\exp\{2\pi i f t\}\mathrm{d}f$ is then given by

$$y(t) = \int_{-\infty}^{\infty} \hat{x}(f)\hat{h}(f)\exp\{2\pi i f t\}\mathrm{d}f. \tag{9.4}$$

We see from (9.4) that the Fourier transform of the output $y(t)$ is $\hat{y}(f) = \hat{x}(f)\hat{h}(f)$. From the convolution theorem, this means that

$$y(t) = \int_{-\infty}^{\infty} x(\tau)h(t-\tau)\mathrm{d}\tau, \tag{9.5}$$

where $h(t) = \int_{-\infty}^{\infty} \hat{h}(f)\exp\{2\pi i f t\}\mathrm{d}f$ is the inverse Fourier transform of $\hat{h}(f)$. Since the physical input and output must be real, $\hat{x}(f) = \hat{x}^*(-f)$ and $\hat{y}(f) = \hat{y}^*(-f)$. It is then necessary that $\hat{h}(f) = \hat{h}^*(-f)$ also.

The channel in this fixed-location free-space example is thus a conventional linear-time-invariant (LTI) system with impulse response $h(t)$ and system function $\hat{h}(f)$.

For the special case where the the combined antenna pattern $\alpha(\theta, \psi, f)$ is real and independent of frequency (at least over the frequency range of interest), we see that $\hat{h}(f)$ is a complex exponential[3] in f and thus $h(t)$ is $(\alpha/r)\delta(t - r/c)$, where δ is the Dirac delta function. From (9.5), $y(t)$ is then given by

$$y(t) = \frac{\alpha}{r}x\left(t - \frac{r}{c}\right).$$

If $\hat{h}(f)$ is other than a complex exponential, then $h(t)$ is not an impulse, and $y(t)$ becomes a nontrivial filtered version of $x(t)$ rather than simply an attenuated and delayed version. From (9.4), however, $y(t)$ only depends on $\hat{h}(f)$ over the frequency band where $\hat{x}(f)$ is nonzero. Thus it is common to model $\hat{h}(f)$ as a complex exponential (and thus $h(t)$ as a scaled and shifted Dirac delta function) whenever $\hat{h}(f)$ is a complex exponential over the frequency band of use.

We will find in what follows that linearity is a good assumption for all the wireless channels to be considered, but that time invariance does not hold when either the antennas or reflecting objects are in relative motion.

[3] More generally, $\hat{h}(f)$ is a complex exponential if $|\alpha|$ is independent of f and $\angle\alpha$ is linear in f.

9.2.2 Free space, moving antenna

Continue to assume a fixed antenna transmitting into free space, but now assume that the receiving antenna is moving with constant velocity v in the direction of increasing distance from the transmitting antenna. That is, assume that the receiving antenna is at a moving location described as $\boldsymbol{d}(t) = (r(t), \theta, \psi)$ with $r(t) = r_0 + vt$. In the absence of the receiving antenna, the electric field at the moving point $\boldsymbol{d}(t)$, in response to an input $\exp(2\pi i f t)$, is given by (9.1) as follows:

$$E(f, t, \boldsymbol{d}(t)) = \frac{\alpha_s(\theta, \psi, f) \exp\{2\pi i f(t - r_0/c - vt/c)\}}{r_0 + vt}. \tag{9.6}$$

We can rewrite $f(t - r_0/c - vt/c)$ as $f(1 - v/c)t - fr_0/c$. Thus the sinusoid at frequency f has been converted to a sinusoid of frequency $f(1 - v/c)$; there has been a *Doppler shift* of $-fv/c$ due to the motion of $\boldsymbol{d}(t)$.[4] Physically, each successive crest in the transmitted sinusoid has to travel a little further before it is observed at this moving observation point.

Placing the receiving antenna at $\boldsymbol{d}(t)$, the waveform at the terminals of the receiving antenna, in response to $\exp(2\pi i f t)$, is given by

$$\frac{\alpha(\theta, \psi, f) \exp\{2\pi i [f(1 - v/c)t - fr_0/c]\}}{r_0 + vt}, \tag{9.7}$$

where $\alpha(\theta, \psi, f)$ is the product of the transmitting and receiving antenna patterns.

This channel cannot be represented as an LTI channel since the response to a sinusoid is not a sinusoid of the same frequency. The channel is still linear, however, so it is characterized as a *linear time-varying channel*. Linear time-varying channels will be studied in Section 9.3, but, first, several simple models will be analyzed where the received electromagnetic wave also includes reflections from other objects.

9.2.3 Moving antenna, reflecting wall

Consider Figure 9.2, in which there is a fixed antenna transmitting the sinusoid $\exp(2\pi i f t)$. There is a large, perfectly reflecting, wall at distance r_0 from the transmitting antenna. A vehicle starts at the wall at time $t = 0$ and travels toward the sending antenna at velocity v. There is a receiving antenna on the vehicle whose distance from the sending antenna at time $t > 0$ is then given by $r_0 - vt$.

In the absence of the vehicle and receiving antenna, the electric field at $r_0 - vt$ is the sum of the free-space waveform and the waveform reflected from the wall. Assuming

[4] Doppler shifts of electromagnetic waves follow the same principles as Doppler shifts of sound waves. For example, when an airplane flies overhead, the noise from it appears to drop in frequency as it passes by.

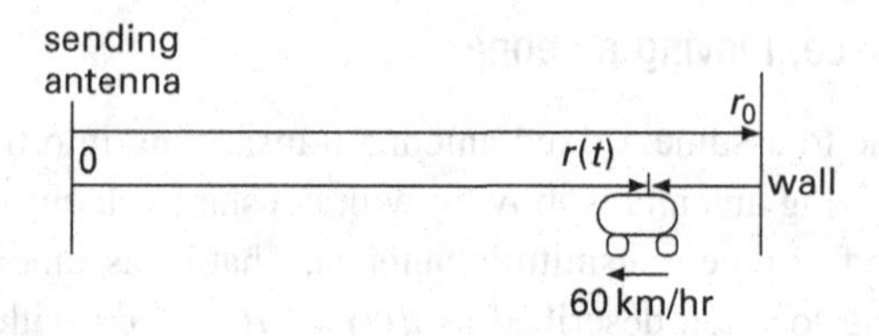

Figure 9.2. Two paths from a stationary antenna to a moving antenna. One path is direct and the other reflects off a wall.

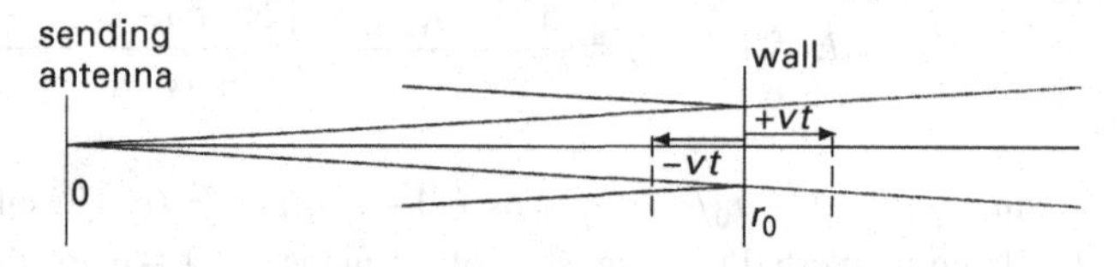

Figure 9.3. Relation of reflected wave to the direct wave in the absence of a wall.

that the wall is very large, the reflected wave at $r_0 - vt$ is the same (except for a sign change) as the free-space wave that would exist on the opposite side of the wall in the absence of the wall (see Figure 9.3). This means that the reflected wave at distance $r_0 - vt$ from the sending antenna has the intensity and delay of a free-space wave at distance $r_0 + vt$. The combined electric field at $\boldsymbol{d}(t)$ in response to the input $\exp(2\pi i f t)$ is then given by

$$E(f, t, \boldsymbol{d}(t)) = \frac{\alpha_s(\theta, \psi, f)\exp\{2\pi i f[t - (r_0 - vt)/c]\}}{r_0 - vt} - \frac{\alpha_s(\theta, \psi, f)\exp\{2\pi i f[t - (r_0 + vt)/c]\}}{r_0 + vt}. \tag{9.8}$$

Including the vehicle and its antenna, the signal at the antenna terminals, say $y(t)$, is again the electric field at the antenna as modified by the receiving antenna pattern. Assume for simplicity that this pattern is identical in the directions of the direct and the reflected wave. Letting α denote the combined antenna pattern of transmitting and receiving antenna, the received signal is then given by

$$y_f(t) = \frac{\alpha \exp\{2\pi i f[t - \frac{r_0 - vt}{c}]\}}{r_0 - vt} - \frac{\alpha \exp\{2\pi i f[t - \frac{r_0 + vt}{c}]\}}{r_0 + vt}. \tag{9.9}$$

In essence, this approximates the solution of Maxwell's equations using an approximate method called *ray tracing*. The approximation comes from assuming that the wall is infinitely large and that both fields are ideal far fields.

The first term in (9.9), the direct wave, is a sinusoid of frequency $f(1+v/c)$; its magnitude is slowly increasing in t as $1/(r_0-vt)$. The second is a sinusoid of frequency $f(1-v/c)$; its magnitude is slowly decreasing as $1/(r_0+vt)$. The combination of the two frequencies creates a beat frequency at fv/c. To see this analytically, assume initially that t is very small so the denominator of each term above can be approximated as r_0. Then, factoring out the common terms in the above exponentials, $y_f(t)$ is given by

$$y_f(t) \approx \frac{\alpha \exp\{2\pi i f[t-\frac{r_0}{c}]\}\,(\exp\{2\pi i f v t/c\} - \exp\{-2\pi i f v t/c\})}{r_0}$$
$$= \frac{2i\,\alpha \exp\{2\pi i f[t-\frac{r_0}{c}]\}\ \sin\{2\pi f v t/c\}}{r_0}. \tag{9.10}$$

This is the product of two sinusoids, one at the input frequency f, which is typically on the order of gigahertz, and the other at the Doppler shift fv/c, which is typically 500 Hz or less.

As an example, if the antenna is moving at $v = 60$ km/hr and if $f = 900$ MHz, this beat frequency is $fv/c = 50$ Hz. The sinusoid at f has about 1.8×10^7 cycles for each cycle of the beat frequency. Thus $y_f(t)$ looks like a sinusoid at frequency f whose amplitude is sinusoidally varying with a period of 20 ms. The amplitude goes from its maximum positive value to 0 in about 5 ms. Viewed another way, the response alternates between being unfaded for about 5 ms and then faded for about 5 ms. This is called *multipath fading*. Note that in (9.9) the response is viewed as the sum of two sinusoids, each of different frequency, while in (9.10) the response is viewed as a single sinusoid of the original frequency with a time-varying amplitude. These are just two different ways to view essentially the same waveform.

It can be seen why the denominator term in (9.9) was approximated in (9.10). When the difference between two paths changes by a quarter wavelength, the phase difference between the responses on the two paths changes by $\pi/2$, which causes a very significant change in the overall received amplitude. Since the carrier wavelength is very small relative to the path lengths, the time over which this phase change is significant is far smaller than the time over which the denominator changes significantly. The phase changes are significant over millisecond intervals, whereas the denominator changes are significant over intervals of seconds or minutes. For modulation and detection, the relevant time scales are milliseconds or less, and the denominators are effectively constant over these intervals.

The reader might notice that many more approximations are required with even very simple wireless models than with wired communication. This is partly because the standard linear time-invariant assumptions of wired communication usually provide straightforward models, such as the system function in (9.3). Wireless systems are usually time-varying, and appropriate models depend very much on the time scales of interest. For wireless systems, making the appropriate approximations is often more important than subsequent manipulation of equations.

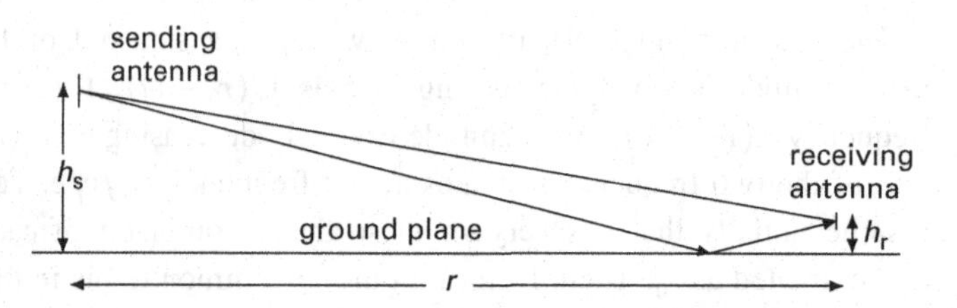

Figure 9.4. Two paths between antennas above a ground plane. One path is direct and the other reflects off the ground.

9.2.4 Reflection from a ground plane

Consider a transmitting antenna and a receiving antenna, both above a plane surface such as a road (see Figure 9.4). If the angle of incidence between antenna and road is sufficiently small, then a dielectric reflects most of the incident wave, with a sign change. When the horizontal distance r between the antennas becomes very large relative to their vertical displacements from the ground plane, a very surprising thing happens. In particular, the difference between the direct path length and the reflected path length goes to zero as r^{-1} with increasing r.

When r is large enough, this difference between the path lengths becomes small relative to the wavelength c/f of a sinusoid at frequency f. Since the sign of the electric field is reversed on the reflected path, these two waves start to cancel each other out. The combined electric field at the receiver is then attenuated as r^{-2}, and the received power goes down as r^{-4}. This is worked out analytically in Exercise 9.3. What this example shows is that the received power can decrease with distance considerably faster than r^{-2} in the presence of reflections. This particular geometry leads to an attenuation of r^{-4} rather than multipath fading.

The above example is only intended to show how attenuation can vary other than with r^{-2} in the presence of reflections. Real road surfaces are not perfectly flat and behave in more complicated ways. In other examples, power attenuation can vary with r^{-6} or even decrease exponentially with r. Also, these attenuation effects cannot always be cleanly separated from multipath effects.

A rapid decrease in power with increasing distance is helpful in one way and harmful in another. It is helpful in reducing the interference between adjoining cells, but is harmful in reducing the coverage of cells. As cellular systems become increasingly heavily used, however, the major determinant of cell size is the number of cell phones in the cell. The size of cells has been steadily decreasing in heavily used areas, and one talks of micro-cells and pico-cells as a response to this effect.

9.2.5 Shadowing

Shadowing is a wireless phenomenon similar to the blocking of sunlight by clouds. It occurs when partially absorbing materials, such as the walls of buildings, lie between the sending and receiving antennas. It occurs both when cell phones are inside buildings and when outside cell phones are shielded from the base station by buildings or other structures.

The effect of shadow fading differs from multipath fading in two important ways. First, shadow fades have durations on the order of multiple seconds or minutes. For this reason, shadow fading is often called slow fading and multipath fading is called fast fading. Second, the attenuation due to shadowing is exponential in the width of the barrier that must be passed through. Thus the overall power attenuation contains not only the r^{-2} effect of free-space transmission, but also the exponential attenuation over the depth of the obstructing material.

9.2.6 Moving antenna, multiple reflectors

Each example with two paths above has used ray tracing to calculate the individual response from each path and then added those responses to find the overall response to a sinusoidal input. An arbitrary number of reflectors may be treated the same way. Finding the amplitude and phase for each path is, in general, not a simple task. Even for the very simple large wall assumed in Figure 9.2, the reflected field calculated in (9.9) is valid only at small distances from the wall relative to the dimensions of the wall. At larger distances, the total power reflected from the wall is proportional both to r_0^{-2} and the cross section of the wall. The portion of this power reaching the receiver is proportional to $(r_0 - r(t))^{-2}$. Thus the power attenuation from transmitter to receiver (for the reflected wave at large distances) is proportional to $[r_0(r_0 - r(t)]^{-2}$ rather than to $[2r_0 - r(t)]^{-2}$. This shows that ray tracing must be used with some caution. Fortunately, however, linearity still holds in these more complex cases.

Another type of reflection is known as scattering and can occur in the atmosphere or in reflections from very rough objects. Here the very large set of paths is better modeled as an integral over infinitesimally weak paths rather than as a finite sum.

Finding the amplitude of the reflected field from each type of reflector is important in determining the coverage, and thus the placement, of base stations, although ultimately experimentation is necessary. Jakes (1974) considers these questions in much greater detail, but this would take us too far into electromagnetic theory and too far away from questions of modulation, detection, and multiple access. Thus we now turn our attention to understanding the nature of the aggregate received waveform, given a representation for each reflected wave. This means modeling the input/output behavior of a channel rather than the detailed response on each path.

9.3 Input/output models of wireless channels

This section discusses how to view a channel consisting of an arbitrary collection of J electromagnetic paths as a more abstract input/output model. For the reflecting wall example, there is a direct path and one reflecting path, so $J = 2$. In other examples, there might be a direct path along with multiple reflected paths, each coming from a separate reflecting object. In many cases, the direct path is blocked and only indirect paths exist.

In many physical situations, the important paths are accompanied by other insignificant and highly attenuated paths. In these cases, the insignificant paths are omitted from the model and J denotes the number of remaining significant paths.

As in the examples of Section 9.2, the J significant paths are associated with attenuations and delays due to path lengths, antenna patterns, and reflector characteristics. As illustrated in Figure 9.5, the signal at the receiving antenna coming from path j in response to an input $\exp(2\pi i f t)$ is given by

$$\frac{\alpha_j \exp\{2\pi i f[t - \frac{r_j(t)}{c}]\}}{r_j(t)}.$$

The overall response at the receiving antenna to an input $\exp(2\pi i f t)$ is then

$$y_f(t) = \sum_{j=1}^{J} \frac{\alpha_j \exp\{2\pi i f[t - \frac{r_j(t)}{c}]\}}{r_j(t)}. \tag{9.11}$$

For the example of a perfectly reflecting wall, the combined antenna gain α_1 on the direct path is denoted as α in (9.9). The combined antenna gain α_2 for the reflected path is $-\alpha$ because of the phase reversal at the reflector. The path lengths are $r_1(t) = r_0 - vt$ and $r_2(t) = r_0 + vt$, making (9.11) equivalent to (9.9) for this example.

For the general case of J significant paths, it is more convenient and general to replace (9.11) with an expression explicitly denoting the complex attenuation $\beta_j(t)$ and delay $\tau_j(t)$ on each path:

$$y_f(t) = \sum_{j=1}^{J} \beta_j(t) \exp\{2\pi i f[t - \tau_j(t)]\}, \tag{9.12}$$

$$\beta_j(t) = \frac{\alpha_j(t)}{r_j(t)}, \qquad \tau_j(t) = \frac{r_j(t)}{c}. \tag{9.13}$$

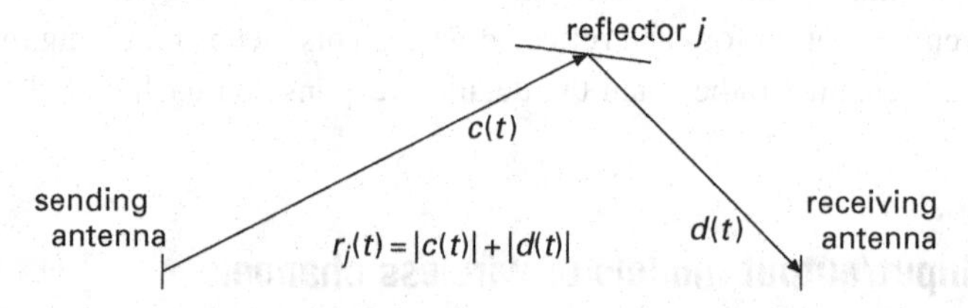

Figure 9.5. The reflected path above is represented by a vector $c(t)$ from sending antenna to reflector j and a vector $d(t)$ from reflector to receiving antenna. The path length $r_j(t)$ is the sum of the *lengths* $|c(t)|$ and $|d(t)|$. The complex function $\alpha_j(t)$ is the product of the transmitting antenna pattern in the direction toward the reflector, the loss and phase change at the reflector, and the receiver pattern in the direction from the reflector.

Equation (9.12) can also be used for arbitrary attenuation rates rather than just the $1/r^2$ power loss assumed in (9.11). By factoring out the term $\exp\{2\pi i f t\}$, (9.12) can be rewritten as follows:

$$y_f(t) = \hat{h}(f,t)\exp\{2\pi i f t\}, \qquad \text{where} \quad \hat{h}(f,t) = \sum_{j=1}^{J} \beta_j(t)\exp\{-2\pi i f \tau_j(t)\}. \quad (9.14)$$

The function $\hat{h}(f,t)$ is similar to the system function $\hat{h}(f)$ of a linear-time-invariant (LTI) system except for the variation in t. Thus $\hat{h}(f,t)$ is called the *system function* for the linear-time-varying (LTV) system (i.e. channel) above.

The path attenuations $\beta_j(t)$ vary slowly with time and frequency, but these variations are negligibly slow over the time and frequency intervals of concern here. Thus a simplified model is often used in which each attenuation is simply a constant β_j. In this simplified model, it is also assumed that each path delay is changing at a constant rate, $\tau_j(t) = \tau_j^{\mathrm{o}} + \tau_j' t$. Thus $\hat{h}(f,t)$ in the simplified model is given by

$$\hat{h}(f,t) = \sum_{j=1}^{J} \beta_j \exp\{-2\pi i f \tau_j(t)\} \qquad \text{where} \quad \tau_j(t) = \tau_j^0 + \tau_j' t. \quad (9.15)$$

This simplified model was used in analyzing the reflecting wall. There, $\beta_1 = -\beta_2 = \alpha/r_0$, $\tau_1^0 = \tau_2^0 = r_0/c$, and $\tau_1' = -\tau_2' = -v/c$.

9.3.1 The system function and impulse response for LTV systems

The LTV system function $\hat{h}(f,t)$ in (9.14) was defined for a multipath channel with a finite number of paths. A simplified model was defined in (9.15). The system function could also be generalized in a straightforward way to a channel with a continuum of paths. More generally yet, if $y_f(t)$ is the response to the input $\exp\{2\pi i f t\}$, then $\hat{h}(f,t)$ is defined as $\hat{y}_f(t)\exp\{-2\pi i f t\}$.

In this subsection, $\hat{h}(f,t)\exp\{2\pi i f t\}$ is taken to be the response to $\exp\{2\pi i f t\}$ for each frequency f. The objective is then to find the response to an arbitrary input $x(t)$. This will involve generalizing the well known impulse response and convolution equation of LTI systems to the LTV case.

The key assumption in this generalization is the linearity of the system. That is, if $y_1(t)$ and $y_2(t)$ are the responses to $x_1(t)$ and $x_2(t)$, respectively, then $\alpha_1 y_1(t) + \alpha_2 y_2(t)$ is the response to $\alpha_1 x_1(t) + \alpha_2 x_2(t)$. This linearity follows from Maxwell's equations.[5]

Using linearity, the response to a superposition of complex sinusoids, say $x(t) = \int_{-\infty}^{\infty} \hat{x}(f)\exp\{2\pi i f t\}\mathrm{d}f$, is given by

$$y(t) = \int_{-\infty}^{\infty} \hat{x}(f)\hat{h}(f,t)\exp(2\pi i f t)\mathrm{d}f. \quad (9.16)$$

[5] Nonlinear effects can occur in high-power transmitting antennas, but we ignore that here.

There is a temptation here to imitate the theory of LTI systems blindly and to confuse the Fourier transform of $y(t)$, namely $\hat{y}(f)$, with $\hat{x}(f)\hat{h}(f,t)$. This is wrong both logically and physically. It is wrong logically because $\hat{x}(f)\hat{h}(f,t)$ is a function of t and f, whereas $\hat{y}(f)$ is a function only of f. It is wrong physically because Doppler shifts cause the response to $\hat{x}(f)\exp(2\pi i f t)$ to contain multiple sinusoids around f rather than a single sinusoid at f. From the receiver's viewpoint, $\hat{y}(f)$ at a given f depends on $\hat{x}(\tilde{f})$ over a range of $\tilde{f}$ around f.

Fortunately, (9.16) can still be used to derive a very satisfactory form of impulse response and convolution equation. Define the *time-varying impulse response* $h(\tau,t)$ as the inverse Fourier transform (in the time variable τ) of $\hat{h}(f,t)$, where t is viewed as a parameter. That is, for each $t \in \mathbb{R}$,

$$h(\tau,t) = \int_{-\infty}^{\infty} \hat{h}(f,t)\exp(2\pi i f\tau)\mathrm{d}f; \qquad \hat{h}(f,t) = \int_{-\infty}^{\infty} h(\tau,t)\exp(-2\pi i f\tau)\mathrm{d}\tau. \quad (9.17)$$

Intuitively, $\hat{h}(f,t)$ is regarded as a conventional LTI system function that is slowly changing with t, and $h(\tau,t)$ is regarded as a channel impulse response (in τ) that is slowly changing with t. Substituting the second part of (9.17) into (9.16), we obtain

$$y(t) = \int_{-\infty}^{\infty} \hat{x}(f)\left[\int_{-\infty}^{\infty} h(\tau,t)\exp[2\pi i f(t-\tau)]\mathrm{d}\tau\right]\mathrm{d}f.$$

Interchanging the order of integration,[6]

$$y(t) = \int_{-\infty}^{\infty} h(\tau,t)\left[\int_{-\infty}^{\infty} \hat{x}(f)\exp[2\pi i f(t-\tau)]\mathrm{d}f\right]\mathrm{d}\tau.$$

Identifying the inner integral as $x(t-\tau)$, we get the *convolution equation for LTV filters*:

$$y(t) = \int_{-\infty}^{\infty} x(t-\tau)h(\tau,t)\mathrm{d}\tau. \quad (9.18)$$

This expression is really quite nice. It says that the effects of mobile transmitters and receivers, arbitrarily moving reflectors and absorbers, and all of the complexities of solving Maxwell's equations, finally reduce to an input/output relation between transmit and receive antennas which is simply represented as the impulse response of an LTV channel filter. That is, $h(\tau,t)$ is the response at time t to an impulse at time $t-\tau$. If $h(\tau,t)$ is a constant function of t, then $h(\tau,t)$, as a function of τ, is the conventional LTI impulse response.

This derivation applies for both real and complex inputs. The actual physical input $x(t)$ at bandpass must be real, however, and, for every real $x(t)$, the corresponding output $y(t)$ must also be real. This means that the LTV impulse response $h(\tau,t)$ must also be real. It then follows from (9.17) that $\hat{h}(-f,t) = \hat{h}^*(f,t)$, which defines $\hat{h}(-f,t)$ in terms of $\hat{h}(f,t)$ for all $f > 0$.

[6] Questions about convergence and interchange of limits will be ignored in this section. This is reasonable since the inputs and outputs of interest should be essentially time and frequency limited to the range of validity of the simplified multipath model.

There are many similarities between the results for LTV filters and the conventional results for LTI filters. In both cases, the output waveform is the convolution of the input waveform with the impulse response; in the LTI case, $y(t) = \int x(t-\tau)h(\tau)\mathrm{d}\tau$, whereas, in the LTV case, $y(t) = \int x(t-\tau)h(\tau, t)\mathrm{d}\tau$. In both cases, the system function is the Fourier transform of the impulse response; for LTI filters, $h(\tau) \leftrightarrow \hat{h}(f)$, and, for LTV filters, $h(\tau, t) \leftrightarrow \hat{h}(f, t)$; i.e., for each t, the function $\hat{h}(f, t)$ (as a function of f) is the Fourier transform of $h(\tau, t)$ (as a function of τ). The most significant difference is that $\hat{y}(f) = \hat{h}(f)\,\hat{x}(f)$ in the LTI case, whereas, in the LTV case, the corresponding statement says only that $y(t)$ is the inverse Fourier transform of $\hat{h}(f, t)\hat{x}(f)$.

It is important to realize that the Fourier relationship between the time-varying impulse response $h(\tau, t)$ and the time-varying system function $\hat{h}(f, t)$ is valid for any LTV system and does not depend on the simplified multipath model of (9.15). This simplified multipath model is valuable, however, in acquiring insight into how multipath and time-varying attenuation affect the transmitted waveform.

For the simplified model of (9.15), $h(\tau, t)$ can be easily derived from $\hat{h}(f, t)$ as follows:

$$\hat{h}(f, t) = \sum_{j=1}^{J} \beta_j \exp\{-2\pi i f \tau_j(t)\} \quad \leftrightarrow \quad h(\tau, t) = \sum_j \beta_j \delta\{\tau - \tau_j(t)\}, \tag{9.19}$$

where δ is the Dirac delta function. Substituting (9.19) into (9.18) yields

$$y(t) = \sum_j \beta_j x(t - \tau_j(t)). \tag{9.20}$$

This says that the response at time t to an arbitrary input is the sum of the responses over all paths. The response on path j is simply the input, delayed by $\tau_j(t)$ and attenuated by β_j. Note that both the delay and attenuation are evaluated at the time t at which the *output* is being measured.

The idealized nonphysical impulses in (9.19) arise because of the tacit assumption that the attenuation and delay on each path are independent of frequency. It can be seen from (9.16) that $\hat{h}(f, t)$ affects the output only over the frequency band where $\hat{x}(f)$ is nonzero. If frequency independence holds over this band, it does no harm to assume it over all frequencies, leading to the above impulses. For typical relatively narrow-band applications, this frequency independence is usually a reasonable assumption.

Neither the general results about LTV systems nor the results for the multipath models of (9.14) and (9.15) provide much immediate insight into the nature of fading. Sections 9.3.2 and 9.3.3 look at this issue, first for sinusoidal inputs, and then for general narrow-band inputs.

9.3.2 Doppler spread and coherence time

Assuming the simplified model of multipath fading in (9.15), the system function $\hat{h}(f, t)$ can be expressed as follows:

$$\hat{h}(f, t) = \sum_{j=1}^{J} \beta_j \exp\{-2\pi i f(\tau_j' t + \tau_j^0)\}.$$

The rate of change of delay, τ_j', on path j is related to the Doppler shift on path j at frequency f by $\mathcal{D}_j = -f\tau_j'$, and thus $\hat{h}(f, t)$ can be expressed directly in terms of the Doppler shifts:

$$\hat{h}(f, t) = \sum_{j=1}^{J} \beta_j \exp\{2\pi i(\mathcal{D}_j t - f\tau_j^0)\}.$$

The response to an input $\exp\{2\pi i f t\}$ is then

$$y_f(t) = \hat{h}(f, t)\exp\{2\pi i f t\} = \sum_{j=1}^{J} \beta_j \exp\{2\pi i(f + \mathcal{D}_j)t - f\tau_j^0\}. \tag{9.21}$$

This is the sum of sinusoids around f ranging from $f + \mathcal{D}_{\min}$ to $f + \mathcal{D}_{\max}$, where $\mathcal{D}_{\min}$ is the smallest of the Doppler shifts and $\mathcal{D}_{\max}$ is the largest. The terms $-2\pi i f\tau_j^0$ are simply phases.

The Doppler shifts $\mathcal{D}_j$ can be positive or negative, but can be assumed to be small relative to the transmission frequency f. Thus $y_f(t)$ is a narrow-band waveform whose bandwidth is the spread between $\mathcal{D}_{\min}$ and $\mathcal{D}_{\max}$. This spread, given by

$$\mathcal{D} = \max_j \mathcal{D}_j - \min_j \mathcal{D}_j, \tag{9.22}$$

is defined as the *Doppler spread* of the channel. The Doppler spread is a function of f (since all the Doppler shifts are functions of f), but it is usually viewed as a constant since it is approximately constant over any given frequency band of interest.

As shown above, the Doppler spread is the bandwidth of $y_f(t)$, but it is now necessary to be more specific about how to define fading. This will also lead to a definition of the *coherence time* of a channel.

The fading in (9.21) can be brought out more clearly by expressing $\hat{h}(f, t)$ in terms of its magnitude and phase, i.e. as $|\hat{h}(f, t)|\, e^{i\angle\hat{h}(f,t)}$. The response to $\exp\{2\pi i f t\}$ is then given by

$$y_f(t) = |\hat{h}(f, t)| \exp\{2\pi i f t + i\angle\hat{h}(f, t)\}. \tag{9.23}$$

This expresses $y_f(t)$ as an amplitude term $|\hat{h}(f, t)|$ times a phase modulation of magnitude 1. This amplitude term $|\hat{h}(f, t)|$ is now defined as the *fading amplitude* of the channel at frequency f. As explained above, $|\hat{h}(f, t)|$ and $\angle\hat{h}(f, t)$ are slowly varying with t relative to $\exp\{2\pi i f t\}$, so it makes sense to view $|\hat{h}(f, t)|$ as a slowly varying envelope, i.e. a fading envelope, around the received phase-modulated sinusoid.

The fading amplitude can be interpreted more clearly in terms of the response $\Re[y_f(t)]$ to an actual real input sinusoid $\cos(2\pi f t) = \Re[\exp(2\pi i f t)]$. Taking the real part of (9.23), we obtain

$$\Re[y_f(t)] = |\hat{h}(f, t)| \cos[2\pi f t + \angle\hat{h}(f, t)].$$

The waveform $\Re[y_f(t)]$ oscillates at roughly the frequency f inside the slowly varying limits $\pm|\hat{h}(f, t)|$. This shows that $|\hat{h}(f, t)|$ is also the envelope, and thus the fading amplitude, of $\Re[y_f(t)]$ (at the given frequency f). This interpretation will be extended later to narrow-band inputs around the frequency f.

We have seen from (9.21) that $\mathcal{D}$ is the bandwidth of $y_f(t)$, and it is also the bandwidth of $\Re[y_f(t)]$. Assume initially that the Doppler shifts are centered around 0, i.e. that $\mathcal{D}_{\max} = -\mathcal{D}_{\min}$. Then $\hat{h}(f,t)$ is a baseband waveform containing frequencies between $-\mathcal{D}/2$ and $+\mathcal{D}/2$. The envelope of $\Re[y_f(t)]$, namely $|\hat{h}(f,t)|$, is the magnitude of a waveform baseband limited to $\mathcal{D}/2$. For the reflecting wall example, $\mathcal{D}_1 = -\mathcal{D}_2$, the Doppler spread is $\mathcal{D} = 2\mathcal{D}_1$, and the envelope is $|\sin[2\pi(\mathcal{D}/2)t]|$.

More generally, the Doppler shifts might be centered around some nonzero Δ defined as the midpoint between $\min_j \mathcal{D}_j$ and $\max_j \mathcal{D}_j$. In this case, consider the frequency-shifted system function $\hat{\psi}(f,t)$ defined as

$$\hat{\psi}(f,t) = \exp(-2\pi \mathrm{i} t\Delta)\hat{h}(f,t) = \sum_{j=1}^{J} \beta_j \exp\{2\pi \mathrm{i} t(\mathcal{D}_j - \Delta) - 2\pi \mathrm{i} f\tau_j^0\}. \tag{9.24}$$

As a function of t, $\hat{\psi}(f,t)$ has bandwidth $\mathcal{D}/2$. Since

$$|\hat{\psi}(f,t)| = |\mathrm{e}^{-2\pi \mathrm{i}\Delta t}\,\hat{h}(f,t)| = |\hat{h}(f,t)|,$$

the envelope of $\Re[y_f(t)]$ is the same as[7] the magnitude of $\hat{\psi}(f,t)$, i.e. the magnitude of a waveform baseband limited to $\mathcal{D}/2$. Thus this limit to $\mathcal{D}/2$ is valid independent of the Doppler shift centering.

As an example, assume there is only one path and its Doppler shift is $\mathcal{D}_1$. Then $\hat{h}(f,t)$ is a complex sinusoid at frequency $\mathcal{D}_1$, but $|\hat{h}(f,t)|$ is a constant, namely $|\beta_1|$. The Doppler spread is 0, the envelope is constant, and there is no fading. As another example, suppose the transmitter in the reflecting wall example is moving away from the wall. This decreases both of the Doppler shifts, but the difference between them, namely the Doppler spread, remains the same. The envelope $|\hat{h}(f,t)|$ then also remains the same. Both of these examples illustrate that it is the *Doppler spread* rather than the individual Doppler shifts that controls the envelope.

Define the *coherence time* $\mathcal{T}_{\mathrm{coh}}$ of the channel to be[8]

$$\mathcal{T}_{\mathrm{coh}} = \frac{1}{2\mathcal{D}}. \tag{9.25}$$

This is one-quarter of the wavelength of $\mathcal{D}/2$ (the maximum frequency in $\hat{\psi}(f,t)$) and one-half the corresponding sampling interval. Since the envelope is $|\hat{\psi}(f,t)|$, $\mathcal{T}_{\mathrm{coh}}$ serves as a crude order-of-magnitude measure of the typical time interval for the envelope to change significantly. Since this envelope is the fading amplitude of the channel at frequency f, $\mathcal{T}_{\mathrm{coh}}$ is fundamentally interpreted as the order-of-magnitude

[7] Note that $\hat{\psi}(f,t)$, as a function of t, is baseband limited to $\mathcal{D}/2$, whereas $\hat{h}(f,t)$ is limited to frequencies within $\mathcal{D}/2$ of Δ and $\hat{y}_f(t)$ is limited to frequencies within $\mathcal{D}/2$ of $f+\Delta$. It is rather surprising initially that all these waveforms have the same envelope. We focus on $\hat{\psi}(f,t) = \mathrm{e}^{-2\pi \mathrm{i} f\Delta}\hat{h}(f,t)$ since this is the function that is baseband limited to $\mathcal{D}/2$. Exercises 6.17 and 9.5 give additional insight and clarifying examples about the envelopes of real passband waveforms.

[8] Some authors define $\mathcal{T}_{\mathrm{coh}}$ as $1/(4\mathcal{D})$ and others as $1/\mathcal{D}$; these have the same order-of-magnitude interpretations.

duration of a fade at f. Since $\mathcal{D}$ is typically less than 1000 Hz, $\mathcal{T}_{\text{coh}}$ is typically greater than 0.5 ms.

Although the rapidity of changes in a baseband function cannot be specified solely in terms of its bandwidth, high-bandwidth functions tend to change more rapidly than low-bandwidth functions; the definition of coherence time captures this loose relationship. For the reflecting wall example, the envelope goes from its maximum value down to 0 over the period $\mathcal{T}_{\text{coh}}$; this is more or less typical of more general examples.

Crude though $\mathcal{T}_{\text{coh}}$ might be as a measure of fading duration, it is an important parameter in describing wireless channels. It is used in waveform design, diversity provision, and channel measurement strategies. Later, when stochastic models are introduced for multipath, the relationship between fading duration and $\mathcal{T}_{\text{coh}}$ will become sharper.

It is important to realize that Doppler shifts are linear in the input frequency, and thus Doppler spread is also. For narrow-band inputs, the variation of Doppler spread with frequency is insignificant. When comparing systems in different frequency bands, however, the variation of $\mathcal{D}$ with frequency is important. For example, a system operating at 8 GHz has a Doppler spread eight times that of a 1 GHz system, and thus a coherence time one-eighth as large; fading is faster, with shorter fade durations, and channel measurements become outdated eight times as fast.

9.3.3 Delay spread and coherence frequency

Another important parameter of a wireless channel is the spread in delay between different paths. The *delay spread* $\mathcal{L}$ is defined as the difference between the path delay on the longest significant path and that on the shortest significant path. That is,

$$\mathcal{L} = \max_j[\tau_j(t)] - \min_j[\tau_j(t)].$$

The difference between path lengths is rarely greater than a few kilometers, so $\mathcal{L}$ is rarely more than several microseconds. Since the path delays, $\tau_j(t)$, are changing with time, $\mathcal{L}$ can also change with time, so we focus on $\mathcal{L}$ at some given t. Over the intervals of interest in modulation, however, $\mathcal{L}$ can usually be regarded as a constant.[9]

A closely related parameter is the *coherence frequency* of a channel. It is defined as follows:[10]

$$\mathcal{F}_{\text{coh}} = \frac{1}{2\mathcal{L}}. \tag{9.26}$$

The coherence frequency is thus typically greater than 100 kHz. This section shows that $\mathcal{F}_{\text{coh}}$ provides an approximate answer to the following question: if the channel is badly faded at one frequency f, how much does the frequency have to be changed to

---

[9] For the reflecting wall example, the path lengths are $r_0 - vt$ and $r_0 + vt$, so the delay spread is $\mathcal{L} = 2vt/c$. The change with t looks quite significant here, but at reasonable distances from the reflector the change is small relative to typical intersymbol intervals.

[10] $\mathcal{F}_{\text{coh}}$ is sometimes defined as $1/\mathcal{L}$ and sometimes as $1/(4\mathcal{L})$; the interpretation is the same.

find an unfaded frequency? We will see that, to a very crude approximation, f must be changed by $\mathcal{F}_{\text{coh}}$.

The analysis of the parameters $\mathcal{L}$ and $\mathcal{F}_{\text{coh}}$ is, in a sense, a time/frequency dual of the analysis of $\mathcal{D}$ and $\mathcal{T}_{\text{coh}}$. More specifically, the fading envelope of $\Re[y_f(t)]$ (in response to the input $\cos(2\pi ft)$) is $|\hat{h}(f,t)|$. The analysis of $\mathcal{D}$ and $\mathcal{T}_{\text{coh}}$ concern the variation of $|\hat{h}(f,t)|$ with t. That of $\mathcal{L}$ and $\mathcal{F}_{\text{coh}}$ concern the variation of $|\hat{h}(f,t)|$ with f.

In the simplified multipath model of (9.15), $\hat{h}(f,t) = \sum_j \beta_j \exp\{-2\pi i f\tau_j(t)\}$. For fixed t, this is a weighted sum of J complex sinusoidal terms in the variable f. The "frequencies" of these terms, viewed as functions of f, are $\tau_1(t), \ldots, \tau_J(t)$. Let τ_{mid} be the midpoint between $\min_j \tau_j(t)$ and $\max_j \tau_j(t)$ and define the function $\hat{\eta}(f,t)$ as follows:

$$\hat{\eta}(f,t) = e^{2\pi i f\tau_{\text{mid}}}\,\hat{h}(f,t) = \sum_j \beta_j \exp\{-2\pi i f[\tau_j(t) - \tau_{\text{mid}}]\}. \tag{9.27}$$

The shifted delays, $\tau_j(t) - \tau_{\text{mid}}$, vary with j from $-\mathcal{L}/2$ to $+\mathcal{L}/2$. Thus $\hat{\eta}(f,t)$, as a function of f, has a "baseband bandwidth"[11] of $\mathcal{L}/2$. From (9.27), we see that $|\hat{h}(f,t)| = |\hat{\eta}(f,t)|$. Thus the envelope $|\hat{h}(f,t)|$, as a function of f, is the magnitude of a function "baseband limited" to $\mathcal{L}/2$.

It is then reasonable to take one-quarter of a "wavelength" of this bandwidth, i.e. $\mathcal{F}_{\text{coh}} = 1/(2\mathcal{L})$, as an order-of-magnitude measure of the required change in f to cause a significant change in the envelope of $\Re[y_f(t)]$.

The above argument relating $\mathcal{L}$ to $\mathcal{F}_{\text{coh}}$ is virtually identical to that relating $\mathcal{D}$ to $\mathcal{T}_{\text{coh}}$. The interpretations of $\mathcal{T}_{\text{coh}}$ and $\mathcal{F}_{\text{coh}}$ as order-of-magnitude approximations are also virtually identical. The duality here, however, is between the t and f in $\hat{h}(f,t)$ rather than between time and frequency for the actual transmitted and received waveforms. The envelope $|\hat{h}(f,t)|$ used in both of these arguments can be viewed as a short-term time average in $|\Re[y_f(t)]|$ (see Exercise 9.6(b)), and thus $\mathcal{F}_{\text{coh}}$ is interpreted as the frequency change required for significant change in this short-term time average rather than in the response itself.

One of the major issues faced by wireless communication is how to spread an input signal or codeword over time and frequency (within the available delay and frequency constraints). If a signal is essentially contained both within a time interval $\mathcal{T}_{\text{coh}}$ and a frequency interval $\mathcal{F}_{\text{coh}}$, then a single fade can bring the entire signal far below the noise level. If, however, the signal is spread over multiple intervals of duration $\mathcal{T}_{\text{coh}}$ and/or multiple bands of width $\mathcal{F}_{\text{coh}}$, then a single fade will affect only one portion of the signal. Spreading the signal over regions with relatively independent fading is called *diversity*, which is studied later. For now, note that the parameters $\mathcal{T}_{\text{coh}}$ and $\mathcal{F}_{\text{coh}}$ tell us how much spreading in time and frequency is required for using such diversity techniques.

In earlier chapters, the receiver timing was delayed from the transmitter timing by the overall propagation delay; this is done in practice by timing recovery at the receiver.

[11] In other words, the inverse Fourier transform, $h(\tau - \tau_{\text{mid}}, t)$ is nonzero only for $|\tau - \tau_{\text{mid}}| \le \mathcal{L}/2$.

Timing recovery is also used in wireless communication, but since different paths have different propagation delays, timing recovery at the receiver will approximately center the path delays around 0. This means that the offset τ_{mid} in (9.27) becomes zero and the function $\hat{\eta}(f,t) = \hat{h}(f,t)$. Thus $\hat{\eta}(f,t)$ can be omitted from further consideration, and it can be assumed, without loss of generality, that $h(\tau,t)$ is nonzero only for $|\tau| \le L/2$.

Next, consider fading for a narrow-band waveform. Suppose that $x(t)$ is a transmitted real passband waveform of bandwidth W around a carrier f_c. Suppose moreover that $\mathsf{W} \ll \mathcal{F}_{\text{coh}}$. Then $\hat{h}(f,t) \approx \hat{h}(f_c,t)$ for $f_c - \mathsf{W}/2 \le f \le f_c + \mathsf{W}/2$. Let $x^+(t)$ be the positive frequency part of $x(t)$, so that $\hat{x}^+(f)$ is nonzero only for $f_c - \mathsf{W}/2 \le f \le f_c + \mathsf{W}/2$. The response $y^+(t)$ to $x^+(t)$ is given by (9.16) as $y^+(t) = \int_{f\ge 0} \hat{x}(f)\hat{h}(f,t)e^{2\pi i f t}\,df$, and is thus approximated as follows:

$$y^+(t) \approx \int_{f_c-\mathsf{W}/2}^{f_c+\mathsf{W}/2} \hat{x}(f)\hat{h}(f_c,t)e^{2\pi i f t}\,df = x^+(t)\hat{h}(f_c,t).$$

Taking the real part to find the response $y(t)$ to $x(t)$ yields

$$y(t) \approx |\hat{h}(f_c,t)|\,\Re[x^+(t)e^{i\angle\hat{h}(f_c,t)}]. \tag{9.28}$$

In other words, for narrow-band communication, the effect of the channel is to cause fading with envelope $|\hat{h}(f_c,t)|$ and with phase change $\angle\hat{h}(f_c,t)$. This is called *flat fading* or *narrow-band fading*. The coherence frequency $\mathcal{F}_{\text{coh}}$ defines the boundary between flat and nonflat fading, and the coherence time $\mathcal{T}_{\text{coh}}$ gives the order-of-magnitude duration of these fades.

The flat-fading response in (9.28) looks very different from the general response in (9.20) as a sum of delayed and attenuated inputs. The signal bandwidth in (9.28), however, is so small that, if we view $x(t)$ as a modulated baseband waveform, that baseband waveform is virtually constant over the different path delays. This will become clearer in Section 9.4.

9.4 Baseband system functions and impulse responses

The next step in interpreting LTV channels is to represent the above bandpass system function in terms of a baseband equivalent. Recall that for any complex waveform $u(t)$, baseband limited to $\mathsf{W}/2$, the modulated real waveform $x(t)$ around carrier frequency f_c is given by

$$x(t) = u(t)\exp\{2\pi i f_c t\} + u^*(t)\exp\{-2\pi i f_c t\}.$$

Assume in what follows that $f_c \gg \mathsf{W}/2$.

In transform terms, $\hat{x}(f) = \hat{u}(f - f_c) + \hat{u}^*(-f + f_c)$. The positive-frequency part of $x(t)$ is simply $u(t)$ shifted up by f_c. To understand the modulation and demodulation in simplest terms, consider a baseband complex sinusoidal input $e^{2\pi i f t}$ for $f \in [-\mathsf{W}/2, \mathsf{W}/2]$ as it is modulated, transmitted through the channel, and demodulated (see Figure 9.6). Since the channel may be subject to Doppler shifts, the recovered

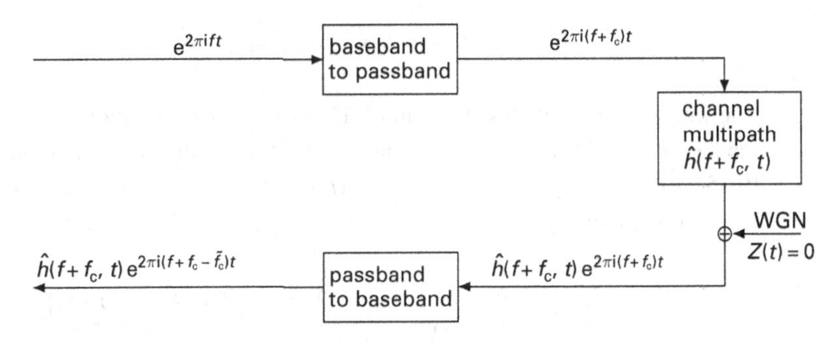

Figure 9.6. Complex baseband sinusoid, as it is modulated to passband, passed through a multipath channel, and demodulated without noise. The modulation is around a carrier frequency f_c and the demodulation is in general at another frequency $\tilde{f}_c$.

carrier, $\tilde{f}_c$, at the receiver might be different than the actual carrier f_c. Thus, as illustrated, the positive-frequency channel output is $y_f(t) = \hat{h}(f+f_c, t)e^{2\pi i(f+f_c)t}$ and the demodulated waveform is $\hat{h}(f+f_c, t)e^{2\pi i(f+f_c-\tilde{f}_c)t}$.

For an arbitrary baseband-limited input, $u(t) = \int_{-W/2}^{W/2} \hat{u}(f)e^{2\pi i f t}\,df$, the positive-frequency channel output is given by superposition:

$$y^+(t) = \int_{-W/2}^{W/2} \hat{u}(f)\hat{h}(f+f_c, t)e^{2\pi i(f+f_c)t}\,df.$$

The demodulated waveform, $v(t)$, is then $y^+(t)$ shifted down by the recovered carrier $\tilde{f}_c$, i.e.

$$v(t) = \int_{-W/2}^{W/2} \hat{u}(f)\hat{h}(f+f_c, t)e^{2\pi i(f+f_c-\tilde{f}_c)t}\,df.$$

Let Δ be the difference between recovered and transmitted carrier,[12] i.e. $\Delta = \tilde{f}_c - f_c$. Thus,

$$v(t) = \int_{-W/2}^{W/2} \hat{u}(f)\hat{h}(f+f_c, t)e^{2\pi i(f-\Delta)t}\,df. \tag{9.29}$$

The relationship between the input $u(t)$ and the output $v(t)$ at baseband can be expressed directly in terms of a baseband system function $\hat{g}(f, t)$ defined as

$$\hat{g}(f, t) = \hat{h}(f+f_c, t)e^{-2\pi i \Delta t}. \tag{9.30}$$

Then (9.29) becomes

$$v(t) = \int_{-W/2}^{W/2} \hat{u}(f)\hat{g}(f, t)e^{2\pi i f t}\,df. \tag{9.31}$$

This is exactly the same form as the passband input–output relationship in (9.16). Letting $g(\tau, t) = \int \hat{g}(f, t)e^{2\pi i f \tau}\,df$ be the LTV baseband impulse response, the same argument as used to derive the passband convolution equation leads to

[12] It might be helpful to assume $\Delta = 0$ on a first reading.

$$v(t) = \int_{-\infty}^{\infty} u(t-\tau)g(\tau, t)\mathrm{d}\tau. \tag{9.32}$$

The interpretation of this baseband LTV convolution equation is the same as that of the passband LTV convolution equation, (9.18). For the simplified multipath model of (9.15), $\hat{h}(f,t) = \sum_{j=1}^{J} \beta_j \exp\{-2\pi i f \tau_j(t)\}$ and thus, from (9.30), the baseband system function is given by

$$\hat{g}(f,t) = \sum_{j=1}^{J} \beta_j \exp\{-2\pi i(f+f_c)\tau_j(t) - 2\pi i \Delta t\}. \tag{9.33}$$

We can separate the dependence on t from that on f by rewriting this as follows:

$$\hat{g}(f,t) = \sum_{j=1}^{J} \gamma_j(t) \exp\{-2\pi i f \tau_j(t)\}, \tag{9.34}$$

where $\gamma_j(t) = \beta_j \exp\{-2\pi i f_c \tau_j(t) - 2\pi i \Delta t\}$. Taking the inverse Fourier transform for fixed t, the LTV baseband impulse response is given by

$$g(\tau,t) = \sum_j \gamma_j(t)\delta\{\tau - \tau_j(t)\}. \tag{9.35}$$

Thus the impulse response at a given receive-time t is a sum of impulses, the jth of which is delayed by $\tau_j(t)$ and has an attenuation and phase given by $\gamma_j(t)$. Substituting this impulse response into the convolution equation, the input–output relation is given by

$$v(t) = \sum_j \gamma_j(t) u(t - \tau_j(t)).$$

This baseband representation can provide additional insight about Doppler spread and coherence time. Consider the system function in (9.34) at $f = 0$ (i.e. at the passband carrier frequency). Letting $\mathcal{D}_j$ be the Doppler shift at f_c on path j, we have $\tau_j(t) = \tau_j^0 - \mathcal{D}_j t / f_c$. Then

$$\hat{g}(0,t) = \sum_{j=1}^{J} \gamma_j(t), \qquad \text{where} \quad \gamma_j(t) = \beta_j \exp\{2\pi i[\mathcal{D}_j - \Delta]t - 2\pi i f_c \tau_j^0\}.$$

The carrier recovery circuit estimates the carrier frequency from the received sum of Doppler-shifted versions of the carrier, and thus it is reasonable to approximate the shift in the recovered carrier by the midpoint between the smallest and largest Doppler shift. Thus $\hat{g}(0,t)$ is the same as the frequency-shifted system function $\hat{\psi}(f_c, t)$ of (9.24). In other words, the frequency shift Δ, which was introduced in (9.24) as a mathematical artifice, now has a physical interpretation as the difference between f_c and the recovered carrier $\tilde{f}_c$. We see that $\hat{g}(0,t)$ is a waveform with bandwidth $\mathcal{D}/2$, and that $\mathcal{T}_{\mathrm{coh}} = 1/(2\mathcal{D})$ is an order-of-magnitude approximation to the time over which $\hat{g}(0,t)$ changes significantly.

Next consider the baseband system function $\hat{g}(f,t)$ at baseband frequencies other than 0. Since $\mathsf{W} \ll f_c$, the Doppler spread at $f_c + f$ is approximately equal to that at f_c, and thus $\hat{g}(f, t)$, as a function of t for each $f \leq \mathsf{W}/2$, is also approximately baseband limited to $\mathcal{D}/2$ (where $\mathcal{D}$ is defined at $f = f_c$).

Finally, consider flat fading from a baseband perspective. Flat fading occurs when $\mathsf{W} \ll \mathcal{F}_{\text{coh}}$, and in this case[13] $\hat{g}(f, t) \approx \hat{g}(0, t)$. Then, from (9.31),

$$v(t) = \hat{g}(0, t)u(t). \tag{9.36}$$

In other words, the received waveform, in the absence of noise, is simply an attenuated and phase-shifted version of the input waveform. If the carrier recovery circuit also recovers phase, then $v(t)$ is simply an attenuated version of $u(t)$. For flat fading, then, $\mathcal{T}_{\text{coh}}$ is the order-of-magnitude interval over which the ratio of output to input can change significantly.

In summary, this section has provided both a passband and a baseband model for wireless communication. The basic equations are very similar, but the baseband model is somewhat easier to use (although somewhat more removed from the physics of fading). The ease of use comes from the fact that all the waveforms are slowly varying and all are complex. This can be seen most clearly by comparing the flat-fading relations, (9.28) for passband and (9.36) for baseband.

9.4.1 A discrete-time baseband model

This section uses the sampling theorem to convert the above continuous-time baseband channel to a discrete-time channel. If the baseband input $u(t)$ is band limited to $\mathsf{W}/2$, then it can be represented by its T-spaced samples, $T = 1/\mathsf{W}$, as $u(t) = \sum_\ell u_\ell \operatorname{sinc}(t/T - \ell)$, where $u_\ell = u(\ell T)$. Using (9.32), the baseband output is given by

$$v(t) = \sum_\ell u_\ell \int g(\tau, t)\operatorname{sinc}(t/T - \tau/T - \ell)\mathrm{d}\tau. \tag{9.37}$$

The sampled outputs, $v_m = v(mT)$, at multiples of T are then given by[14]

$$v_m = \sum_\ell u_\ell \int g(\tau, mT)\operatorname{sinc}(m - \ell - \tau/T)\mathrm{d}\tau \tag{9.38}$$

$$= \sum_k u_{m-k} \int g(\tau, mT)\operatorname{sinc}(k - \tau/T)\mathrm{d}\tau, \tag{9.39}$$

[13] There is an important difference between saying that the Doppler spread at frequency $f + f_c$ is close to that at f_c and saying that $\hat{g}(f, t) \approx \hat{g}(0, t)$. The first requires only that W be a relatively small fraction of f_c, and is reasonable even for $\mathsf{W} = 100$ MHz and $f_c = 1$ GHz, whereas the second requires $\mathsf{W} \ll \mathcal{F}_{\text{coh}}$, which might be on the order of hundreds of kilohertz.

[14] Due to Doppler spread, the bandwidth of the output $v(t)$ can be slightly larger than the bandwidth $\mathsf{W}/2$ of the input $u(t)$. Thus the output samples v_m do not fully represent the output waveform. However, a QAM demodulator first generates each output signal v_m corresponding to the input signal u_m, so these output samples are of primary interest. A more careful treatment would choose a more appropriate modulation pulse than a sinc function and then use some combination of channel estimation and signal detection to produce the output samples. This is beyond our current interest.

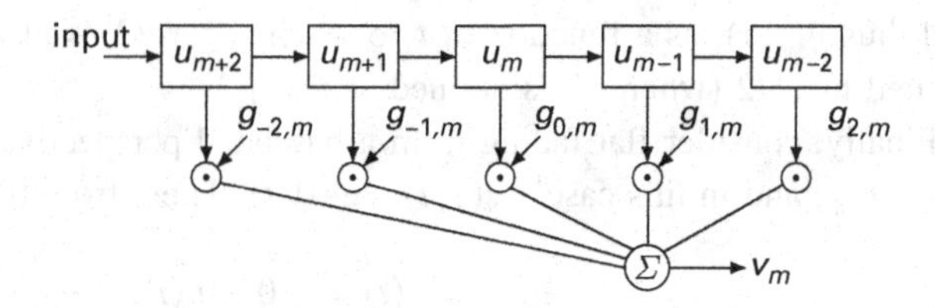

Figure 9.7. Time-varying discrete-time baseband channel model. Each unit of time a new input enters the shift register and the old values shift right. The channel taps also change, but slowly. Note that the output timing here is offset from the input timing by two units.

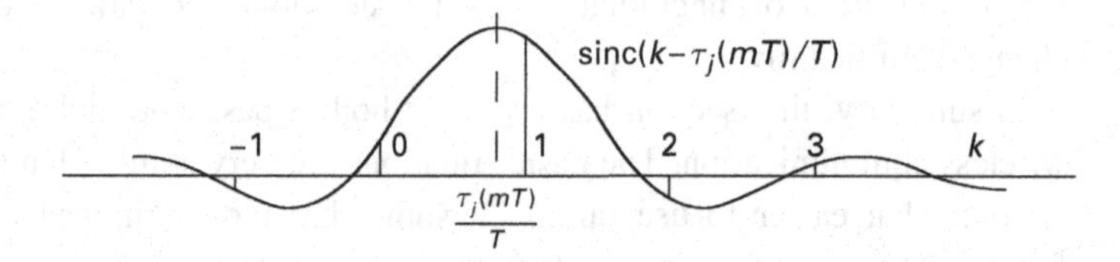

Figure 9.8. This shows $\mathrm{sinc}(k - \tau_j(mt)/T)$, as a function of k, marked at integer values of k. In the illustration, $\tau_j(mt)/T) = 0.8$. The figure indicates that each path contributes primarily to the tap or taps closest to the given path delay.

where $k = m - \ell$. By labeling the above integral as $g_{k,m}$, (9.39) can be written in the following discrete-time form

$$v_m = \sum_k g_{k,m}\, u_{m-k}, \qquad \text{where} \quad g_{k,m} = \int g(\tau, mT)\mathrm{sinc}(k - \tau/T)\mathrm{d}\tau. \tag{9.40}$$

In discrete-time terms, $g_{k,m}$ is the response at mT to an input sample at $(m-k)T$. We refer to $g_{k,m}$ as the kth (complex) channel filter tap at discrete output time mT. This discrete-time filter is represented in Figure 9.7. As discussed later, the number of channel filter taps (i.e. different values of k) for which $g_{k,m}$ is significantly nonzero is usually quite small. If the kth tap is unchanging with m for each k, then the channel is linear-time-invariant. If each tap changes slowly with m, then the channel is called *slowly time-varying*. Cellular systems and most wireless systems of current interest are slowly time-varying.

The filter tap $g_{k,m}$ for the simplified multipath model is obtained by substituting (9.35), i.e. $g(\tau, t) = \sum_j \gamma_j(t)\, \delta\{\tau - \tau_j(t)\}$, into the second part of (9.40), yielding

$$g_{k,m} = \sum_j \gamma_j(mT)\, \mathrm{sinc}\left[k - \frac{\tau_j(mT)}{T}\right]. \tag{9.41}$$

The contribution of path j to tap k can be visualized from Figure 9.8. If the path delay equals kT for some integer k, then path j contributes only to tap k, whereas if the path delay lies between kT and $(k+1)T$, it contributes to several taps around k and $k+1$.

The relation between the discrete-time and continuous-time baseband models can be better understood by observing that when the input is baseband-limited to $\mathsf{W}/2$, then the baseband system function $\hat{g}(f, t)$ is irrelevant for $f > \mathsf{W}/2$. Thus an equivalent filtered system function $\hat{g}_{\mathsf{W}}(f, t)$ and impulse response $g_{\mathsf{W}}(\tau, t)$ can be defined by filtering out the frequencies above $\mathsf{W}/2$, i.e.

$$\hat{g}_{\mathsf{W}}(f, t) = \hat{g}(f, t)\ \mathrm{rect}(f/\mathsf{W}); \qquad g_{\mathsf{W}}(\tau, t) = g(\tau, t) * \mathsf{W}\ \mathrm{sinc}(\tau \mathsf{W}). \tag{9.42}$$

Comparing this with the second half of (9.40), we see that the tap gains are simply scaled sample values of the filtered impulse response, i.e.

$$g_{k,m} = T g_{\mathsf{W}}(kT, mT). \tag{9.43}$$

For the simple multipath model, the filtered impulse response replaces the impulse at $\tau_j(t)$, by a scaled sinc function centered at $\tau_j(t)$, as illustrated in Figure 9.8.

Now consider the number of taps required in the discrete-time model. The delay spread, $\mathcal{L}$, is the interval between the smallest and largest path delay,[15] and thus there are about $\mathcal{L}/T$ taps close to the various path delays. There are a small number of additional significant taps corresponding to the decay time of the sinc function. In the special case where $\mathcal{L}/T$ is much smaller than 1, the timing recovery will make all the delay terms close to 0 and the discrete-time model will have only one significant tap. This corresponds to the flat-fading case we looked at earlier.

The coherence time $\mathcal{T}_{\mathrm{coh}}$ provides a sense of how fast the individual taps $g_{k,m}$ are changing with respect to m. If a tap $g_{k,m}$ is affected by only a single path, then $|g_{k,m}|$ will be virtually unchanging with m, although $\angle g_{k,m}$ can change according to the Doppler shift. If a tap is affected by several paths, then its magnitude can fade at a rate corresponding to the spread of the Doppler shifts affecting that tap.

9.5 Statistical channel models

Section 9.4.1 created a discrete-time baseband fading channel in which the individual tap gains $g_{k,m}$ in (9.41) are scaled sums of the attenuation and smoothed delay on each path. The physical paths are unknown at the transmitter and receiver, however, so from an input/output viewpoint, it is the tap gains themselves[16] that are of primary interest. Since these tap gains change with time, location, bandwidth, carrier frequency, and

[15] Technically, $\mathcal{L}$ varies with the output time t, but we generally ignore this since the variation is slow and $\mathcal{L}$ has only an order-of-magnitude significance.

[16] Many wireless channels are characterized by a very small number of significant paths, and the corresponding receivers track these individual paths rather than using a receiver structure based on the discrete-time model. The discrete-time model is, nonetheless, a useful conceptual model for understanding the statistical variation of multiple paths.

other parameters, a statistical characterization of the tap gains is needed in order to understand how to communicate over these channels. This means that each tap gain $g_{k,m}$ should be viewed as a sample value of a random variable $G_{k,m}$.

There are many approaches to characterizing these tap-gain random variables. One would be to gather statistics over a very large number of locations and conditions and then model the joint probability densities of these random variables according to these measurements, and do this conditionally on various types of locations (cities, hilly areas, flat areas, highways, buildings, etc.) Many such experiments have been performed, but the results provide more detail than is desirable to achieve an initial understanding of wireless issues.

Another approach, which is taken here and in virtually all the theoretical work in the field, is to choose a few very simple probability models that are easy to work with, and then use the results from these models to gain insight about actual physical situations. After presenting the models, we discuss the ways in which the models might or might not reflect physical reality. Some standard results are then derived from these models, along with a discussion of how they might reflect actual performance.

In the Rayleigh tap-gain model, the real and imaginary parts of all the tap gains are taken to be zero-mean jointly Gaussian random variables. Each tap gain $G_{k,m}$ is thus a complex Gaussian random variable, which is further assumed to be circularly symmetric, i.e. to have iid real and imaginary parts. Finally, it is assumed that the probability density of each $G_{k,m}$ is the same for all m. We can then express the probability density of $G_{k,m}$ as follows:

$$f_{\Re(G_{k,m}),\Im(G_{k,m})}(g_{\text{re}}, g_{\text{im}}) = \frac{1}{2\pi\sigma_k^2} \exp\left\{\frac{-g_{\text{re}}^2 - g_{\text{im}}^2}{2\sigma_k^2}\right\}, \tag{9.44}$$

where σ_k^2 is the variance of $\Re(G_{k,m})$ (and thus also of $\Im(G_{k,m})$) for each m. We later address how these rvs are related between different m and k.

As shown in Exercise 7.1, the magnitude $|G_{k,m}|$ of the kth tap is a *Rayleigh* rv with density given by

$$f_{|G_{k,m}|}(|g|) = \frac{|g|}{\sigma_k^2} \exp\left\{\frac{-|g|^2}{2\sigma_k^2}\right\}. \tag{9.45}$$

This model is called the *Rayleigh fading* model. Note from (9.44) that the model includes a uniformly distributed phase that is independent of the Rayleigh distributed amplitude. The assumption of uniform phase is quite reasonable, even in a situation with only a small number of paths, since a quarter-wavelength at cellular frequencies is only a few inches. Thus, even with fairly accurately specified path lengths, we would expect the phases to be modeled as uniform and independent of each other. This would also make the assumption of independence between tap-gain phase and amplitude reasonable.

The assumption of Rayleigh distributed amplitudes is more problematic. If the channel involves scattering from a large number of small reflectors, the central limit theorem would suggest a jointly Gaussian assumption for the tap gains,[17] thus making (9.44) reasonable. For situations with a small number of paths, however, there is no good justification for (9.44) or (9.45).

There is a frequently used alternative model in which the line of sight path (often called a *specular* path) has a known large magnitude, and is accompanied by a large number of independent smaller paths. In this case, $g_{k,m}$, at least for one value of k, can be modeled as a sample value of a complex Gaussian rv with a mean (corresponding to the specular path) plus real and imaginary iid fluctuations around the mean. The magnitude of such a rv has a *Rician* distribution. Its density has quite a complicated form, but the error probability for simple signaling over this channel model is quite simple and instructive.

The preceding paragraphs make it appear as if a model is being constructed for some known number of paths of given character. Much of the reason for wanting a statistical model, however, is to guide the design of transmitters and receivers. Having a large number of models means investigating the performance of given schemes over all such models, or measuring the channel, choosing an appropriate model, and switching to a scheme appropriate for that model. This is inappropriate for an initial treatment, and perhaps inappropriate for design, returning us to the Rayleigh and Rician models. One reasonable point of view here is that these models are often poor approximations for individual physical situations, but when averaged over all the physical situations that a wireless system must operate over, they make more sense.[18] At any rate, these models provide a number of insights into communication in the presence of fading.

Modeling each $g_{k,m}$ as a sample value of a complex rv $G_{k,m}$ provides part of the needed statistical description, but this is not the only issue. The other major issue is how these quantities vary with time. In the Rayleigh fading model, these rvs have zero mean, and it will make a great deal of difference to useful communication techniques if the sample values can be estimated in terms of previous values. A statistical quantity that models this relationship is known as the *tap-gain correlation function*, $R(k, n)$. It is defined as

$$R(k, n) = \mathsf{E}[G_{k,m} G^*_{k,m+n}]. \tag{9.46}$$

This gives the autocorrelation function of the sequence of complex random variables, modeling each given tap k as it evolves in time. It is tacitly assumed that this is not a function of time m, which means that the sequence $\{G_{k,m}; m \in \mathbb{Z}\}$ for each k is assumed to be wide-sense stationary. It is also assumed that, as a random variable,

[17] In fact, much of the current theory of fading was built up in the 1960s when both space communication and military channels of interest were well modeled as scattering channels with a very large number of small reflectors.

[18] This is somewhat oversimplified. As shown in Exercise 9.9, a random choice of a small number of paths from a large possible set does not necessarily lead to a Rayleigh distribution. There is also the question of an initial choice of power level at any given location.

$G_{k,m}$ is independent of $G_{k',m'}$ for all $k \neq k'$ and all m, m'. This final assumption is intuitively plausible[19] since paths in different ranges of delay contribute to $G_{k,m}$ for different values of k.

The tap-gain correlation function is useful as a way of expressing the statistics for how tap gains change, given a particular bandwidth W. It does not address the questions comparing different bandwidths for communication. If we visualize increasing the bandwidth, several things happen. First, since the taps are separated in time by $1/\mathsf{W}$, the range of delay corresponding to a single tap becomes narrower. Thus there are fewer paths contributing to each tap, and the Rayleigh approximation can, in many cases, become poorer. Second, the sinc functions of (9.41) become narrower, so the path delays spill over less in time. For this same reason, $R(k, 0)$ for each k gives a finer grained picture of the amount of power being received in the delay window of width k/W. In summary, as this model is applied to larger W, more detailed statistical information is provided about delay and correlation at that delay, but the information becomes more questionable.

In terms of $R(k, n)$, the multipath spread $\mathcal{L}$ might be defined as the range of kT over which $R(k, 0)$ is significantly nonzero. This is somewhat preferable to the previous "definition" in that the statistical nature of $\mathcal{L}$ becomes explicit and the reliance on some sort of stationarity becomes explicit. In order for this definition to make much sense, however, the bandwidth W must be large enough for several significant taps to exist.

The coherence time $\mathcal{T}_{\text{coh}}$ can also be defined more explicitly as nT for the smallest value of $n > 0$ for which $R(0, n)$ is significantly different from $R(0, 0)$. Both these definitions maintain some ambiguity about what "significant" means, but they face the reality that $\mathcal{L}$ and $\mathcal{T}_{\text{coh}}$ should be viewed probabilistically rather than as instantaneous values.

9.5.1 Passband and baseband noise

The preceding statistical channel model focuses on how multiple paths and Doppler shifts can affect the relationship between input and output, but the noise and the interference from other wireless channels have been ignored. The interference from other users will continue to be ignored (except for regarding it as additional noise), but the noise will now be included.

Assume that the noise is WGN with power $\mathsf{W}N_0$ over the bandwidth W. The earlier convention will still be followed of measuring both signal power and noise power at baseband. Extending the deterministic baseband input/output model $v_m = \sum_k g_{k,m} u_{m-k}$ to include noise as well as randomly varying gap gains, we obtain

$$V_m = \sum_k G_{k,m} U_{m-k} + Z_m, \tag{9.47}$$

[19] One could argue that a moving path would gradually travel from the range of one tap to another. This is true, but the time intervals for such changes are typically large relative to the other intervals of interest.

where $\ldots, Z_{-1}, Z_0, Z_1, \ldots$ is a sequence of iid circularly symmetric complex Gaussian random variables. Assume also that the inputs, the noise, and the tap gains at a given time are statistically independent of each other.

The assumption of WGN essentially means that the primary source of noise is at the receiver or is radiation impinging on the receiver that is independent of the paths over which the signal is being received. This is normally a very good assumption for most communication situations. Since the inputs and outputs here have been modeled as samples at rate W of the baseband processes, we have $\mathsf{E}[|U_m|^2] = P$, where P is the baseband input power constraint. Similarly, $\mathsf{E}[|Z_m|^2] = N_0\mathsf{W}$. Each complex noise rv is thus denoted as $Z_m \sim \mathcal{CN}(0, WN_0)$.

The channel tap gains will be normalized so that $V'_m = \sum_k G_{k,m} U_{m-k}$ satisfies $\mathsf{E}[|V'_m|^2] = P$. It can be seen that this normalization is achieved by

$$\mathsf{E}\left[\sum_k |G_{k,0}|^2\right] = 1. \tag{9.48}$$

This assumption is similar to our earlier assumption for the ordinary (nonfading) WGN channel that the overall attenuation of the channel is removed from consideration. In other words, both here and there we are defining signal power as the power of the received signal in the absence of noise. This is conventional in the communication field and allows us to separate the issue of attenuation from that of coding and modulation.

It is important to recognize that this assumption cannot be used in a system where feedback from receiver to transmitter is used to alter the signal power when the channel is faded.

There has always been a certain amount of awkwardness about scaling from baseband to passband, where the signal power and noise power each increase by a factor of 2. Note that we have also gone from a passband channel filter $\hat{H}(f, t)$ to a baseband filter $\hat{G}(f, t)$ using the same convention as used for input and output. It is not difficult to show that if this property of treating signals and channel filters identically is preserved, and the convolution equation is preserved at baseband and passband, then losing a factor of 2 in power is inevitable in going from passband to baseband.

9.6 Data detection

A reasonable approach to detection for wireless channels is to measure the channel filter taps as they evolve in time, and to use these measured values in detecting data. If the response can be measured accurately, then the detection problem becomes very similar to that for wireline channels, i.e. detection in WGN.

Even under these ideal conditions, however, there are a number of problems. For one thing, even if the transmitter has perfect feedback about the state of the channel, power control is a difficult question; namely, how much power should be sent as a function of the channel state?

For voice, maintaining both voice quality and small constant delay is important. This leads to a desire to send information at a constant rate, which in turn leads to increased transmission power when the channel is poor. This is very wasteful of power, however, since common sense says that if power is scarce and delay is unimportant, then the power and transmission rate should be *decreased* when the channel is poor.

Increasing power when the channel is poor has a mixed impact on interference between users. This strategy maintains equal received power at a base station for all users in the cell corresponding to that base station. This helps reduce the effect of multiaccess interference within the same cell. The interference between neighboring cells can be particularly bad, however, since fading on the channel between a cell phone and its base station is not highly correlated with fading between that cell phone and another base station.

For data, delay is less important, so data can be sent at high rate when the channel is good and at low rate (or zero rate) when the channel is poor. There is a straightforward information-theoretic technique called water filling that can be used to maximize overall transmission rate at a given overall power. The scaling assumption that we made above about input and output power must be modified for all of these issues of power control.

An important insight from this discussion is that the power control used for voice should be very different from that for data. If the same system is used for both voice and data applications, then the basic mechanisms for controlling power and rate should be very different for the two applications.

In this section, power control and rate control are not considered, and the focus is simply on detecting signals under various assumptions about the channel and the state of knowledge at the receiver.

9.6.1 Binary detection in flat Rayleigh fading

Consider a very simple example of communication in the absence of channel measurement. Assume that the channel can be represented by a single discrete-time complex filter tap $G_{0,m}$, which we abbreviate as G_m. Also assume Rayleigh fading; i.e., the probability density of the magnitude of each G_m is given by

$$f_{|G_m|}(|g|) = 2|g|\exp\{-|g|^2\}, \qquad |g| \geq 0; \tag{9.49}$$

or, equivalently, the density of $\gamma = |G_m|^2 \geq 0$ is given by

$$f(\gamma) = \exp(-\gamma), \qquad \gamma \geq 0. \tag{9.50}$$

The phase is uniform over $[0, 2\pi)$ and independent of the magnitude. Equivalently, the real and imaginary parts of G_m are iid Gaussian, each with variance $1/2$. The Rayleigh fading has been scaled in this way to maintain equality between the input power, $\mathsf{E}[|U_m|^2]$, and the output signal power, $\mathsf{E}[|U_m|^2\,|G_m|^2]$. It is assumed that U_m and G_m are independent, i.e. that feedback is not used to control the input power as a function of the fading. For the time being, however, the dependence between the taps G_m at different times m is not relevant.

This model is called *flat* fading for the following reason. A single-tap discrete-time model, where $v(mT) = g_{0,m}u(mT)$, corresponds to a continuous-time baseband model for which $g(\tau, t) = g(0, t)\ \mathrm{sinc}(\tau/T)$. Thus the baseband system function for the channel is given by $\hat{g}(f, t) = g_0(t)\ \mathrm{rect}(fT)$. Thus the fading is constant (i.e. flat) over the baseband frequency range used for communication. When more than one tap is required, the fading varies over the baseband region. To state this another way, the flat-fading model is appropriate when the coherence frequency is greater than the baseband bandwidth.

Consider using binary antipodal signaling with $U_m = \pm a$ for each m. Assume that $\{U_m;\ m \in \mathbb{Z}\}$ is an iid sequence with equiprobable use of plus and minus a. This signaling scheme fails completely, even in the absence of noise, since the phase of the received symbol is uniformly distributed between 0 and 2π under each hypothesis, and the received amplitude is similarly independent of the hypothesis. It is easy to see that phase modulation is similarly flawed. In fact, signal structures must be used in which either different symbols have different magnitudes, or, alternatively, successive signals must be dependent.[20]

Next consider a form of binary pulse-position modulation where, for each pair of time-samples, one of two possible signal pairs, $(a, 0)$ or $(0, a)$, is sent. This has the same performance as a number of binary orthogonal modulation schemes such as minimum shift keying (see Exercise 8.16), but is simpler to describe in discrete time. The output is then given by

$$V_m = U_m G_m + Z_m, \qquad m = 0, 1, \tag{9.51}$$

where, under one hypothesis, the input signal pair is $\boldsymbol{U} = (a, 0)$, and under the other hypothesis $\boldsymbol{U} = (0, a)$. The noise samples $\{Z_m;\ m \in \mathbb{Z}\}$ are iid circularly symmetric complex Gaussian random variables, $Z_m \sim \mathcal{CN}(0, N_0 W)$. Assume for now that the detector looks only at the outputs V_0 and V_1.

Given $\boldsymbol{U} = (a, 0)$, $V_0 = aG_0 + Z_0$ is the sum of two independent complex Gaussian random variables, the first with variance $a^2/2$ per dimension and the second with variance $N_0 W/2$ per dimension. Thus, given $\boldsymbol{U} = (a, 0)$, the real and imaginary parts of V_0 are independent, each $\mathcal{N}(0, a^2/2 + N_0 W/2)$. Similarly, given $\boldsymbol{U} = (a, 0)$, the real and imaginary parts of $V_1 = Z_1$ are independent, each $\mathcal{N}(0, N_0 W/2)$. Finally, since the noise variables are independent, V_0 and V_1 are independent (given $\boldsymbol{U} = (a, 0)$). The joint probability density[21] of (V_0, V_1) at (v_0, v_1), conditional on hypothesis $\boldsymbol{U} = (a, 0)$, is therefore given by

$$f_0(v_0, v_1) = \frac{1}{(2\pi)^2 (a^2/2 + WN_0/2)(WN_0/2)} \exp\left\{ -\frac{|v_0|^2}{a^2 + WN_0} - \frac{|v_1|^2}{WN_0} \right\}, \tag{9.52}$$

[20] For example, if the channel is slowly varying, differential phase modulation, where data are sent by the difference between the phase of successive signals, could be used.

[21] V_0 and V_1 are complex rvs, so the probability density of each is defined as probability per unit area in the real and complex plane. If V_0 and V_1 are represented by amplitude and phase, for example, the densities are different.

where f_0 denotes the conditional density given hypothesis $\boldsymbol{U} = (a, 0)$. Note that the density in (9.52) depends only on the magnitude and not the phase of v_0 and v_1. Treating the alternative hypothesis in the same way, and letting f_1 denote the conditional density given $\boldsymbol{U} = (0, a)$, we obtain

$$f_1(v_0, v_1) = \frac{1}{(2\pi)^2(a^2/2 + \mathsf{W}N_0/2)(\mathsf{W}N_0/2)} \exp\left\{ -\frac{|v_0|^2}{\mathsf{W}N_0} - \frac{|v_1|^2}{a^2 + \mathsf{W}N_0} \right\}. \tag{9.53}$$

The log likelihood ratio is then given by

$$\text{LLR}(v_0, v_1) = \ln\left\{ \frac{f_0(v_0, v_1)}{f_1(v_0, v_1)} \right\} = \frac{[|v_0|^2 - |v_1|^2]a^2}{(a^2 + \mathsf{W}N_0)(\mathsf{W}N_0)}. \tag{9.54}$$

The maximum likelihood (ML) decision rule is therefore to decode $\tilde{\boldsymbol{U}} = (a, 0)$ if $|v_0|^2 \geq |v_1|^2$ and decode $\tilde{\boldsymbol{U}} = (0, a)$ otherwise. Given the symmetry of the problem, this is certainly no surprise. It may, however, be somewhat surprising that this rule does not depend on any possible dependence between G_0 and G_1.

Next consider the ML probability of error. Let $X_m = |V_m|^2$ for $m = 0, 1$. The probability densities of $X_0 \geq 0$ and $X_1 \geq 0$, conditioning on $\boldsymbol{U} = (a, 0)$ throughout, are then given by

$$f_{X_0}(x_0) = \frac{1}{a^2 + \mathsf{W}N_0} \exp\left\{ -\frac{x_0}{a^2 + \mathsf{W}N_0} \right\}; \qquad f_{X_1}(x_1) = \frac{1}{\mathsf{W}N_0} \exp\left\{ -\frac{x_1}{\mathsf{W}N_0} \right\}.$$

Then, $\Pr(X_1 > x) = \exp(-x/\mathsf{W}N_0)$ for $x \geq 0$, and therefore

$$\begin{aligned} \Pr(X_1 > X_0) &= \int_0^\infty \frac{1}{a^2 + \mathsf{W}N_0} \exp\left\{ -\frac{x_0}{a^2 + \mathsf{W}N_0} \right\} \exp\left\{ -\frac{x_0}{\mathsf{W}N_0} \right\} \mathrm{d}x_0 \\ &= \frac{1}{2 + a^2/\mathsf{W}N_0}. \end{aligned} \tag{9.55}$$

Since $X_1 > X_0$ is the condition for an error when $\boldsymbol{U} = (a, 0)$, this is $\Pr(e)$ under the hypothesis $\boldsymbol{U} = (a, 0)$. By symmetry, the error probability is the same under the hypothesis $\boldsymbol{U} = (0, a)$, so this is the unconditional probability of error.

The mean signal power is $a^2/2$ since half the inputs have a square value a^2 and half have value 0. There are $\mathsf{W}/2$ binary symbols per second, so E_b, the energy per bit, is a^2/W. Substituting this into (9.55), we obtain

$$\Pr(e) = \frac{1}{2 + E_b/N_0}. \tag{9.56}$$

This is a very discouraging result. To get an error probability $\Pr(e) = 10^{-3}$ would require $E_b/N_0 \approx 1000$ (30 dB). Stupendous amounts of power would be required for more reliable communication.

After some reflection, however, this result is not too surprising. There is a constant signal energy E_b per bit, independent of the channel response G_m. The errors generally occur when the sample values $|g_m|^2$ are small, i.e. during fades. Thus the damage here is caused by the combination of fading and constant signal power. This result, and the result to follow, make it clear that to achieve reliable communication, it is necessary either to have diversity and/or coding between faded and unfaded parts of the channel, or to use channel measurement and feedback to control the signal power in the presence of fades.

9.6.2 Noncoherent detection with known channel magnitude

Consider the same binary pulse position modulation of Section 9.6.1, but now assume that G_0 and G_1 have the same magnitude, and that the sample value of this magnitude, say g, is a fixed parameter that is known at the receiver. The phase ϕ_m of G_m, $m = 0, 1$, is uniformly distributed over $[0, 2\pi)$ and is unknown at the receiver. The term noncoherent detection is used for detection that does not make use of a recovered carrier phase, and thus applies here. We will see that the joint density of ϕ_0 and ϕ_1 is immaterial. Assume the same noise distribution as before. Under hypothesis $\boldsymbol{U} = (a, 0)$, the outputs V_0 and V_1 are given by

$$V_0 = ag\exp\{\mathrm{i}\phi_0\} + Z_0, \qquad V_1 = Z_1 \qquad \text{(under } \boldsymbol{U} = (a, 0)). \tag{9.57}$$

Similarly, under $\boldsymbol{U} = (0, a)$,

$$V_0 = Z_0, \qquad V_1 = ag\exp\{\mathrm{i}\phi_1\} + Z_1 \qquad \text{(under } \boldsymbol{U} = (0, a)). \tag{9.58}$$

Only V_0 and V_1, along with the fixed channel magnitude g, can be used in the decision, but it will turn out that the value of g is not needed for an ML decision. The channel phases ϕ_0 and ϕ_1 are not observed and cannot be used in the decision.

The probability density of a complex rv is usually expressed as the joint density of the real and imaginary parts, but here it is more convenient to use the joint density of magnitude and phase. Since the phase ϕ_0 of $ag\exp\{\mathrm{i}\phi_0\}$ is uniformly distributed, and since Z_0 is independent with uniform phase, it follows that V_0 has uniform phase; i.e., $\angle V_0$ is uniform conditional on $\boldsymbol{U} = (a, 0)$. The magnitude $|V_0|$, conditional on $\boldsymbol{U} = (a, 0)$, is a Rician rv which is independent of ϕ_0, and therefore also independent of $\angle V_0$. Thus, conditional on $\boldsymbol{U} = (a, 0)$, V_0 has independent phase and amplitude, and uniformly distributed phase.

Similarly, conditional on $\boldsymbol{U} = (0, a)$, $V_0 = Z_0$ has independent phase and amplitude, and uniformly distributed phase. What this means is that both the hypothesis and $|V_0|$ are statistically independent of the phase $\angle V_0$. It can be seen that they are also statistically independent of ϕ_0.

Using the same argument on V_1, we see that both the hypothesis and $|V_1|$ are statistically independent of the phases $\angle V_1$ and ϕ_1. It should then be clear that $|V_0|$, $|V_1|$, and the hypothesis are independent of the phases $(\angle V_0, \angle V_1, \phi_0, \phi_1)$. This means that the sample values $|v_0|^2$ and $|v_1|^2$ are sufficient statistics for choosing between the hypotheses $\boldsymbol{U} = (a, 0)$ and $\boldsymbol{U} = (0, a)$.

Given the sufficient statistics $|v_0|^2$ and $|v_1|^2$, we must determine the ML detection rule, again assuming equiprobable hypotheses. Since v_0 contains the signal under hypothesis $U = (a, 0)$, and v_1 contains the signal under hypothesis $U = (0, a)$, and since the problem is symmetric between $U = (a, 0)$ and $U = (0, a)$, it appears obvious that the ML detection rule is to choose $U = (a, 0)$ if $|v_0|^2 > |v_1|^2$ and to choose $U = (0, a)$ otherwise. Unfortunately, to show this analytically it seems necessary to calculate the likelihood ratio. Appendix 9.11 gives this likelihood ratio and calculates the probability of error. The error probability for a given g is derived there as

$$\Pr(e) = \frac{1}{2}\exp\left(-\frac{a^2g^2}{2\mathsf{W}N_0}\right). \tag{9.59}$$

The mean received baseband signal power is $a^2g^2/2$ since only half the inputs are used. There are $\mathsf{W}/2$ bits per second, so $E_b = a^2g^2/\mathsf{W}$. Thus, this probability of error can be expressed as

$$\Pr(e) = \frac{1}{2}\exp\left(-\frac{E_b}{2N_0}\right) \qquad \text{(noncoherent)}. \tag{9.60}$$

It is interesting to compare the performance of this noncoherent detector with that of a coherent detector (i.e. a detector such as those in Chapter 8 that use the carrier phase) for equal-energy orthogonal signals. As seen in (8.27), the error probability in the latter case is given by

$$\Pr(e) = Q\left(\sqrt{\frac{E_b}{N_0}}\right) \approx \sqrt{\frac{N_0}{2\pi E_b}}\exp\left(-\frac{E_b}{2N_0}\right) \qquad \text{(coherent)}. \tag{9.61}$$

Thus both expressions have the same exponential decay with E_b/N_0 and differ only in the coefficient. The error probability with noncoherent detection is still substantially higher[22] than with coherent detection, but the difference is nothing like that in (9.56). More to the point, if E_b/N_0 is large, we see that the additional energy per bit required in noncoherent communication to make the error probability equal to that of coherent communication is very small. In other words, a small increment in dB corresponds to a large decrease in error probability. Of course, with noncoherent detection, we also pay a 3 dB penalty for not being able to use antipodal signaling.

Early telephone-line modems (in the 1200 bps range) used noncoherent detection, but current high-speed wireline modems generally track the carrier phase and use coherent detection. Wireless systems are subject to rapid phase changes because of the transmission medium, so noncoherent techniques are still common there.

It is even more interesting to compare the noncoherent result here with the Rayleigh fading result. Note that both use the same detection rule, and thus knowledge of the magnitude of the channel strength at the receiver in the Rayleigh case would not reduce the error probability. As shown in Exercise 9.11, if we regard g as a sample value of

---

[22] As an example, achieving $\Pr(e) = 10^{-6}$ with noncoherent detection requires E_b/N_0 to be 26.24, which would yield $\Pr(e) = 1.6 \times 10^{-7}$ with coherent detection. However, it would require only about 0.5 dB of additional power to achieve that lower error probability with noncoherent detection.

a rv that is known at the receiver, and average over the result in (9.59), then the error probability is the same as that in (9.56).

The conclusion from this comparison is that the real problem with binary communication over flat Rayleigh fading is that when the signal is badly faded, there is little hope for successful transmission using a fixed amount of signal energy. It has just been seen that knowledge of the fading amplitude at the receiver does not help. Also, as seen in the second part of Exercise 9.11, using power control at the transmitter to maintain a fixed error probability for binary communication leads to infinite average transmission power. The only hope, then, is either to use variable rate transmission or to use coding and/or diversity. In this latter case, knowledge of the fading magnitude will be helpful at the receiver in knowing how to weight different outputs in making a block decision.

Finally, consider the use of only V_0 and V_1 in binary detection for Rayleigh fading and noncoherent detection. If there are no inputs other than the binary input at times 0 and 1, then all other outputs can be seen to be independent of the hypothesis and of V_0 and V_1. If there are other inputs, however, the resulting outputs can be used to measure both the phase and amplitude of the channel taps.

The results in Sections 9.6.1 and 9.6.2 apply to any pair of equal-energy baseband signals that are orthogonal in the sense that both the real and imaginary parts of one waveform are orthogonal to both the real and imaginary parts of the other. For this more general result, however, we must assume that G_m is constant over the range of m used by the signals.

9.6.3 Noncoherent detection in flat Rician fading

Flat Rician fading occurs when the channel can be represented by a single tap and one path is significantly stronger than the other paths. This is a reasonable model when a line-of-sight path exists between transmitter and receiver, accompanied by various reflected paths. Perhaps more importantly, this model provides a convenient middle ground between a large number of weak paths, modeled by Rayleigh fading, and a single path with random phase, modeled in Section 9.6.2. The error probability is easy to calculate in the Rician case, and contains the Rayleigh case and known magnitude case as special cases. When we study diversity, the Rician model provides additional insight into the benefits of diversity.

As with Rayleigh fading, consider binary pulse-position modulation where $\boldsymbol{U} = \boldsymbol{u}^0 = (a, 0)$ under one hypothesis and $\boldsymbol{U} = \boldsymbol{u}^1 = (0, a)$ under the other hypothesis. The corresponding outputs are then given by

$$V_0 = U_0 G_0 + Z_0 \qquad \text{and} \quad V_1 = U_1 G_1 + Z_1.$$

Using noncoherent detection, ML detection is the same for Rayleigh, Rician, or deterministic channels; i.e., given sample values v_0 and v_1 at the receiver,

$$|v_0|^2 \underset{\tilde{U}=\boldsymbol{u}^1}{\overset{\tilde{U}=\boldsymbol{u}^0}{\gtrless}} |v_1|^2. \tag{9.62}$$

The magnitude of the strong path is denoted by $\bar{g}$ and the collective variance of the weaker paths is denoted by σ_g^2. Since only the magnitudes of v_0 and v_1 are used in detection, the phases of the tap gains G_0 and G_1 do not affect the decision, so the tap gains can be modeled as $G_0 \sim G_1 \sim \mathcal{CN}(\bar{g}, \sigma_g^2)$. This is explained more fully, for the known magnitude case, in Appendix 9.11.

From the symmetry between the two hypotheses, the error probability is clearly the same for both. Thus the error probability will be calculated conditional on $\boldsymbol{U} = \boldsymbol{u}^0$. All of the following probabilities and probability densities are assumed to be conditional on $\boldsymbol{U} = \boldsymbol{u}^0$. Under this conditioning, the real and imaginary parts of V_0 and V_1 are independent and characterized by

$$V_{0,\mathrm{re}} \sim \mathcal{N}(a\bar{g}, \sigma_0^2), \qquad V_{0,\mathrm{im}} \sim \mathcal{N}(0, \sigma_0^2),$$

$$V_{1,\mathrm{re}} \sim \mathcal{N}(0, \sigma_1^2), \qquad V_{1,\mathrm{im}} \sim \mathcal{N}(0, \sigma_1^2),$$

where

$$\sigma_0^2 = \frac{WN_0 + a^2\sigma_g^2}{2}; \qquad \sigma_1^2 = \frac{WN_0}{2}. \tag{9.63}$$

Observe that $|V_1|^2$ is an exponentially distributed rv and for any $x \geq 0$, $\Pr(|V_1|^2 \geq x) = \exp(-x/2\sigma_1^2)$. Thus the probability of error, conditional on $|V_0|^2 = x$, is $\exp(-x/2\sigma_1^2)$. The unconditional probability of error (still conditioning on $\boldsymbol{U} = \boldsymbol{u}^0$) can then be found by averaging over $\boldsymbol{V}_0$.

$$\Pr(e) = \int_{-\infty}^{\infty}\int_{-\infty}^{\infty} \frac{1}{2\pi\sigma_0^2} \exp\left[-\frac{(v_{0,\mathrm{re}} - a\bar{g})^2}{2\sigma_0^2} - \frac{v_{0,\mathrm{im}}^2}{2\sigma_0^2}\right] \exp\left[-\frac{v_{0,\mathrm{re}}^2 + v_{0,\mathrm{im}}^2}{2\sigma_1^2}\right] \mathrm{d}v_{0,\mathrm{re}}\,\mathrm{d}v_{0,\mathrm{im}}.$$

Integrating this over $v_{0,\mathrm{im}}$, we obtain

$$\Pr(e) = \sqrt{\frac{2\pi\sigma_0^2\sigma_1^2}{\sigma_0^2 + \sigma_1^2}} \int_{-\infty}^{\infty} \frac{1}{2\pi\sigma_0^2} \exp\left[-\frac{(v_{0,\mathrm{re}} - a\bar{g})^2}{2\sigma_0^2} - \frac{v_{0,\mathrm{re}}^2}{2\sigma_1^2}\right] \mathrm{d}v_{0,\mathrm{re}}.$$

This can be integrated by completing the square in the exponent, resulting in

$$\frac{\sigma_1^2}{\sigma_0^1 + \sigma_1^2} \exp\left[-\frac{a^2\bar{g}^2}{2(\sigma_0^2 + \sigma_1^2)}\right].$$

Substituting the values for σ_0 and σ_1 from (9.63), the result is as follows:

$$\Pr(e) = \frac{1}{2 + a^2\sigma_g^2/WN_0} \exp\left[-\frac{\bar{g}_2 a^2}{2WN_0 + a^2\sigma_g^2}\right].$$

Finally, the channel gain should be normalized so that $\bar{g}^2 + \sigma_g^2 = 1$. Then E_b becomes a^2/W and

$$\Pr(e) = \frac{1}{2 + E_\mathrm{b}\sigma_g^2/N_0} \exp\left[-\frac{\bar{g}^2 E_\mathrm{b}}{2N_0 + E_\mathrm{b}\sigma_g^2}\right]. \tag{9.64}$$

In the Rayleigh fading case, $\overline{g} = 0$ and $\sigma_g^2 = 1$, simplifying $\Pr(e)$ to $1/(2+E_b/N_0)$, agreeing with the result derived earlier. For the fixed amplitude case, $\overline{g} = 1$ and $\sigma_g^2 = 0$, reducing $\Pr(e)$ to $(1/2)\exp(-E_b/2N_0)$, again agreeing with the earlier result.

It is important to realize that this result does not depend on the receiver knowing that a strong path exists, since the detection rule is the same for noncoherent detection whether the fading is Rayleigh, Rician, or deterministic. The result says that, with Rician fading, the error probability can be much smaller than with Rayleigh. However, if $\sigma_g^2 > 0$, the exponent approaches a constant with increasing E_b, and $\Pr(e)$ still goes to 0 with $(E_b/N_0)^{-1}$. What this says, then, is that this slow approach to zero error probability with increasing E_b cannot be avoided by a strong specular path, but only by the lack of an arbitrarily large number of arbitrarily weak paths. This is discussed further in Section 9.8.

9.7 Channel measurement

This section introduces the topic of dynamically measuring the taps in the discrete-time baseband model of a wireless channel. Such measurements are made at the receiver based on the received waveform. They can be used to improve the detection of the received data, and, by sending the measurements back to the transmitter, to help in power and rate control at the transmitter.

One approach to channel measurement is to allocate a certain portion of each transmitted packet for that purpose. During this period, a known *probing sequence* is transmitted and the receiver uses this known sequence either to estimate the current values for the taps in the discrete-time baseband model of the channel or to measure the actual paths in a continuous-time baseband model. Assuming that the actual values for these taps or paths do not change rapidly, these estimated values can then help in detecting the remainder of the packet.

Another technique for channel measurement is called a *rake receiver*. Here the detection of the data and the estimation of the channel are done together. For each received data symbol, the symbol is detected using the previous estimate of the channel and then the channel estimate is updated for use on the next data symbol.

Before studying these measurement techniques, it will be helpful to understand how such measurements will help in detection. In studying binary detection for flat-fading Rayleigh channels, we saw that the error probability is very high in periods of deep fading, and that these periods are frequent enough to make the overall error probability large even when E_b/N_0 is large. In studying noncoherent detection, we found that the ML detector does not use its knowledge of the channel strength, and thus, for binary detection in flat Rayleigh fading, knowledge at the receiver of the channel strength is not helpful. Finally, we saw that when the channel is good (the instantaneous E_b/N_0 is high), knowing the phase at the receiver is of only limited benefit.

It turns out, however, that binary detection on a flat-fading channel is very much a special case, and that channel measurement can be very helpful at the receiver both for nonflat fading and for larger signal sets such as coded systems. Essentially, when the receiver observation consists of many degrees of freedom, knowledge of the channel helps the detector weight these degrees of freedom appropriately.

Feeding channel measurement information back to the transmitter can be helpful in general, even in the case of binary transmission in flat fading. The transmitter can then send more power when the channel is poor, thus maintaining a constant error probability,[23] or can send at higher rates when the channel is good. The typical round-trip delay from transmitter to receiver in cellular systems is usually on the order of a few microseconds or less, whereas typical coherence times are on the order of 100 ms or more. Thus feedback control can be exercised within the interval over which a channel is relatively constant.

9.7.1 The use of probing signals to estimate the channel

Consider a discrete-time baseband channel model in which the channel, at any given output time m, is represented by a given number k_0 of randomly varying taps, $G_{0,m}, \ldots, G_{k_0-1,m}$. We will study the estimation of these taps by the transmission of a probing signal consisting of a known string of input signals. The receiver, knowing the transmitted signals, estimates the channel taps. This procedure has to be repeated at least once for each coherence-time interval.

One simple (but not very good) choice for such a known signal is to use an input of maximum amplitude, say a, at a given epoch, say epoch 0, followed by zero inputs for the next $k_0 - 1$ epochs. The received sequence over the corresponding k_0 epochs in the absence of noise is then $(ag_{0,0}, ag_{1,1}, \ldots, ag_{k_0-1,k_0-1})$. In the presence of sample values $z_0, z_1, \ldots$ of complex discrete-time WGN, the output $\boldsymbol{v} = (v_0, \ldots, v_{k_0-1})^{\mathsf{T}}$ from time 0 to $k_0 - 1$ is given by

$$\boldsymbol{v} = (ag_{0,0} + z_0,\ ag_{1,1} + z_1,\ \ldots,\ ag_{k_0-1,k_0-1} + z_{k_0-1})^{\mathsf{T}}.$$

A reasonable estimate of the kth channel tap, $0 \le k \le k_0 - 1$, is then

$$\tilde{g}_{k,k} = \frac{v_k}{a}. \tag{9.65}$$

The principles of estimation are quite similar to those of detection, but are not essential here. In detection, an observation (a sample value v of a rv or vector V) is used to select a choice $\tilde{u}$ from the possible sample values of a discrete rv U (the hypothesis). In estimation, a sample value v of V is used to select a choice $\tilde{g}$ from the possible sample values of a continuous rv G. In both cases, the likelihoods $f_{V|U}(v|u)$ or $f_{V|G}(v|g)$ are assumed to be known and the a-priori probabilities $p_U(u)$ or $f_G(g)$ are assumed to be known.

Estimation, like detection, is concerned with determining and implementing reasonable rules for estimating g from v. A widely used rule is the *maximum likelihood*

[23] Exercise 9.11 shows that this leads to infinite expected power on a pure flat-fading Rayleigh channel, but in practice the very deep fades that require extreme instantaneous power simply lead to outages.

(ML) rule. This chooses the estimate $\tilde{g}$ to be the value of g that maximizes $f_{V|G}(v|g)$. The ML rule for estimation is the same as the ML rule for detection. Note that the estimate in (9.65) is a ML estimate.

Another widely used estimation rule is *minimum mean-square error* (MMSE) estimation. The MMSE rule chooses the estimate $\tilde{g}$ to be the mean of the a-posteriori probability density $f_{G|V}(g|v)$ for the given observation v. In many cases, such as where G and V are jointly Gaussian, this mean is the same as the value of g which maximizes $f_{G|V}(g|v)$. Thus the MMSE rule is somewhat similar to the MAP rule of detection theory.

For detection problems, the ML rule is usually chosen when the a-priori probabilities are all the same, and in this case ML and MAP are equivalent. For estimation problems, ML is more often chosen when the a-priori probability density is unknown. When the a-priori density is known, the MMSE rule typically has a strictly smaller mean-square-estimation error than the ML rule.

For the situation at hand, there is usually very little basis for assuming any given model for the channel taps (although Rayleigh and Rician models are frequently used in order to have something specific to discuss). Thus the ML estimate makes considerable sense and is commonly used. Since the channel changes very slowly with time, it is reasonable to assume that the measurement in (9.65) can be used at any time within a given coherence interval. It is also possible to repeat the above procedure several times within one coherence interval. The multiple measurements of each channel filter tap can then be averaged (corresponding to ML estimation based on the multiple observations).

The problem with the single-pulse approach above is that a peak constraint usually exists on the input sequence; this is imposed both to avoid excessive interference to other channels and also to simplify implementation. If the square of this peak constraint is little more than the energy constraint per symbol, then a long input sequence with equal energy in each symbol will allow much more signal energy to be used in the measurement process than the single-pulse approach. As seen in what follows, this approach will then yield more accurate estimates of the channel response than the single-pulse approach.

Using a predetermined antipodal *pseudo-noise* (PN) input sequence $\boldsymbol{u} = (u_1, \ldots, u_n)^{\mathsf{T}}$ is a good way to perform channel measurements with such evenly distributed energy.[24] The components $u_1, \ldots, u_n$ of $\boldsymbol{u}$ are selected to be $\pm a$, and the desired property is that the covariance function of $\boldsymbol{u}$ approximates an impulse. That is, the sequence is chosen to satisfy

$$\sum_{m=1}^{n} u_m u_{m+k} \approx \begin{cases} a^2 n; & k = 0 \\ 0; & k \neq 0 \end{cases} \quad = \quad a^2 n \delta_k, \tag{9.66}$$

[24] This approach might appear to be an unimportant detail here, but it becomes more important for the rake receiver to be discussed shortly.

where u_m is taken to be 0 outside of $[1, n]$. For long PN sequences, the error in this approximation can be viewed as additional but negligible noise. The implementation of such vectors (in binary rather than antipodal form) is discussed at the end of this subsection.

An almost obvious variation on choosing $\boldsymbol{u}$ to be an antipodal PN sequence is to choose it to be complex with antipodal real and imaginary parts, i.e. to be a 4-QAM sequence. Choosing the real and imaginary parts to be antipodal PN sequences and also to be approximately uncorrelated, (9.66) becomes

$$\sum_{m=1}^{n} u_m u_{m+k}^* \approx 2a^2 n\delta_k. \tag{9.67}$$

The QAM form spreads the input measurement energy over twice as many degrees of freedom for the given n time units, and is thus usually advantageous. Both the antipodal and the 4-QAM form, as well as the binary version of the antipodal form, are referred to as PN sequences. The QAM form is assumed in what follows, but the only difference between (9.66) and (9.67) is the factor of 2 in the covariance. It is also assumed for simplicity that (9.66) is satisfied with equality.

The condition (9.67) (with equality) states that $\boldsymbol{u}$ is orthogonal to each of its time shifts. This condition can also be expressed by defining the *matched filter* sequence for $\boldsymbol{u}$ as the sequence $\boldsymbol{u}^\dagger$, where $u_j^\dagger = u_{-j}^*$. That is, $\boldsymbol{u}^\dagger$ is the complex conjugate of $\boldsymbol{u}$ reversed in time. The convolution of $\boldsymbol{u}$ with $\boldsymbol{u}^\dagger$ is then $\boldsymbol{u} * \boldsymbol{u}^\dagger = \sum_m u_m u_{k-m}^\dagger$. The covariance condition in (9.67) (with equality) is then equivalent to the convolution condition:

$$\boldsymbol{u} * \boldsymbol{u}^\dagger = \sum_{m=1}^{n} u_m u_{k-m}^\dagger = \sum_{m=1}^{n} u_m u_{m-k}^* = 2a^2 n\delta_k. \tag{9.68}$$

Let the complex-valued rv $G_{k,m}$ be the value of the kth channel tap at time m. The channel output at time m for the input sequence $\boldsymbol{u}$ (before adding noise) is the convolution

$$V_m' = \sum_{k=0}^{n-1} G_{k,m} u_{m-k}. \tag{9.69}$$

Since $\boldsymbol{u}$ is zero outside of the interval $[1, n]$, the noise-free output sequence $\boldsymbol{V}'$ is zero outside of $[1, n+k_0-1]$. Assuming that the channel is random but unchanging during this interval, the kth tap can be expressed as the complex rv G_k. Correlating the channel output with $u_1^*, \ldots, u_n^*$ results in the covariance at each epoch j given by

$$C_j' = \sum_{m=-j+1}^{-j+n} V_m' u_{m+j}^* = \sum_{m=-j+1}^{-j+n} \sum_{k=0}^{n-1} G_k u_{m-k} u_{m+j}^* \tag{9.70}$$

$$= \sum_{k=0}^{n-1} G_k (2a^2 n)\delta_{j+k} = 2a^2 n G_{-j}. \tag{9.71}$$

Thus the result of correlation, in the absence of noise, is the set of channel filter taps, scaled and reversed in time.

It is easier to understand this by looking at the convolution of $\boldsymbol{V}'$ with $\boldsymbol{u}^\dagger$. That is,

$$\boldsymbol{V}' * \boldsymbol{u}^\dagger = (\boldsymbol{u} * \boldsymbol{G}) * \boldsymbol{u}^\dagger = (\boldsymbol{u} * \boldsymbol{u}^\dagger) * \boldsymbol{G} = 2a^2 n\boldsymbol{G}.$$

This uses the fact that convolution of sequences (just like convolution of functions) is both associative and commutative. Note that the result of convolution with the matched filter is the time reversal of the result of correlation, and is thus simply a scaled replica of the channel taps. Finally note that the matched filter $\boldsymbol{u}^\dagger$ is zero outside of the interval $[-n, -1]$. Thus if we visualize implementing the measurement of the channel using such a discrete filter, we are assuming (conceptually) that the receiver time reference lags the transmitter time reference by at least n epochs.

With the addition of noise, the overall output is $\boldsymbol{V} = \boldsymbol{V}' + \boldsymbol{Z}$, i.e. the output at epoch m is $V_m = V'_m + Z_m$. Thus the convolution of the noisy channel output with the matched filter $\boldsymbol{u}^\dagger$ is given by

$$\boldsymbol{V} * \boldsymbol{u}^\dagger = \boldsymbol{V}' * \boldsymbol{u}^\dagger + \boldsymbol{Z} * \boldsymbol{u}^\dagger = 2a^2 n\boldsymbol{G} + \boldsymbol{Z} * \boldsymbol{u}^\dagger. \tag{9.72}$$

After dividing by $2a^2n$, the kth component of this vector equation is

$$\frac{1}{2a^2n}\sum_m V_m u^\dagger_{k-m} = G_k + \Psi_k, \tag{9.73}$$

where Ψ_k is defined as the complex rv

$$\Psi_k = \frac{1}{2a^2n}\sum_m Z_m u^\dagger_{k-m}. \tag{9.74}$$

This estimation procedure is illustrated in Figure 9.9.

Assume that the channel noise is WGN so that the discrete-time noise variables $\{Z_m\}$ are circularly symmetric $\mathcal{CN}(0, \mathsf{W}N_0)$ and iid, where $\mathsf{W}/2$ is the baseband bandwidth.[25]

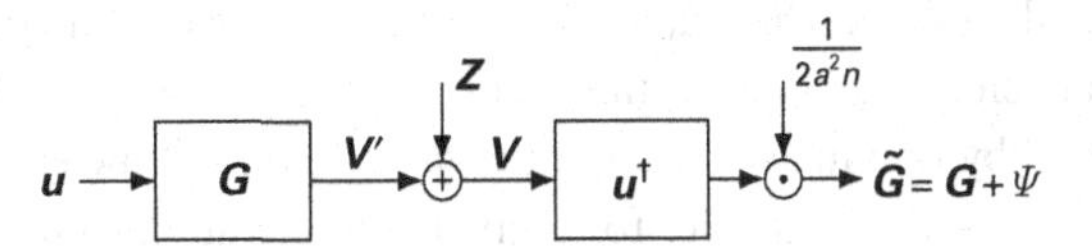

Figure 9.9. Channel measurement using a filter matched to a PN input. We have assumed that $\boldsymbol{G}$ is nonzero only in the interval $[0, k_0 - 1]$ so the output is observed only in this interval. Note that the component $\boldsymbol{G}$ in the output is the response of the matched filter to the input $\boldsymbol{u}$, whereas Ψ is the response to $\boldsymbol{Z}$.

[25] Recall that these noise variables are samples of white noise filtered to $\mathsf{W}/2$. Thus their mean-square value (including both real and imaginary parts) is equal to the bandlimited noise power $N_0\mathsf{W}$. Viewed alternatively, the sinc functions in the orthogonal expansion have energy $1/\mathsf{W}$, so the variance of each real and imaginary coefficient in the noise expansion must be scaled up by W from the noise energy $N_0/2$ per degree of freedom.

Since $\boldsymbol{u}$ is orthogonal to each of its time shifts, its matched filter vector $\boldsymbol{u}^\dagger$ must have the same property. It then follows that

$$\mathsf{E}[\Psi_k \Psi_i^*] = \frac{1}{4a^4 n^2} \sum_m \mathsf{E}[|Z_m|^2] u_{k-m}^\dagger (u_{i-m}^\dagger)^* = \frac{N_0 \mathsf{W}}{2a^2 n} \delta_{k-i}. \tag{9.75}$$

The random variables $\{\Psi_k\}$ are jointly Gaussian from (9.74) and uncorrelated from (9.75), so they are independent Gaussian rvs. It is a simple additional exercise to show that each Ψ_k is circularly symmetric, i.e. $\Psi_k \sim \mathcal{CN}(0, N_0\mathsf{W}/2a^2 n)$.

Going back to (9.73), it can be seen that for each k, $0 \le k \le k_0 - 1$, the ML estimate of G_k from the observation of $G_k + \Psi_k$ is given by

$$\tilde{G}_k = \frac{1}{2a^2 n} \sum_m V_m u_{k-m}^\dagger.$$

It can also be shown that this is the ML estimate of G_k from the entire observation $\boldsymbol{V}$, but deriving this would take us too far afield. From (9.73), the error in this estimate is Ψ_k, so the mean-squared error in the real part of this estimate, and similarly in the imaginary part, is given by $\mathsf{W}N_0/(4a^2 n)$.

By increasing the measurement length n or by increasing the input magnitude a, we can make the estimate arbitrarily good. Note that the mean-squared error is independent of the fading variables $\{G_k\}$; the noise in the estimate does not depend on how good or bad the channel is. Finally observe that the energy in the entire measurement signal is $2a^2 n\mathsf{W}$, so the mean-squared error is inversely proportional to the measurement-signal energy.

What is the duration over which a channel measurement is valid? Fortunately, for most wireless applications, the coherence time $\mathcal{T}_{\text{coh}}$ is many times larger than the delay spread, typically on the order of hundreds of times larger. This means that it is feasible to measure the channel and then use those measurements for an appreciable number of data symbols. There is, of course, a tradeoff, since using a long measurement period n leads to an accurate measurement, but uses an appreciable part of $\mathcal{T}_{\text{coh}}$ for measurement rather than data. This tradeoff becomes less critical as the coherence time increases.

One clever technique that can be used to increase the number of data symbols covered by one measurement interval is to do the measurement in the middle of a data frame. It is also possible, for a given data symbol, to interpolate between the previous and the next channel measurement. These techniques are used in the popular GSM cellular standard. These techniques appear to increase delay slightly, since the early data in the frame cannot be detected until after the measurement is made. However, if coding is used, this delay is necessary in any case. We have also seen that one of the primary purposes of measurement is for power/rate control, and this clearly cannot be exercised until after the measurement is made.

The above measurement technique rests on the existence of PN sequences which approximate the correlation property in (9.67). Pseudo-noise sequences (in binary form) are generated by a procedure very similar to that by which output streams are generated in a convolutional encoder. In a convolutional encoder of constraint

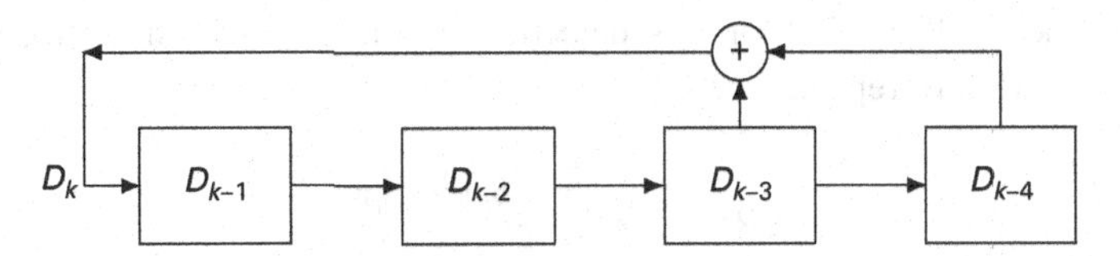

Figure 9.10. Maximal-length shift register with $n = 4$ stages and a cycle of length $2^n - 1$ that cycles through all states except the all-zero state.

length n, each bit in a given output stream is the mod-2 sum of the current input and some particular pattern of the previous n inputs. Here there are no inputs, but instead the output of the shift register is fed back to the input as shown in Figure 9.10.

By choosing the stages that are summed mod 2 in an appropriate way (denoted a *maximal-length shift register*), any nonzero initial state will cycle through all possible $2^n - 1$ nonzero states before returning to the initial state. It is known that maximal-length shift registers exist for all positive integers n.

One of the nice properties of a maximal-length shift register is that it is linear (over mod-2 addition and multiplication). That is, let $\boldsymbol{y}$ be the sequence of length $2^n - 1$ bits generated by the initial state $\boldsymbol{x}$, and let $\boldsymbol{y}'$ be that generated by the initial state $\boldsymbol{x}'$. Then it can be seen with a little thought that $\boldsymbol{y} \oplus \boldsymbol{y}'$ is generated by $\boldsymbol{x} \oplus \boldsymbol{x}'$. Thus the difference between any two such cycles started in different initial states contains 1 in 2^{n-1} positions and 0 in the other $2^{n-1} - 1$ positions. In other words, the set of cycles forms a binary simplex code.

It can be seen that any nonzero cycle of a maximal-length shift register has an almost ideal correlation with a cyclic shift of itself. Here, however, it is the correlation over a single period, where the shifted sequence is set to 0 outside of the period, that is important. There is no guarantee that such a correlation is close to ideal, although these shift register sequences are usually used in practice to approximate the ideal.

9.7.2 Rake receivers

A rake receiver is a type of receiver that combines channel measurement with data reception in an iterative way. It is primarily applicable to spread spectrum systems in which the input signals are pseudo-noise (PN) sequences. It is, in fact, just an extension of the PN measurement technique described in Section 9.7.1. Before describing the rake receiver, it will be helpful to review binary detection, assuming that the channel is perfectly known and unchanging over the duration of the signal.

Let the input $\boldsymbol{U}$ be one of the two signals $\boldsymbol{u}^0 = (u_1^0, \ldots, u_n^0)^\mathsf{T}$ and $\boldsymbol{u}^1 = (u_1^1, \ldots, u_n^1)^\mathsf{T}$. Denote the known channel taps as $\boldsymbol{g} = (g_0, \ldots, g_{k_0-1})^\mathsf{T}$. Then the channel output, before the addition of white noise, is either $\boldsymbol{u}^0 * \boldsymbol{g}$, which we denote by $\boldsymbol{b}_0$, or $\boldsymbol{u}^1 * \boldsymbol{g}$, which we denote by $\boldsymbol{b}_1$. These convolutions are contained within the interval $[1, n+k_0-1]$. After the addition of WGN, the output is either $\boldsymbol{V} = \boldsymbol{b}_0 + \boldsymbol{Z}$ or $\boldsymbol{V} = \boldsymbol{b}_1 + \boldsymbol{Z}$. The detection problem is to decide, from observation of $\boldsymbol{V}$, which of these two possibilities is more likely. The LLR for this detection problem is shown in Section 8.3.4 to be given by (8.28), repeated here:

$$\begin{aligned}\mathrm{LLR}(\boldsymbol{v}) &= \frac{-\|\boldsymbol{v}-\boldsymbol{b}_0\|^2 + \|\boldsymbol{v}-\boldsymbol{b}_1\|^2}{N_0} \\ &= \frac{2\Re(\langle \boldsymbol{v}, \boldsymbol{b}_0\rangle) - 2\Re(\langle \boldsymbol{v}, \boldsymbol{b}_1\rangle) - \|\boldsymbol{b}_0\|^2 + \|\boldsymbol{b}_1\|^2}{N_0}.\end{aligned} \tag{9.76}$$

It is shown in Exercise 9.17 that if $\boldsymbol{u}^0$ and $\boldsymbol{u}^1$ are ideal PN sequences, i.e. sequences that satisfy (9.68), then $\|\boldsymbol{b}_0\|^2 = \|\boldsymbol{b}_1\|^2$. The ML test then simplifies as follows:

$$\Re(\langle \boldsymbol{v}, \boldsymbol{u}^0 * \boldsymbol{g}\rangle) \underset{\tilde{U}=\boldsymbol{u}^1}{\overset{\tilde{U}=\boldsymbol{u}^0}{\gtrless}} \Re(\langle \boldsymbol{v}, \boldsymbol{u}^1 * \boldsymbol{g}\rangle). \tag{9.77}$$

Finally, for $i = 0, 1$, the inner product $\langle \boldsymbol{v}, \boldsymbol{u}^i * \boldsymbol{g}\rangle$ is simply the output at epoch 0 when $\boldsymbol{v}$ is the input to a filter matched to $\boldsymbol{u}^i * \boldsymbol{g}$. The filter matched to $\boldsymbol{u}^i * \boldsymbol{g}$, however, is just the filter matched to $\boldsymbol{u}^i$ convolved with the filter matched to $\boldsymbol{g}$. The block diagram for this is shown in Figure 9.11.

If the signals $\boldsymbol{u}^0$ and $\boldsymbol{u}^1$ are PN sequences, there is a great similarity between Figures 9.9 and 9.11. In particular, if $\boldsymbol{u}^0$ is sent, then the output of the matched filter $(\boldsymbol{u}^0)^\dagger$, i.e. the first part of the lower matched filter, will be $2a^2n\boldsymbol{g}$ in the absence of noise. Note that $\boldsymbol{g}$ is a vector, meaning that the noise-free output at epoch k is $2a^2ng_k$ Similarly, if $\boldsymbol{u}^1$ is sent, then the noise-free output of the first part of the upper matched filter, at epoch k, will be a^2ng_k. The decision is made at receiver time 0 after the sequence $2a^2n\boldsymbol{g}$, along with noise, passes through the unrealizable filter $\boldsymbol{g}^\dagger$. These unrealizable filters are made realizable by the delay in receiver timing relative to transmitter timing.

Under the assumption that a correct decision is made, an estimate can also be made of the channel filter $\boldsymbol{g}$. In particular, if the decision is $\tilde{U} = \boldsymbol{u}^0$, then the outputs of the first part of the lower matched filter, at receiver times $-k_0+1$ to 0, will be scaled noisy versions of g_0 to g_{k_0-1}. Instead of using these outputs as a ML estimate of the filter taps, they must be combined with earlier estimates, constantly updating the current estimate

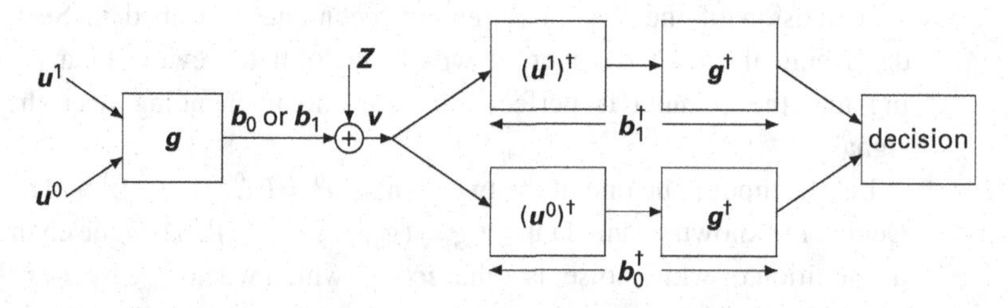

Figure 9.11. Detection for binary signals passed through a known filter $\boldsymbol{g}$. The real parts of the inputs entering the decision box at epoch 0 are compared; $\tilde{U} = \boldsymbol{u}^0$ if the real part of the lower input is larger, and $\tilde{U} = \boldsymbol{u}^1$ is chosen otherwise.

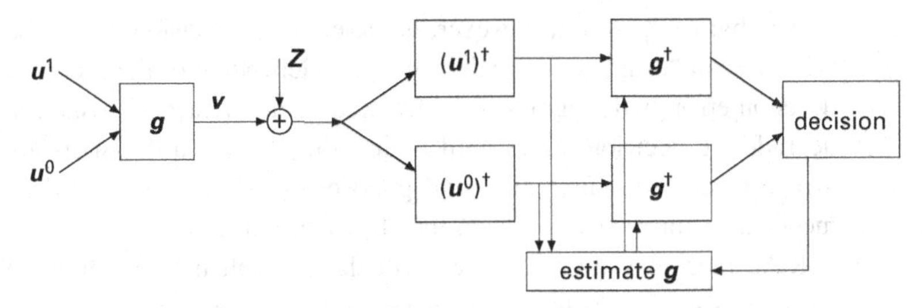

Figure 9.12. Rake receiver. If $\tilde{U} = \boldsymbol{u}^0$, the corresponding k_0 outputs from the matched filter $(\boldsymbol{u}^0)^\dagger$ are used to update the estimate of $\boldsymbol{g}$ (and thus the taps of each matched filter $\boldsymbol{g}^\dagger$). Alternatively, if $\tilde{U} = \boldsymbol{u}^1$, the output from the matched filter $(\boldsymbol{u}^1)^\dagger$ is used. These updated matched filters $\boldsymbol{g}^\dagger$ are then used, with the next block of outputs from $(\boldsymbol{u}^0)^\dagger$ and $(\boldsymbol{u}^1)^\dagger$, to make the next decision, and so forth for subsequent estimates and decisions.

each n epochs. This means that if the coherence time is long, then the filter taps will change very slowly in time, and the continuing set of channel estimates, one each n sample times, can be used to continually improve and track the channel filter taps.

Note that the decision in Figure 9.11 was based on knowledge of $\boldsymbol{g}$ and thus knowledge of the matched filter $\boldsymbol{g}^\dagger$. The ability to estimate $\boldsymbol{g}$ as part of the data detection thus allows us to improve the estimate $\boldsymbol{g}^\dagger$ at the same time as making data decisions. When $\tilde{U} = \boldsymbol{u}^i$ (and the decision is correct), the outputs of the matched filter $(\boldsymbol{u}^i)^\dagger$ provide an estimate of $\boldsymbol{g}$, and thus allow $\boldsymbol{g}^\dagger$ to be updated. The combined structure for making decisions and estimating the channel is called a *rake receiver* and is illustrated in Figure 9.12.

The rake receiver structure can only be expected to work well if the coherence time of the channel includes many decision points. That is, the updated channel estimate made on one decision can only be used on subsequent decisions. Since the channel estimates made at each decision epoch are noisy, and since the channel changes very slowly, the estimate $\hat{\boldsymbol{g}}$ made at one decision epoch will only be used to make a small change to the existing estimate.

A rough idea of the variance in the estimate of each tap g_k can be made by continuing to assume that decisions are made correctly. Assuming as before that the terms in the input PN sequences have magnitude a, it can be seen from (9.75) that for each signaling interval of n samples, the variance of the measurement noise (in each of the real and imaginary directions) is $\mathsf{W}N_0/(4a^2n)$. There are roughly $\mathcal{T}_{\text{coh}}\mathsf{W}/n$ signaling intervals in a coherence-time interval, and we can approximate the estimate of g_k as the average of those measurements. This reduces the measurement noise by a factor of $\mathcal{T}_{\text{coh}}\mathsf{W}/n$, reducing the variance of the measurement error[26] to $N_0/(4a^2\mathcal{T}_{\text{coh}})$.

[26] The fact that the variance of the measurement error does not depend on W might be surprising. The estimation error per discrete epoch 1/W is $\mathsf{W}N_0/(4a^2\mathcal{T}_{\text{coh}})$, which increases with W, but the number of measurements per second increases in the same way, leading to no overall variation with W. Since the number of taps is increasing with W, however, the effect of estimation errors increases with W. However, this assumes a model in which there are many paths with propagation delays within 1/W of each other, and this is probably a poor assumption when W is large.

An obvious question, however, is the effect of decision errors. Each decision error generates an "estimate" of each g_k that is independent of the true g_k. Clearly, too many decision errors will degrade the estimated value of each g_k, which in turn will further degrade the decision errors until both estimations and decisions are worthless. Thus a rake receiver requires an initial good estimate of each g_k and also requires some mechanism for recovering from the above catastrophe.

Rake receivers are often used with larger alphabets of input PN sequences, and the analysis of such nonbinary systems is the same as for the binary case above. For example, the IS95 cellular standard to be discussed later uses spread spectrum techniques with a bandwidth of 1.25 MHz. In this system, a signal set of 64 orthogonal signal waveforms is used with a 64-ary rake receiver. In that example, however, the rake receiver uses noncoherent techniques.

Usually, in a rake system, the PN sequences are chosen to be mutually orthogonal, but this is not really necessary. So long as each signal is a PN sequence with the appropriate autocorrelation properties, the channel estimation will work as before. The decision element for the data, of course, must be designed for the particular signal structure. For example, we could even use binary antipodal signaling, given some procedure to detect if the channel estimates become inverted.

9.8 Diversity

Diversity has been mentioned several times in the previous sections as a way to reduce error probabilities at the receiver. Diversity refers to a rather broad set of techniques, and the model of the last two sections must be generalized somewhat.

The first part of this generalization is to represent the baseband modulated waveform as an orthonormal expansion $u(t) = \sum_k u_k \phi_k(t)$ rather than the sinc expansion of Sections 9.6 and 9.7. For the QAM-type systems in these sections, this is a somewhat trivial change. The modulation pulse $\mathrm{sinc}(\mathsf{W}t)$ is normalized to $\mathsf{W}^{-1/2}\ \mathrm{sinc}(\mathsf{W}t)$. With this normalization, the noise sequence $Z_1, Z_2, \ldots$ becomes $Z_k \sim \mathcal{CN}(0, N_0)$ for $k \in \mathbb{Z}^+$, and the antipodal input signal $\pm a$ satisfies $a^2 = E_{\mathrm{b}}$.

Before discussing other changes in the model, we give a very simple example of diversity using the tapped gain model of Section 9.5.

Example 9.8.1 Consider a Rayleigh fading channel modeled as a two-tap discrete-time baseband model. The input is a discrete-time sequence U_m and the output is a discrete-time complex sequence described, as illustrated in Figure 9.13, by

$$V_m = G_{0,m}U_m + G_{1,m}U_{m-1} + Z_m.$$

For each m, $G_{0,m}$ and $G_{1,m}$ are iid circularly symmetric complex Gaussian rvs with $G_{0,m} \sim \mathcal{CN}(0, 1/2)$. This satisfies the condition $\sum_k \mathsf{E}[|G_k|^2] = 1$ given in (9.48). The correlation of $G_{0,m}$ and $G_{1,m}$ with m is immaterial here, and can be assumed uncorrelated. Assume that the sequence Z_m is a sequence of iid circularly symmetric complex Gaussian rvs, $Z_m \sim \mathcal{CN}(0, N_0)$.

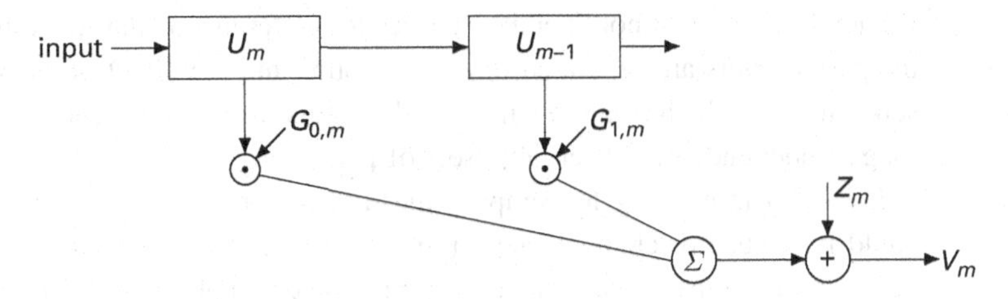

Figure 9.13. Two-tap discrete-time Rayleigh fading model.

Assume that a single binary digit is sent over this channel, sending either $\boldsymbol{u}^0 = (\sqrt{E_b}, 0, 0, 0)$ or $\boldsymbol{u}^1 = (0, 0, \sqrt{E_b}, 0)$, each with equal probability. The input for the first hypothesis is at epoch 0 and for the second hypothesis at epoch 2, thus allowing a separation between the responses from the two hypotheses.

Conditional on $\boldsymbol{U} = \boldsymbol{u}^0$, it can be seen that $V_0 \sim \mathcal{CN}(0, E_b/2 + N_0)$, where the signal contribution to V_0 comes through the first tap. Similarly, $V_1 \sim \mathcal{CN}(0, E_b/2 + N_0)$, with the signal contribution coming through the second tap. Given $\boldsymbol{U} = \boldsymbol{u}^0$, $V_2 \sim \mathcal{CN}(0, N_0)$ and $V_3 \sim \mathcal{CN}(0, N_0)$. Since the noise variables and the two gains are independent, it can be seen that $V_0, \ldots, V_3$ are independent conditional on $\boldsymbol{U} = \boldsymbol{u}^0$. The reverse situation occurs for $\boldsymbol{U} = \boldsymbol{u}^1$, with $V_m \sim \mathcal{CN}(0, E_b/2 + N_0)$ for $m = 2, 3$ and $V_m \sim \mathcal{CN}(0, N_0)$ for $m = 0, 1$.

Since $\angle V_m$ for $0 \le m \le 3$ are independent of the hypothesis, it can be seen that the energy in the set of received components, $X_m = |V_m|^2$, $0 \le m \le 3$, forms a sufficient statistic. Under hypothesis $\boldsymbol{u}^0$, X_0 and X_1 are exponential rvs with mean $E_b/2 + N_0$, and X_2 and X_3 are exponential with mean N_0; all are independent. Thus the probability densities of X_0 and X_1 (given $\boldsymbol{u}^0$) are given by $\alpha e^{-\alpha x}$ for $x \ge 0$, where $\alpha = 1/(N_0 + E_b/2)$. Similarly, the probability densities of X_2 and X_3 are given by $\beta e^{-\beta x}$ for $x \ge 0$, where $\beta = 1/N_0$. The reverse occurs under hypothesis $\boldsymbol{u}^1$.

The LLR and the probability of error (under ML detection) are then evaluated in Exercise 9.13 to be as follows:

$$\text{LLR}(\boldsymbol{x}) = (\beta - \alpha)(x_0 + x_1 - x_2 - x_3);$$

$$\Pr(e) = \frac{3\alpha^2\beta + \alpha^3}{(\alpha + \beta)^3} = \frac{4 + 3E_b/2N_0}{(2 + E_b/2N_0)^3}.$$

Note that as E_b/N_0 becomes large, the error probability approaches 0 as $(E_b/N_0)^{-2}$ instead of $(E_b/N_0)^{-1}$, as with flat Raleigh fading. This is a good example of diversity; errors are caused by high fading levels, but with two independent taps there is a much higher probability that one or the other has reasonable strength.

Note that multiple physical transmission paths give rise both to multipath fading and to diversity; the first usually causes difficulties and the second usually ameliorates those difficulties. It is important to understand what the difference is between them.

If the input bandwidth is chosen to be half as large as in Example 9.8.1, then the two-tap model would essentially become a one-tap model; this would lead to flat

Rayleigh fading and no diversity. The major difference is that with the two-tap model, the path outputs are separated into two groups and the effect of each can be observed separately. With the one-tap model, the paths are all combined, since there are no longer independently observable sets of paths.

It is also interesting to compare the diversity receiver above with a receiver that could make use of channel measurements. If the tap values were known, then a ML detector would involve a matched filter on the channel taps, as in Figure 9.12. In terms of the particular input in Example 9.8.1, this would weight the outputs from the two channel taps according to the magnitudes of the taps, whereas the diversity receiver above weights them equally. In other words, the diversity detector above does not do quite the right thing given known tap values, but it is certainly a large improvement over narrow-band transmission.

The type of diversity used above is called time diversity since it makes use of the delay between different sets of paths. The analysis hides a major part of the benefit to be gained by time diversity. For example, in the familiar reflecting wall example, there are only two paths. If the signal bandwidth is large enough that the response comes on different taps (or if the receiver measures the time delay on each path), then the fading will be eliminated.

It appears that many wireless situations, particularly those in cellular and local area networks, contain a relatively small number of significant coherent paths, and if the bandwidth is large enough to resolve these paths then the gain is far greater than that indicated in Example 9.8.1.

The diversity receiver can be generalized to other discrete models for wireless channels. For example, the frequency band could be separated into segments separated by the coherence frequency, thus getting roughly independent fading in each and the ability to separate the outputs in each of those bands. Diversity in frequency is somewhat different than diversity in time, since it does not allow the resolution of paths of different delays.

Another way to achieve diversity is through multiple antennas at the transmitter and receiver. Note that multiple antennas at the receiver allow the full received power available at one antenna to be received at each antenna, rather than splitting the power as occurs with time diversity or frequency diversity. For all of these more general ways to achieve diversity, the input and output should obviously be represented by the appropriate orthonormal expansions to bring out the diversity terms.

The two-tap example can be easily extended to an arbitrary number of taps. Assume the model of Figure 9.13, modified to have L taps, $G_{0,m}, \dots, G_{L-1,m}$, satisfying $G_{k,m} \sim \mathcal{CN}(0, 1/L)$ for $0 \le k \le L-1$. The input is assumed to be either $\boldsymbol{u}^0 = (\sqrt{E_b}, 0, \dots, 0)$ or $\boldsymbol{u}^1 = (0, \dots, 0, \sqrt{E_b}, 0, \dots, 0)$, where each of these $2L$-tuples has 0s in all but one position, namely position 0 for $\boldsymbol{u}^0$ and position L for $\boldsymbol{u}^1$. The energy in the set of received components, $X_m = |V_m|^2$, $0 \le m \le 2L-1$, forms a sufficient statistic for the same reason as in the dual diversity case. Under hypothesis $\boldsymbol{u}^0$, $X_0, \dots, X_{L-1}$ are exponential rvs with density $\alpha \exp(-\alpha x)$, where $\alpha = 1/(N_0 + E_b/L)$. Similarly, $X_L, \dots, X_{2L-1}$ are exponential rvs with density $\beta \exp(-\beta x)$. All are conditionally independent given $\boldsymbol{u}^0$. The reverse is true given hypothesis $\boldsymbol{u}^1$.

It can be seen that the ML detection rule is to choose $\boldsymbol{u}^0$ if $\sum_{m=0}^{L-1} X_m \geq \sum_{m=L}^{2L-1} X_m$ and to choose $\boldsymbol{u}^1$ otherwise. Exercise 9.14 then shows that the error probability is given by

$$\Pr(e) = \sum_{\ell=L}^{2L-1} \binom{2L-1}{\ell} p^{\ell}(1-p)^{2L-1-\ell},$$

where $p = \alpha/(\alpha+\beta)$. Substituting in the values for α and β, this becomes

$$\Pr(e) = \sum_{\ell=L}^{2L-1} \binom{2L-1}{\ell} \frac{(1+E_b/LN_0)^{2L-1-\ell}}{(2+E_b/LN_0)^{2L-1}}. \tag{9.78}$$

It can be seen that the dominant term in this sum is $\ell = L$. For any given L, then, the probability of error decreases with E_b as E_b^{-L}. At the same time, however, if L is increased for a given E_b, then eventually the probability of error starts to increase and approaches 1/2 asymptotically. In other words, increased diversity can decrease error probability up to a certain point, but then further increased diversity, for fixed E_b, is counter-productive.

If one evaluates (9.78) as a function of E_b/N_0 and L, one finds that $\Pr(e)$ is minimized for large but fixed E_b/N_0 when L is on the order of $0.3E_b/N_0$. The minimum is quite broad, but too much diversity does not help. The situation remains essentially the same with channel measurement. Here the problem is that, when the available energy is spread over too many degrees of freedom, there is not enough energy per degree of freedom to measure the channel.

The preceding discussion assumed that each diversity path is Rayleigh, but we have seen that, with time diversity, the individual paths might become separable, thus allowing much lower error probability than if the taps remain Rayleigh. Perhaps at this point we are trying to model the channel too accurately. If a given transmitter and receiver design is to be used over a broad set of different channel behaviors, then the important question is the fraction of behaviors over which the design works acceptably. This question ultimately must be answered experimentally, but simple models such as Rayleigh fading with diversity provide some insight into what to expect.

9.9 CDMA: the IS95 standard

In this section, IS95, one of the major classes of cellular standards, is briefly described. This system has been selected both because it is conceptually more interesting, and because most newer systems are focusing on this approach. This standard uses spread spectrum, which is often known by the name CDMA (code division multiple access). There is no convincing proof that spread spectrum is inherently superior to other approaches, but it does have a number of inherent engineering advantages over traditional narrow-band systems. Our main purpose, however, is to get some insight into how a major commercial cellular network system deals with some of the issues we have been discussing. The discussion here focuses on the issues arising with voice transmission.

IS95 uses a frequency band from 800 to 900 megahertz (MHz). The lower half of this band is used for transmission from cell phones to base station (the uplinks), and the upper half is used for base station to cell phones (the downlinks). There are multiple subbands[27] within this band, each 1.25 MHz wide. Each base station uses each of these subbands, and multiple cell phones within a cell can share the same subband. Each downlink subband is 45 MHz above the corresponding uplink subband. The transmitted waveforms are sufficiently well filtered at both the cell phones and the base stations so that they do not interfere appreciably with reception on the opposite channel.

The other two major established cellular standards use TDMA (time-division multiple access). The subbands are more narrow in TDMA, but only one cell phone uses a subband at a time to communicate with a given base station. In TDMA, there is little interference between different cell phones in the same cell, but considerable interference between cells. CDMA has more interference between cell phones in the same cell, but less between cells. A high-level block diagram for the parts of a transmitter is given in Figure 9.14.

The receiver, at a block level viewpoint (see Figure 9.15), performs the corresponding receiver functions in reverse order. This can be viewed as a layered system, although the choice of function in each block is somewhat related to that in the other blocks.

Section 9.9.1 discusses the voice compression layer; Sections 9.9.2 and 9.9.3 discuss the channel coding layer; and Sections 9.9.4 and 9.9.5 discuss the modulation layer. The voice compression and channel coding are quite similar in each of the standards, but the modulation is very different.

9.9.1 Voice compression

The voice waveform, in all of these standards, is first segmented into 20 ms increments. These segments are long enough to allow considerable compression, but short enough

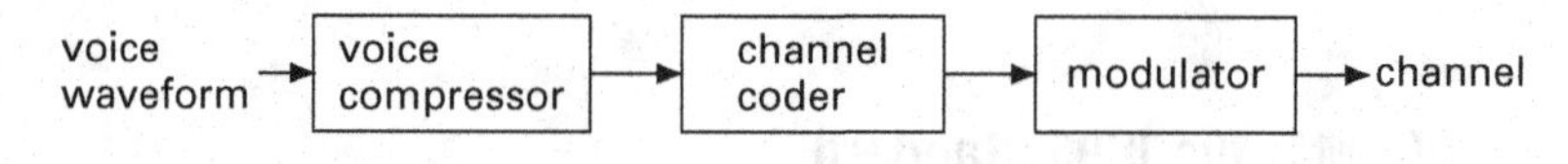

Figure 9.14. High-level block diagram of transmitters.

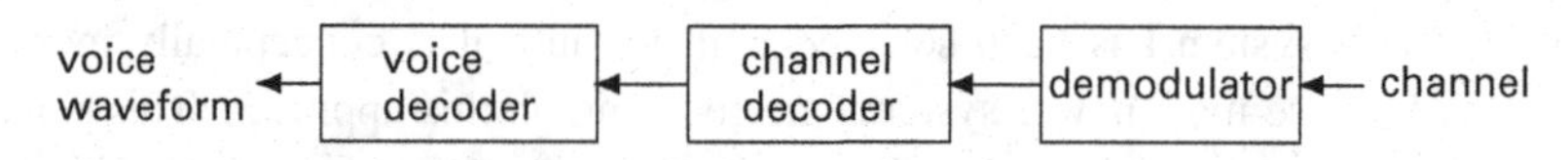

Figure 9.15. High-level block diagram of receiver.

[27] It is common in the cellular literature to use the word channel for a particular frequency subband; we will continue to use the word channel for the transmission medium connecting a particular transmitter and receiver. Later we use the words multiaccess channel to refer to the uplinks for multiple cell phones in the same cell.

to cause relatively little delay. In IS95, each 20 ms segment is encoded into 172 bits. The digitized voice rate is then $8600 = 172/0.02$ bits per second (bps). Voice compression has been an active research area for many years. In the early days, voice waveforms, which lie in a band from about 400–3200 Hz, were simply sampled at 8000 times a second, corresponding to a 4 kHz band. Each sample was then quantized to 8 bits for a total of 64 000 bps. Achieving high-quality voice at 8600 bps is still a moderate challenge today and requires considerable computation.

The 172 bits per 20 ms segment from the compressor is then extended by 12 bits per segment for error detection. This error detection is unrelated to the error correction algorithms to be discussed later, and is simply used to detect when those systems fail to correct the channel errors. Each of these 12 bits is a parity check (i.e. a modulo-2 sum) of a prespecified set of the data bits. Thus, it is very likely, when the channel decoder fails to decode correctly, that one of these parity checks will fail to be satisfied. When such a failure occurs, the corresponding frame is mapped into 20 ms of silence, thus avoiding loud squawking noises under bad channel conditions.

Each segment of $172 + 12$ bits is then extended by 8 bits, all set to 0. These bits are used as a terminator sequence for the convolutional code to be described shortly. With the addition of these bits, each 20 ms segment generates 192 bits, so this overhead converts the rate from 8600 to 9600 bps. The timing everywhere else in the transmitter and receiver is in multiples of this bit rate. In all the standards, many overhead items creep in, each performing small but necessary functions, but each increasing the overall required bit rate.

9.9.2 Channel coding and decoding

The channel encoding and decoding use a convolutional code and a Viterbi decoder. The convolutional code has rate 1/3, thus producing three output bits per input bit, and mapping the 9600 bps input into a 28.8 kbps output. The choice of rate is not very critical, since it involves how much coding is done here and how much is done later as part of the modulation proper. The convolutional encoder has a constraint length of 8, so each of the three outputs corresponding to a given input depends on the current input plus the eight previous inputs. There are then $2^8 = 256$ possible states for the encoder, corresponding to the possible sets of values for the previous eight inputs.

The complexity of the Viterbi algorithm is directly proportional to the number of states, so there is a relatively sharp tradeoff between complexity and error probability. The fact that decoding errors are caused primarily by more fading than expected (either a very deep fade that cannot be compensated by power control or by an inaccurate channel measurement), suggests that increasing the constraint length from 8 to 9 would, on the one hand be somewhat ineffective, and, on the other hand, double the decoder complexity.

The convolutional code is terminated at the end of each voice segment, thus turning the convolutional encoder into a block code of block length 576 and rate $1/3$, with 192 input bits per segment. As mentioned in Section 9.9.1, these 192 bits include 8 bits to terminate the code and return it to state 0. Part of the reason for this termination is the

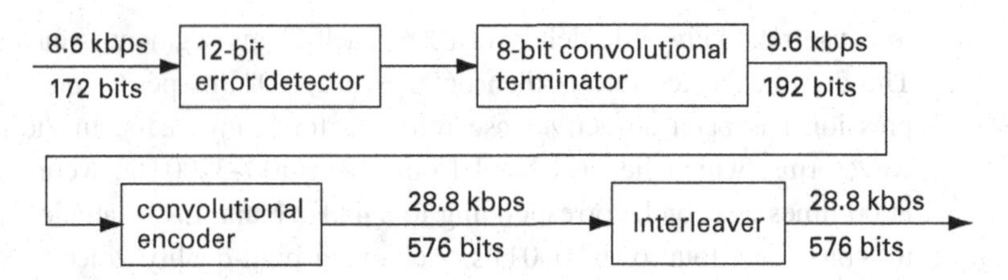

Figure 9.16. Block diagram of channel encoding, giving the number of bits in a segment (and the corresponding bit rate in kbps) at each stage

requirement for small delay, and part is the desire to prevent a fade in one segment from causing errors in multiple voice segments (the failure to decode correctly in one segment makes decoding in the next segment less reliable in the absence of this termination).

When a Viterbi decoder makes an error, it is usually detectable from the likelihood ratios in the decoder, so the 12-bit overhead for error detection could probably have been avoided. Many such tradeoffs between complexity, performance, and overhead must be made in both standards and products.

The decoding uses soft decisions from the output of the demodulator. The ability to use likelihood information (i.e. soft decisions) from the demodulator is one reason for the use of convolutional codes and Viterbi decoding. Viterbi decoding uses this information in a natural way, whereas, for some other coding and decoding techniques, this can be unnatural and difficult. All of the major standards use convolutional codes, terminated at the end of each voice segment, and decode with the Viterbi algorithm. It is worth noting that channel measurements are useful in generating good likelihood inputs to the Viterbi decoder.

The final step in the encoding process is to interleave the 576 output bits from the encoder corresponding to a given voice segment. Correspondingly, the first step in the decoding process is to de-interleave the bits (actually the soft decisions) coming out of the demodulator. It can be seen without analysis that if the noise coming into a Viterbi decoder is highly correlated, then the Viterbi decoder, with its short constraint length, is more likely to make a decoding error than if the noise is independent. Section 9.9.3 will show that the noise from the demodulator is in fact highly correlated, and thus the interleaving breaks up this correlation. Figure 9.16 summarizes this channel encoding process.

9.9.3 Viterbi decoding for fading channels

In order to provide some sense of why the above convolutional code with Viterbi decoding will not work very well if the coding is followed by straightforward binary modulation, suppose the pulse-position modulation of Section 9.6.1 is used and the channel is represented by a single tap with Rayleigh fading. The resulting bandwidth is well within typical values of $\mathcal{F}_{\text{coh}}$, so the single-tap model is reasonable. The coherence time is typically at least 1 ms, but, in the absence of moving vehicles, it could easily be more than 20 ms.

This means that an entire 20 ms segment of voice could easily be transmitted during a deep fade, and the convolutional encoder, even with interleaving within that 20 ms would not be able to decode successfully. If the fading is much faster, the Viterbi decoder, with likelihood information on the incoming bits, would probably work fairly successfully, but that is not something that can be relied upon.

There are only three remedies for this situation. One is to send more power when the channel is faded. As shown in Exercise 9.11, however, if the input power compensates completely for the fading (i.e. the input power at time m is $1/|g_m|^2$), then the expected input power is infinite. This means that, with finite average power, deep fades for prolonged periods cause outages.

The second remedy is diversity, in which each codeword is spread over enough coherence bandwidths or coherence-time intervals to achieve averaging over the channel fades. Using diversity over several coherence-time intervals causes delays proportional to the coherence-time, which is usually unacceptable for voice. Diversity can be employed by using a bandwidth larger than the coherence frequency (this can be done using multiple taps in the tapped delay line model or multiple frequency bands).

The third remedy is the use of variable rate transmission. This is not traditional for voice, since the voice encoding traditionally produces a constant rate stream of input bits into the channel, and the delay constraint is too stringent to queue this input and transmit it when the channel is good. It would be possible to violate the source/channel separation principle and have the source produce "important bits" at one rate and "unimportant bits" at another rate. Then, when the channel is poor, only the important bits would be transmitted. Some cellular systems, particularly newer ones, have features resembling this.

For data, however, variable rate transmission is very much a possibility since there is usually not a stringent delay requirement. Thus, data can be transmitted at high rate when the channel is good and at low or zero rate when the channel is poor. Newer systems also take advantage of this possibility.

9.9.4 Modulation and demodulation

The final part of the high-level block diagram of the IS95 transmitter is to modulate the output of the interleaver before channel transmission. This is where spread spectrum comes in, since this 28.8 kbps data stream is now spread into a 1.25 MHz bandwidth. The bandwidth of the corresponding received spread waveform will often be broader than the coherence frequency, thus providing diversity protection against deep fades. A rake receiver will take advantage of this diversity. Before elaborating further on these diversity advantages, the mechanics of the spreading is described.

The first step of the modulation is to segment the interleaver output into strings of length 6, and then map each successive 6 bit string into a 64 bit binary string. The mapping maps each of the 64 strings of length 6 into the corresponding row of the

H_6 Hadamard matrix described in Section 8.6.1. Each row of this Hadamard matrix differs from each other row in 32 places, and each row, except the all-zero row, contains exactly 32 1s and 32 0s. It is thus a binary orthogonal code.

Suppose the selected word from this code is mapped into a PAM sequence by the 2-PAM map $\{0, 1\} \longrightarrow \{+a, -a\}$. These 64 sequences of binary antipodal values are called *Walsh functions*. The symbol rate coming out of this 6 bit to 64 bit mapping is $(64/6) \cdot 28\,800 = 307\,200$ symbols per second.

To get some idea of why these Walsh functions are used, let $x_1^k, \ldots, x_{64}^k$ be the kth Walsh function, amplified by a factor a, and consider this as a discrete-time baseband input. For simplicity, assume flat fading with a single-channel tap of amplitude g. Suppose that baseband WGN of variance $N_0/2$ (per real and imaginary part) is added to this sequence, and consider detecting which of the 64 Walsh functions was transmitted. Let E_s be the expected received energy for each of the Walsh functions. The noncoherent detection result from (9.59) shows that the probability that hypothesis j is more likely than k, given that $x^k(t)$ is transmitted, is $1/2\exp[-E_s/2N_0]$. Using the union bound over the 63 possible incorrect hypotheses, the probability of error, using noncoherent detection and assuming a single-tap channel filter, is

$$\Pr(e) \leq \frac{63}{2}\exp\left[\frac{-E_s}{2N_0}\right]. \tag{9.79}$$

The probability of error is not the main subject of interest here, since the detector output comprises soft decisions that are then used by the Viterbi decoder. However, the error probability lets us understand the rationale for using such a large signal set with orthogonal signals.

If coherent detection were used, the analogous union bound on error probability would be $63Q(\sqrt{E_s/N_0})$. As discussed in Section 9.6.2, this goes down exponentially with E_s in the same way as (9.79), but the coefficient is considerably smaller. However, the number of additional dB required using noncoherent detection to achieve the same $\Pr(e)$ as coherent detection decreases almost inversely with the exponent in (9.79). This means that by using a large number of orthogonal functions (64 in this case), we make the exponent in (9.79) large in magnitude, and thus approach (in dB terms) what could be achieved by coherent detection.

The argument above is incomplete, because E_s is the transmitted energy per Walsh function. However, six binary digits are used to select each transmitted Walsh function. Thus, E_b in this case is $E_s/6$, and (9.79) becomes

$$\Pr(e) \leq 63\exp(-3E_b/N_0). \tag{9.80}$$

This large signal set also avoids the 3 dB penalty for orthogonal signaling rather than antipodal signaling that we have seen for binary signal sets. Here the cost of orthogonality essentially lies in using an orthogonal code rather than the corresponding biorthogonal code with 7 bits of input and 128 codewords,[28] i.e. a factor of 6/7 in rate.

[28] This biorthogonal code is called a (64, 7, 32) Reed–Muller code in the coding literature.

A questionable issue here is that two codes (the convolutional code as an outer code, followed by the Walsh function code as an inner code) are used in place of a single code. There seems to be no clean analytical way of showing that this choice makes good sense over all choices of single or combined codes. On the other hand, each code is performing a rather different function. The Viterbi decoder is eliminating the errors caused by occasional fades or anomalies, and the Walsh functions allow noncoherent detection and also enable a considerable reduction in error probability because of the large orthogonal signal sets rather than binary transmission.

The modulation scheme in IS95 next spreads the above Walsh functions into an even wider bandwidth transmitted signal. The stream of binary digits out of the Hadamard encoder[29] is combined with a pseudo-noise (PN) sequence at a rate of 1228.8 kbps, i.e. four PN bits for each signal bit. In essence, each bit of the 307.2 kbps stream out of the Walsh encoder is repeated four times (to achieve the 1228.8 kbps rate) and is then added mod-2 to the PN sequence. This further spreading provides diversity over the available 1.25 MHz bandwidth.

The constraint length here is $n = 42$ binary digits, so the period of the cycle is $2^{42} - 1$ (about 41 days). We can ignore the difference between simplex and orthogonal, and simply regard each cycle as orthogonal to each other cycle. Since the cycle is so long, however, it is better simply to approximate each cycle as a sequence of iid binary digits. There are several other PN sequences used in the IS95 standard, and this one, because of its constraint length, is called the "long PN sequence." PN sequences have many interesting properties, but for us it is enough to view them as iid but also known to the receiver.

The initial state of the long PN sequence is used to distinguish between different cell phones, and in fact this initial state is the only part of the transmitter system that is specific to a particular cell phone.

The resulting binary stream, after adding the long PN sequence, is at a rate of 1.2288 Mbps. This stream is duplicated into two streams prior to being quadrature modulated onto a cosine and sine carrier. The cosine stream is added mod-2 to another PN-sequence (called the in-phase or I-PN) sequence at rate 1.2288 Mbps, and the sine stream is added mod-2 to another PN sequence called the quadrature, or Q-PN, sequence. The I-PN and Q-PN sequences are the same for all cell phones and help in demodulation.

The final part of modulation is for the two binary streams to go through a 2-PAM map into digital streams of $\pm a$. Each of these streams (over blocks of 256 bits) maintains the orthogonality of the 64 Walsh functions. Each of these streams is then passed through a baseband filter with a sharp cutoff at the Nyquist bandwidth of 614.4 kHz. This is then quadrature modulated onto the carrier with a bandwidth of 614.4 kHz above and below the carrier, for an overall bandwidth of 1.2288 MHz. Note that almost all the modulation operation here is digital, with only the final filter and

[29] We visualized mapping the Hadamard binary sequences by a 2-PAM map into Walsh functions for simplicity. For implementation, it is more convenient to maintain binary (0,1) sequences until the final steps in the modulation process are completed.

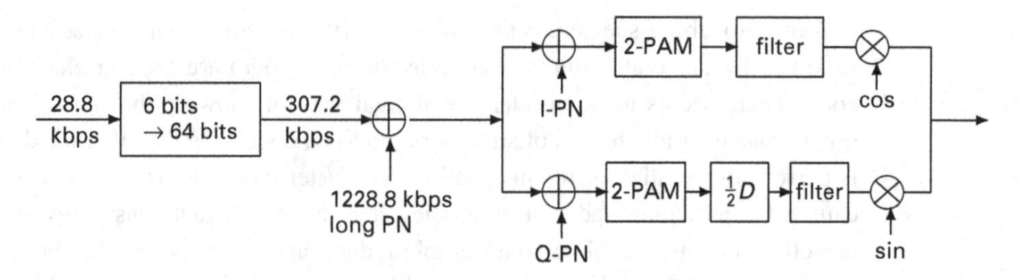

Figure 9.17. Block diagram of source and channel encoding.

modulation being analog. The question of what should be done digitally and what in analog form (other than the original binary interface) is primarily a question of ease of implementation. A block diagram of the modulator is shown in Figure 9.17.

Next consider the receiver. The fixed PN sequences that have been added to the Walsh functions do not alter the orthogonality of the signal set, which now consists of 64 functions, each of length 256 and each (viewed at baseband) containing both a real and imaginary part. The received waveform, after demodulation to baseband and filtering, is passed through a rake receiver similar to the one discussed earlier. The rake receiver here has a signal set of 64 signals rather than 2. Also, the channel here is viewed not as taps at the sampling rate, but rather as three taps at locations dynamically moved to catch the major received paths. As mentioned before, the detection is noncoherent rather than coherent.

The output of the rake receiver is a likelihood value for each of the 64 hypotheses. This is then converted into a likelihood value for each of the 6 bits in the inverse of the 6 bit to 64 bit Hadamard code map.

One of the reasons for using an interleaver between the convolutional code and the Walsh function encoder is now apparent. After the Walsh function detection, the errors in the string of 6 bits from the detection circuit have highly correlated errors. The Viterbi decoder does not work well with bursts of errors, so the interleaver spreads these errors out, allowing the Viterbi decoder to operate with noise that is relatively independent from bit to bit.

9.9.5 Multiaccess interference in IS95

A number of cell phones will use the same 1.2288 MHz frequency band in communicating with the same base station, and other nearby cell phones will also use the same band in communicating with their base stations. We now want to understand what kind of interference these cell phones cause for each other. Consider the detection process for any given cell phone and the effect of the interference from the other cell phones.

Since each cell phone uses a different phase of the long PN sequence, the PN sequences from the interfering cell phones can be modeled as random iid binary streams. Since each of these streams is modeled as iid, the mod-2 addition of the PN stream and the data is still an iid stream of binary digits. If the filter used before transmission is very sharp (which it is, since the 1.2288 MHz bands are quite close together), the Nyquist pulses can be approximated by sinc pulses. It also makes sense to model the sample clock of each interfering cell phone as being uniformly distributed. This means that the interfering cell phones can be modeled as being wide-sense stationary with a flat spectrum over the 1.2288 MHz band.

The more interfering cell phones there are in the same frequency band, the more interference there is, but also, since these interfering signals are independent of each other, we can invoke the central limit theorem to see that this aggregate interference will be approximately Gaussian.

To get some idea of the effect of the interference, assume that each interfering cell phone is received at the same baseband energy per information bit given by E_{b}. Since there are 9600 information bits per second entering the encoder, the power in the interfering waveform is then $9600E_{\mathrm{b}}$. This noise is evenly spread over 2 457 600 dimensions per second, so is $(4800/2.4576\times 10^6)E_{\mathrm{b}} = E_{\mathrm{b}}/512$ per dimension. Thus the noise per dimension is increased from $N_0/2$ to $(N_0/2+kE_{\mathrm{b}}/512)$, where k is the number of interferers. With this change, (9.80) becomes

$$\Pr(e) \leq \frac{63}{2}\exp\left[\frac{-3E_{\mathrm{b}}}{N_0+kE_{\mathrm{b}}/256}\right]. \tag{9.81}$$

In reality, the interfering cell phones are received with different power levels, and because of this the system uses a fairly elaborate system of power control to attempt to equalize the received powers of the cell phones being received at a given base station. Those cell phones being received at other base stations presumably have lower power at the given base station, and thus cause less interference. It can be seen that with a large set of interferers, the assumption that they form a Gaussian process is even better than with a single interferer.

The factor of 256 in (9.81) is due to the spreading of the waveforms (sending them in a bandwidth of 1.2288 MHz rather than in a narrow band. This spreading, of course, is also the reason why appreciable numbers of other cell phones must use the same band. Since voice users are typically silent half the time while in a conversation, and the cell phone need send no energy during these silent periods, the number of tolerable interferers is doubled.

The other types of cellular systems (GSM and TDMA) attempt to keep the interfering cell phones in different frequency bands and time slots. If successful, this is, of course, preferable to CDMA, since there is then no interference rather than the limited interference in (9.81). The difficulty with these other schemes is that frequency slots and time slots must be reused by cell phones going to other cell stations (although preferably not by cell phones connected with neighboring cell stations). The need to avoid slot re-use between neighboring cells leads to very complex algorithms for allocating re-use patterns between cells, and these algorithms cannot make use of the factor of 2 due to users being quiet half the time.

Because these transmissions are narrow-band, when interference occurs, it is not attenuated by a factor of 256 as in (9.81). Thus the question boils down to whether it is preferable to have a large number of small interferers or a small number of larger interferers. This, of course, is only one of the issues that differ between CDMA systems and narrow-band systems. For example, narrow-band systems cannot make use of rake receivers, although they can make use of many techniques developed over the years for narrow-band transmission.

9.10 Summary of wireless communication

Wireless communication differs from wired communication primarily in the time-varying nature of the channel and the interference from other wireless users. The time-varying nature of the channel is the more technologically challenging of the two, and has been the primary focus of this chapter.

Wireless channels frequently have multiple electromagnetic paths of different lengths from transmitter to receiver, and thus the receiver gets multiple copies of the transmitted waveform at slightly different delays. If this were the only problem, then the channel could be represented as a linear-time-invariant (LTI) filter with the addition of noise, and this could be treated as a relatively minor extension to the nonfiltered channels with noise studied in earlier chapters.

The problem that makes wireless communication truly different is the fact that the different electromagnetic paths are also sometimes moving with respect to each other, thus giving rise to different Doppler shifts on different paths.

Section 9.3 showed that these multiple paths with varying Doppler shifts lead to an input/output model which, in the absence of noise, is modeled as a linear-time-varying (LTV) filter $h(\tau, t)$, which is the response at time t to an impulse τ seconds earlier. This has a time-varying system function $\hat{h}(f, t)$ which, for each fixed t, is the Fourier transform of $h(\tau, t)$. These LTV filters behave in a somewhat similar fashion to the familiar LTI filters. In particular, the channel input $x(t)$ and noise-free output $y(t)$ are related by the convolution equation, $y(t) = \int h(\tau, t)x(t-\tau)\mathrm{d}\tau$. Also, $y(t)$, for each fixed t, is the inverse Fourier transform of $\hat{x}(f)\hat{h}(f, t)$. The major difference is that $\hat{y}(f)$ is not equal to $\hat{x}(f)\hat{h}(f, t)$ unless $\hat{h}(f, t)$ is nonvarying in t.

The major parameters of a wireless channel (at a given carrier frequency f_c) are the Doppler spread $\mathcal{D}$ and the time spread $\mathcal{L}$. The Doppler spread is the difference between the largest and smallest significant Doppler shift on the channel (at f_c). It was shown to be twice the bandwidth of $|\hat{h}(f_c, t)|$ viewed as a function of t. Similarly, $\mathcal{L}$ is the time spread between the longest and shortest multipath delay (at a fixed output time t_0). It was shown to be twice the "bandwidth" of $|\hat{h}(f, t_0)|$ viewed as a function of f.

The coherence-time $\mathcal{T}_{\mathrm{coh}}$ and coherence frequency $\mathcal{F}_{\mathrm{coh}}$ were defined as $\mathcal{T}_{\mathrm{coh}} = 1/2\mathcal{D}$ and $\mathcal{F}_{\mathrm{coh}} = 1/2\mathcal{L}$. Qualitatively, these parameters represent the duration of multipath fades in time and the duration over frequency, respectively. Fades, as their name suggests, occur gradually, both in time and frequency, so these parameters represent duration only in an order-of-magnitude sense.

As shown in Section 9.4, these bandpass models of wireless channels can be converted to baseband models and then converted to discrete-time models. The relation between the bandpass and baseband model is quite similar to that for nonfading channels. The discrete-time model relies on the sampling theorem, and, while mathematically correct, can somewhat distort the view of channels with a small number of paths, sometimes yielding only one tap, and sometimes yielding many more taps than paths. Nonetheless, this model is so convenient for acquiring insight about wireless channels that it is widely used, particularly among those who dislike continuous-time models.

Section 9.5 then breaks the link with electromagnetic models and views the baseband tapped delay line model probabilistically. At the same time, WGN is added. A one-tap model corresponds to situations where the transmission bandwidth is narrow relative to the coherence frequency $\mathcal{F}_{\text{coh}}$ and multitap models correspond to the opposite case. We generally model the individual taps as being Rayleigh faded, corresponding to a large number of small independent paths in the corresponding delay range. Several other models, including the Rician model and noncoherent deterministic model, were analyzed, but physical channels have such variety that these models only provide insight into the types of behavior to expect. The modeling issues are quite difficult here, and our point of view has been to analyze the consequences of a few very simple models.

Consistent with the above philosophy, Section 9.6 analyzes a single-tap model with Rayleigh fading. The classical Rayleigh fading error probability, using binary orthogonal signals and no knowledge of the channel amplitude or phase, is calculated to be $1/[2+\mathsf{E}_{\text{b}}/N_0]$. The classical error probability for noncoherent detection, where the receiver knows the channel magnitude but not the phase, is also calculated and compared with the coherent result as derived for nonfaded channels. For large E_{b}/N_0, the results are very similar, saying that knowledge of the phase is not very important in that case. However, the noncoherent detector does not use the channel magnitude in detection, showing that detection in Rayleigh fading would not be improved by knowledge of the channel magnitude.

The conclusion from this study is that reasonably reliable communication for wireless channels needs diversity or coding or needs feedback with rate or power control. With Lth-order diversity in Rayleigh fading, it was shown that error probability tends to 0 as $(E_{\text{b}}/4N_0)^{-L}$ for large E_{b}/N_0. If the magnitude of the various diversity paths are known, then the error probability can be made still smaller.

Knowledge of the channel as it varies can be helpful in two ways. One is to reduce the error probability when coding and/or diversity are used, and the other is to exercise rate control or power control at the transmitter. Section 9.7 analyzes various channel measurement techniques, including direct measurement by sending known probing sequences and measurement using rake receivers. These are both widely used and effective tools.

Finally, all of the above analysis and insight about wireless channels is brought to bear in Section 9.9, which describes the IS95 CDMA cellular system. In fact, this section illustrates most of the major topics throughout this text.

9.11 Appendix: Error probability for noncoherent detection

Under hypothesis $\boldsymbol{U}=(a,0)$, $|V_0|$ is a Rician random variable R which has the density[30]

$$f_R(r) = \frac{r}{\mathsf{W}N_0/2}\exp\left\{-\frac{r^2+a^2g^2}{\mathsf{W}N_0}\right\} I_0\left(\frac{rag}{\mathsf{W}N_0/2}\right), \qquad r \geq 0, \tag{9.82}$$

where I_0 is the modified Bessel function of zeroth order. Conditional on $\boldsymbol{U}=(0,a)$, $|V_1|$ has the same density, so the likelihood ratio is given by

$$\frac{f[(|v_0|,|v_1|)\,|\,\boldsymbol{U}=(a,0)]}{f[(|v_0|,|v_1|)\,|\,\boldsymbol{U}=(0,a)]} = \frac{I_0(2|v_0|ag/\mathsf{W}N_0)}{I_0(2|v_1|ag/\mathsf{W}N_0)}. \tag{9.83}$$

It can be shown that I_0 is monotonic increasing in its argument, which verifies that the ML decision rule is to choose $\boldsymbol{U}=(a,0)$ if $|v_0|>|v_1|$ and $\boldsymbol{U}=(0,a)$ otherwise.

By symmetry, the probability of error is the same for either hypothesis, and is given by

$$\Pr(e) = \Pr\left\{|V_0|^2 \leq |V_1|^2 \mid \boldsymbol{U}=(a,0)\right\} = \Pr\left\{|V_0|^2 > |V_1|^2 \mid \boldsymbol{U}=(0,a)\right\}. \tag{9.84}$$

This can be calculated by straightforward means without any reference to Rician rvs or Bessel functions. We calculate the error probability, conditional on hypothesis $U=(a,0)$, and do this by returning to rectangular coordinates. Since the results are independent of the phase ϕ_i of G_i for $i=0$ or 1, we will simplify our notation by assuming $\phi_0=\phi_1=0$.

Conditional on $\boldsymbol{U}=(a,0)$, $|V_1|^2$ is just $|Z_1|^2$. Since the real and imaginary parts of Z_1 are iid Gaussian with variance $\mathsf{W}N_0/2$ each, $|Z_1|^2$ is exponential with mean $\mathsf{W}N_0$. Thus, for any $x \geq 0$,

$$\Pr(|V_1|^2 \geq x \mid \boldsymbol{U}=(a,0)) = \exp\left(-\frac{x}{\mathsf{W}N_0}\right). \tag{9.85}$$

Next, conditional on hypothesis $\boldsymbol{U}=(a,0)$ and $\phi_0=0$, we see from (9.57) that $V_0 = ag+Z_0$. Letting $V_{0,\mathrm{re}}$ and $V_{0,\mathrm{im}}$ be the real and imaginary parts of V_0, the probability density of $V_{0,\mathrm{re}}$ and $V_{0,\mathrm{im}}$, given hypothesis $\boldsymbol{U}=(a,0)$ and $\phi_0=0$, is given by

$$f(v_{0,\mathrm{re}}, v_{0,\mathrm{im}} \mid \boldsymbol{U}=(a,0)) = \frac{1}{2\pi\mathsf{W}N_0/2}\exp\left(-\frac{[v_{0,\mathrm{re}}-ag]^2+v_{0,\mathrm{im}}^2}{\mathsf{W}N_0}\right). \tag{9.86}$$

We now combine (9.85) and (9.86). All probabilities below are implicitly conditioned on hypothesis $\boldsymbol{U}=(a,0)$ and $\phi_0=0$. For a given observed pair $v_{0,\mathrm{re}}, v_{0,\mathrm{im}}$, an error will be made if $|V_1|^2 \geq v_{0,\mathrm{re}}^2+v_{0,\mathrm{im}}^2$. Thus,

[30] See, for example, Proakis (2000, p. 304).

$$\Pr(e) = \iint f(v_{0,\text{re}}, v_{0,\text{im}} \mid \boldsymbol{U} = (a, 0)) \Pr(|V_1|^2 \geq v_{0,\text{re}}^2 + v_{0,\text{im}}^2) dv_{0,\text{re}}\, dv_{0,\text{im}}$$

$$= \iint \frac{1}{2\pi W N_0/2} \exp\left(-\frac{(v_{0,\text{re}} - ag)^2 + v_{0,\text{im}}^2}{W N_0}\right) \exp\left(-\frac{v_{0,\text{re}}^2 + v_{0,\text{im}}^2}{W N_0}\right) dv_{0,\text{re}}\, dv_{0,\text{im}}.$$

The following equations combine these exponentials, "complete the square," and recognize the result as simple Gaussian integrals:

$$\Pr(e) = \iint \frac{1}{2\pi W N_0/2} \exp\left(-\frac{2v_{0,\text{re}}^2 - 2agv_{0,\text{re}} + a^2g^2 + 2v_{0,\text{im}}^2}{W N_0}\right) dv_{0,\text{re}}\, dv_{0,\text{im}}$$

$$= \frac{1}{2} \iint \frac{1}{2\pi W N_0/4} \exp\left(-\frac{(v_{0,\text{re}} - (1/2)ag)^2 + v_{0,\text{im}}^2 + (1/4)a^2g^2}{W N_0/2}\right) dv_{0,\text{re}}\, dv_{0,\text{im}}$$

$$= \frac{1}{2} \exp\left(-\frac{a^2 g}{2W N_0}\right) \iint \frac{1}{2\pi W N_0/4} \exp\left(-\frac{(v_{0,\text{re}} - (1/2)ag)^2 + v_{0,\text{im}}^2}{W N_0/2}\right) dv_{0,\text{re}}\, dv_{0,\text{im}}.$$

Integrating the Gaussian integrals, we obtain

$$\Pr(e) = \frac{1}{2} \exp\left(-\frac{a^2 g^2}{2W N_0}\right). \tag{9.87}$$

9.12 Exercises

9.1 (a) Equation (9.6) is derived under the assumption that the motion is in the direction of the line of sight from sending antenna to receiving antenna. Find this field under the assumption that there is an arbitrary angle ϕ between the line of sight and the motion of the receiver. Assume that the time range of interest is small enough that changes in (θ, ψ) can be ignored.

(b) Explain why, and under what conditions, it is reasonable to ignore the change in (θ, ψ) over small intervals of time.

9.2 Equation (9.10) is derived by an assumption to (9.9). Derive an exact expression for the received waveform $y_f(t)$ starting with (9.9). [Hint. Express each term in (9.9) as the sum of two terms, one the approximation used in (9.10) and the other a correction term.] Interpret your result.

9.3 (a) Let r_1 be the length of the direct path in Figure 9.4. Let r_2 be the length of the reflected path (summing the path length from the transmitter to ground plane and the path length from ground plane to receiver). Show that as r increases, $r_2 - r_1$ is asymptotically equal to b/r for some constant r; find the value of b. [Hint. Recall that for x small, $\sqrt{1+x} \approx (1 + x/2)$ in the sense that $[\sqrt{1+x} - 1]/x \to 1/2$ as $x \to 0$.]

(b) Assume that the received waveform at the receiving antenna is given by

$$E_r(f,t) = \frac{\Re\left[\alpha \exp\{2\pi i[ft - fr_1/c]\right]}{r_1} - \frac{\Re\left[\alpha \exp\{2\pi i[ft - fr_2/c]\right]}{r_2}. \quad (9.88)$$

Approximate the denominator r_2 by r_1 in (9.88) and show that $E_r \approx \beta/r^2$ for r^{-1} much smaller than c/f. Find the value of β.

(c) Explain why this asymptotic expression remains valid without first approximating the denominator r_2 in (9.88) by r_1.

9.4 Evaluate the channel output $y(t)$ for an arbitrary input $x(t)$ when the channel is modeled by the multipath model of (9.14). [Hint. The argument and answer are very similar to that in (9.20), but you should think through the possible effects of time-varying attenuations $\beta_j(t)$.]

9.5 (a) Consider a wireless channel with a single path having a Doppler shift $\mathcal{D}_1$. Assume that the response to an input $\exp\{2\pi i f t\}$ is $y_f(t) = \exp\{2\pi i t(f + \mathcal{D}_1)\}$. Evaluate the Doppler spread $\mathcal{D}$ and the midpoint between minimum and maximum Doppler shifts Δ. Evaluate $\hat{h}(f,t)$, $|\hat{h}(f,t)|$, $\hat{\psi}(f,t)$, and $|\hat{\psi}(f,t)|$ for $\hat{\psi}$ in (9.24). Find the envelope of the output when the input is $\cos(2\pi f t)$.

(b) Repeat part (a) where $y_f(t) = \exp\{2\pi i t(f + \mathcal{D}_1)\} + \exp\{2\pi i t f\}$.

9.6 (a) Bandpass envelopes. Let $y_f(t) = e^{2\pi i f t}\hat{h}(f,t)$ be the response of a multipath channel to the input $e^{2\pi i f t}$ and assume that f is much larger than any of the channel Doppler shifts. Show that the envelope of $\Re[y_f(t)]$ is equal to $|y_f(t)|$.

(b) Find the power $(\Re[y_f(t)])^2$ and consider the result of lowpass filtering this power waveform. Interpret this filtered waveform as a short-term time-average of the power and relate the square root of this time-average to the envelope of $\Re[y_f(t)]$.

9.7 Equations (9.34) and (9.35) give the baseband system function and impulse response for the simplified multipath model. Rederive those formulas using the slightly more general multipath model of (9.14) where each attenuation β_j can depend on t but not f.

9.8 It is common to define Doppler spread for passband communication as the Doppler spread at the carrier frequency and to ignore the change in Doppler spread over the band. If f_c is 1 GHz and W is 1 mHz, find the percentage error over the band in making this approximation.

9.9 This illustrates why the tap gain corresponding to the sum of a large number of potential independent paths is not necessarily well approximated by a Gaussian distribution. Assume there are N possible paths and each appears independently with probability $2/N$. To make the situation as simple as possible, suppose that if path n appears, its contribution to a given random tap gain, say $G_{0,0}$, is equiprobably ± 1, with independence between paths. That is,

$$G_{0,0} = \sum_{n=1}^{N} \theta_n \phi_n,$$

where $\phi_1, \phi_2, \ldots, \phi_N$ are iid rvs taking on the value 1 with probability $2/N$ and taking on the value 0 otherwise and $\theta_1, \ldots, \theta_N$ are iid and equiprobably ± 1.

(a) Find the mean and variance of $G_{0,0}$ for any $N \geq 1$ and take the limit as $N \to \infty$.
(b) Give a common sense explanation of why the limiting rv is not Gaussian. Explain why the central limit theorem does not apply here.
(c) Give a qualitative explanation of what the limiting distribution of $G_{0,0}$ looks like. If this sort of thing amuses you, it is not hard to find the exact distribution.

9.10 Let $\hat{g}(f, t)$ be the baseband equivalent system function for a linear time-varying filter, and consider baseband inputs $u(t)$ limited to the frequency band $(-\mathsf{W}/2, \mathsf{W}/2)$. Define the baseband-limited impulse response $g(\tau, t)$ by

$$g(\tau, t) = \int_{-\mathsf{W}/2}^{\mathsf{W}/2} \hat{g}(f, t) \exp\{2\pi i f \tau\} \mathrm{d}f.$$

(a) Show that the output $v(t)$ for input $u(t)$ is given by

$$v(t) = \int_{\tau} u(t - \tau) g(\tau, t) \mathrm{d}\tau.$$

(b) For the discrete-time baseband model of (9.41), find the relationship between $g_{k,m}$ and $g(k/\mathsf{W}, m/\mathsf{W})$. [Hint. It is a very simple relationship.]
(c) Let $G(\tau, t)$ be a rv whose sample values are $g(\tau, t)$ and define

$$\mathcal{R}(\tau, t') = \frac{1}{\mathsf{W}} \mathsf{E}\{G(\tau, t) G^*(\tau, t + t')\}.$$

What is the relationship between $\mathcal{R}(\tau, t')$ and $R(k, n)$ in (9.46)?
(d) Give an interpretation to $\int_{\tau} \mathcal{R}(\tau, 0) \mathrm{d}\tau$ and indicate how it might change with W. Can you explain, from this, why $\mathcal{R}(t, \tau)$ is defined using the scaling factor W?

9.11 (a) Average over gain in the noncoherent detection result in (9.59) to rederive the Rayleigh fading error probability.
(b) Assume narrow-band fading with a single tap G_m. Assume that the sample value of the tap magnitude, $|g_m|$, is measured perfectly and fed back to the transmitter. Suppose that the transmitter, using pulse-position modulation, chooses the input magnitude dynamically so as to maintain a constant received signal to noise ratio. That is, the transmitter sends $a/|g_m|$ instead of a. Find the expected transmitted energy per binary digit.

9.12 Consider a Rayleigh fading channel in which the channel can be described by a single discrete-time complex filter tap G_m. Consider binary communication where, for each pair of time-samples, one of two equiprobable signal pairs is sent, either (a, a) or $(a, -a)$. The output at discrete times 0 and 1 is given by

$$V_m = U_m G + Z_m; \quad m = 0, 1.$$

The magnitude of G has density $f(|g|) = 2|g|\exp\{-|g|^2\}$; $|g| \geq 0$. Note that G is the same for $m = 0, 1$ and is independent of Z_0 and Z_1, which in turn are iid circularly symmetric Gaussian with variance $N_0/2$ per real and imaginary part. Explain your answers in each part.

(a) Consider the noise transformation

$$Z_0' = \frac{Z_1 + Z_0}{\sqrt{2}}; \qquad Z_1' = \frac{Z_1 - Z_0}{\sqrt{2}}.$$

Show that Z_0' and Z_1' are statistically independent and give a probabilistic characterization of them.

(b) Let

$$V_0' = \frac{V_1 + V_0}{\sqrt{2}}; \qquad V_1' = \frac{V_1 - V_0}{\sqrt{2}}.$$

Give a probabilistic characterization of (V_0', V_1') under $\boldsymbol{U} = (a, a)$ and under $\boldsymbol{U} = (a, -a)$.

(c) Find the log likelihood ratio $\Lambda(v_0', v_1')$ and find the MAP decision rule for using v_0', v_1' to choose $\tilde{\boldsymbol{U}} = (a, a)$ or $(a, -a)$.

(d) Find the probability of error using this decision rule.

(e) Is the pair V_0, V_1 a function of V_0', V_1'? Why is this question relevant?

9.13 Consider the two-tap Rayleigh fading channel of Example 9.8.1. The input $\boldsymbol{U} = U_0, U_1, \ldots$ is one of two possible hypotheses, either $\boldsymbol{u}^0 = (\sqrt{E_b}, 0, 0, 0)$ or $\boldsymbol{u}^1 = (0, 0, \sqrt{E_b}, 0)$ where $U_\ell = 0$ for $\ell \geq 4$ for both hypotheses. The output is a discrete-time complex sequence $\boldsymbol{V} = V_0, V_1, \ldots$, given by

$$V_m = G_{0,m} U_m + G_{1,m} U_{m-1} + Z_m.$$

For each m, $G_{0,m}$ and $G_{1,m}$ are iid and circularly symmetric complex Gaussian rvs with $G_{0,m} \sim \mathcal{CN}(0, 1/2)$ for m both 0 and 1. The correlation of $G_{0,m}$ and $G_{1,m}$ with m is immaterial, and can be assumed uncorrelated. Assume that the sequence $Z_m \sim \mathcal{CN}(0, N_0)$ is a sequence of iid circularly symmetric complex Gaussian rvs. The signal, the noise, and the channel taps are all independent. As explained in the example, the energy vector $\boldsymbol{X} = (X_0, X_1, X_2, X_3)^{\mathsf{T}}$, where $X_m = |V_m|^2$ is a sufficient statistic for the hypotheses $\boldsymbol{u}^0$ and $\boldsymbol{u}^1$. Also, as explained there, these energy variables are independent and exponential given the hypothesis. More specifically, define $\alpha = 1/(E_b/2 + N_0)$ and $\beta = 1/N_0$. Then, given $\boldsymbol{U} = \boldsymbol{u}^0$, the variables X_0 and X_1 each have the density $\alpha e^{-\alpha x}$ and $\boldsymbol{X}_2$ and $\boldsymbol{X}_3$ each have the density $\beta e^{-\beta x}$, all for $x \geq 0$. Given $\boldsymbol{U} = \boldsymbol{u}^1$, these densities are reversed.

(a) Give the probability density of $\boldsymbol{X}$ conditional on $\boldsymbol{u}^0$.

(b) Show that the log likelihood ratio is given by

$$\mathrm{LLR}(\boldsymbol{x}) = (\beta - \alpha)(x_0 + x_1 - x_2 - x_3).$$

(c) Let $Y_0 = X_0 + X_1$ and let $Y_1 = X_2 + X_3$. Find the probability density and the distribution function for Y_0 and Y_1 conditional on $\boldsymbol{u}^0$.

(d) Conditional on $\boldsymbol{U} = \boldsymbol{u}^0$, observe that the probability of error is the probability that Y_1 exceeds Y_0. Show that this is given by

$$\Pr(e) = \frac{3\alpha^2\beta + \alpha^3}{(\alpha+\beta)^3} = \frac{4 + 3E_b/2N_0}{(2 + E_b/2N_0)^3},$$

[Hint. To derive the second expression, first convert the first expression to a function of β/α. Recall that $\int_0^\infty e^{-y}\, dy = \int_0^\infty ye^{-y}\, dy = 1$ and $\int_0^\infty y^2e^{-y}\, dy = 2$.]

(e) Explain why the assumption that $G_{k,i}$ and $G_{k,j}$ are uncorrelated for $i \neq j$ was not needed.

9.14 (Lth-order diversity) This exercise derives the probability of error for Lth-order diversity on a Rayleigh fading channel. For the particular model described at the end of Section 9.8, there are L taps in the tapped delay line model for the channel. Each tap k multiplies the input by $G_{k,m} \sim \mathcal{CN}(0, 1/L)$, $0 \leq k \leq L-1$. The binary inputs are $\boldsymbol{u}^0 = (\sqrt{E_b}, 0, \ldots, 0)$ and $\boldsymbol{u}^1 = (0, \ldots, 0, \sqrt{E_b}, 0, \ldots, 0)$, where $\boldsymbol{u}^0$ and $\boldsymbol{u}^1$ contain the signal at times 0 and L, respectively.

The complex received signal at time m is $V_m = \sum_{k=0}^{L-1} G_{k,m} U_{m-k} + Z_m$ for $0 \leq m \leq 2L-1$, where $Z_m \sim \mathcal{CN}(0, N_0)$ is independent over time and independent of the input and channel tap gains. As shown in Section 9.8, the set of energies $X_m = |V_m|^2$, $0 \leq m \leq 2L-1$, are conditionally independent, given either $\boldsymbol{u}^0$ or $\boldsymbol{u}^1$, and constitute a sufficient statistic for detection; the ML detection rule is to choose $\boldsymbol{u}^0$ if $\sum_{m=1}^{L-1} X_m \geq \sum_{m=L}^{2L-1} X_m$ and $\boldsymbol{u}^1$ otherwise. Finally, conditional on $\boldsymbol{u}^0$, $X_0, \ldots, X_{L-1}$ are exponential with mean $N_0 + E_b/L$. Thus for $0 \leq m < L$, X_m has the density $\alpha \exp(-\alpha X_m)$, where $\alpha = 1/(N_0 + E_b/L)$. Similarly, for $L \leq m < 2L$, X_m has the density $\beta \exp(-\beta X_m)$, where $\beta = 1/N_0$.

(a) The following parts of the exercise demonstrate a simple technique to calculate the probability of error $\Pr(e)$ conditional on either hypothesis. This is the probability that the sum of L iid exponential rvs of rate α is less than the sum of L iid exponential rvs of rate $\beta = N_0$. View the first sum, i.e. $\sum_{m=0}^{L-1} X_m$ (given $\boldsymbol{u}_0$), as the time of the Lth arrival in a Poisson process of rate α and view the second sum, $\sum_{m=L}^{2L-1} X_m$, as the time of the Lth arrival in a Poisson process of rate β (see Figure 9.18). Note that the notion of time here has nothing to do with the actual detection problem and is strictly a mathematical artifice for viewing the problem in terms of Poisson processes. Show that $\Pr(e)$ is the probability that, out of the first $2L-1$ arrivals in the combined Poisson process above, at least L of those arrivals are from the first process.

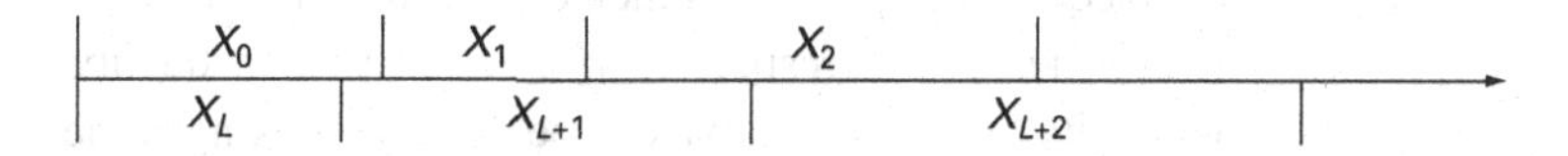

Figure 9.18. Poisson process with interarrival times $\{X_k; 0 \leq k < L\}$, and another with interarrival times $\{X_{L+\ell}; 0 \leq \ell < L\}$. The combined process can be shown to be a Poisson process of rate $\alpha + \beta$.

(b) Each arrival in the combined Poisson process is independently drawn from the first process with probability $p = \alpha/(\alpha+\beta)$ and from the second process with probability $p = \beta/(\alpha+\beta)$. Show that

$$\Pr(e) = \sum_{\ell=L}^{2L-1} \binom{2L-1}{\ell} p^{\ell}(1-p)^{2L-1-\ell}.$$

(c) Express the result in (b) in terms of α and β and then in terms of E_b/LN_0.

(d) Use the result in (b) to recalculate $\Pr(e)$ for Rayleigh fading without diversity (i.e. with $L = 1$). Use it with $L = 2$ to validate the answer in Exercise 9.13.

(e) Show that $\Pr(e)$ for very large E_b/N_0 decreases with increasing L as $[E_b/(4N_0)]^L$.

(f) Show that $\Pr(e)$ for Lth-order diversity (using ML detection as above) is *exactly* the same as the probability of error that would result by using $(2L-1)$-order diversity, making a hard decision on the basis of each diversity output, and then using majority rule to make a final decision.

9.15 Consider a wireless channel with two paths, both of equal strength, operating at a carrier frequency f_c. Assume that the baseband equivalent system function is given by

$$\hat{g}(f,t) = 1 + \exp\{i\phi\}\exp[-2\pi i(f+f_c)\tau_2(t)]. \tag{9.89}$$

(a) Assume that the length of path 1 is a fixed value r_0 and the length of path 2 is $r_0 + \Delta r + vt$. Show (using (9.89)) that

$$\hat{g}(f,t) \approx 1 + \exp\{i\psi\}\exp\left[-2\pi i\left(\frac{f\Delta r}{c} + \frac{f_c vt}{c}\right)\right]. \tag{9.90}$$

Explain what the parameter ψ is in (9.90); also explain the nature of the approximation concerning the relative values of f and f_c.

(b) Discuss why it is reasonable to define the multipath spread $\mathcal{L}$ here as $\Delta r/c$ and to define the Doppler spread $\mathcal{D}$ as $f_c v/c$.

(c) Assume that $\psi = 0$, i.e. that $\hat{g}(0,0) = 2$. Find the smallest $t > 0$ such that $\hat{g}(0,t) = 0$. It is reasonable to denote this value t as the coherence-time $\mathcal{T}_{coh}$ of the channel.

(d) Find the smallest $f > 0$ such that $\hat{g}(f,0) = 0$. It is reasonable to denote this value of f as the coherence frequency $\mathcal{F}_{coh}$ of the channel.

9.16 Union bound. Let $E_1, E_2, \ldots, E_k$ be independent events each with probability p.

(a) Show that $\Pr(\cup_{j=1}^k E_j) = 1-(1-p)^k$.

(b) Show that $pk - (pk)^2/2 \le \Pr(\cup_{j=1}^k E_j) \le pk$. [Hint. One approach is to demonstrate equality at $p = 0$ and then demonstrate the inequality for the derivative of each term with respect to p. For the first inequality, demonstrating the inequality for the derivative can be done by looking at the second derivative.]

9.17 (a) Let $\boldsymbol{u}$ be an ideal PN sequence, satisfying $\sum_\ell u_\ell u^*_{\ell+k} = 2a^2 n\delta_k$. Let $\boldsymbol{b} = \boldsymbol{u} * \boldsymbol{g}$ for some channel tap gain $\boldsymbol{g}$. Show that $\|\boldsymbol{b}\|^2 = \|\boldsymbol{u}\|^2\|\boldsymbol{g}\|^2$. [Hint. One approach is to convolve $\boldsymbol{b}$ with its matched filter $\boldsymbol{b}^{\dagger}$.] Use the commutativity of convolution along with the properties of $\boldsymbol{u} * \boldsymbol{u}^{\dagger}$.

(b) If $\boldsymbol{u}^0$ and $\boldsymbol{u}^1$ are each ideal PN sequences as in part (a), show that $\boldsymbol{b}_0 = \boldsymbol{u}^0 * \boldsymbol{g}$ and $\boldsymbol{b}_1 = \boldsymbol{u}^1 * \boldsymbol{g}$ satisfy $\|\boldsymbol{b}_0\|^2 = \|\boldsymbol{b}_1\|^2$.

9.18 This exercise explores the difference between a rake receiver that estimates the analog baseband channel and one that estimates a discrete-time model of the baseband channel. Assume that the channel is estimated perfectly in each case, and look at the resulting probability of detecting the signal incorrectly.

We do this, somewhat unrealistically, with a 2-PAM modulator sending $\operatorname{sinc}(t)$ given $H = 0$ and $-\operatorname{sinc}(t)$ given $H = 1$. We assume a channel with two paths having an impulse response $\delta(t) - \delta(t - \varepsilon)$, where $0 < \varepsilon \ll 1$. The received waveform, after demodulation from passband to baseband, is given by

$$V(t) = \pm[\operatorname{sinc}(t) - \operatorname{sinc}(t - \varepsilon)] + Z(t),$$

where $Z(t)$ is WGN of spectral density $N_0/2$. We have assumed for simplicity that the phase angles due to the demodulating carrier are 0.

(a) Describe the ML detector for the analog case where the channel is perfectly known at the receiver.

(b) Find the probability of error $\Pr(e)$ in terms of the energy of the low-pass received signal, $E = \|\operatorname{sinc}(t) - \operatorname{sinc}(t - \varepsilon)\|^2$.

(c) Approximate E by using the approximation $\operatorname{sinc}(t - \varepsilon) \approx \operatorname{sinc}(t) - \varepsilon \operatorname{sinc}'(t)$. [Hint. Recall the Fourier transform pair $u'(t) \leftrightarrow 2\pi i f \hat{u}(f)$.]

(d) Next consider the discrete-time model where, since the multipath spread is very small relative to the signaling interval, the discrete channel is modeled with a single tap g. The sampled output at epoch 0 is $\pm g[1 - \operatorname{sinc}(-\varepsilon)] + Z(0)$. We assume that $Z(t)$ has been filtered to the baseband bandwidth $\mathsf{W} = 1/2$. Find the probability of error using this sampled output as the observation and assuming that g is known.

(e) The probability of error for both the result in (d) and the result in (b) and (c) approach 1/2 as $\varepsilon \to 0$. Contrast the way in which each result approaches 1/2.

(f) Try to explain why the discrete approach is so inferior to the analog approach here. [Hint. What is the effect of using a single-tap approximation to the sampled lowpass channel model?]

References

Bertsekas, D. and Gallager, R. G. (1992). *Data Networks*, 2nd edn (Englewood Cliffs, NJ: Prentice-Hall).

Bertsekas, D. and Tsitsiklis, J. (2002). *An Introduction to Probability Theory* (Belmont, MA: Athena).

Carleson, L. (1966). "On convergence and growth of partial sums of Fourier series," *Acta Mathematica* **116**, 135–157.

Cover, T. M. and Thomas, J. A. (2006). *Elements of Information Theory*, 2nd edn (New York: Wiley).

Feller, W. (1968). *An Introduction to Probability Theory and its Applications*, vol. 1 (New York: Wiley).

Feller, W. (1971). *An Introduction to Probability Theory and its Applications*, vol. 2 (New York: Wiley).

Forney, G. D. (2005). *Principles of Digital Communication II*, MIT Open Course Ware, http://ocw.mit.edu/OcwWeb/Electrical-Engineering-and-Computer-Science/6-451Spring-2005/CourseHome/index.htm

Gallager, R. G. (1968). *Information Theory and Reliable Communication* (New York: Wiley).

Gallager, R. G. (1996). *Discrete Stochastic Processes* (Dordrecht: Kluwer Academic Publishers).

Goldsmith, A. (2005). *Wireless Communication* (New York: Cambridge University Press).

Gray, R. M. (1990). *Source Coding Theory* (Dordrecht: Kluwer Academic Publishers).

Hartley, R. V. L. (1928). "Transmission of information," *Bell Syst. Tech. J.* **7**, 535.

Haykin, S. (2002). *Communication Systems* (New York: Wiley).

Huffman, D. A. (1952). "A method for the construction of minimum redundancy codes," *Proc. IRE* **40**, 1098–1101.

Jakes, W. C. (1974). *Microwave Mobile Communications* (New York: Wiley).

Lin, S. and Costello, D. J. (2004). *Error Control Coding*, 2nd edn (Englewood Cliffs, NJ: Prentice-Hall).

Lloyd, S. P. (1982). "Least squares quantization in PCM," *IEEE Trans. Inform. Theory* **IT-28** (2), 129–136.

Kraft, L. G. (1949). "A device for quantizing, grouping, and coding amplitude modulated pulses," M. S. Thesis, Department of Electrical Engineering MIT, Cambridge, MA.

Max, J. (1960). "Quantization for minimum distortion," *IRE Trans. Inform. Theory* **IT-6** (2), 7–12.

Nyquist, H. (1928). "Certain topics in telegraph transmission theory," *Trans. AIEE* **47**, 627–644.

Paley, R. E. A. C. and Wiener, N. (1934). "Fourier transforms in the complex domain," Colloquium Publications, vol. 19. (New York: American Mathematical Society).

Proakis, J. G. (2000). *Digital Communications*, 4th edn (New York: McGraw-Hill).

Proakis, J. G. and Salehi, M. (1994). *Communication Systems Engineering* (Englewood Cliffs, NJ: Prentice-Hall).

Pursley, M. (2005). *Introduction to Digital Communications* (Englewood Cliffs, NJ: Prentice-Hall).

Ross, S. (1994). *A First Course in Probability*, 4th edn (New York: Macmillan & Co.).

Ross, S. (1996). *Stochastic Processes*, 2nd edn (New York: Wiley and Sons).

Rudin, W. (1966). *Real and Complex Analysis* (New York: McGraw-Hill).

Shannon, C. E. (1948). "A mathematical theory of communication," *Bell Syst. Tech. J.* **27**, 379–423, 623–656. Available on the web at http://cm.bell-labs.com/cm/ms/what/shannonday/paper.html

Shannon, C. E. (1956). "The zero-error capacity of a noisy channel," *IRE Trans. Inform. Theory* **IT-2**, 8–19.

Slepian, D. and Pollak, H. O. (1961), "Prolate spheroidal waveforms, Fourier analysis, and uncertainty–I," *Bell Syst. Tech. J.* **40**, 43–64.

Steiner, M. (1994). "The strong simplex conjecture is false," *IEEE Trans. Inform. Theory* **IT-25**, 721–731.

Tse, D. and Viswanath, P. (2005). *Fundamentals of Wireless Communication* (New York: Cambridge University Press).

Viterbi, A. J. (1995). *CDMA: Principles of Spread Spectrum Communications* (Reading, MA: Addison-Wesley).

Wilson, S. G. (1996). *Digital Modulation and Coding* (Englewood Cliffs, NJ: Prentice-Hall).

Wozencraft, J. M. and Jacobs, I. M. (1965). *Principles of Communication Engineering* (New York: Wiley).

Wyner, A. and Ziv, J. (1994). "The sliding window Lempel–Ziv algorithm is asymptotically optimal," *Proc. IEEE* **82**, 872–877.

Ziv, J. and Lempel, A. (1977). "A universal algorithm for sequential data compression," *IEEE Trans. Inform. Theory* **IT-83**, 337–343.

Ziv, J. and Lempel, A. (1978). "Compression of individual sequences via variable-rate coding," *IEEE Trans. Inform. Theory* **IT-24**, 530–536.

Index

Printed in the United States
By Bookmasters